High-Temperature Superconductors and Novel Inorganic Materials

NATO Science Series

A Series presenting the results of activities sponsored by the NATO Science Committee. The Series is published by IOS Press and Kluwer Academic Publishers, in conjunction with the NATO Scientific Affairs Division.

General Sub-Series

A. Life Sciences	IOS Press
B. Physics	Kluwer Academic Publishers
C. Mathematical and Physical Sciences	Kluwer Academic Publishers
D. Behavioural and Social Sciences	Kluwer Academic Publishers
E. Applied Sciences	Kluwer Academic Publishers
F. Computer and Systems Sciences	IOS Press

Partnership Sub-Series

1. Disarmament Technologies	Kluwer Academic Publishers
2. Environmental Security	Kluwer Academic Publishers
3. High Technology	Kluwer Academic Publishers
4. Science and Technology Policy	IOS Press
5. Computer Networking	IOS Press

The Partnership Sub-Series incorporates activities undertaken in collaboration with NATO's Partners in the Euro-Atlantic Partnership Council – countries of the CIS and Central and Eastern Europe – in Priority Areas of concern to those countries.

NATO-PCO-DATA BASE

The NATO Science Series continues the series of books published formerly in the NATO ASI Series. An electronic index to the NATO ASI Series provides full bibliographical references (with keywords and/or abstracts) to more than 50000 contributions from international scientists published in all sections of the NATO ASI Series.
Access to the NATO-PCO-DATA BASE is possible via CD-ROM "NATO-PCO-DATA BASE" with user-friendly retrieval software in English, French and German (© WTV GmbH and DATAWARE Technologies Inc. 1989).

The CD-ROM of the NATO ASI Series can be ordered from: PCO, Overijse, Belgium.

3. High Technology – Vol. 62

High-Temperature Superconductors and Novel Inorganic Materials

edited by

G. Van Tendeloo
University of Antwerp,
Belgium

E. V. Antipov
Moscow State University,
Russia

and

S. N. Putilin
Moscow State University,
Russia

Springer-Science+Business Media, B.V.

Proceedings of the NATO Advanced Workshop on
High-Temperature Superconductors and Novel Inorganic Materials Engineering MSU-HTSC V
Moscow, Russia
March 24–29, 1998

A C.I.P. Catalogue record for this book is available from the Library of Congress.

ISBN 978-0-7923-5346-1 ISBN 978-94-011-4732-3 (eBook)
DOI 10.1007/978-94-011-4732-3

Printed on acid-free paper

TABLE OF CONTENTS

STRUCTURAL ASPECTS HIGH Tc MATERIALS

PHYSICAL AND CHEMICAL ASPECTS OF HIGH Tc MATERIALS

NOVEL INORGANIC MATERIALS

Preface

The discovery of high temperature superconductivity in 1986 stimulated an enormous research activity around the world in physics, chemistry as well as in materials science. The synthesis, the analysis and the understanding of superconducting Cu-based mixed oxides are difficult scientific challenges. Moreover, the fabrication of superconducting ceramics and of thin films and devices poses new technological problems. Actually, the complexity of these materials is one of the main reasons of their relatively slow appearence on the world market. A successful research in the field of High-Tc superconductivity strongly demands a deep cooperation between scientists from various fields. This is exactly why High-Tc superconductivity became a crystallization center or a nucleus for scientific cooperation of researchers from various fields and from different countries.

The numerous international conferences on High-Tc materials often unify physicist, chemists and materials scientists, theoreticians as well as experimentalists, aiming to discuss and to find the optimum solution for important problems in this field. This idea was the reason why the Department of Inorganic Chemistry of the Moscow State University organized in 1989 the 1-st International Workshop "Chemistry and Technology of High-Tc materials MSU-HTSC-I". These workshops, organised every other year, allowed to establish and develop scientific cooperation between Western and Russian scientists. In 1998 the 5-th International Workshop on "High Temperature Superconductors and Novel Inorganic Materials Engineering - MSU-HTSC-V" was organized. This Workshop was supported by the NATO and organized in the framework of the NATO Advanced Research Workshop Program. The significant support from NATO as well as from Moscow State University, the Russian Fund for Basic Research and the Russian State Program on Superconductivity allowed to organise the present Workshop at a distinguished level with about 300 participants from 22 countries. For the first time the topic of the Workshop was extended beyond High Tc superconductivity and included also other advanced oxide materials, mainly colossal magnetoresistance (CMR) materials which are closely related to the ceramic superconducting materials. The study of these materials makes use of similar synthetic, crystallographic and characterisation techniques.

The scientific program of the Workshop, as well as the manuscripts collected in this book, represent the main topics discussed in the Workshop.and can be subdivided in two main sections:

1. Advanced oxide materials:
   High-Tc superconductors (new materials, structural and physical properties, powders and bulk materials, thin films)
   Giant Magnetoresistance materials, Ferroelectrics, Ionic Conductors etc.

2. Processing Chemistry and Engineering (self-organization phenomena, MOCVD, new synthetic routes and precursors, melt processing etc.)

The scientific program consisted of 2 plenary lectures, 46 oral presentations and 254 posters. Russian participants delivered 17 oral talks and 200 posters. An important feature of the Workshop was the active participation of a large number of students and young researchers. The special committee awarded 5 young researchers for the excellent presentation of their results: A. Molodyk, V. Kuzmenko and V. Petrykin (Moscow State University, Russia), G. Nilsson (Stockholm University, Sweden) and J. Hadermann (Antwerp University, Belgium).

The editors of this book would like to express our deep appreciation for the help and cooperation of the researchers from the Department of Chemistry of the Moscow State University: Dr. Oleg D'yachenko, Dr. Artyom Abakumov, Dr. Yuri Metlin, Dr. Elena Eremina and Prof. Andrei Kaul.

Gustaaf Van Tendeloo
EMAT, University of Antwerp (RUCA), Belgium
gvt@ruca.ua.ac.be

Evgeny Antipov
Dpt. of Chemistry, Moscow State University, Russia
antipov@inorg.chem.msu.su

Sergei Putilin
Dpt. of Chemistry, Moscow State University, Russia
putilin@inorg.chem.msu.su

RECENT ADVANCES IN FABRICATION OF HIGH-T_c SUPERCONDUCTORS FOR ELECTRIC POWER APPLICATIONS

U. (BALU) BALACHANDRAN
Energy Technology Division, Argonne National Laboratory
9700 South Cass Avenue, Argonne, IL 60439

1. Introduction

The U.S. Department of Energy (DOE) supports an applied superconductivity program entitled "Superconductivity Program for Electric Power Systems." Activities within this program contribute to development of the high-temperature superconductor (HTS) technology needed for industry to proceed with the commercial development of electric power applications such as motors, generators, transformers, transmission cables, and current limiters. Research is conducted in three categories: wire development, systems technology development, and Superconductivity Partnership Initiative (SPI). Wire development activities are devoted to improving the critical current density (J_c) of short-length HTS wires, whereas systems technology development focuses on fabrication of long-length wires, coils, and on magnets. The SPI activities are aimed at development of prototype products.

A J_c of 5×10^4 A/cm^2 at 77 K has been obtained in short-length (≈4 cm) powder-in-tube (PIT)-processed Bi-2223 tapes. The J_c value increased to $\approx 2 \times 10^5$ A/cm^2 at 77 K in short-length samples when an Ag wire was introduced into the Ag tube while preparing the billet. Enhancements in J_c were made through increased packing density of the precursor powder, improved mechanical deformation, and adjustments to cooling rate. Ag-clad Bi-2223 tapes of lengths >1 km with $J_c \approx 15$ kA/cm^2 are being manufactured routinely. Due to the poor flux pinning characteristics of Bi-2223, this material has been limited to applications that operate at 35 K in a magnetic field of ≈1 T. However, because Y-123 performs better than Bi-2223 in a magnetic field, efforts are currently underway to further its development. A $J_c \approx 10^6$ A/cm^2 at 77 K has been reported for short-length Y-123 films on a polycrystalline metallic substrate with buffer layers. A 2.4 kV/2.2 kA current limiter, a 286 hp motor, and a 50-m-long cable have already been demonstrated under the SPI activities. A current overview and recent progress in HTS tape and application development will be presented in this paper.

G. Van Tendeloo et al. (eds.), High-Temperature Superconductors and Novel Inorganic Materials, 1–9.

2. Applications Development

Development of HTS technology will make possible the design and production of smaller, lighter, and more efficient electric power devices such as motors, generators, transformers, cables, and fault current limiters. As shown in Fig. 1, over the 1986-96 decade several potential applications have been envisioned for HTSs [1].

To expedite the commercialization HTS technology, DOE has formed multidisciplinary teams comprising national laboratories, industry, utilities, and universities. The national laboratories and universities play a key role in developing the fundamental technology necessary for HTS applications, while industry and utilities are responsible for development, cost-benefit analysis, and commercialization of various products. The SPI project, which began in 1994, is a flagship program in the nation's effort to commercialite HTS technologies. The SPI program consists of a vertically integrated team of utilities, system manufacturers, HTS component manufacturers, and national laboratories. Phase I of the SPI consisted of four industry-led teams focusing on four HTS products.

The first team was led by Lockheed-Martin Corporation and included American Superconductor Corp. (ASC, which has since withdrawn), Los Alamos National Laboratory (LANL), and Southern California Edison (SCE). The team is developing superconductor fault current limiters, which have an estimated global market of $20 billion. A sudden surge in power can render equipment or a system ineffective, but traditional approaches of using circuit breakers with higher fault current ratings to curtail such power surges have proved to be expensive and inefficient. A fault current limiter that was developed with HTSs can reduce the current to a fraction of its peak value in <8 ms, thereby mitigating the effects of short circuits or power fluctuations. The Lockheed-Martin-led team successfully demonstrated a 2.4 kV, 2.2-kA HTS fault current limiter in 1995. This experimental prototype is being scaled up in the Phase II (1996-98) SPI project to a precommercial unit rated at 15 kV and 20 kA, and Intermagnetics General Corp. (IGC) has joined the team for the Phase II activitiy [2].

The second SPI team, selected to develop superconducting generators to lower the cost of electricity and serve a worldwide market estimated at $30 billion, was led by General Electric Company (GE) and included the Electric Power Research Institute (EPRI), IGC, New York State Energy Authority, New York State Institute for Superconductivity, Niagara Mohawk Power, and Argonne, Los Alamos, and Oak Ridge National Laboratories (ANL, LANL, ORNL). This team focused on the conceptual design and assessment of a 100-MVA generator and the development of an HTS racetrack coil suitable for use in a full-scale generator. Ag-sheathed multifilamentary Bi-2223 superconductor tapes for the racetrack coil was manufactured by IGC. The coil, assembled with ≈3.5 km of Bi-2223 tape, achieved 34 A at 25 K, which corresponds to ≈40,000 ampere turns and is considered adequate for an HTS generator.

The third SPI team was led by Reliance Electric Co. and included ASC, Centerior Energy, EPRI, and Sandia National Laboratories. This team is working on a project to develop superconducting motors. Large industrial motors consume nearly 30% of the electricity generated in the U.S., so the development of a more efficient motor would

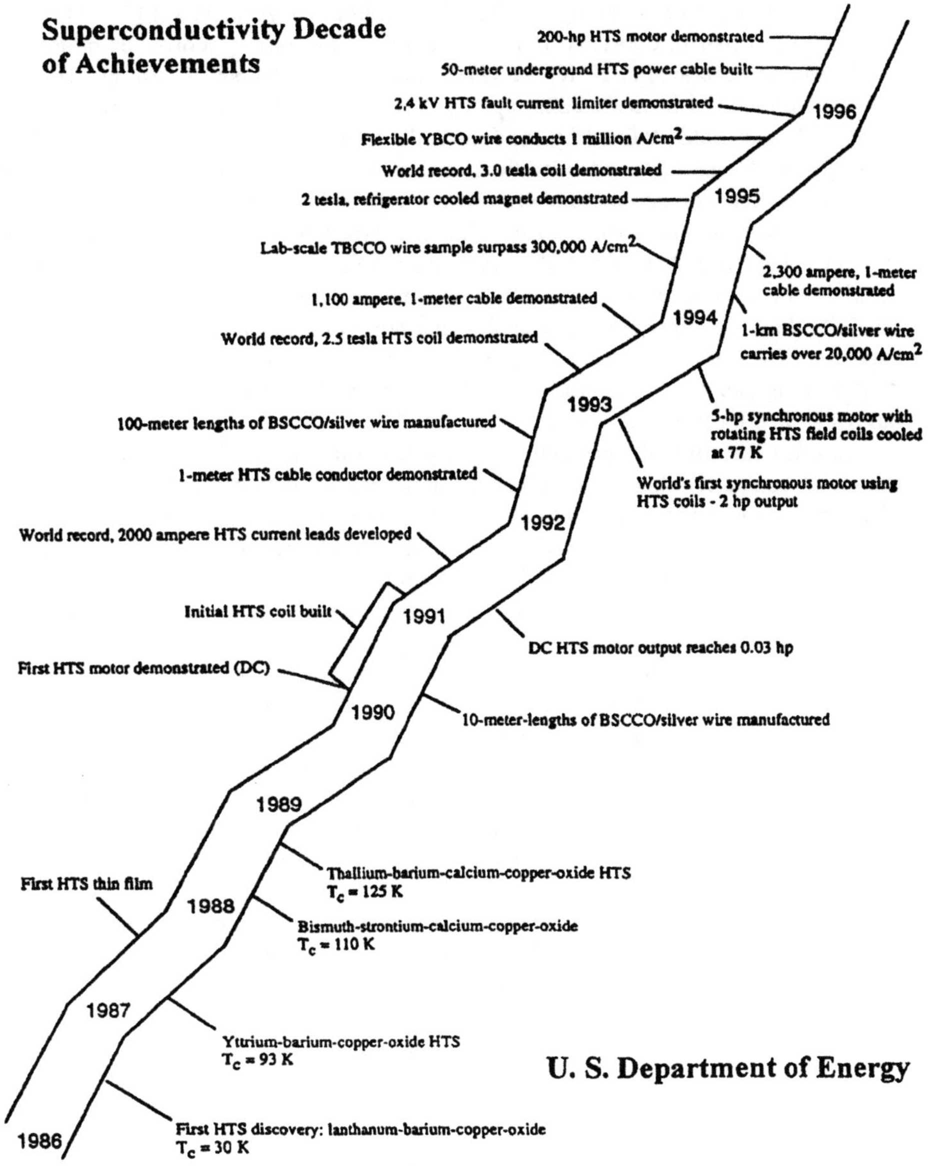

Figure 1. Progress made, 1986-1996, in developing several potential bulk and microelectronic applications (ref. #1).

provide a competitive advantage in the $300 million per year market [3]. This team successfully tested in February 1996 a four-pole, 1800-rpm synchronous motor with HTS windings operated at 27 K. The HTS motor had an output of ≈28 hp. The HTS coils, manufactured from Bi-2223 tape by ASC, achieved currents of 100 A, and the combined peak field attained by the four coils during the testing was nearly 1 T [4]. In August 1997, the DOE extended the Reliance-led motor program into a Phase II SPI effort to develop a precommercial prototype of a 5000 hp HTS motor.

The fourth team's goal is to develop two types of HTS power cable systems. Pirelli Cables North America, ASC, EPRI, and LANL are at work on retrofit cables and coaxial cables. The retrofit cables, as the name suggests, will be drawn into existing conduits that will replace copper cables that are worn out or exhibit reduced power capacity. The coaxial cables will be used primarily for underground installations [5]. A 50-m-long flexible conductor assembly was manufactured by Pirelli with Bi-2223 tapes supplied by ASC and was tested at 3300 A DC.

Besides the four SPI projects, industry-led teams have also started projects on transformers and flywheels, as well as one new transmission cable design. Waukesha Electric Systems has teamed with IGC, Rochester Gas and Electric Corp., Rensselaer Polytechnic Institute, and ORNL to develop a HTS transformer. This team has conducted a series of reference designs concentrating on a 30-MVA, 138-kV/13.8 kV HTS transformer. This rating is representative of a medium-power transformer class foreseen as comprising about 50% of all U.S. power transformer sales in the next two decades [6]. One major advantage of the HTS transformer is elimination of the fire-hazardous oil used in conventional transformers. In the HTS transformer, the only substance present in large volume is nonflammable and environmentally benign liquid nitrogen. The Waukesha-led team has focused on the design of a low-loss conductor and winding configuration and has come up with a proprietary design for its HTS transformer. At present, this team is working on the demonstration of a 1-MVA cryocooled, 13.8 kV/6.9-kV single-phase transformer that uses a relatively low-cost surface-coated Bi-2212 conductor.

Commonwealth Research Corporation, a subsidiary of Commonwealth Edison, has joined ANL in developing flywheels with HTS bearings for energy storage applications. Other groups active in the flywheel project in the U.S. include the University of Houston and Boeing Co. Southwire Company, the largest U.S. cable manufacturer, has teamed with ORNL, ANL, and IGC to develop a new transmission cable design. The Southwire-led team plans to build a 33-m-long, cryogenic-dielectric, three-phase, 12.4 kV, 1250 A system by 1999. This team has already fabricated and tested several 1-m-long conductors carrying up to 2 kA at 77 K.

In Japan, Tokyo Electric Power Co. (TEPCO) is working with Sumitomo Electric Industries (SEI) and Furukawa on developing a 66 kV, 1000 MVA HTS cold-dielectric cable system with the ultimate goal of deploying it around Tokyo to meet the city's growing needs. The TEPCO team has tested a 50-m-long conductor carrying 2 kA AC [7]. In Europe, Siemens plans to develop a 100-m-long prototype cable by 1999. Siemens and Hydro Quebec are collaborating to develop fault current limiters based on two different concepts (screened core and resistive type). Asea Brown Boveri (ABB) is developing an inductive fault current limiter and a transformer. ABB connected the world's first operating HTS transformer to the Geneva (Switzerland) Power Supply network in March 1997. This three-phase transformer has a rating of 630 KVA and converts power from 18.7 kV into 420 V, using Bi-2223 windings cooled in liquid nitrogen. This unit is now undergoing long-term performance monitoring under actual power grid conditions. ABB is planning two commercial 10-MVA transformers in 2001. ABB has also reported endurance and performance test results for a 1 MVA HTS fault current limiter, installed in November 1996 at the Kraftwerk am Loentsch hydroelectric power plant in Switzerland. This fault current limiter uses bulk Bi-2212 rings and displayed smooth current limitation without high overvoltages when tested with a prospective short-circuit fault current of 60 KVA. Endurance tests indicated no major problems after six months of operation and, according to ABB, refilling of the nitrogen tank was the only significant maintenance required during the testing period. ABB is beginning to focus its efforts on developing a 10-MVA unit with closed-cycle cooling.

3. HTS Wire Development

Most prototype products currently use either bismuth-strontium-calcium-copper oxide (BSCCO) or yttrium-barium-copper oxide (YBCO) superconductors. Current transport in HTSs is controlled by intra- and intergranular J_C. High intragranular J_C values have been achieved in HTSs. Because the current flows by a percolative process, intergranular J_C is the main factor that controls the overall current transport property of a polycrystalline superconductor. A grain with high intragranular J_C will not be of much use if it is poorly coupled to the adjacent grain. Therefore, phase purity, grain alignment, and grain connectivity are very important in the fabrication of HTSs. For successful applications, HTSs must be fabricated into long-length conductors that exhibit high-J_C values and mechanical reliability. However, the processing of HTSs (which are brittle oxide ceramics) into more reliable and robust forms is an extremely challenging and formidable task. The groups at Vacuumschmelze and SEI were the first to develop the powder-in-tube (PIT) technique for the fabrication of HTSs [8,9] In this approach, various ingredients for the precursor powder are mixed and treated so they react with each other. The resultant powder is then packed into a silver or silver alloy tube that is then swaged and drawn through a series of dies to form a silver-sheathed wire ≈2 mm in diameter. These wires are then rolled to form a tape with a final thickness of ≈0.1 mm. Multifilament conductors are fabricated by restacking monofilament wires in a larger silver tube and then using a similar deformation sequence. The final step involves thermomechanical treatment that consists of a series of heat treatments and mechanical deformation steps [10-13]. During this stage, the precursor powder reacts to form the desired superconducting phases.

Tapes made by the PIT technique have a desirable geometry in which a brittle superconductor oxide is surrounded by a metallic sheath that protects the superconductor core from chemical, thermal, and mechanical abrasion. The main advantage of the PIT technique is that the metallurgical techniques employed in the fabrication of the tapes are all well established and can therefore be readily scaled up for mass production of long-length conductors. Companies such as IGC, ASC, and SEI have been successful in making kilometer-long high-quality mono- and multifilament conductors. Recently, ASC reported a self-field J_c of ≈7 x 10^4 A/cm^2 in short-length rolled multifilamentary tapes [14]. A J_c of 1.5 x 10^4 A/cm^2 has also been achieved in a 1.26-km-long multifilament conductor fabricated by IGC [15].

Several research groups have reported that most of the current in the BSCCO tape flows exclusively in a thin layer of highly aligned Bi-2223 grains near the silver sheath. Existence of such a layer has been confirmed by microslice experiments, high-resolution electron microscopy, and magneto-optical imaging [16-18]. The high-current superconducting layers are generally ≈2.3 μm thick and have been shown to support a transport current with a J_c value >10^5 A/cm^2 at 77 K [16,19]. Transport J_c values of tapes with identical superconductor cross-sectional areas, but differing Ag/Bi-2223 interfacial lengths, confirm the importance of the interfacial region [20]: the critical current was shown to be proportional to the Ag/Bi-2223 interface perimeter length (IPL), expressed as a linear function. These results imply that <10% of the Bi-2223 superconductor transports the vast majority of supercurrent in Ag-clad Bi-2223 tapes. Thus, J_c values can be increased by microstructural design through optimization of the IPL.

Efforts to enhance J_c by increasing the Ag/Bi-2223 interfacial area continue. Fabrication of multifilamentary tapes achieves this goal, but, in general, the areal fraction of Ag increases for such tapes [21]. An alternative approach is to incorporate Ag wires into a Bi-2223 core. Initial work focused on use of a single Ag wire. The J_c value increased to ≈2 x 10^5 A/cm^2 at 77 K in short-length samples by this approach [19]. In addition to offering possibilities for improved transport J_c, significant enhancement of bend-strain tolerance with a wire-in-tube approach has been reported [22]. This duplex-core work has recently been extended to a two-step process in which many fine Ag wires are coated with Bi-2223 precursor powder and then loaded into an Ag tube. Conventional PIT working then produces a tape with a very large Ag/Bi-2223 interfacial area [23]. To date, up to 600 Ag wires, coated with Bi-2223 precursor, have been loaded into a single Ag tube and processed into tapes. Despite the smaller cross-sectional area of the superconducting core, transport J_c values are now greater than those of corresponding monofilament tapes [23].

Because BSCCO exhibits poor flux-pinning characteristics, applications of this superconductor have been limited to those areas where operations are conducted at 35 K and in a magnetic field <1 T. Because YBCO behaves better than BSCCO in a magnetic field, effort is currently underway to use it for conductor development. In 1991, Iijima and collaborators [24] at Fujikura Ltd. successfully developed a thick film of YBCO. The film was deposited by laser ablation on a nickel alloy substrate with a buffer layer of yttria-stabilized-zirconia (YSZ). The buffer layer was obtained by the ion-beam-assisted deposition (IBAD) technique. The group achieved a critical current density of 1.13 x 10^6 A/cm^2 at 77 K and zero field in short-length samples, and a J_c of 2.1 x 10^5 A/cm^2 in a 70-cm-long tape. SEI reported a J_c of 3 x 10^5 A/cm^2 in a 10-cm-long YBCO film on a Hastelloy substrate with a YSZ buffer layer. The YSZ layer was produced on the substrate by an inclined substrate deposition technique. The research group at LANL, working on the IBAD technique, reported a YBCO film

≈1.2 μm thick and ≈2 cm long with $J_c \approx 1 \times 10^6$ A/cm^2 [25]. The group reported very little loss in J_c when a field of 19 T was applied parallel to the tape surface. In a field of 5 T applied normal to the plane of the tape, J_c was 1×10^5 A/cm^2 at 77 K. ORNL has produced epitaxial YBCO films with J_c in the range of 1-3 x 10^6 A/cm^2 at 77 K [26]. The YBCO film was obtained by using laser ablation on rolling-assisted biaxially textured substrates. Oxide buffer layers were deposited on textured nickel substrates by three techniques: laser ablation, electron-beam evaporation, and sputtering. A team consisting of researchers from the University of Munich, University of Göttingen, and Siemens have deposited YBCO films using the coevaporation technique on substrates as large as 20 x 20 cm. They reported a J_c of 0.7-2 x 10^6 A/cm^2 on small regions of 20 x 20 cm film [27]. These results are encouraging because they promise to extend the range of HTS applications, especially in the presence of high magnetic fields and temperatures.

4. Conclusions

Significant progress has been made in the development of HTSs for various applications: some applications have already made significant strides in the marketplace, while others are still in the developmental stages. For successful electric power applications, it is very important that the HTS be fabricated into long-length conductors that exhibit desired superconducting and mechanical properties. Several parameters of the PIT technique must be carefully controlled to obtain the desired properties. Long lengths of Bi-2223 tapes with respectable superconducting properties have been fabricated by a carefully designed thermomechanical treatment process. A 1-MVA capacity fault current limiter, a 286-hp motor, 630-kVA transformers, and a 50-m-long conductor, all using HTSs, have already been demonstrated. While the use of HTS devices in the electric utility area has clear advantages, impediments to successful commercialization remain. Issues such as AC losses, conductor cost, and reliable superconducting joints must be addressed. The cost of HTS conductors are still quite high, and significant R&D effort must be focused on this issue. The general acceptance of HTS power equipment will ultimately be based on system performance, reliability and maintenance, efficiency, and installed cost relative to those of conventional technologies.

5. Acknowledgments

This work is supported by the U.S. Department of Energy (DOE), Energy Efficiency and Renewable Energy, as part of a DOE program to develop electric power technology, under Contract W-31-109-Eng-38.

6. References

1. U.S. Department of Energy, Office of Energy Efficiency and Renewable Energy, Office of Utility Technology, 1996.

2. Leung, E. (1997) Surge protection for power grids, *IEEE Spectrum* **34,** 26-30.
3. Ott, K.D. (1994) Superconductors for our energy future, *Supercond. Ind.* **7**, 10.
4. Blaugher, R.D. (1997) Low-calorie, high-energy generators and motors, *IEEE Spectrum* **34**, 36-42.
5. HTS Cable Wires, News Release, American Superconductor Corp., Westborough, MA, march 1996.
6. Mehta, S.P., Aversa, N., and Walker, M.S. (1997) Transforming transformers, *IEEE Spectrum* **34**, 43-49.
7. Schiff, N. and Schempp, E. (1994) TEPCO demonstrated AC transmission cable, *Supercond. Ind.* **7**, 14-15.
8. Tenbrink, J., Wilhelm, M., Heine, K., and Krauth, H. (1991) Development of high-T_c superconductor wires for magnet applications, *IEEE Trans. Mag.* **27**, 1239.
9. Sato, K., Hikata, T., Mukai, H., Ueyama, M., Shibuta, N., Kato, T., Masuda, T., Nagata, M., Iwata, K., and Mitui, T. (1991) High-J_c silver-sheathed Bi-based superconducting wires, *IEEE Trans. Mag.* **27**, 1231.
10. Sandhage, K.H., Riley, G.N. Jr., and Carter, W.L. (1991) Critical issues in the OPIT processing of high-J_c BSCCO superconductors,*J. Metals* **43**, 21.
11. Haldar, P. and Motowidlo, L.R. (1992) Processing high critical current density Bi-2223 wires and tapes, *J. Metals* **44**, 54.
12. Balachandran, U., Iyer, A.N., Haldar, P., and Motowidlo, L.R. (1994) Recent advances in bismuth-based superconductors, *J. Metals* **46**, 23.
13. Luo, J.S., Merchant, N., Escorcia-Aparicio, E., Maroni, V.A., Gruen, D.M., Tani, B.S., Riley, G.N., Jr., and Carter, W.L. (1994) Composition and microstructural evolution of nonsuperconducting phases in silver-clad $(Bi,Pb)_2Sr_2Ca_2Cu_3O_x$ composite conductors, *J. Mater. Res.* **9**, 3059.
14. Riley, G.N., Jr. (1992) BSCCO development at ASC, presented at DOE Wire Workshop, St. Petersburg, FL.
15. Balachandran, U., Iyer, A.N., Jammy, R., Haldar, P., Hoehn, J.G., Jr., and Suenaga, M. (1995) Processing and properties of long-lengths of Ag-clad BSCCO superconductors and high-T_c magnets, in Advances in Superconductivity VII, eds. K. Yamafuji and T. Movishita (Springer-Verlag, Tokyo, 1995), p. 781.
16. Lelovic, M., Krishnaraj, P., Eror, N.G., Iyer, A.N., and Balachandran, U. (1995) Minimum critical current density of 10^5 A/cm^2 at 77 K in the thin layer of $Bi_{1.8}Pb_{0.4}Sr_{2.0}Ca_{2.2}Cu_{3.0}O_y$ superconductor near the Ag in Ag-sheathed tape, *Physica C* **242**, 246.
17. Feng, Y. and Larbalestier, D. (1994) Texture relationships and interface structure in Ag-sheathed Bi(Pb)-Sr-Ca-Cu-O superconducting tapes, *Interface Science* **1**, 401.
18. Larbalestier, D., Cai, X.Y., Edelman, H., Field, M.B., Feng, Y., Parrell, J., Pashitski, A., and Polyanskii, A. (1994) Visualizing current flow in high-T_c superconductors, *J. Metals* **46**, 20.

19. Lelovic, M., Krishnaraj, P., Eror, N.G., and Balachandran, U (1995) Magnetization current of a thin layer of $Bi_{1.9}Pb_{0.4}Sr_{2.0}Ca_{2.2}Cu_{3.0}O_y$ near silver in Ag-sheathed BSCCO-2223 tapes, *Supercond. Sci. Technol.* **8**, 336.
20. Prorok, B.C., Lelovic, M., Eror, N.G., Deis, T.A., Krishnaraj, P., Iyer, A.N., and Balachandran, U. (1997) Improvements in critical current density for $Bi_{1.8}Pb_{0.4}Sr_2Ca_2Cu_3O_x$ (2223) OPIT tapes, *Adv. Cryo. Eng.* **42A**, 739.
21. Wang, W.G., Horvat, J., Liu, H.K., and Dou, S.X. (1996) The effect of mechanical deformation on silver-core interface and critical current density in Ag-Bi-2223 single and multifilament tapes, *Supercond. Sci. Technol.* **9**, 875.
22. Vasanthamohan, N. and Singh, J.P. (1997) Enhancement of critical current density and strain tolerance in Ag-sheathed BSCCO (2223) tapes by continuous silver reinforcement, *Supercond. Sci. Technol.* **10,** 113.
23. Dorris, S.E., Ashcom, N., Truchan, T., Vasanthamohan, V, Burlone, D.A., and Woolf, L.D. (1997) to be published in Proc. Supercond. Symp., 99th Ann. Mtg. Am. Ceram. Soc., Cincinnati, May 4-7, 1997.
24. Iijima, Y., Tanabe, N., Kohno, O., and Ikeno, Y. (1992) In-plane aligned $YBa_2Cu_3O_{7-x}$ thin films deposited on polycrystalline metallic substrates, *Appl. Phys. Lett.* **60**, 769.
25. Peterson, D.E. (1995) YBCO conductor development at LANL, in Proc. Superconductivity Program for Electric Power Systems, 1995 Annual Peer Review Mtgs., U.S. Department of Energy, Alexandria, VA, Aug. 1-2, 1995.
26. Goyal, A. (1997) High-J_c YBCO films on rolling-assisted biaxially textured substrates (RABiTS), presented at 8th U.S.-Japan Workshop on High-Temperature Superconductors, Tallahassee, FL, Dec. 7-10, 1997.
27. Kinder, H. (1997) Thermal coevaporation of YBCO on large area substrates, presented at MRS Spring Mtg., San Francisco, CA, March 31-April 4, 1997.

THE LOCAL STRUCTURE OF YBCO BASED MATERIALS BY TEM.

G. VAN TENDELOO[1], O.I. LEBEDEV[1], K. VERBIST[1], A.M.ABAKUMOV[1,2], R. V. SHPANCHENKO[2], E. V. ANTIPOV[2], D. H. A. BLANK[3].
[1] EMAT, University of Antwerp (RUCA), Groenenborgerlaan 171, B-2020 Antwerp, Belgium
[2] Department of Chemistry, Moscow State University, 119899 Moscow, Russia
[3] Dept. Applied Physics, University of Twente, POB 217, 7500 AE Enschede, The Netherlands

1. ABSTRACT

Transmission electron microscopy (TEM) allows to identify and determine the structure of new phases as well as to investigate the defect structure of ramp-type Josephson junctions. A F doped $YBa_2Cu_3O_{6+\epsilon}$ high temperature superconductor (Tc = 94K) has been prepared from strongly reduced $YBa_2Cu_3O_{6+e}$ using XeF_2 as a fluorination agent. HREM and microanalysis unambiguously demonstrated that fluorene enters the structure at the position of the Cu(1) planes. The structure of the fully fluorinated phase $YBa_2Cu_3O_6F_2$ was determined. Strongly reduced $Y_2Ba_4Cu_7O_{14.09}$ was also synthesized using the same method. For a high level of fluorination, the intercalation of fluorine is accompanied by a significant structural rearrangement and leads to an increase of T_c from 30K to 62K. HREM reveals that the structure with a complete octahedral arrangement of the copper atoms only exists in limited regions; the *c* parameter in these areas amounts to 53.5Å.

Ramp-type Josephson junctions (JJ) with highly doped $PrBa_2Cu_{3-x}Ga_xO_{7-\delta}$ barrier layers (x = 0.7 or x = 1) show a Ga-rich intergrowth and Ga-diffusion in the ion-milled $SrTiO_3$ substrate. The Ga-segregation is responsible for the deviating electrical behaviour as expected from low doping levels (x = 0.1-0.4). The Ga-diffusion in the ion-milled substrate and possibly in the base electrode may hamper JJ fabrication using substituted $REBa_2Cu_3O_{7-\delta}$ materials.

2. INTRODUCTION.

After the discovery in 1987 of the $YBa_2Cu_3O_{7-\delta}$ superconductor, several groups reported the enhancement of T_c (up to 150K) for the 123 phase with partial replacement of oxygen by fluorine [1-3]. The main -confirmed- results in the field of fluorination can be summarized as follows:

- For the partially reduced ($\delta \leq 0.35$) $YBa_2Cu_3O_{7-\delta}$ compound, the F incorporation occurs in the Cu-1 layer and fluorine preferentially occupies vacant sites along the *a*-direction [4,5]. This process does not alter the symmetry; it causes a decrease of the *c* and *b* parameters while the *a* parameter increases. For $YBa_2Cu_3O_{6.7}F_{0.2}$ an enhancement of Tc from 65 K to 90 K is reported [5], but at a higher level of fluorination superconductivity is suppressed [4].

G. Van Tendeloo et al. (eds.), High-Temperature Superconductors and Novel Inorganic Materials, 11–19.

- Fluorination of highly reduced non-superconducting $YBa_2Cu_3O_{7-\delta}$ ($\delta \geq 0.8$) preserves the tetragonal symmetry with an increase of the *a*-parameter, while the *c*-parameter decreases again [6-8]. This fluorine up-take induces superconductivity with $T_c \approx$ 55-60K.

Fluorine together with the standard YBCO cations forms thermodynamically very stable phases (BaF_2, LnOF etc) and high temperature synthesis results in the formation of these. Therefore a soft-chemistry technique is more favourable to obtain fluorinated YBCO. XeF_2 may be successfully used as a fluorinating reagent at temperatures below 300°C [9]. In the case of 123-YBCO an anion deficient starting material is definitely needed since anion exchange in oxidised YBCO demands higher temperatures leading to the formation of thermodynamically stable admixtures.

Also $Y_2Ba_4Cu_7O_{14+x}$ (247), which is closely related to the 123 compound has been fluorinated. Its ideal crystal structure can be described as an ordered alternation of blocks comprising single CuO chains (123-units) and double CuO chains (124-units). 247 exhibits a superconducting transition with T_c ranging from 4K to 95K, depending on the oxygen content [10,11]. By analogy to the stoichiometric 124 compound, the double CuO chains remain fully oxidised and oxygen non-stoichiometry is connected to the presence of 123 blocks [12-15]. One may therefore expect that 123 and 247 compounds under fluorination should exhibit a similar behavior.

3. EXPERIMENTAL

$BaCO_3$, Y_2O_3, CuO and XeF_2 are taken as starting materials. The reduced $YBa_2Cu_3O_{6.11}$ was prepared by annealing $YBa_2Cu_3O_{6.95}$ pellets in an Ar flow under controled partial oxygen pressure at 780°C for 10 h. Subsequent operations were carried out in a dry box in dried N_2 atmosphere. 0.4 g of YBCO was mixed with 0.07 g of XeF_2 and intimately grinded. Syntheses were carried out in Ni-crucibles placed in a hermetically closed steel or Ni-container. Time and annealing temperature were varied from 3 h till 70 h and from 100°C to 350°C [16]. Phase compositions and lattice parameters of the compounds were determined by X-ray powder diffraction. The formal copper valence in the initial and fluorinated samples was determined by iodometric titration. AC susceptibility measurements were performed in the temperature range 12-100K. Electron diffraction (ED) and high resolution electron microscopy (HREM) and microanalysis (EDX) were done using different microscopes at the EMAT laboratory.

4. RESULTS

4.1.X-RAY STUDY AND MAGNETIC MEASUREMENTS.

The fluorination of $YBa_2Cu_3O_{6.95}$ did not lead to any significant structural changes or to any change on the T_c value [16]. No noticeable insertion of fluorine into the structure or a replacement of O-atoms by F-atoms was found.

For the fluorination of $YBa_2Cu_3O_{6.11}$ significant structural changes were detected for samples synthesised above 200°C. The increase of temperature and time of fluorination led to a dramatic decrease of the intensity of reflections with $l \neq 0$. Apart from these sharp reflections a few very broad *hkl* ($l \neq 0$) reflections could be attributed to the presence of related phases having a variable *c*-parameter, ranging from 11.6Å to 13Å. This suggests a well established in-plane order, but a strong disorder along *c*. The change of phase composition is also prominent from the T_c dependence on the fluorination temperature (Fig.1). The initial tetragonal phase is known to be non-superconducting, but annealing at 250°C during 1 day already reveals a superconducting transition. The highest value of T_c = 94K was found for 50 h at 350°C.

For the 247 compound, the X-ray patterns of fluorinated samples also exhibit significant changes upon fluorination. Some samples also contained two phases with a 247-type structure. Like in the 123 compound, the formation of a fluorinated phase is accompanied by a drastical change of the intensities and a broadening of the reflections in the X-ray pattern. An increase of the fluorination reaction temperature to 350 ^{o}C leads to decomposition of the 247 compound. Both initial reduced $Y_2Ba_4Cu_7O_{14.09}$ and oxidized $Y_2Ba_4Cu_7O_{14.92}$ samples exhibit superconducting transitions at 30K and 80K respectively. The estimated superconducting volume fraction is about 15%.

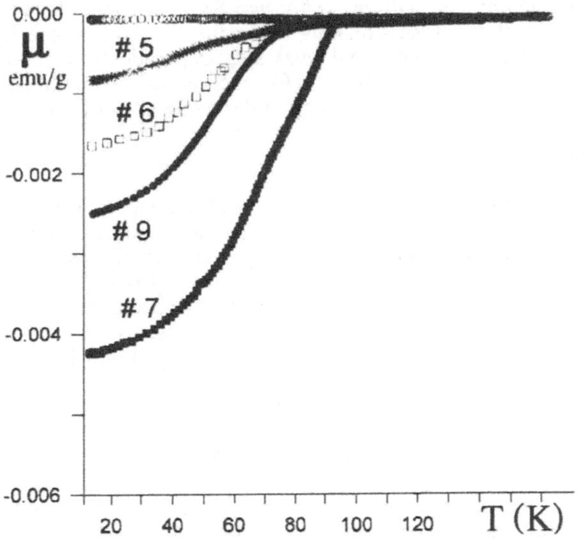

Fig. 1. Temperature dependence of magnetic susceptibility for $YBa_2Cu_3O_{6.11}$ oxide before and after fluorination under different treatment conditions.

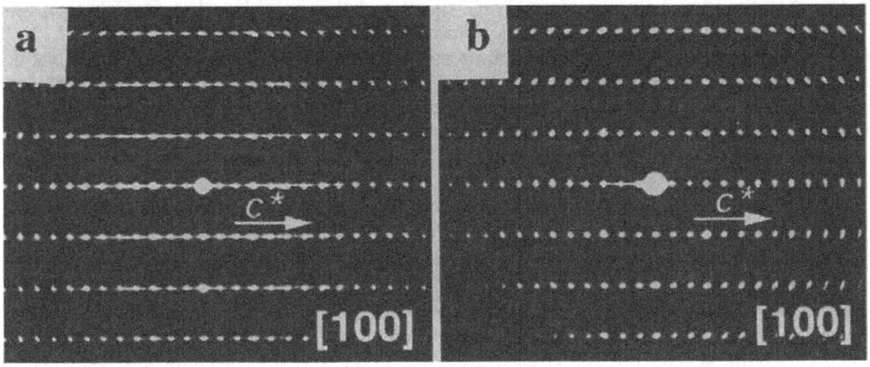

Fig. 2 [100] electron diffraction patterns of the F2 phase (a) and the F1 phase (b).

4.2. ELECTRON DIFFRACTION

The [001]* diffraction pattern of fluorinated YBCO shows a tetragonal symmetry with a = 3.86 Å. The [$hk0$]* sections however show a different aspect: two different patterns are observed. Some of the patterns (fig. 2b) can be indexed in a tetragonal cell with $a = 3.869$ Å and $c \approx 11.66$ Å. The sharp reflections as well as the absence of streaks reveal a perfectly crystalline structure. We call this phase F1. The [100]* zone of fig. 2a on the other hand exhibits pronounced diffuse streaks along the c^* axis. A detailed examination of such patterns indicated the periodic presence along c* of extra intensity maxima. These reflections are compatible with a new unit cell with lattice parameters $a = 3.869$ Å and $c \approx 13$ Å. This phase we will call the F2-phase. It is clear however that, contrary to the F1 phase, the F2 phase does contain a large number of planar defects, inducing streaking of the reflections along c* [17].

All diffraction patterns of the 247 fluorinated compound can be indexed in an A-centered orthorhombic unit cell with a=3.86Å, b=3.88Å and c=51.9Å. Both [010]* and [100]* patterns exhibit diffuse streaking along the c^* axis. Some ED patterns along the [100]* axis also show additional weak streaks parallel to the c^* direction at the position $h, k+1/2, l$. The [001]* pattern for such areas also exhibits $0,k+1/2,0$ superstructure reflections, resulting in a doubling of the b -parameter [18].

4.3. EDX ANALYSIS

The fluorine content of the individual crystallites was investigated by EDX analysis inside the electron microscope and related to the crystal structure (F1 or F2). The analysis revealed the presence of fluorine in all crystallites, F1 as well as F2. The fluorine distribution is not homogenous and may be deviate strongly for different areas of one crystallite. A quantitative analysis of the fluorine content however is not straightforward, because of its low Z-number. We have used here the F:O intensity peaks ratio as a qualitative estimation for the fluorine content. This ratio for F1 cystallites is systematically lower than that for crystallites with show F2 type ED patterns. Similar results hold for the 247 compound.

4.4. HIGH RESOLUTION ELECTRON MICROSCOPY STUDY

[100] HREM images are able to stress the difference between the F1 and the F2 phase. The F1 phase shows a contrast very similar to that of the undoped YBCO. The F2 phase, with an enlarged c-parameter, exist as an intergrowth within the F1 matrix (Fig. 3) or occurs as an isolated defect with a limited extension.. Using the F1 phase as an internal standard, we obtain a c-parameter of 13.0-13.2 Å for the F2 phase. In the F2 phase, the distance between the Cu1-O layer and the neighbouring BaO layers has drastically increased, pointing towards a fluorine uptake within the Cu1 plane.

In order to determine the stacking sequence for this new structure we performed image simulations for different models based on partially and fully fluorinated structures. For the F2-phase we assumed that in the Cu(1)-plane all oxygen has been replaced by fluorine, leading to a structural formula $YBa_2Cu_3O_6F_2$. The atomic coordinates were deduced according to the increased unit cell, taking into account a change of the Cu1-O(apical) separation due to a Jahn-Teller effect. The Cu2-Cu2 distance was kept unaltered with respect to the undoped YBCO. A structural model for the new $YBa_2Cu_3O_6F_2$ compound is shown in Fig. 4. Calculated images for different thicknesses and defocus values show a good agreement with experimental images.

Direct images indicate a highly interweaving of the F1 and F2 phase. The F2 phase never exists over large areas but mainly occurs as bands within an F1 matrix. These bands are more or less parallel to the (001) planes but there is no rigid interface. The extra wide separation between the Cu1-O layer and the surrounding BaO layers makes that the atom planes are no longer straight but become wavy when a "pancake" of fluorine is intercalated.

Within a well ordered 247 matrix, containing a high F-content, numerous pancake type defects were observed; as indicated by arrows in Fig. 5. The contrast inside the defects significantly differs from the matrix contrast: the dots imaging Ba atoms become particularly bright. This contrast variation is accompanied by a significant increase of the distance between the Cu1-layer and the neighboring Ba-O layers in the block with the 123-type structure. As a result the local value of the *c*-parameter increases to 53.5Å. At the same time the brightness of Cu1 dots becomes similar to that of copper atoms neighbouring the Y-layer. This can be understood by the formation of a similar equatorial environment for these copper atoms.

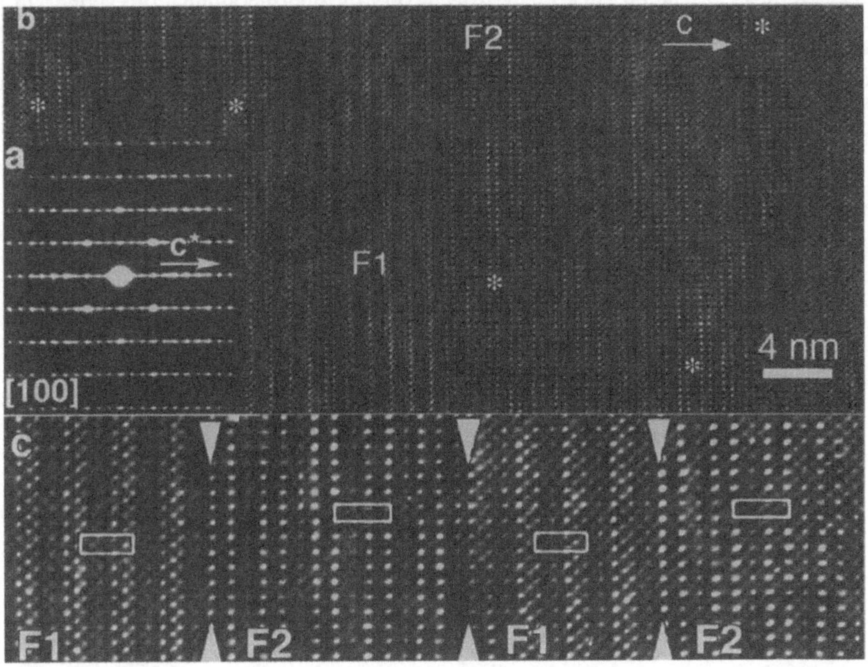

Fig. 3 .HREM image along [100], showing alternation of F1 and F2 phases. Corresponding unit cells are marked.

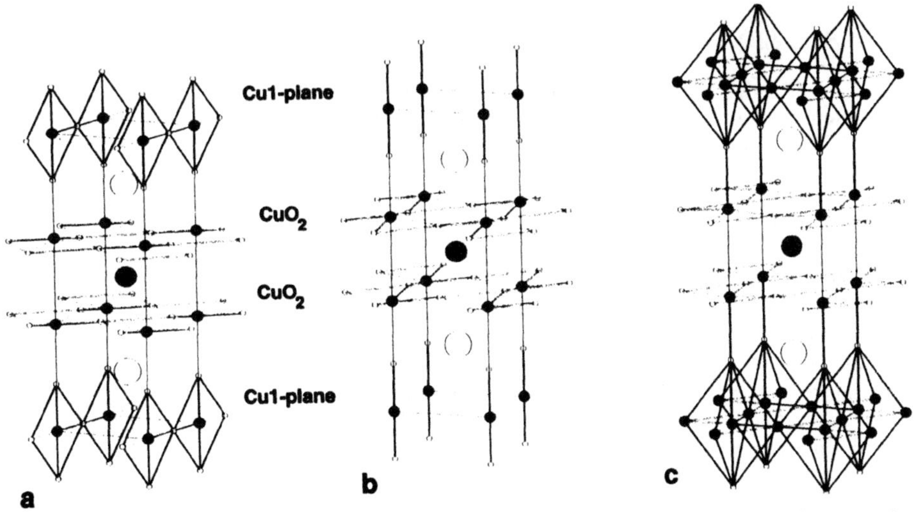

Fig. 4. Crystal structures of orthorhombic (a) and tetragonal (b) YBCO and fully fluorinated $YBa_2Cu_3O_6F_2$ (c).

The variation of the c-parameter along the a-axis appears to be related to the occupation of anion positions within the Cu1 layers by fluorine atoms. We assume that this leads to an elongation of the apical Cu1-O distances followed by a shift of the Ba atoms away from the Cu1 plane. A model based on a unit cell with c=53.5Å and full occupation of the anion sites in the Cu1 layers by fluorine atoms corresponds to a composition $Y_2Ba_4Cu_7O_{14}F_2$.. The comparison of the images calculated for different thickness and defocus values with the experimental ones shows that a model with partial occupation of the anion sites in the Cu1 layers corresponds to the perfect structure while the model with a full occupation is in good agreement with the specific contrast observed in the local defects of fig. 5.

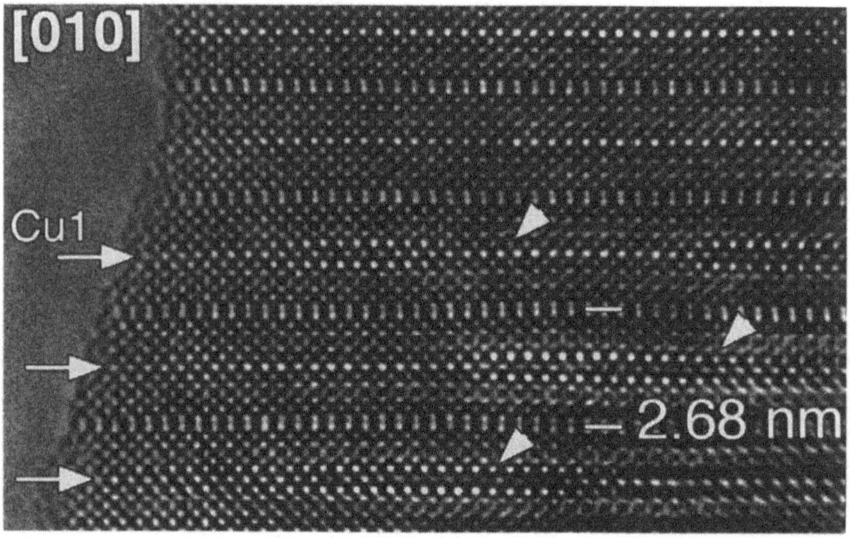

Fig.5 [010] HREM image showing the appearance of a fully fluorinated 247 phase as defects (marked by arrows) in the Cu1 layers.

5. DISCUSSION

The insertion of fluorene atoms into the YBCO structure may occur at two different positions. The most probable is an occupation of the positions in the *xy0* plain (Cu1-layer). Up to 1 fluorene atom per unit cell (F1-phase) may either occupy one of the two empty positions in the Cu1-plane and form a square Cu1 coordination polyhedron or may be statistically distributed on the 1/2,0,0 and 0,1/2,0 sites. If the fluorine content is more than one atom per unit cell (F2-phase), the extra atoms will occupy the remaining sites in the Cu1 plane. The presence of two F atoms in this plane results in the formation of an octahedral coordination for the Cu1 atoms. Such arrangement will introduce a distortion of the octahedra due to a Jahn-Teller effect and increase the Cu1-Cu2 separation. One may expect an increase of the Cu1-O distance with 0.5-0.7Å which leads to an increased c-parameter for the F2-phase of 12.8-13.2Å. Such intercalation of F_x ($x > 1$) into the *xy0* plane will also maintain the tetragonal symmetry. The fully fluorination will result in an oxidation of the Cu atoms and keep the superconducting CuO_2 layers unaltered. The structural transformation associated with this fluorination is reproduced in Fig. 4.

The composition of the new F2 phase is close to $YBa_2Cu_3O_6F_2$, which should have an ordered arrangement of fluorine atoms in the Cu-1 plane. HREM images demonstrate without reasonable doubt that fluorine has been successfully introduced into the YBCO structure with the formation of a new superconducting material.

For the $Y_2Ba_4Cu_7O_{14}$ compound, built up alternatively of blocks of 124 and blocks of 123, the situation is very similar. Assuming a constant anion stoichiometry for the 124 block, there are again two possible positions for insertion of fluorine atoms into the 123 blocks. Full occupation of both positions in the Cu1 plane results in the formation of an octahedral coordination for the Cu1 atoms. Also here it will introduce a Jahn-Teller effect and a significant increase of the apical Cu1-O1 separation, which will result in an increase of the *c*-parameter for the 247 phase to about 53.5Å. The fluorine incorporation increases the Cu valence from +2.03 to +2.11 and T_c from 30K to 62K.

The present investigation shows that the results of fluorination of strongly reduced 123 and 247 phases by XeF_2 are quite similar. For both phases the insertion of a fluorine-oxidizer leads to an increase of the formal copper valence, followed by an increase of T_c. For both structures the fluorine enters into the anion-deficient Cu1 layers, resulting in the formation of an octahedral arrangement of Cu atoms with a significant elongation of the apical Cu-O distances. However, in contrast to the fully fluorinated tetragonal $YBa_2Cu_3O_6F_2$ phase, the 247 phase remains orthorhombic, even for a high fluorination degree since the double CuO chains are unaffected by the fluorination.

6. TEM OF RAMP TYPE JOSEPHSON JUNCTIONS (JJ).

The ramp-type JJ geometry allows adjustment of the junction properties by changing the barrier thickness and/or the barrier material itself. The choice of the barrier material is of great importance to ensure good quality JJ, where the interface resistance should not dominate over the barrier properties. The substitution of the Cu in the chain layer by Ga, Co and Nd in $PrBa_2Cu_3O_{7-\delta}$ (PrBCO) results in an increase in the bulk resistivities [19-20]. $PrBa_2Cu_{3-x}Ga_xO_{7-\delta}$ (PrBCGaO) as barrier layer is expected to produce junctions with a higher normal state resistance than non-substituted PrBCO.

All layers of the JJ have been deposited by off-axis rf magnetron sputtering on (001) $SrTiO_3$ substrate [21]. The base electrode (DyBCO) and the separating layer (PrBCO) have been deposited *in situ*. The ramps are structured by an Ar ion beam source using a photoresist mask. After stripping the photoresist, the ramp surface is cleaned by ion milling. Without breaking the vacuum the sample is transported into the sputtering chamber to achieve re-oxygenation of the surfaces. Subsequently, the barrier layer (PrBCGaO) and the top electrode (DyBCO) are deposited.

The effect of Ga-substitution in the PrBCO barrier layer on the junction parameters, is an increase in the resistivity up to an order of magnitude, whereas the influence on Jc is negligible. However, this trend is not continued for the higher substitution levels. No increase in resistivity for x = 0.7 and 1.0 compared to x = 0.4 was observed, leading to disappointing I_cR_n products.

The junctions discussed here have a ramp angle of 20° to 28° and a barrier layer thickness of 10 to 30 nm. All layers grow epitaxially with the c-axis perpendicular to the substrate. The HREM image of Fig. 6 shows part of a PrBCGaO barrier layer (x= 0.7) grown on the ion-milled substrate. The barrier layer contains isolated unit cells with c-axis lengths of approximately c_i = 0.8 nm, different from c_{123} (double arrows in fig. 6). They are perfectly intergrown and often occur periodically within the 123 structure. The intergrowth occurs only in the $PrBa_2Cu_{2.3}Ga_{0.7}O_{7-\delta}$ barrier layer, not in the DyBCO bottom or top layer nor in the PrBCO separating layer. This intergrowth is interpreted as the occurrence of a full Ga-plane. A model for the Ga-containing intergrowth layer is based on $PrSrGaCuO_5$ (fig.6). The tetrahedral oxygen coordination of Ga induces shifts of the Ga and O atoms away from the ideal perovskite positions.

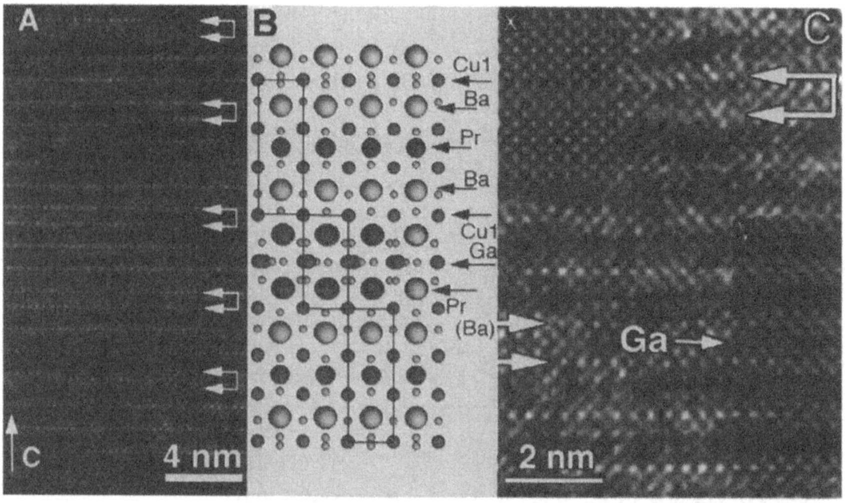

Fig.6 : (a) Overview of a 30 nm $PrBa_2Cu_{2.3}Ga_{0.7}O_{7-\delta}$ barrier layer on an ion-milled $SrTiO_3$ substrate. The non 123 unit cells are indicated are by double arrows and arrange locally in a stacking sequence $[c_{123}]_3[c_i]_1$. (b) A model of the Ga-intergrowth cell. (c) HREM image together with computer simulated image.

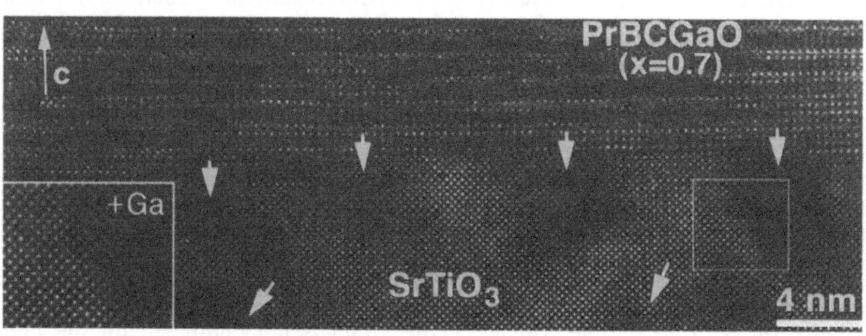

Fig. 7 : [001] HREM micrograph of the interface between $PrBa_2Cu_{2.3}Ga_{0.7}O_{7-\delta}$ and ion-milled $SrTiO_3$. Dark contrast areas in the substrate are indicated by arrows. The inset shows the distorted lattice image at a higher magnification.

Fig. 7 shows the interface region between the barrier layer and the ion-milled SrTiO3 substrate. In the upper most 10 nm of the substrate pronounced black patches, indicated by arrows, are observed. The SrTiO3 lattice in the dark contrast areas is distorted. This specific contrast exists in all parts of the SrTiO3 substrate, which have been exposed to Ar milling to create the ramp-edges,. Several possible origins for this contrast variation such as ion milling artifacts, oxygen depletion or interdiffusion of either photoresist or gallium from the barrier layer have been considered. We conclude that Ga-diffusion in the surface region of the ion-milled $SrTiO_3$ substrate is the origin of the local structural distortions [22]. This can be understood if we assume that lattice

imperfections, such as oxygen vacancies or point defects, created during ion-milling, facilitate the diffusion of gallium in the interface region of the substrate.
In conclusion we may say that the behaviour of the junction parameters at high Ga-substitution levels is better understood due to a TEM microstructural characterisation. The microstructure of $PrBa_2Cu_{3-x}Ga_xO_{7-\delta}$ (x = 0.7 or x = 1) thin barrier layers is clearly different from the one in bulk material. A Ga-rich intergrowth was detected in the barrier layers together with the diffusion of Ga in the ion-milled substrate. Thus two distinct mechanisms of Ga segregation in thin barrier layers are operative and render high Ga-substitution levels ineffective to enhance the normal state resistivity.

ACKNOWLEDGEMENT
This research was performed within the framework of IUAP 4/10.

REFERENCES

1. Ovshinsky S R, Young R T, Allred D D, De Maggio G, and Van der Leeden G A (1987) Superconductivity at 155 K. *Phys. Rev. Lett.* **58** : 2579 -
2. Cirillo K M, Wright J C, Seuntjens J, Daeumling M, and Larbalestier D C (1988) Fluorine oxidation of high-temperature superconductors. *Solid State Comm.* **66** : 1237 - .
3. Vanderah T A, Decker D L, Harris D C, and Chamberland B L (1989) Synthesis and characterization of fluoride-substituted $RBa_2Cu_3O_7$: R=Y, Eu, La *Mat. Res. Bull.* **24:** 121-129 .
4. LaGraff J R, E C Behrman, J A T Taylor, F J Rotella, J D Jorgensen, L Q Wang, and P G Mattocks (1989) Magnetic and structure properties of $YBa_2Cu_3O_{7-x}F_y$ as prepared by a NF_3 gas anneal. *Phys.Rev. B* **39** : 347 -
5. Perrin C, Dinia A, Peña O, Sergent M, Burlet P, and Rossat-Mignod J (1989) Fluorination of $YBa_2Cu_3O_{6.7}$: powder neutron diffraction determination of fluorine sites and their influence on the superconducting properties. *Solid State Comm.* **76** : 401 -
6. C. Perrin, O. Peña, M. Sergent, P. Christensen, G. Fonteneau, J. Lucas, *Supercond. Sci. Technol.* **2** (1989) 35
7. MacManus J J L, Fray D J, and Evatts J E (1991) Fluorination of Y1Ba2Cu3O7-x by a solid state electrochemical method. *Physica C* **184** : 172 -184.
8. Sacuto A, Julien C, Shchukin V A, Perrin C, and Mokhtari M (1995) Charge transfer in $YBa_2Cu_3O_6$ doped with fluorine: Infrared reflectance and Raman scattering studies. *Phys. Rev. B* **52** : 7619 -
9. E. I. Ardashnikova, S. V. Lubarsky, D. I. Denisenko, R. V. Shpanchenko, E. V. Antipov and G. Van Tendeloo, *Physica C* **253**, 59 (1995).
10. J. Karpinski , S. Rusiecki, B. Bucher, E. Kaldis, E. Jilek, Physika C 161 (1989) 618
11. J.-Y. Genoud, T. Graf, G. Triscone, A. Junod, J. Muller, Physica C 192 (1992) 137
12. H. Shwer, E. Kaldis, J. Karpinski, C. Rossel, Physica C 211 (1993) 165
13. A. Hewat P. Fisher, E. Kaldis, J. Karpinski, S. Rusiecki, E. Jilek, Physica C 167 (1990) 579
14. P. Bordet, C. Chaillout, J. Chevanas, J. Hodeau, M. Marezio, J. Karpinski, E. Kaldis, Nature 334 (1988) 596
15. T. Krekels, G. Van Tendeloo, S. Amelinckx, J. Karpinski, S. Rusieski, E. Kaldis, E. Jilek, Physica C 178 (1991) 383
16. R.V. Shpanchenko, M.G. Rozova, A.M. Abakumov, E.I. Ardashnikova, M.L. Kovba, S.N. Putilin, E.V. Antipov, O.I. Lebedev, G. Van Tendeloo, Physica C 280 (1997) 272
17. G. Van Tendeloo, O. Lebedev, R. Shpanchenko, E. Antipov, J. Electron Microscopy 1 (1997) 23
18. A.M.Abakumov, M.G.Rozova, R.V.Shpanchenko, M.L.Kovba, S.N.Putilin, E.V.Antipov, O.I. Lebedev, G. Van Tendeloo, E.M.Kopnin, J.Karpinski, Physica C (1998) in the press.
19. Y. Xu, W. Guan, *Physica C* **206**, 59 (1993).
20. G. Koren, E. Polturak, G. M. Reisner, B. Fisher, L. Patlagan, *Phys ica C* **225**, 21 (1994).
21. M. A. J. Verhoeven, G. J. Gerritsma, H. Rogalla, A. A. Golubov, *IEEE Trans. Appl. Supercond.* **5**, 2095 (1995).
22. K. Verbist, O.I. Lebedev, G. Van Tendeloo, M.A.J. Verhoeven, A.J.H.M. Rijnders, D.H.A. Blank and H. Rogalla. Appl. Phys. Lett.**70**, (1997)1167-1169.

HIGH PRESSURE CRYSTAL GROWTH AND PROPERTIES OF Hg-SUPERCONDUCTORS AND QUASI-ONE-DIMENSIONAL CUPRATES

J.KARPINSKI, H.SCHWER, R.MOLINSKI, G.I.MEIJER, E.KOPNIN, M.ANGST, S.BÜCHI, *Laboratorium für Festkörperphysik ETH 8093-Zürich*, J.HOFER, *Physik-Instytut der Universität Zürich*
V.ALYOSHIN, *Chemical Department, Moscow State University*,
R.PUZNIAK, A.WISNIEWSKI, *Institute of Physics, PAS Warsaw*

1. Introduction

Various aspects of application of high gas pressure to the growth of single crystals and polycrystalline samples will be presented. Hydrostatic pressure can change properties of materials during crystallization or annealing due to: a) structural transitions at high pressure, b) reduced evaporation of volatile components, c) enhanced pressure activity (fugacity) of a gaseous component of the system [1,2].
Examples of a) are the spin ladder $Sr_{n-1}Cu_{n+1}O_{2n}$ or infinite layer $Ca_xA_{1-x}CuO_2$ compounds. Examples of b) are $HgBa_2Ca_{n-1}Cu_nO_{2n+2+\delta}$ (Hg-12(n-1)n) compounds which melt incongruently and have high decomposition pressures of HgO, Hg, and O_2. Examples of c) are $A_{1-x}CuO_2$ (where A=Sr,Ca,Ba) infinite chain compounds.
Hg-1223 compounds exhibit the highest T_c, but magnetic flux-pinning properties of these materials are not as good as of some other high-T_c superconductors (e.g. $YBa_2Cu_3O_{7-x}$). Several authors reported recently significant improvement of the flux pinning when Hg is partially replaced by Re in ceramic samples [3]. However, due to high anisotropy of HTSC and additional phases in ceramic samples only measurements on single crystals are reliable for determination true intrinsic properties of these compounds.
By application of a high O_2 pressure during synthesis it was possible to modify CuO_2 planes and obtain quasi-one-dimensional $Sr_{0.73}CuO_2$ [4,5], $Ca_{0.83}CuO_2$ [6] and $Ba_{0.67}CuO_2$ compounds containing only CuO_2 infinite chains. We obtained also single crystals of the three leg ladder $(Sr,Ca)_4Cu_6O_{10}$ [7], as well as of the infinite layer $(Sr,Ca)CuO_2$ compounds [1,2].

2. Crystal growth of $HgBa_2Ca_{n-1}Cu_nO_{2n+2+\delta}$

In Fig.1, curve (a) shows the partial pressures of Hg and O_2 versus temperature in the closed crucible containing Hg-1201 sample. When the temperature increases above the melting point, partial pressures can reach values corresponding to those of the volatile chemical component of the Hg-O system. Curves (c), (d) and (e) in Fig.1 present the temperature dependent partial pressures caused by decomposition of HgO.

G. Van Tendeloo et al. (eds.), High-Temperature Superconductors and Novel Inorganic Materials, 21–26.

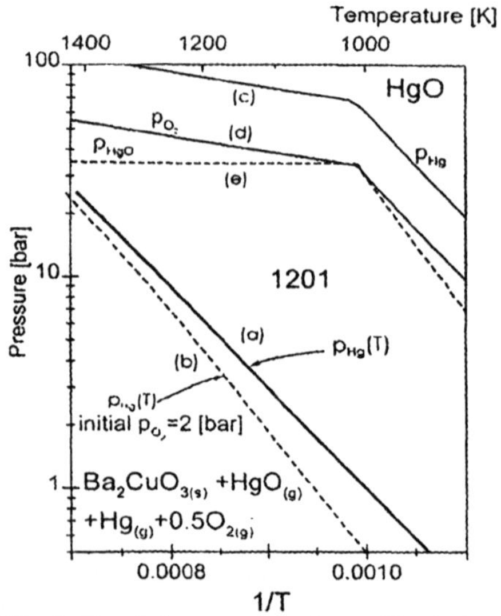

Since the total pressure of the volatile components: P_{O_2}, P_{Hg} and P_{HgO} exceeds 100 bar encapsulation of the sample with a high hydrostatic inert gas pressure was used to prevent decomposition. All experiments were performed at Ar pressures of 10-15 kbar. Due to the high density of Ar gas at this pressure ($>1 g/cm^3$), evaporation of Hg was strongly supressed. As a flux we used $BaCuO_2$-CuO-Ag_2O mixture or PbO. PbO flux leads to a partial substitution of Pb for Hg. The crystallization temperature is 930°C<T<1070°C. We obtained crystals of Hg-12(n-1)n for n=1-6.

Fig.1. Partial pressures above Hg-1201: (a) closed ampoule with Hg-1201 sample, (b) the same as (a) with initial P_{O_2}=2 bar, (c),(d),(e): partial pressures created in a closed crucible of a free volume 3.2 cm^3 by a decomposition of 0.87g HgO. Thermodynamical data are from Ref.[8].

3. Influence of Re- and oxygen doping on the irreversibility line of Hg-1223

Hg-1223 exhibit the highest Tc (135 K) achieved up to now, but the flux-pinning properties of these materials are worse than for some other high-T_c superconductors (e.g. $YBa_2Cu_3O_{6+x}$). Chemical substitutions can enhance the effectiveness of the pinning through reducing anisotropy by decreasing of the blocking layer thickness or creating points defects acting as pinning centers. Rhenium can substitute Hg up to the maximum concentration of 25%. It is coordinated octahedrally by oxygen [9] and forms short bonds leading to a decrease of the lattice parameters (Fig.6). Figure 2 shows the temperature dependence of the irreversibility field $H_{irr}(T)$ for H||c for three crystals, the first underdoped Hg-1223 with T_c=120K, the second almost optimally

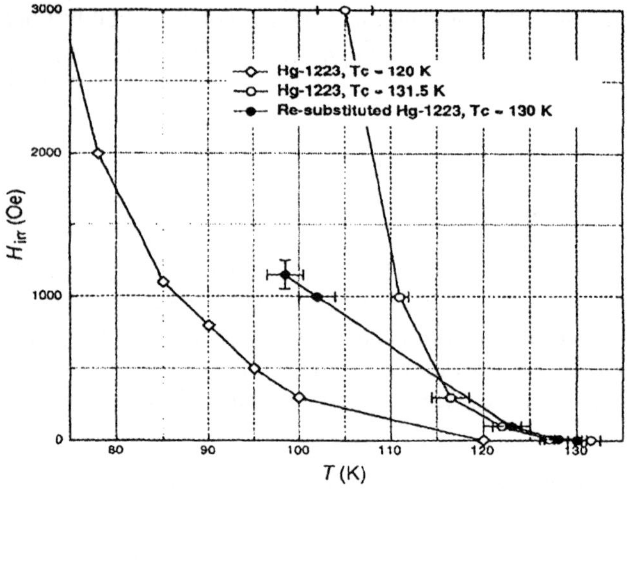

Fig.2. Comparison of the irreversibility lines for Hg-1223 single crystals with various oxygen content and Re-doping.

doped Hg-1223 with T_c=131K and the third Hg-1223 doped with 20%Re (T_c=130K). The irreversibility field of the Re substituted HgRe-1223 compound is lower than the one of optimally doped Hg-1223. It can also be seen clearly that the oxygen doping level has a much greater impact on the position of the irreversibility line than the Re substitution. Thus, the possible metallization of the blocking layer by Re substitution alone does not seem to extend the irreversibility region significantly. Rather, a shortening of the blocking layer thickness (e.g. by substitution of Ba by Sr) seems to achieve this goal

4. Doping dependence of the structure and superconducting parameters in $HgBa_2CuO_{4+\delta}$ single crystals

$HgBa_2CuO_{4+\delta}$ crystals were grown from PbO flux or from a stoichiometric melt. Undoped crystals have an almost perfect structure with a full occupation (97%) of the Hg sites and a maximum T_c=96 K (Fig.3). PbO flux leads to Pb substitution for Hg up to 10%. X-ray investigations of single crystal show, that oxygen and mercury atoms are shifted from their ideal positions and oxygen atoms are distributed in a ring-like configuration around the (1/2,1/2,0) sites [2]. This is due to the formation of short Pb-O bonds (2.1 A). Pb doping has consequences for the physical properties: decreases T_c to 81 K and changes the pinning properties of the crystals. Torque measurements indicate a strong fish-tail effect for Pb-doped crystal.

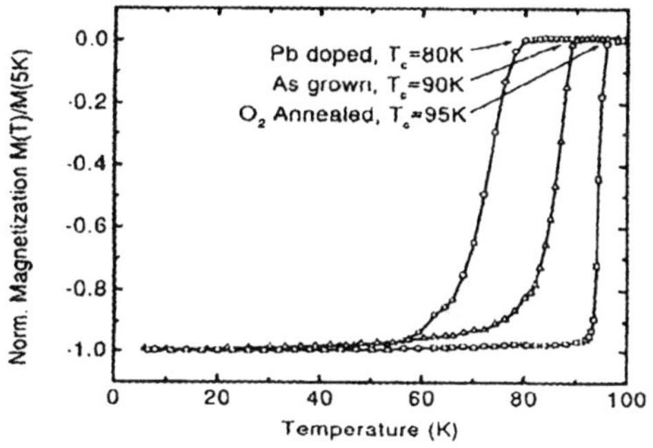

Fig.3. DC magnetization of undoped and Pb doped Hg-1201 single crystals (ZF-cooling).

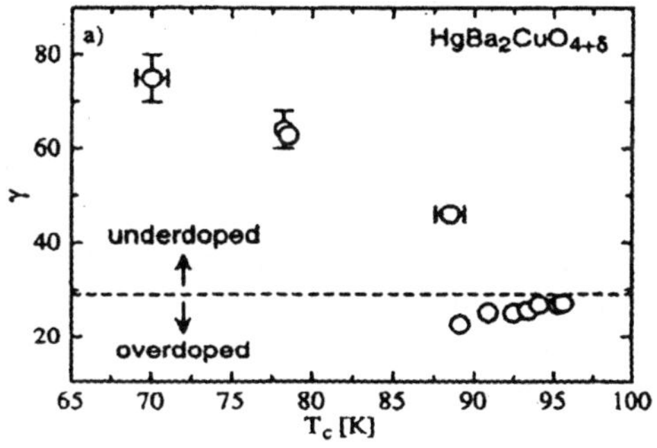

Fig.4. Anisotropy of Hg-1201 single crystals as a function of critical temperature.

We have investigated the influence of oxygen content on superconducting parameters of the single crystals [10]: the anisotropy γ (Fig.4), the penetration depth λ_{ab} and the coherence length ξ_{ab}. The anisotropy γ increases monotonically with decreasing δ, suggesting a smooth 3D-2D crossover in the superconducting properties on reducing of the oxygen content. The value of γ=23 for overdoped crystals (T_c=89K), and γ=27 for optimally doped crystals (T_c=95.6K), increases to γ=72 for underdoped crystals (T_c=79K). The in-plane penetration depth λ_{ab} shows a minimum at optimal doping.

5. New structures of single crystals of the Hg-Re-Ba-Ca-Cu-O system

Single crystals of two new types of superstructures were obtained upon Re doping and analyzed with X-ray single crystal diffraction.

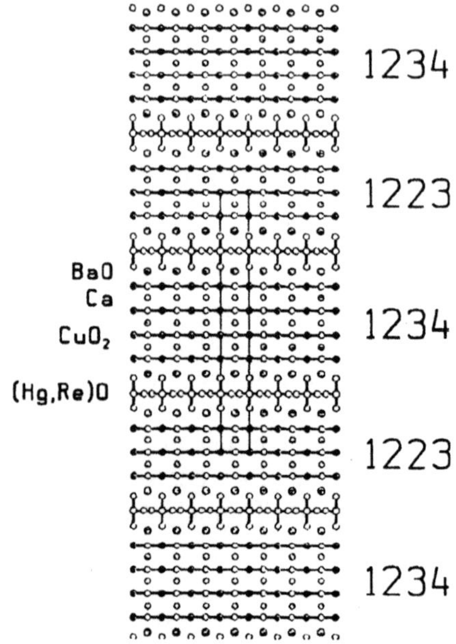

Fig.5. Structure of new $(Hg,Re)_2Ba_4Ca_5Cu_7O_{18+\delta}$ compound.

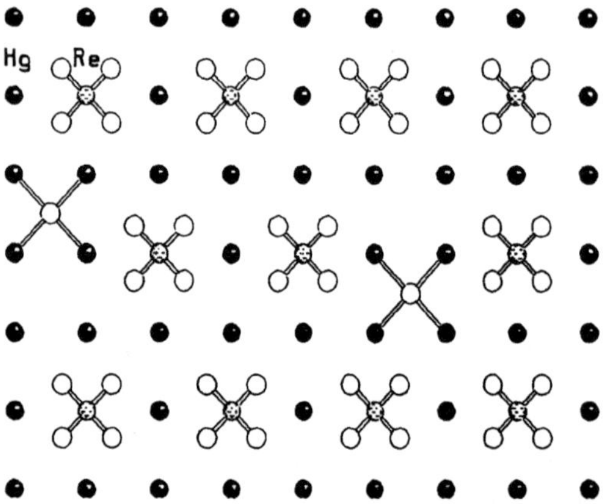

Fig.6. (Hg,Re)O layer in $Hg_xRe_{1-x}Ba_2Ca_{n-1}Cu_nO_{2n+2+\delta}$ single crystals.

Rhenium is coordinated octahedrally in Hg-based cuprates and forms short Re - O bonds (< 2 Å). It substitutes mercury and therefore lattice parameters are decreased upon Re doping. The maximum substitution of Re for Hg is 25 % because higher Re concentrations would cause short O - O distances which are not possible.

5.1. $Hg_xRe_{1-x}Ba_2Ca_5Cu_6O_{14+\delta}$ (1256)

The single crystals of the six layer $Hg_xRe_{1-x}Ba_2Ca_5Cu_6O_{14+\delta}$ (HgRe-1256) compound have transition temperatures up to T_c = 98 K. Rhenium is introduced at the Hg site of the structure and coordinated octahedrally by oxygen atoms in agreement with studies of Chmaissem et al. [9]. Re - O bondlengths (1.86 Å) in the basal z = 0 plane are much shorter than the corresponding Hg - O distances (2.725 Å). Hg and Re atoms cannot be distinguished with X-rays, but oxygen atoms bonded to Re can easily be located in the Fourier maps. The oxygen content in the basal plane was 0.88 at (x, x, 0) bonded to 0.22 Re, and 0.12 at (½, ½, 0) connected with Hg. Strong difference electron densities (ΔF) at (0, 0, 0.126), (½, ½, 0.021), and (½, ½, 0.231) are attributed to the presence of stacking faults [2]. Their number increases with the number of CuO_2 layers in the structure and is about 6 % in the case of (HgRe)1256. In contrast to undoped members of the 12(n-1)n family which show a strong Hg deficiency, the total amount of Hg + Re is 0.97 in these crystals leading to a composition of $Hg_{0.75}Re_{0.22}Ba_2Ca_5Cu_6O_{15}$.

5.2. $(Hg,Re)_2Ba_4Ca_5Cu_7O_{18+\delta}$ (2457)

HgRe-2457 is a new structure in the Hg-Re-Ba-Ca-Cu-O system, based on the Hg-12(n-1)n family. It has a supercell period of c = 34.377 Å and consists of an alternating stacking of 1223 and 1234 units in c-direction (Fig.5). Until now, intergrowth structures in the Hg based series have been only observed in a small submicron scale or as a defect [11,12]. X-ray precession photographs of single crystals show that such an intergrowth structure may form a real, bulk phase. The onset of T_c is at 93K.
Results of single crystal structure analysis show that there is no excess oxygen at (½, ½, z) in the (Hg,Re)O plane. This is not surprising because Hg is substituted to the maximum level of 25 % by ReO_4 and there is no more space left for additional excess oxygen. Based on the ideal 2457 model strong difference electron density peaks were identified at (0, 0, 0.32), (½, ½, 0.24), and (½, ½, 0.40). This feature can be explained by a disturbance of the stacking sequence by a double 3 or 4 layer unit, like ...3-4-3-4-3-4-4-3-4-... ; a further ΔF peak at (0, 0, 0.193), is due to triple 3 or 4 layer units like ...3-4-3-4-4-4-3-4-.... Again, the sum of Hg and Re atoms is close to the ideal value like in HgRe-1256, leading to a crystal composition of $Hg_{1.49}Re_{0.5}Ba_4Ca_5Cu_7O_{20}$. The observation of this rather complex 3-4 layer intergrowth structure makes it plausible, that at least one simpler 2-3 layer intergrowth phase exists.

6. Magnetic properties of quasi-one-dimensional $Sr_{0.73}CuO_2$, $Ca_{0.83}CuO_2$ and $Ba_{0.67}CuO_2$

Magnetic properties of low-dimensional materials have attracted much attention in the field of solid state physics. This is because of the wide variety of interesting effects observed in solids with reduced dimensionality. Spin-Peierls transition and dimerisation in spin 1/2 Heisenberg chains and Haldane spin gaps in integer spin chains are a few examples. Also the intermediate region is nowadays a subject of a debate. By linking together 1D spin 1/2 chains one can construct the so-called spin ladder compounds. For these compounds, with an even number of legs superconductivity was predicted when the ladders become lightly doped. Superconductivity is found in such a compound containing the two leg spin ladder phase and 1D Cu-O chains. We investigate magnetic properties of the new low-dimensional Cu-O spin systems which have been synthesized at high oxygen pressure [4,5,6]. These materials, $Sr_{0.73}CuO_2$, $Ca_{0.83}CuO_2$ and $Ba_{0.67}CuO_2$ consist of 1D Cu-O spin chains separated by Sr, Ca or Ba planes and are incommensurate in chain direction.(Fig.7) The Cu atoms in the chain are connected through two O sites by about 90 degree bonds. Our 1D compounds are insulating and characterized by anti-ferromagnetic spin 1/2 Heisenberg chains. In the low temperature limit we observe a broad maximum in the susceptibility data which might originate from a dimerization gap in the spin chain. Surprisingly, these quasi-1D materials exhibits magnetic ordering at $T \leq 12$ K despite their substantial hole doping. Weak-ferromagnetic ordering is found for the Sr and antiferromagnetic for Ca and Ba compounds (Fig.8). We have performed elastic scattering neutron measurements on $Ca_{0.83}CuO_2$ and $Sr_{0.73}CuO_2$ at the SINQ neutron source (PSI Villigen) which reveal that the ordering is of long range character. This finding makes $Ca_{0.83}CuO_2$ and $Sr_{0.73}CuO_2$ the first highly hole doped materials with long range order.

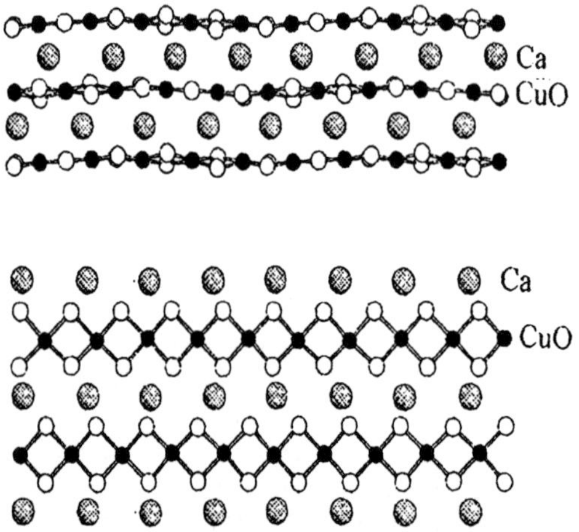

Fig.7. Structure of the infinite-chain $A_{1-x}CuO_2$ compounds (A=Sr, Ca, Ba).

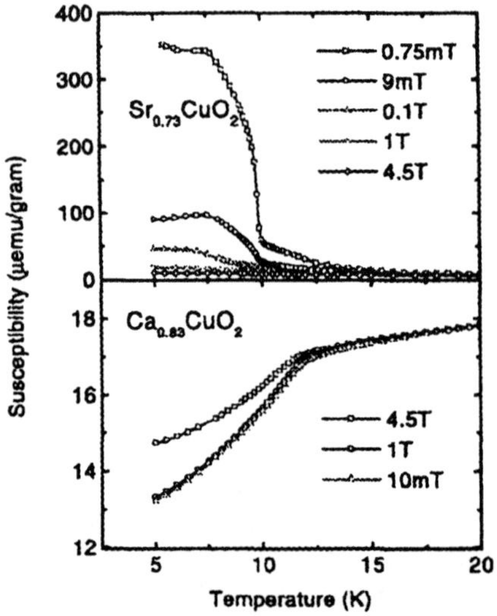

Fig.8. Susceptibility of $A_{1-x}CuO_2$ for A = Sr and Ca. $Sr_{0.73}CuO_2$ shows weak ferromagnetic transition at T=10K, $Ca_{0.83}CuO_2$ antiferromagnetic at T=12K

Acknowledgments: This work was supported by the Swiss National Science Foundation and Polish Government Agency KBN (Contract No.722P039509).

References

[1] J.Karpinski et al., in "Studies of High Temperature Superconductors" Vol. 24, Nova Science Publ., ed. A. Narlikar 1997, 165.
[2] H.Schwer, J.Karpinski, ibid, 49.
[3] J.Shimoyama et al., Physica C 235-240, (1994), 2795.
[4] J.Karpinski, H.Schwer, G.I.Meijer, K.Conder, E.M.Kopnin, Physica C 274 (97) 99.
[5] A.Shengelaya et al. Phys. Rev. Letters, in press.
[6] G.I.Meijer et al., Europhysics Letters. in press.
[7] H.Schwer et al., J.Solid State Chem. 134, (1997) 427-430
[8] V.A.Alyoshin, D.A.Mikhailova, E.V.Antipov, Physica C 271 (1996) 197
[9] O. Chmaissem et al., Physica C 292, (1997), 305.
[10] J.Hofer, J.Karpinski, M.Willemin, G.I.Meijer, E.Kopnin, R.Molinski, H.Schwer, C.Rossel, H Keller, Physica C 297, (1998), 103.
[11] M.Cantoni, H.-U.Nissen, A.Schilling, H.R.Ott, Physica C 235, (1994), 899.
[12] Y.S.Yao et al., Physica C 282-287, (1997), 897.

POLYMERIZABLE COMPLEX SYNTHESIS OF MULTI-COMPONENT OXIDES INCLUDING HIGH-TEMPERATURE SUPERCONDUCTING OXIDES

M. KAKIHANA* and M. OSADA
Materials and Structures Laboratory, Tokyo Institute of Technology, Nagatsuta 4259, Yokohama 226-8503, Japan
S. TOCHIHARA, H. YASUOKA and H. MAZAKI
Department of Mathematics and Physics, National Defense Academy, 1-10-20 Hashirimizu, Yokosuka 238, Japan

Abstract: The "polymerizable complex (**PC**)" technique (first documented as a patent by Pechini in 1967) offers a convenient means for fabricating complex multi-component oxides. The method is based on the formation of a polymer resin precursor in which various metal ions can be uniformly distributed keeping their initial stoichiometric ratio. The approach allows for the development of multicomponent systems with very complex compositions. Feasibility of the **PC** method with respect to homogeneous distribution of dopants within the lattice was demonstrated on the ceramic synthesis of plycrystalline compounds, $Y_{1-x}Ca_xBa_2Cu_{2.8}Co_{0.2}O_{6+d}$ (x=0-0.3). T_c determined precisely from the complex magnetic susceptibility measurements was systematically increased from 57 K for the x=0 sample up to 89 K for the x=0.3 sample. The restoration of T_c with Ca was interpreted as a consequence of reintroduction of holes on the $Cu(2)O_2$ planes; which was supported from the Raman scattering measurements, where specific modes involving atoms in the $Cu(2)O_2$ layers decrease in intensity with Ca indicating an increased metallic screening in the Ca-rich samples.

1. Introduction

The occurrence of high-T_c superconductivity in the series of cuprate based oxides that have been discovered for the last one decade has stimulated intense research activities worldwide aiming at elucidating the origin of the superconductivity at high temperatures. Current research is directed both to the precise characterization of the fundamental physics behind the superconducting oxides and to the development of new and useful synthetic methods. Clearly the understanding of the unique properties of these compounds, as well as their technological application, depends crucially upon the progress of the synthetic method [1].

The so-called "polymerizable complex (**PC**)" technique [1], that is a subject of the present paper, is known originally as the Pechini patent [2], and the **PC** method has

* To whom all correspondence should be addressed. E-mail: kakihana@rlem.titech.ac.jp

G. Van Tendeloo et al. (eds.), High-Temperature Superconductors and Novel Inorganic Materials, 27–32.

received considerable attention among other solution-based ceramic processing techniques as it offers a more convenient and easier means for fabricating multi-component oxides. The **PC** technique is based on the formation of a polymer resin precursor in which various metal ions can be uniformly distributed keeping their initial stoichiometric ratio. It is probable that low cation mobility in the highly viscous polymer network causes little segregation of metal ions during the pyrolysis. It should be stressed here that the approach allows particularly for the development of multicomponent systems with complex compositions, like high-T_c superconducting oxides [3-8]. Because of this unique character, the **PC** technique has to be considered as one of preferable means in preparing samples of solid solutions, that are subsequently used to study doping effects for which a uniform distribution of the dopant through the lattice is strictly required.

We therefore demonstrate in this paper the advantages of the **PC** method in the synthesis of solid solutions with complicated compositions. A series of solid solutions with the chemical form of $Y_{1-x}Ca_xBa_2Cu_{2.8}Co_{0.2}O_{6+d}$ (x=0, 0.1, 0.2, 0.3) were then chosen as a focus of the study, as their T_c's are expected to increase systematically with Ca only when dopants (Ca and Co in this case) are homogeneously distributed within the lattice.

2. Experimental

2.1 Synthesis

A flow-chart for the synthesis of our $Y_{1-x}Ca_xBa_2Cu_{2.8}Co_{0.2}O_{6+d}$ samples is summarized in Fig.1. Stoichiometric amounts of $Y_{1.5}(CO_3)_{1.5}\cdot 1.5H_2O$, $CaCO_3$, $BaCO_3$, $CuCO_3\cdot Cu(OH)_2$ and $Co(CO_3)_{1.5}$ were dissolved in a solution containing anhydrous citric acid (CA) and ethylene glycol (EG). The molar ratio of [CA] : [EG] : [Y+Ca+Ba+Cu+Co] was set to 5 : 25 : 1. This solution was then heated at 110-130 °C to promote the ester reaction between CA and EG. As the solution became concentrated, it became highly viscous indicating the formation of a polymeric gel. Noteworthy is that neither colloid nor sol has been formed during the polymerization. The viscous polymeric product was heat-treated at about 350 °C to remove residual solvent and organics, and the resulting powder was used as a precursor for $Y_{1-x}Ca_xBa_2Cu_{2.8}Co_{0.2}O_{6+d}$. The powder precursor was calcined at 940 °C for 18 h in O_2. The powder was then sintered in pellet form at 950 °C in O_2 for 15 h, and the temperature was decreased to 400 °C where the sample was annealed in O_2 for 12 h.

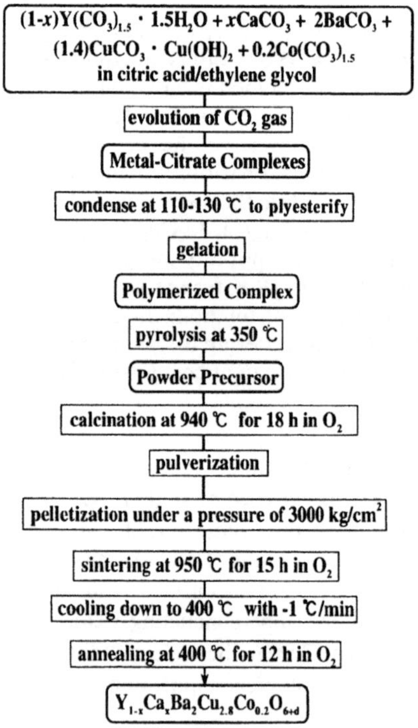

Fig.1 Synthesis of $Y_{1-x}Ca_xBa_2Cu_{2.8}Co_{0.2}O_{6+d}$ by the **PC** method

2.2 Characterization

Phase identification was carried out by use of a standard powder X-ray diffractometer (Mac-Science MXP^{3VA}) using CuKα radiation and a triple Raman spectrometer (Atago-Jobin Yvon T64000) equipped with a CCD (Charge Coupled Device) detector. An Ar laser with 514.5 nm wave length was used for excitation of Raman spectra. The superconducting transition of the $Y_{1-x}Ca_xBa_2Cu_{2.8}Co_{0.2}O_{6+d}$ samples was measured in terms of complex magnetic susceptibility, $\chi=\chi' - i\chi''$ using a Hartshorn bridge. A full description of the measuring system for χ was reported in [9].

3. Results and Discussion

3.1 Sample purity

X-ray diffraction (XRD) of our $Y_{1-x}Ca_xBa_2Cu_{2.8}Co_{0.2}O_{6+d}$ samples with x=0.1, 0.2 and 0.3 has shown no impurities (Fig.2); the detection limit being about 1.0 % by weight for CuO, $BaCuO_2$ and Y_2BaCuO_5 in Y123 based superconductors. The high pure nature of our Y123 sample was confirmed by the corresponding Raman spectra (shown later in Fig.4). It should be stressed that Raman spectroscopy is advantageous over XRD with regard to the detection of non-superconducting impurity phases in high-Tc superconductors. This is because the insulating or semi-conducting oxides have much larger Raman scattering cross-sections than the metallic superconducting oxides (e.g. $BaCuO_2$ has ~30 times larger intensities than those of Y123, the detection limit being about 0.1-0.5 wt% [1, 3]).

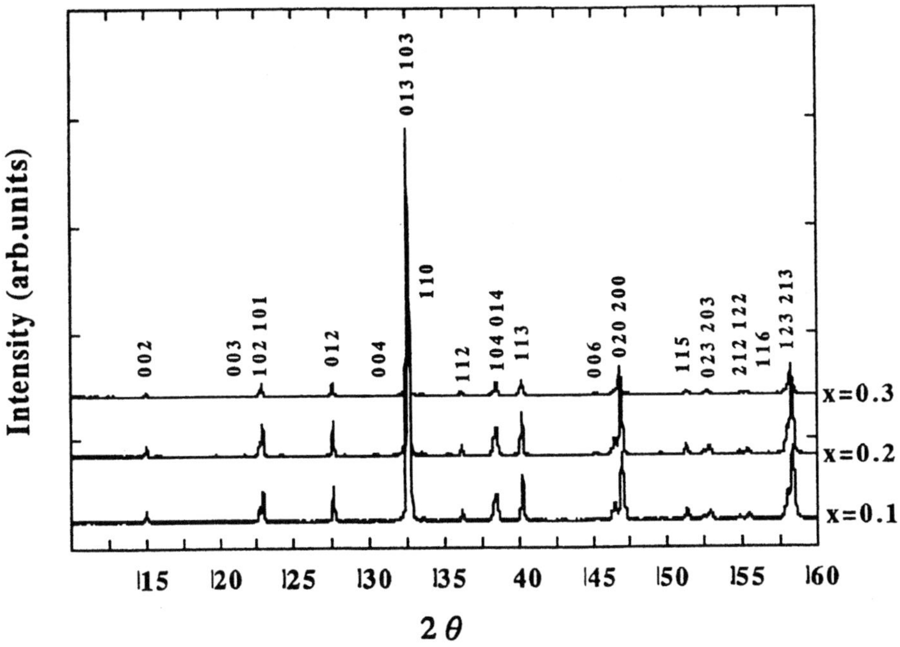

Fig.2 X-ray diffraction patterns of $Y_{1-x}Ca_xBa_2Cu_{2.8}Co_{0.2}O_{6+d}$ polycrystalline samples with x=0.1, 0.2 and 0.3 prepared by the **PC** method.

3.2 Superconductivity

Figure 3 shows a typical result of complex magnetic susceptibility measurements for some of our $Y_{1-x}Ca_xBa_2Cu_{2.8}Co_{0.2}O_{6+d}$ polycrystalline samples with x=0, 0.1 and 0.2, where only the real part, χ', is presented for clarity (the imaginary part, χ'', exhibited a single peak for each sample). The superconducting onset temperature, T_c(onset)s, for x=0, 0.1, 0.2 and 0.3 were, respectively, 57.1, 78.7, 88.6 and 89.1 K. Note that the transition towards the full diamagnetism is rather sharp and quite smooth without any discontinuity during the transition. It is generally acceptable that broadening of the superconducting transition (or transition with discontinuity) reflects the presence of impurity phases and inhomogeneities of the material [1]. The narrow and smooth superconducting transition is therefore attributable to the high purity and compositional uniformity of the sample prepared by our **PC** technique. This is consistent with the conclusion drawn from the XRD and Raman scattering measurements.

It is known that in the system $YBa_2Cu_{3-y}Co_yO_{6+d}$ T_c is almost linearly decreased with Co concentration, and the observed depression of T_c with increasing y can be explained by a reduced number of mobile holes on the $Cu(2)O_2$ planes, which is consistent with reported Hall effect measurements [10] where the number of mobile holes on the $Cu(2)O_2$ conduction planes decreases with increasing Co concentration. The co-substitution of Ca for Y and Co for Cu(1) in $Y_{1-x}Ca_xBa_2Cu_{2.8}Co_{0.2}O_{6+d}$ can introduce additional holes on the $Cu(2)O_2$ planes when compared with the corresponding single doping system $YBa_2Cu_{2.8}Co_{0.2}O_{6+d}$, from which we can expect a significant enhancement of T_c as has been indeed observed (Fig.3). The depression of T_c with Co and the restoration of T_c brought about by extra substitution of Ca for Y come from variations of the hole population between the $Cu(2)O_2$ planes and the $Cu(1)O_y$ chains. An increase in holes on $Cu(2)O_2$ planes for $Y_{1-x}Ca_xBa_2Cu_{2.8}Co_{0.2}O_{6+d}$ with x will be confirmed by Raman scattering, the result of which is shown in the next section.

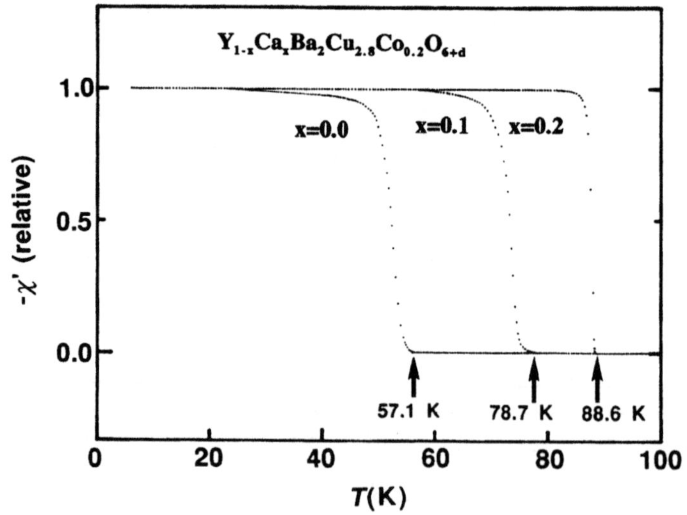

Fig.3 Real (χ') parts of complex magnetic susceptibility of $Y_{1-x}Ca_xBa_2Cu_{2.8}Co_{0.2}O_{6+d}$ with x=0, 0.1 and 0.2 prepared by the **PC** method.

3.3 Raman scattering

In Fig. 4 we show typical Raman spectra of $Y_{1-x}Ca_xBa_2Cu_{2.8}Co_{0.2}O_{7-d}$ with x=0.0, 0.1, 0.2 and 0.3. The incident and scattered lights were polarized along the c axis of the grains. In this configuration A_{1g} modes with displacements along the c axis can be observed. The modes at about 110, 150, 440 and 500 cm^{-1} have previously been assigned to the Ba, Cu(2), O(2)-O(3) in-phase and O(4) vibrational modes, respectively [11], where Cu(2), O(2) and O(3) denote the atoms in the superconducting $Cu(2)O_2$ planes, while O(4) denotes the apical oxygen connecting the $Cu(2)O_2$ plane and the $Cu(1)O_y$ chain. An additional broad feature at 520-620 cm^{-1} has been assigned to the disorder induced Raman scattering related to the Cu(1)-O(1) vibrations which are normally Raman forbidden IR modes [12-14]. Modes involving atoms in the $Cu(2)O_2$ planes are of particular importance, as they are intimately connected with the superconductivity which takes place in the $Cu(2)O_2$ planes. At focus here is that the intensity of the O(2,3) in-phase mode is systematically decreased with increasing Ca content. The effect is most probably direct manifestations of the increase in the number of charge carriers induced by Ca doping. The larger number of holes in the Ca-substituted samples leads to an increased metallic screening within $Cu(2)O_2$ planes, leading to smaller $Cu(2)O_2$ bond polarizabilities and phonon intensities. Similar effects have also been observed for the corresponding vibrations related to the CuO_2 plane in $Bi_2Sr_2Ca_{1-x}Y_xCu_2O_{8+d}$ [15].

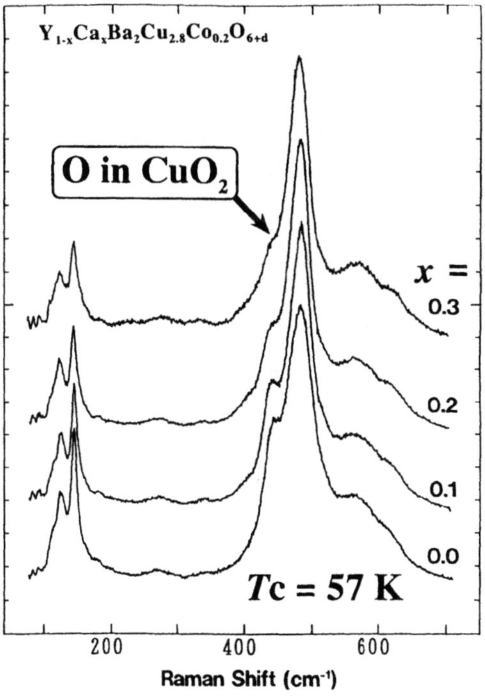

Fig.4 Raman spectra of $Y_{1-x}Ca_xBa_2Cu_{2.8}Co_{0.2}O_{6+d}$ polycrystalline samples with x=0, 0.1, 0.2 and 0.3 prepared by the **PC** method.

4. Conclusion

The **PC** method has been successfully applied to the pure and homogeneous synthesis of solid solutions with the chemical form, $Y_{1-x}Ca_xBa_2Cu_{2.8}Co_{0.2}O_{6+d}$. The observed rise in T_c on going from $YBa_2Cu_{2.8}Co_{0.2}O_{6+d}$ (T_c=57.1 K) to $Y_{0.7}Ca_{0.3}Ba_2Cu_{2.8}Co_{0.2}O_{6+d}$ (T_c=89.1 K) was interpreted as the consequence of reintroduction of holes on the $Cu(2)O_2$ planes. For increasing degree of Ca doping we find that the in-phase oxygen mode within the $Cu(2)O_2$ plane decreases systematically in intensity. We attribute this effect to the increased metallic screening in the $Cu(2)O_2$ planes, that can be expected in the Ca-doped samples with extra holes.

ACKNOWLEDGMENT

I appreciate supports from Dr. H. Fujimori, Mr. Y. Nasuno, Prof. M. Yashima and Prof. M. Yoshimura.

REFERENCES

1. M.Kakihana, *J. Sol-Gel Sci. Tech.* **6**, 7 (1996).
2. M.P.Pechini, *U.S. Patent* No.3, 330, 697, July (1967).
3. M.Kakihana, L.Börjesson, S.Eriksson and P.Svedlindh, *J. Appl. Phys.* **69**, 867 (1991).
4. M.Kakihana, M.Käll, L.Börjesson, H.Mazaki, H.Yasuoka, P.Berastegui, S.Eriksson and L.G. Johansson, *Physica C*, **173**, 377 (1991).
5. P.Berastegui, M.Kakihana, M.Yoshimura, H.Mazaki, H.Yasuoka, L.G. Johansson, S.Eriksson, L.Börjesson and M.Käll, *J. Appl. Phys.* **73**, 2424 (1993).
6. H.Mazaki, M.Kakihana, H.Yasuoka, *Jpn. J. Appl. Phys.* **29**, 38 (1991).
7. M.Kakihana, M.Yoshimura, H.Mazaki, H.Yasuoka and L.Börjesson, *J. Appl. Phys.* **71**, 3904 (1992).
8. H.Mazaki, H.Yasuoka, M.Kakihana, H.Fujimori, M.Yashima and M.Yoshimura, *Physica C*, **246**, 37 (1995).
9. T.Ishida and H.Mazaki, *Phys. Rev. B*, **20**, 131 (1979).
10. J.Clayhold, S.Hagen, Z.Z.Wang, N.P.Ong, J.M.Tarascon and P.Barboux, *Phys. Rev. B*, **39**, 777 (1989).
11. K.F.McCarty, J.Z.Liu, R.N.Shelton and H.B.Radouski, *Phys. Rev. B*, **41**, 8792 (1990).
12. M.Kakihana, L.Börjesson, S.Eriksson, P.Svedlindh and P.Norling, *Phys. Rev. B*, **40**, 6787 (1989).
13. M.Kakihana, S.Eriksson, L.Börjesson, L.G.Johansson, C.Ström and M.Käll, *Phys. Rev. B*, **47**, 5359 (1993).
14. L.Börjesson, L.V.Hong, M.Käll, M.Kakihana and P.Berastegui, *J. Alloys and Compounds*, **195**, 363 (1993).
15. M.Kakihana, M.Osada, M.Käll, L.Börjesson, H.Mazaki, H.Yasuoka, M.Yashima, and M.Yoshimura, *Phys. Rev. B*, **53**, 13797 (1996).

SPRAY PYROLYSIS OF HIGH-TEMPERATURE (SUPER)CONDUCTORS

C.H. CHEN[†], J. SCHOONMAN[†], Y. KAMLAG[†,‡], I. COLBECK[‡]

† *Laboratory for Inorganic Chemistry, Delft University of Technology, Julianalaan 136, 2628 BL Delft, The Netherlands*
‡ *Institute for Environmental Research, Department of Biological Sciences University of Essex, Colchester CO4 3SQ, United Kingdom*

1. Introduction

In recent years, spray pyrolysis has been developed into a powerful tool to synthesize a variety of electroceramic materials including electronic conductors, semiconductors, high T_c superconductors, ionic conductors, and mixed electronic-ionic conductors. The synthesis always starts from aerosol generation of liquid precursors. This aerosol is subsequently directed either into a pyrolysis chamber in order to form powders, or towards a heated substrate in order to form thin films. It bridges gas-phase techniques and liquid-phase techniques, and combines many superb features of the two kinds of techniques having following advantages: (a) simple, cheap equipment, (b) wide choices of precursors, (c) easy composition control, (d) diverse film morphologies possible, and (e) ease of up-scaling. The aerosols can be generated by a variety of sprayers or atomizers, including pneumatic (pressure, two-fluid, nebulizers), ultrasonic, and electrostatic [1]. The aerosols generated by these atomizers differ in droplet size, rate of atomization, and droplet velocity.

Three kinds of spray pyrolysis techniques have been used for the syntheses of electroceramic materials, i.e. Electrostatic Spray Deposition (ESD) for thin film fabrication [2-8], Flame-Assisted Ultrasonic Spray Pyrolysis (FAUSP) for powder production [9] and Thermally- Activated Ultrasonic Spray Pyrolysis (TAUSP) [10,11]

2. Electrostatic Spray Deposition (ESD)

ESD technique was first developed in our laboratory a few years ago for the preparation of thin films of $LiMn_2O_4$ for rechargeable lithium-ion batteries and yttria-stabilized zirconia (YSZ) for solid oxide fuel cells (SOFC) [2,3]. Very recently, it has been further developed and very extensively applied to deposit a wide range of functional ceramic thin films for energy conversion and storage systems, as well as for chemical gas sensors [4-8].

If a liquid is forced to flow through a small metal nozzle which is subjected to an electric field, the liquid will leave the outlet of the nozzle in different forms or modes as a result of different electrohydrodynamic mechanisms. These modes include among others dripping

G. Van Tendeloo et al. (eds.), High-Temperature Superconductors and Novel Inorganic Materials, 33–38.

mode, cone-jet mode, and spindle mode [12,13]. In which mode a spray is operating is dependent on mainly the electric potential applied to the nozzle and on flow rate, electrical conductivity and surface tension of the liquid as well. A spray of a precursor liquid generated electrostatically and used for film deposition is the very base of the ESD technique. For film deposition the cone-jet mode is the most suitable, because it is a continuous mode and the formation of homogeneous fine spray droplets is possible. In this mode, there exits a so-called "Taylor cone" with 49.3° half-angle at the apex of the cone. This cone is extended by a jet which breaks up into spray droplets to generate an aerosol of the precursor liquid. Subsequently, the aerosol is electrohydrodynamically directed to a heated substrate to form a thin film. Depending on mainly the substrate temperature and the composition of the precursor liquid, various film morphlogies can be obtained [5]. A schematic view of a vertical ESD set-up [6] is shown in Fig.1.

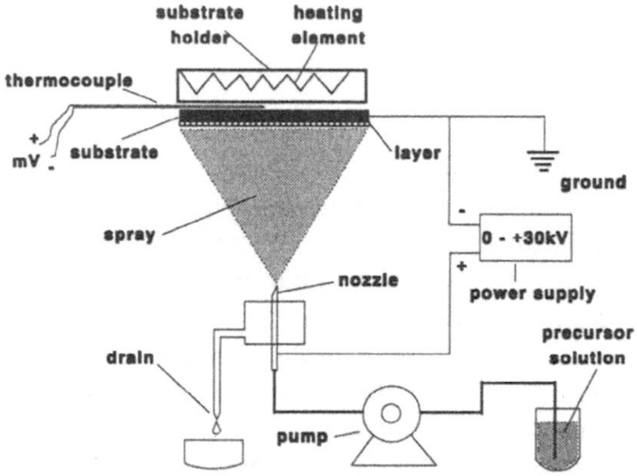

Fig. 1 ESD set-up (vertical configuration) (after [6])

3. Flame-Assisted Ultrasonic Spray Pyrolysis (FAUSP)

Flame spray pyrolysis was developed in the 1980's [14,15]. It operates by injecting the spray of a precursor solution obtained from an aerosol generator into a combustion chamber where the individual droplets are rapidly combusted. With this technique various metal oxide and mixed metal oxide powders can be produced from metal chlorides, sulfates, acetates, and nitrates as precursor. In a conventional flame spray pyrolysis technique, extra fuel gas such as natural gas or hydrogen is introduced in order to generate appropriate high temperatures. However, in our modified flame spray pyrolysis technique, which is referred to as Flame-Assisted Ultrasonic Spray Pyrolysis (FAUSP), flammable alcoholic solutions are used as precursors. Therefore, no extra fuel gas is needed in FAUSP because the spray of the alcoholic solution itself is combusted and a metal oxide powder is thus produced. Many ceramic powders such as lithium cobaltate, titanium dioxide, lead titanate, and tin dioxide have been successfully prepared with this technique in our laboratory. Preliminary results show its versatility to prepare fine powers as well as to deposit films. A schematic representation of a FAUSP set-up is shown in Fig. 2.

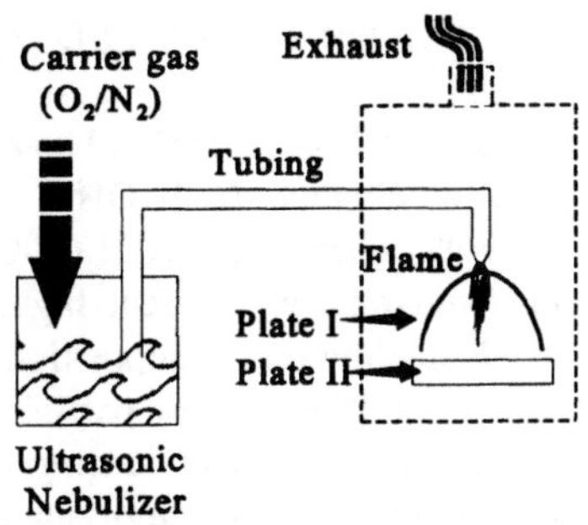

Fig. 2 FAUSP set-up. (Plates I and II are made of stainless steel, acting as powder collectors.)

4. Thermally-Activated Ultrasonic Spray Pyrolysis (TAUSP)

Spray pyrolysis with a furnace as reactor involves chemical reactions between precursor vapors in the gas flow through the furnace. The process involves the atomization of a precursor by an ultrasonic nebulizer (developed at the University of Essex). The droplets are then directed through the furnace by a carrier gas. Inside the furnace the liquid evaporates and intraparticle reactions occur that form the product material which is collected on the cold finger. Thermally activated reactors provide excellent control over temperature and flow, and hence over the product particle characterization. The formation of the particles is well understood. With this technique various materials have been formed like metal oxides, nonoxides ceramics and super conducting materials, the materials produced are only limited by the precursor. The liquid solution can be alcoholic or aqueous which gives this technique an advantage over other spray pyrolysis. A schematic representation of a TAUSP set-up is shown in Fig. 3.

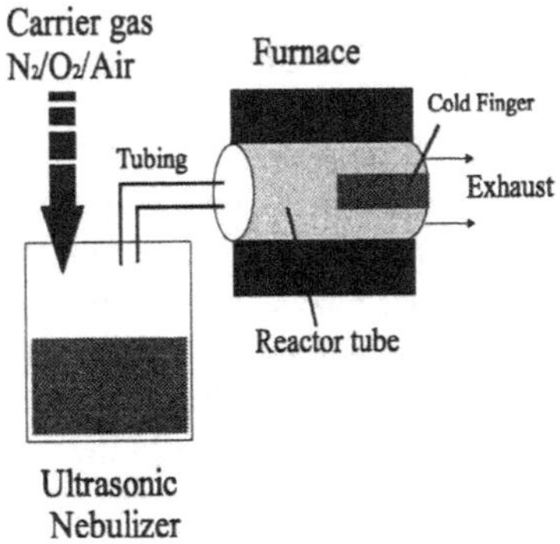

Fig. 3 TAUSP set-up.

5. Examples of applications of ESD, FAUSP and TAUSP

5.1 ELECTRONIC SEMICONDUCTOR TiO_2 BY ESD [8,15]

For the deposition of TiO_2 thin films by ESD, a 0.007 M solution of titanium isopropoxide ($Ti(i\text{-}C_3H_7O)_4$) in 38 vol% ethanol (C_2H_5OH) + 38 vol% butyl carbitol ($C_4H_9OC_2H_4OC_2H_4OH$) + 24 vol% acetic acid (CH_3COOH) was used as precursor. Different morphologies ranging from relatively dense to highly porous including a reticular structure (Fig. 4) were obtained depending on the deposition conditions, mainly substrate temperature and solvent composition. The primary particle size of the films was of the order of a micron for those granular films deposited at temperatures higher than 250°C. X-ray diffraction analysis has confirmed the presence of anatase phase in these films after annealing at temperatures above 500°C.

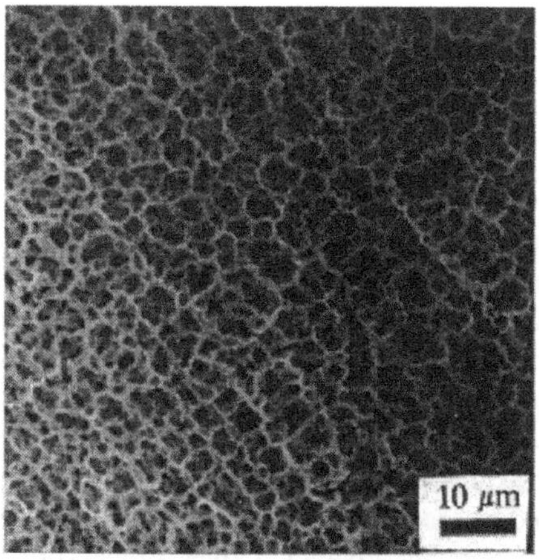

Fig. 4 Reticular TiO_2 film deposited at 240°C from a precursor solution of $Ti(i\text{-}C_3H_7O)_4$.

5.2 IONIC CONDUCTOR ZrO_2 BY FAUSP [9]

To prepare ZrO_2 powders by FAUSP we used 0.2 M zirconium n-propoxide ($Zr(n\text{-}C_3H_7O)_4$) in ethanol as precursor solution. A small quantity of nitric acid was also added to stabilize the solution. The solution was atomized ultrasonically into an aerosol and carried by air through a tube to a glass nozzle. At the outlet of the nozzle the aerosol was ignited and burned to obtain fine powders. The temperature in the flame center was measured to be 850 - 950°C. A typical morphology of these powders is shown in Fig.5. The powder has a narrow particle-size distribution in the micron- or submicron range.

Fig. 5 ZrO_2 powders prepared by FAUSP (after [9]).

5.3 BISMUTH OXIDE SUPERCONDUCTOR BY TAUSP

A high-temperature superconductor powder has been produced by TAUSP under various conditions. An aqueous solution of a nitrate mixture of Bi-Pb-Sr-Ca-Cu was atomized using different carrier gases and temperatures. The aerosol was then dried and decomposed in a furnace and collected on a cold finger. The powder was examined by XRD, SEM and EDX. Unagglomerated spherical particles were obtained at 700°C and 800°C with nitrogen as the carrier gas (Fig.6). At a temperature of 900°C particle agglomeration occurred. The size of the particles was in the range of 100 nm to 1 µm. The powder after TAUSP consists mainly of 2201 phase with calcium separate and a minimum of the 2212 phases at 800°C to 900°C and nitrogen and air as carrier gases. After sintering 16 hours at a temperature of 850°C the 2201 phase with calcium has been transformed to the 2212 phase.

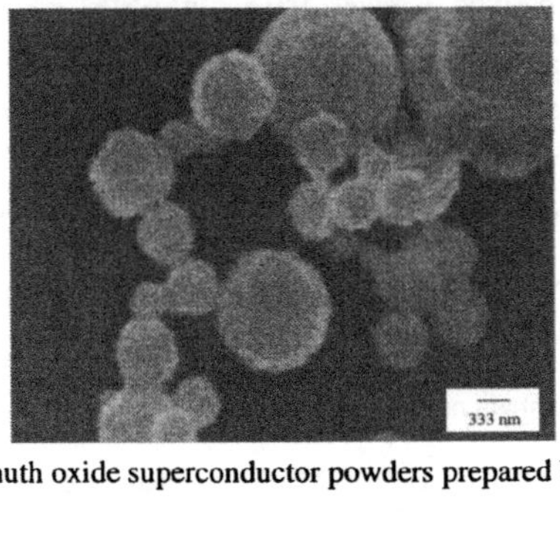

Fig. 6 Bismuth oxide superconductor powders prepared by TAUSP.

6. Summary

We have shown that spray pyrolysis is a powerful tool for the synthesis of electroceramic materials. Three variants of spray pyrolysis techniques, i.e. Electrostatic Spray Deposition (ESD), Flame-Assisted Ultrasonic Spray Pyrolysis (FAUSP) and Thermally-Activated

Ultrasonic Spray Pyrolysis (TAUSP), have been successfully applied to prepare semiconductors such as TiO_2, ionic conductors such as ZrO_2, and bismuth oxide superconductors. The ESD is particularly useful for the fabrication of thin films, while the FAUSP and TAUSP are usually used for the production of fine powders.

Acknowledgements

Mr. Fulong Yuan (Heilongjiang University, P.R. China) is acknowledged for the development of FAUSP. Y. Kamlag is grateful to the European Science Foundation (NANO program) for providing a fellowship which made it possible to establish a co-operation between the two universities. The Foundation for Chemical Research in The Netherlands (SON) under The Netherlands Organization for Scientific Research (NWO) is thanked for financial support of C.h. Chen.

References

1. Messing, G.L., Zhang, S.C., and Jayanthi, G.V. (1993) Ceramic powder synthesis by spray pyrolysis, *J. Am. Ceram. Soc.* **76**, 2707-2726.
2. Van Zomeren, A.A., Kelder, E.M., Marijnissen, J.C.M., and Schoonman, J. (1994) The production of thin films of $LiMn_2O_4$ by electrospraying, *J. Aerosol Sci.* **25**, 1229-1235.
3. Kelder, E.M., Nijs, O.C.J., and Schoonman, J. (1994) Low-temperature synthesis of thin films of YSZ and BaCeO3 using electrostatic spray pyrolysis (ESP), *Solid State Ionics* **68**, 5-7.
4. Chen, C.H., Buysman, A.A.J., Kelder, E.M., and Schoonman, J. (1995) Fabrication of $LiCoO_2$ thin film cathodes for rechargeable lithium battery by electrostatic spray pyrolysis, *Solid State Ionics* **80**, 1-4.
5. Chen, C.H., Kelder, E.M., van der Put, P.J.J.M., and Schoonman, J. (1996) Morphology control of thin films fabricated by electrostatic spray deposition (ESD) technique, *J. Mater. Chem.* **6**, 765-771.
6. Chen, C.H., Kelder, E.M., and Schoonman, J. (1996) Unique porous $LiCoO_2$ thin layers prepared by electrostatic spray deposition, *J. Mater. Sci.* **31**, 5437-5442.
7. Chen, C.H., Nord-Varhaug, K., and Schoonman, J. (1996) Coating of yttria stabilized zirconia (YSZ) thin films on gadolinia doped ceria (GCO) by electrostatic spray deposition (ESD) technique, *J. Mater. Synthesis and Processing* **4**, 189-194.
8. Chen, C.H., Kelder, E.M., and Schoonman, J. (1997) Functional ceramic films with reticular structures prepared by electrostatic spray deposition technique, *J. Electrochem. Soc.* **144**, L289-L291.
9. Yuan, F.L., Chen, C.H., Kelder, E.M., and Schoonman, J. (1998) Spray pyrolysis routes to electroceramic powders and thin films, *Solid State Ionics*, in press.
10. Gurav, A., Kodas, T., Pluym, T., and Xiong, Y. (1994) Aerosol processing of materials, *Aerosol Science and Technology* **19**, 411-452.
11. Gurav, Abhijit S., Kodas, Toivo T., Joutsensaari, J., Kauppinen, E. I., and Zilliacus, R. (1994) Gas-phase particle size distributions and lead loss during spray pyrolysis of (Bi, Pb)-Sr-Ca-Cu-O, *J. Mater. Res.* **10**, 1644-1652.
12. Cloupeau, M. and Prunet-Foch, B. (1994) Electrohydrodynamic spraying functioning modes: a critical review, *J. Aerosol Sci.* **25**, 1021-1036.
13. Grace, J.M. and Marijnissen, J.C.M. (1994) A review of liquid atomization by electrical means, *J. Aerosol Sci.* **25**, 1005-1019.
14. Bickmore, C.R., Waldner, K. F., Treadwell, D. R., and Laine, R. M. (1996) Ultrafine spinel powders by flame spray pyrolysis of a magnesium aluminum double alkoxide, *J. Amer. Ceram. Soc.* **79**, 1419-1423.
15. Helble, J., Moniz,G. A., and Morency, J. R., U. S. Pat. No. 5358696 (1994).

PREPARATION AND CHARACTERISATION OF HTSC POWDERS, SHAPING OF BULK MATERIALS AND THEIR MAGNETIC PROPERTIES

H. ALTENBURG[1], H. DYCK[1], W. JASZCZUK[1], CH. SEEGA[1], C. MAGERKURTH[1], N. MUNSER[1], J. PLEWA[2], M. UELTZEN[2], S. LOZA[1,3], YU.D. TRETYAKOV[3], O. SHLYAKTIN[3], E. KIEFER[4], A. BUEV[5], I.F. KONONYUK[6], V.V. VASHOOK[6]

[1]Fachhochschule Münster - University of Applied Sciences, Stegerwaldstraße 39, D-48565 Steinfurt, Germany
[2]Steinfurter Initiative für Materialforschung e.V. SIMa
[3]Moscow State Universtity, Moscow, Russia
[4]MTD Magnettechnik Deutschland, Dortmund, Germany
[5]Mari University, Yoshkar Ola, Russia
[6]Belarussian Academy of Science, Minsk, Belarus

1. Introduction

The quality of the starting powder is a crucial aspect of the production of superconducting bulk materials. Depending on the application and on the purity, homogeneity, particle size and shape of the powder, different versions of synthesis like ceramic precursor or solution methods were used [1; 2]. If there are great demands on the properties the powder preparation process will be large-scale and costly. This applies to the used superconducting powders of YBCO, NBCO and BSCCO. These powder materials had to be converted into suitable compact forms (e.g. shielding and magnetic levitation elements) for practical usage. Two preparation techniques for bulk materials were mainly applied: first casting and second pressing with following liquid phase sintering or melt texturing. It is necessary to know about the various physical properties of the superconducting monolithic pieces in order to achieve the correct design of the superconducting shields. This work intends to show the application potential of tubular elements made from superconducting granular materials. The aim of this study consists of three parts: firstly to produce superconducting powders and precursors by various methods and the characterisation of them; secondly to shape superconducting hollow cylinders and thirdly to examine their magnetic properties.

2. Powder Preparation

In the case of the *ceramic method* both carbonate ($BaCO_3$, $SrCO_3$, $CaCO_3$) and carbonate free materials ($BaCuO_2$) were applied (TAB. 1). Low carbon content materials were produced by the carbon free method. In the case of $NdBa_2Cu_3O_{7-x}$, the carbon method produces powder with higher critical temperature showing a bigger grain size with

G. Van Tendeloo et al. (eds.), High-Temperature Superconductors and Novel Inorganic Materials, 39–44.

homogeneity at the particle level. This good practical ceramic method yields higher quantities of cuprates with satisfying characteristics, though showing disadvantages like inhomogeneity at the atomic level as well as impurities [1; 5].

The *precursor methods* - reactive mixture process and flame fusion flakes process - were the preferred methods for the preparation of YBCO and NBCO materials (ceramics, cuprates). These precursors reacted and formed a 123 phase at a relatively low temperature (about 850 °C for YBCO), could be sintered rapidly as pure phase and offered high densification. First the reactive mixture of Yttrium ceramics with an approximate stoichiometry for Y:Ba:Cu between the range of 1:2:3 and 21:33:49 were prepared (TAB. 1) and $BaCuO_2$ was used. This compound with a carbon content of 0.8-1.0 % C (low content) was mixed with Y_2O_3 and CuO and homogenised by ball-milling. After a long milling process, this mixture was pre-sintered at 880 °C. The grinding and pre-sintering stages were repeated and so the reactive mixture powder formed a large agglomeration of the resulting compounds $YBa_2Cu_3O_{7-x}$, Y_2BaCuO_5, $BaCuO_2$, and CuO at grain sizes of 10-12 μm. These mixtures were analysed by a Netzsch DTA/TG and showed melting effects at temperatures ranging from 922 to 942 °C [6].

For the reactive mixture of NBCO brown phase ($Nd_4Ba_2Cu_2O_{10}$) was used, however, the molar ratio of Nd:Ba:Cu varied between 9:12:17 and 12:21:31. The mixtures of various small particles of Nd-422 (milled for 0.25 - 25 h) were tested. This compound showed a strong agglomeration of particles after the milling process. In the case of the mixture of Nd compounds the liquid phase formation took place at a temperature range above 1050 °C. Secondly precursor flakes were prepared by a flame fusion process [7]. This newly

TABLE 1. Examples for powder preparation by different methods

Method	Composition / Compounds	C % average	d50 μm	Temperature °C
Ceramic Methods				
- with Carbonate	$YBa_2Cu_3O_{7-x}$	0.175	40-50	850; 950
	$NdBa_2Cu_3O_{7-x}$	0.468	10-30	900; 920; 960
	$Bi_2Sr_2Ca_2Cu_3O_{10+x}$	0.186	40-50	750; 800; 835
	$(Bi;Pb)_2Sr_2Ca_2Cu_3O_{10+x}$	0.014	20-30	750; 840
- without Carbonate	$YBa_2Cu_3O_{7-x}$	0.085	10-20	50; 950
	$NdBa_2Cu_3O_{7-x}$	0.059	10-15	900; 920; 960
Precursor Methods				
- reactive mixture	Y_2O_3+$BaCuO_2$+CuO (*e.g. ratio Y:Ba:Cu, 1:2:3, 9:12:17, 7:11:16)*	0.100	12-15	850; 880
	$Nd_4Ba_2Cu_2O_{10}$+$BaCuO_2$+CuO			
- flakes	Y-Flakes	0.1	1-5	1500
	Nd-Flakes			1500
Solution Methods				
- freeze drying with nitrate	$YBa_2Cu_3O_{7-x}$	0.019	7-10	600; 950
	$NdBa_2Cu_3O_{7-x}$	0.018	5-10	550; 950
- freeze drying with acetate	$YBa_2Cu_3O_{7-x}$	0.011	5-7	600; 950
	$NdBa_2Cu_3O_{7-x}$	0.004	3-5	550; 950
- combustion with citrate	$YBa_2Cu_3O_{7-x}$	0.045	1	900; 920
- sol-gel with citrate	$YBa_2Cu_4O_8$	0.061	2-3	800; 830

developed *flakes precursor process* offers a special possibility of powder preparation. This process is very similar to the reactive mixture process. This reactive mixture was not prepared by mixing various substances but by quenching a crucible-free, overheated melt by means of a special variation of the Verneuil-apparatus (*Fig. 1*). HTSC ceramics were led through the inner chamber of the burner, where the material melts drop by drop in the hydrogen-oxygen flame and then it was quenched between rotating copper rolls. The obtained thin flakes could easily be pulverised so that no further impurification during the milling process occured. Advantages of this technique are: (1) no reaction between the melt and the crucible material; (2) very short melting time and high cooling rate, (3) high homogeneity and distribution of newly produced compounds.

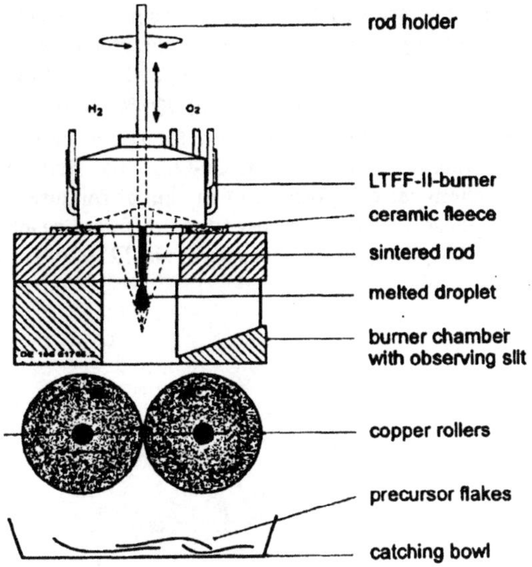

Figure 1. Flakes Precursor Apparatus

Applied YBCO - superconducting ceramics were decomposed into a mixture of Y_2O_3, Y-211, $BaCuO_2$ and $BaCu_2O_2$. Melt textured material made of this mixture showed significantly higher levitation forces in contrast to reference samples of other techniques [5]. Sticks of NBCO-superconductor ceramics with an composition of $Nd_{1.8}Ba_{2.4}Cu_{3.4}O_x$ (molar ratio) have been molten in the same hydrogen-oxygen flame and quenched afterwards. According to x-ray diffraction data the flakes consist of Nd_2O_3, Nd-422, $BaCu_2O_2$, CuO and Cu_2O.

3. Shaping of bulk material and characterisation

Superconducting samples were produced by means of the process of powder compaction followed by a sintering stage at high temperature [8]. Both uniaxial and cold isostatic pressing methods were used for the production of various shapes: hollow cylinders (*Fig. 2*), discs, plates and rings. In general, ceramic powders were compacted in order to produce bulk materials. The quality of the achived bulks, depends on the press ability of the powders and of the press additives. HTSC powder showed no plastic properties and very high pressure (>1 GPa) is required. Big samples could be pressed with special plasticisers under low pressure (<0.1 GPa). The inelastic, frictional particles of HTSC with additive could easily be compacted to homogene green bodies. As the first step of the production process, a plastic paste was made by mixing the superconducting or precursor powders with organic binder (2w.%). As a solvent for the binder diethylketon was used. The YBCO powders with organic binder and solvent were mixed to a slurry in a zirconia ball-mill, thus resulting in a thick paste, which then was ground and dried in an agate mortar at room temperature. After that the material was homogenised by a sifting process. The resulting powder was isostaticly pressed at a pressure of 100 MPa to obtain green ceramic tubes of YBCO and BSCCO. These green samples were treated by sinter or melt-texture process. The typical sinter

process consisted of several steps at maximal temperatures of 950-1000 °C for YBCO and reactive mixtures, 840-850 °C for BSCCO and 1050-1100 °C for Y- and Nd-flakes. All samples were heat-treated in a chamber furnace, where they were slowly heated to about 300 °C to burn out the additives and then they were heated at rates of 60 K/h to the maximum temperature. After that the samples were cooled at a rate of approx. 1 K/h to 900 °C, in order to avoid the "dangerous" temperature at which cracks develop [8]. The YBCO samples were cooled to room temperature in the furnace.
The tubes and plates of YBCO were annealed in an oxygen flow at 450-500 °C for a period of 72 h and cooled slowly in the furnace. In the case of BSCCO there was no „dangerous" temperature with the risk of cracks. It was only important to cool the samples step by step in the furnace. For preparing samples of NBCO the MTG process was applied consisting of several stages: (1) pre-sintering of the pressed pellets at a temperature of 960 °C or 1020 °C, (2) heating the material up to 1150 °C, (3) melting it at this temperature for a short period of time, (4) slow cooling stage for the crystallisation of $NdBa_2Cu_3O_{7-x}$. To study the interdependence of the properties like the chemical composition of the initial mixture, the milling time or the dispersion of Nd-422 phase to MTG process, three different reactive composition with an overall stoichiometry of $Nd_{1,8}Ba_{2,4}Cu_{3,4}O_x$ was used: Nd-123+20 % Nd-422, Nd-422+$BaCuO_2$+CuO and flakes (*Fig. 2*, No. 1, 2, 3).

Figure 2. **Large superconducting tubes of sintered YBCO and MTG processed samples of NBCO (1) Nd-123+20 % Nd-422 (2) Nd-422+$BaCuO_2$+CuO (3) NBCO-flakes**

It can be seen that the size of the crystallites is quite different. In the case of the first two samples it is about 10 mm, the size of the sample prepared by flakes is about 2-3 mm. This can be explained by different crystallisation rates. The mechanism of the crystallisation of Nd-123 is as follows: Nd^{3+} ions of Nd-422 solved in the peritectic melt and then the Nd-123 crystallised out of this melt. So the crystallisation rate of Nd-123 strongly depended on the Nd^{3+} concentration of flux. It is obvious that the particle size of Nd-422 of the flakes sample (*Fig. 2, No.3*) is smaller than of the ceramic samples. That is the reason why the melt saturation by Nd^{3+} is higher in the flakes mixture. Therefore, at the same cooling rate, the crystallisation of Nd-123 of this mixture was quicker and the size of the crystals was smaller. Reducing the cooling rate or lowering the concentration of Nd-422 in the flame fusion mixture may probably lead to bigger crystals. The microstructure in dependence of the melting time of ceramic samples was investigated (*Fig. 3*). As can be seen on the low resolution photos in polarised light the porosity of the flakes sample was first increased due to the oxygen release, but after the melting process the porosity was much lower. In the case of the ceramic sample, under the same sintering conditions, high porosity can be observed. Such a decrease in porosity of the flakes sample may be the result of the low viscosity of the melt. These parts show densities of about 5.2 g/cm^3. Additionally ground Nd - flakes were dry uniaxially pressed at 200 MPa into small cylinders with a diameter of 20 mm and a length of 5 mm and densities of 4.5 g/cm^3. The tubes of YBCO range from 150 to 300 mm in length with an o.d. of 40 mm and an i.d. of 30 mm. In this shaping step, discs with a diameter of 30 mm and a thickness of 5 mm and small rings have also been made by uniaxial pressing. The green samples of YBCO, reactive mixtures, BSCCO show green densities of 2.1, 3.0 and 4.2 g/cm^3, respectively. Parallel to that, BSCCO tubes with bottom were produced by dry isostatic pressing without plasticiser at 1000 MPa. After sintering these pieces were physically and chemically analysed. In general, the sintered tubes are homogenous and faultless. Some samples of YBCO show considerable cracks.

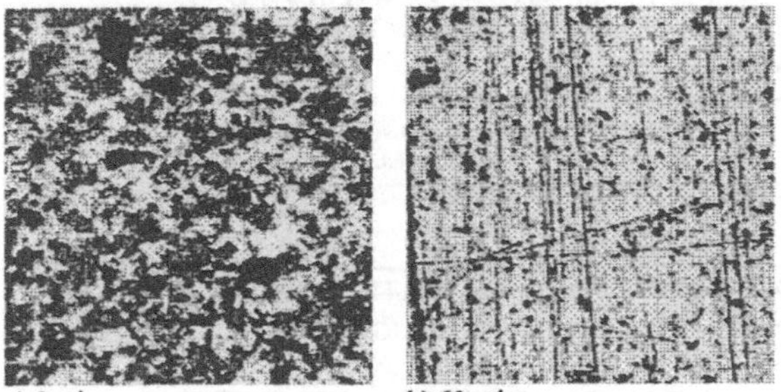

a) 5 min. b) 60 min

Figure 3. **Change of porosity in Nd-flakes samples during peritectic melting. (melting time in minutes)**

4. Magnetic measurement

For investigations of the magnetic shielding properties at a temperature of 77 K a new test bench was developed (*Fig. 4*). Similar studys for different materials and shapes were also done by several groups [3, 4, 9, 10]. The HTSC hollow cylinders or plates were fixed in a liquid nitrogen dewar. A cryogenic Hall probe can be positioned inside the cylinder or directly onto a plate to measure the flux density as a function of the applied dc field. If the magnetic field is weak some of the penetrating flux cannot be measured by the Hall probe because the sensitivity is to low, so SQUID divices are applied. If the external field will be increased beyond a critical value the shielding abruptly breaks down and flux can penetrate the shield as displayed in the hystresis graph in *Fig. 4*. Hollow cylinders (and plates) of BSCCO, NBCO, YBCO and others were tested for homogeneity, density etc. and their electrical and magnetic properties were investigated. At first all test probes were investigated under microscope for presence of fissures and porosity because they dramatically reduce the

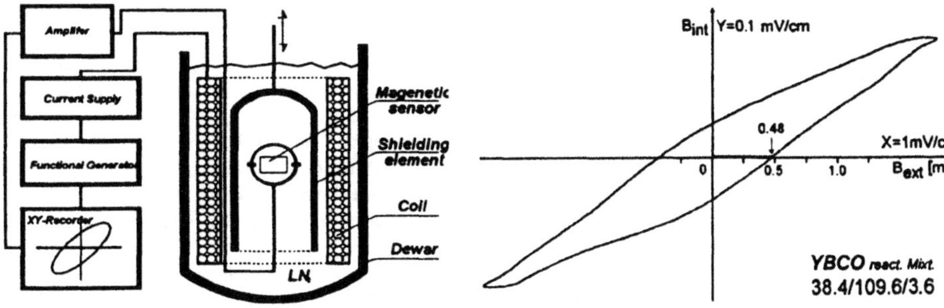

Figure 4. Magnetic shielding measurement system and typical shielding curve of YBCO

shielding properties of HTSC. Critical temperature measurements were carried out on small sections of the test probes to find a difference in the R(T) function of samples with and without structure defects. As the measurements do not show always easy-visible differences between "good" and "bad" samples decisive tests were done with the described shielding measurement system (*Fig. 4)*.

TABLE 2. Hollow cylinders for magnetic shielding (Hall-probe, dc field up to ±11.86 mT) pressed isostatically under 100 MPa with plasticiser Zusoplast®
[1]pressure 1000 MPa without plasticizer, [2]SQUID

Material		Sintering °C	Dimension, mm Ø	L	d	Density g/cm^3	Penetration field, mT
M1	Cryoperm		27.8	118.4	2.1		7.90 (0)[2]
M2	Iron		20.0	140.4	3.0		0
10	YBCO	980. 985	38.8	90.4	4.3	5.44	0.12
12	YBCO	980. 990	37.7	108.4	4.1	5.63	0.77
13	YBCO	990	38.8	109.9	3.6	5.43	0.48
14	BSCCO	840	43.1	131.7	4.2	4.28	3.50
15[1]	BSCCO	840	26.6	106.6	3.5	5.60	6.95 (6.7)[2]
16[1]	BSCCO	840	25.7	106.3	3.3	5.17	2.50
17[1]	BSCCO	850	26.1	104.1	2.7	5.12	7.03
18[1]	BSCCO	840	26.8	72.9	3.6	5.25	5.36
19	NBCO	985	38.1	115.3	3.7	5.68	0.10
K1	14+17						7.96
K2	14+M1						9.60

Acknowledgement

This work is supported by the German Bundesministerium für Bildung und Wissenschaft, Forschung und Technologie (BMBF), the Land NRW and the European Community.

References

1. Altenburg, H., Hauck, J., Rogalla, H.,. Kononyuk, I.F, Tretyakov, Yu.D. (1996) High quality perovskite powder for hightech application in advanced ceramic technology, Periodic Report for INTAS 94-1399.
2. Tretyakov, Yu.D., Oleinikov, N.N., Shlyakhtin, O.A. (1997) *Cryochemical Technology of Advanced Materials*, Chapman & Hall, London,
3. Miller, M., Carroll, T., Soulen, R., Toth, L., Rayne, R., McAlford, N. Saunders, C.S. (1993) Magnetic shielding and noise spectrum of YBCO, BSCCO and BPSCCO tubes, Cryogenics **33**, 180-183
4. Polushkin, V., Buev, A., Koch, H. (1994) Magnetic shielding at liquid nitrogen temperature Applied Superconductors **2**, 597-611,
5. Shlyakhtin, O.A., Vinokurov, A.L., Ischenko, V.V., Altenburg, H., Oleinikov, N.N. (1997) Modification of Bi-2223 precursors by powder engineering methods, Proc. 126th TMS Annual Meeting, Orlando, FL.
6. Plewa, J., DiBenedetto, A., Altenburg, H., Eßer, G., Kugeler, O., Schmitz, G.J. (1997) Thermal investigation of reactive powder mixture as precursors for melt-processing of YBCO, J. Thermal. Anal. **48**, 1011.
7. Ueltzen, M., Seega, Ch., Altenburg, H., Litzkendorf, D., Bruchlos, G., Fischer, K., Görnert, P., (1995) Verfahren zur Herstellung Hochtemperatursupraleitern mit erhöhter Levitationskraft, DE-P 195017552
8. Altenburg, H., Plewa, J., Renzing, C., Jaszczuk, W. (1997) Hochtemperatursupraleitende Hohlzylinder und Verfahren zu ihrer Herstellung, DE-P 19758295.8
9. Itoh, M., Ohyama, T., Minemoto, T., Namuta, K., Hoshino, K. (1992) The improvement in magnetic shielding by use a double-cylinder configuration of high-Tc oxide superconductors J. Phys. D **25**, 1630
10. Willis, J.O., McHenry, M.E., Maley, M.P., Sheinberg, H. (1989) Magnetic shielding by superconducting YBCO hollow cylinders, IEEE Trans. Magn. **25**, 2502-2505

IN SITU MONITORING OF CVD FOR HTS GROWTH

MICHAEL L HITCHMAN, SARKIS H SHAMLIAN, GUGLIELMO G CONDORELLI, FRANCOISE CHABERT-ROCABOIS
Department of Pure and Applied Chemistry,
University of Strathclyde, 295 Cathedral Street,
Glasgow G1 1XL, UK.

1. Introduction

In recent years there has been an increasing awareness of the chemical complexity of CVD processes in general [1, Ch.1, 4] and, in particular, of the rich homogeneous chemistry which can precede homogeneous nucleation. The phenomenon of pre-nucleation reactions is, in fact, likely to be common to most, if not all, CVD processes since before a precursor molecule reaches the reaction surface it will experience a thermal energetic input. In a hot wall reactor the gas phase and the substrate surfaces are all at the same temperature and so molecular reorganisation and/or bond breaking could be expected. In a cold wall reactor, as is often used for the CVD of HTS, precursor molecules will have to traverse a thermal boundary layer close to the substrate surface. The least this will do is to bring about a change in the physical properties of the gas, but in addition thermal molecular modifications are likely. If one is to begin to understand CVD processes in more detail then clearly it is necessary to be able to have some knowledge about this gas phase thermochemistry. For simple molecules *ab initio* calculations can provide both qualitative and quantitative insights into decomposition mechanisms, but for more complex species this is more difficult to achieve theoretically. Therefore there is a need for in situ studies of homogeneous CVD precursor chemistry and in recent years there has been an increasing amount of work done in this area [1, Ch.3]. Here we shall illustrate the usefulness of in situ monitoring for the decomposition of organometallic compounds used in the CVD of HTS.

After the discovery of HTS [2] and a year later [3] of $YBa_2Cu_3O_{7-x}$ (YBCO), the first paper describing the preparation of thin films of YBCO by CVD appeared in 1988 [4]. Subsequently, the discoveries of a HTS containing Bi, Sr, Ca and Cu (BSCCO) and one containing Tl, Ba, Ca, and Cu (TBCCO) in 1988 were rapidly followed by methods describing the CVD of these materials [5, 6]. Since that time there have been several hundred papers on the CVD of YBCO and tens of papers on BSCCO and TBCCO.

At the same time as the research of the new HTS materials has developed there has been intense commercial interest in them. In 1998 it was predicted [7] that the first commercial applications of HTS would be in electronic devices, and this was fulfilled in 1990 with a description of the use of a superconductor transistor in microwave amplifiers, oscillators, phase shifters and signal mixers [8], and with the first semiconductor-CMOS semiconductor circuits being described in 1993 [9]. By the year 2005 it is estimated that the world market for electronic devices based on HTS will be in excess of $300 M [10].

In order to build electronic HTS devices it will be necessary to produce thin films of HTS by techniques which are compatible with semiconductor fabrication lines, which allow close control of layer properties, which produce conformal thin films over large areas and have a high level of intra- and inter-sample uniformity, and which give a high sample throughput. CVD is just such a technique, and in addition it has the features of allowing a range of oxidants to be used (e.g. O_2, N_2O, O_3,) of permitting low temperature deposition (e.g. plasma enhanced CVD, photoCVD), and of maintaining a high partial pressure of oxidant, which is often desirable [11]. However, the C in CVD reminds us that the technique is very dependent on precursor chemistry, deposition chemistry, and reaction parameters, and the CVD of YBCO and other HTS is also a multiparameter problem. Much

G. Van Tendeloo et al. (eds.), High-Temperature Superconductors and Novel Inorganic Materials, 45–50.

of what has been reported in the literature has been of an empirical nature and there is a need for understanding both precursor and deposition chemistry. Some preliminary steps in this direction have been taken in the last few years.

2. Precursors

For the CVD of YBCO, β-diketonates with the general molecular formula shown below have been found to be the most suitable precursors [12, 13].

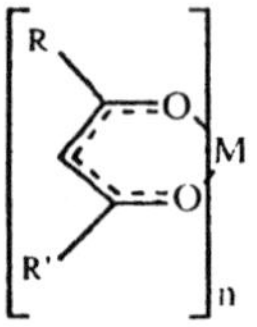

M = Y n = 3
M = Ba n = 2
M = Cu n = 2

For Y and Cu metal complexes of HTMHD, where TMHD = 2,2,6,6-tetramethylheptane-3,5-dione (i.e. R = R' = ^{t}Bu) are frequently used because of their good volatility and stability. For barium, however, no such stable volatile precursor was available [12] until recently. Now several have been reported [14-17]. Little has been reported, though, on the thermal decomposition of any of these compounds, especially in the presence of oxygen which is necessary for the growth of HTS layers. The only studies which have been carried out have been by mass spectrometry for the thermal decomposition of [$Cu(TMHD)_2$] and [$Y(TMHD)_3$] [18-19]. In situ FTIR spectroscopy would be a useful complementary technique since it would probe the decomposition directly in the reactor and at the same time give information about bonding changes. Therefore in order to try and obtain a better understanding of the chemical decomposition of precursors for the MOCVD of HTS thin films [15, 20], we have studied their thermal decomposition in the gas phase in the absence and presence of oxygen under typical deposition conditions [21-23], using both FTIR and mass spectrometry, although here we only describe results for the decomposition of [$Y(TMHD)_3$].

3. Experimental

The experimental arrangement used for the FTIR in this study has already been described elsewhere [21]. Briefly, it consisted of an FTIR spectrometer, a gas-handling facility, a reactor made up of an evaporator, a reactor tube and a thermostatted jacket, and a vacuum system. The yttrium precursor was heated in a stainless steel evaporator and a typical temperature [20] of 145°C was used. The precursor vapour was carried to the reactor by a flow of argon. To prevent condensation of the vapour before it reached the reactor the wall of the jacket was heated to 200°C. The reactor was made of quartz with a diameter of 2.5cm and a total length of 27cm. The temperature range investigated in the reaction zone (15cm) was 200 to 700°C and the total pressure was 1.33×10^3 Pa (10 Torr). Several spectra of the vapour in the decomposition zone were obtained at each temperature for the spectral region extending from 4500 cm^{-1} to 700 cm^{-1}.

Initial attempts to obtain mass spectra of precursors used for the MOCVD of HTS thin films were unsuccessful because a number of serious problems were encountered in introducing the heavy molecular weight complexes into the mass spectrometer. These problems were partly associated with contamination of the quadrupole, but, more significantly, with condensation of the precursors at the inlet of the spectrometer. To overcome the condensation problem a heated box at the entrance zone of the spectrometer for the purpose of maintaining the sampling and analyser heads above 100°C at all times proved to be successful in preventing the orifice becoming blocked. An extension, which was also heated, was built onto this hot box to include a small reactor and a precursor pot, but valve units were eliminated in this extension in order to prevent any long term memory effects. The quartz reactor had a diameter of 0.8 cm and was ca. 45cm long. The precursor vapour was carried to the reactor by a flow of Ar. For experiments with oxygen there was a mixed flow of this gas with Ar. The temperature range investigated in the reaction zone was 160 to 600°C and the total pressure was 3.33×10^2Pa (2.5 Torr). Several spectra were obtained at each temperature for the m/e^{-} range 40 to 640.

4. Results

A typical IR spectrum for [$Y(TMHD)_3$] at 200°C is shown in Fig. 1. It is interesting to note the sensitivity of the FTIR instrument of about 0.01%. Characteristic spectral bands appear at 2967 cm^{-1}, which can be suggested to be due to the $C\text{-}H_x$ stretch of the tertiary butyl groups, at 1571-1556 cm^{-1} arising from C=C or C=O stretching in the rings with the higher frequency corresponding to a C=C stretching [24], and at 1508 cm^{-1} and 1408 cm^{-1} also due to ring C=C stretching. The absorption coefficient (α) was calculated from the relation $\alpha=[log(I_0/I)]/d$ where I_0 is the background intensity, I the signal intensity and d the length of the tube between the two windows. Values of the absorption coefficient normalised to the corresponding mean value obtained at 200°C are plotted as a function of temperature in Fig. 2.

Up to about 600°C and in the absence of oxygen (Fig. 2(a)) no significant change is observed of the relative absorption coefficient with temperature for the absorptions due to vibrations from both the ring and the ring substituent, t-Bu. Above about 600°C the absorptions arising from the C=C stretches begin to decrease slightly more rapidly than the absorptions attributed to the C-H stretch. This probably indicates that the structure of the rings is less stable than the hydrocarbon moiety. The stability of the C-H bonds is not surprising since it is well known [25] that thermal cracking of such bonds requires considerably higher temperatures than those used here. In the presence of oxygen, up to about 400°C again little variation of absorption coefficient with temperature is observed for all the four main bands (Fig. 2(b)). Above ca. 400°C there is a much more dramatic decrease in the relative absorption coefficient with temperature, and above ca. 650°C none of the four initial vibrational modes could be detected. The appearance of a peak in the spectrum at 2356 cm^{-1} due to CO_2 for temperatures >450°C shows that oxidation of the entire organic structure is occurring, although again the hydrocarbon moiety seems to be slightly more resistant to decomposition than the rings.

Fig. 3 shows a typical mass spectrum for [$Y(TMHD)_3$] at 200°C. The four major peaks of highest mass are those at m/e^- of 638, 581, 455 and 183. The first peak corresponds to the single positively charged parent compound [$Y(TMHD)_3$], the second to the parent with the loss of a single t-Bu, the third to the parent with the loss of one ligand (i.e. [$Y(TMHD)_2]^+$), and the fourth to a single ligand moiety - $TMHD^+$. Fig. 4 shows the temperature variation in the absence of oxygen of the four signals mentioned above. Above about 450°C there is clear evidence of decomposition of the parent precursor and of the formation of free ligand, TMHD, with an indication that at 600°C this also starts to break up. All this is consistent with the FTIR data which is summarised at the top of the figure. As noted earlier, these data show no significant variation up to 600°C which indicates that the C=C bonding in the ligand is being retained. Results for the temperature variation of the mass spectrometric signals in the presence of oxygen are shown in Fig. 5. It can be seen that the precursor starts to decompose above about 400°C, but now at the same temperature the TMHD ligand also decomposes. The FTIR data, again summarised at the top of the figure, show the break-up of the TMHD ligand as well. However, whereas the FTIR results indicate a steady decline with temperature of the amount of TMHD in the gas phase, the mass spectrometric data suggest that above 500°C TMHD is being formed faster than it is being oxidised. This effect may be a result of different conditions in the two reactor systems. For example, the space time for the mass spectrometric reactor system was about twice that of the FTIR system, and so there is a correspondingly longer time required to treat a reactor volume of feed in the former case. The FTIR results presented above have shown that the oxidation of the precursor is first order in the precursor itself and in oxygen. If we consider the reaction to occur in consecutive steps of ligand release followed by oxidation, and if the oxidation is slow compared with free ligand formation, then for plug flow systems, such as we have here, the amount of intermediate (i.e. the TMHD ligand) can be higher in the system with the larger space time [26]. This point is being investigated further, but it illustrates the importance of considering reactor design in addition to decomposition kinetics.

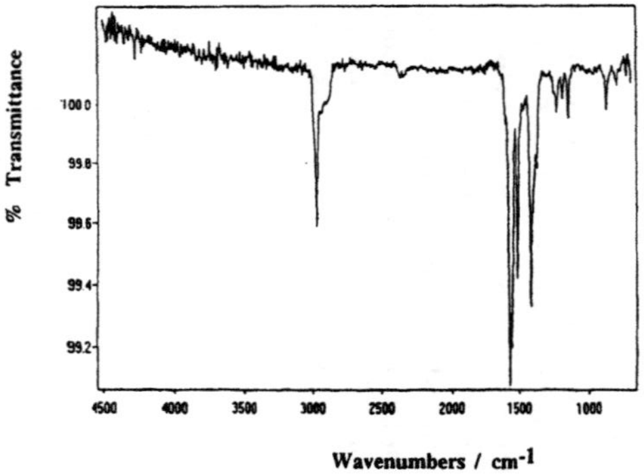

Figure 1 Typical IR spectrum of [Y(TMHD)$_3$] [21]

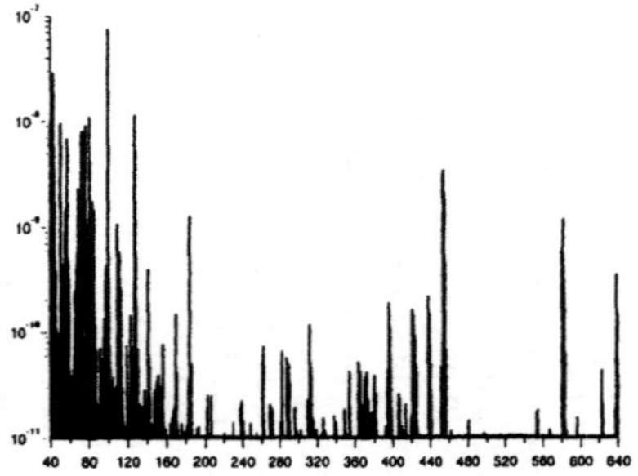

Figure 3 Typical mass spectrum of [Y(TMHD)$_3$] taken at 200°C [22]

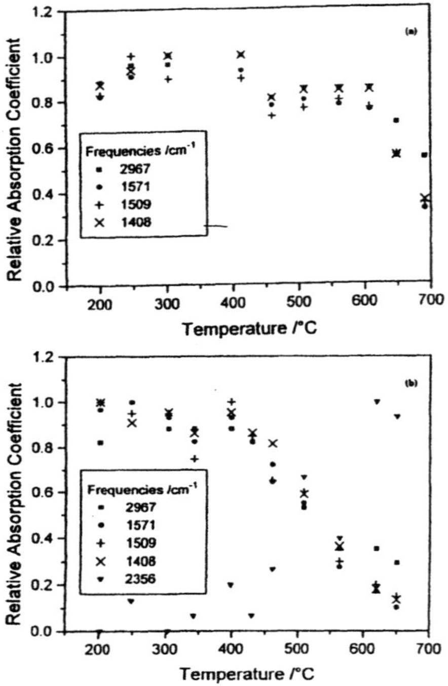

Figure 2 Plots of relative absorption coefficients as a function of temperature for [Y(TMHD)$_3$] vapour under a total pressure of 1.33 x 10^3 Pa (10 Torr) (a) without oxygen and (b) with oxygen [21]

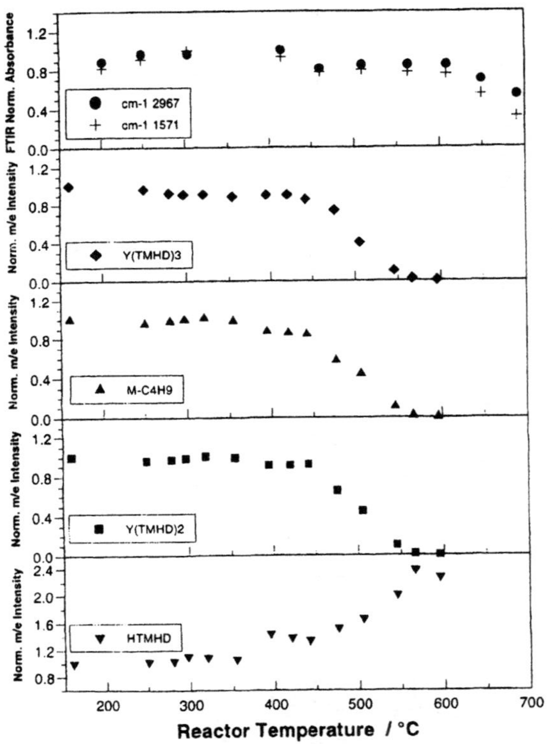

Figure 4 Plots of mass spectral intensities and relative absorption co-efficients as a function of temperature for [Y(TMHD)$_3$] vapour without oxygen [22]

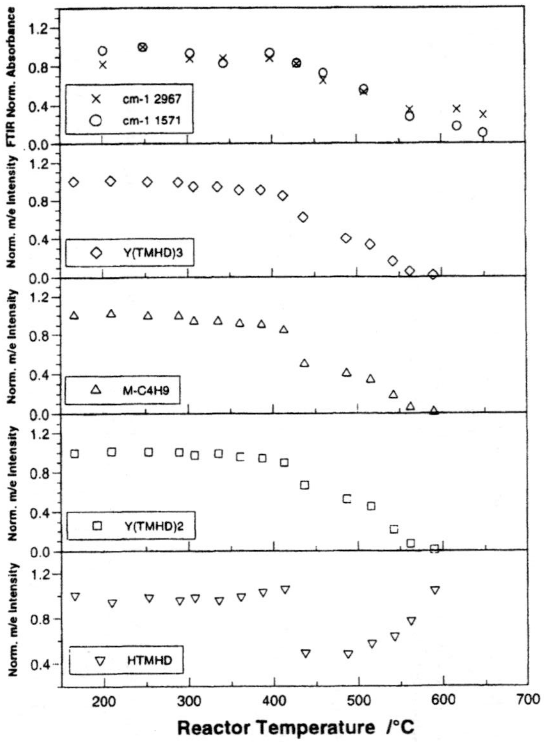

Figure 5 Plots of mass spectral intensities and relative absorption co-efficients as a function of temperature for [Y(TMHD)$_3$] vapour with oxygen [22]

5. Conclusions

The results presented here have shown that insights about the chemistry of the decomposition of precursors in the MOCVD of HTS can be obtained using FTIR and MS analysis. The studies have revealed that the two techniques provide complementary information. Further work is required to investigate the influence of homogeneous chemistry on layer properties.

6. Acknowledgements

This work was partially supported by the CEC under the BRITE/EURAM Programme, Contract No. BREU 0438.

References

1. M.L. Hitchman and K.F. Jensen (eds.), Chemical Vapor Deposition, Academic, London, 1993
2. J.G. Bednorz and K.A. Mueller, Z. Phys. B, 64 (1986) 189
3. M.K. Wu, J.R. Ashburn, C.J. Torng, P.H. Horr, R.L. Meng, L. Gao, Z. J. Huang, Y.Q. Wang and C.W. Chu, Phys. Rev. Letts 58 (1987) 908
4. A.D. Berry, K.D. Gaskill, R.T. Holm, E.J. Cukauskas, R. Kaplan and R.L. Henry, Appl. Phys. Lett. 52 (1988) 1743
5. A.D. Berry, R.T. Holm, E.J. Cukauskas, M. Fatemi, D.K. Gaskill, R. Kaplan and W.B. Fox, J. Crystal Growth 92 (1988) 344
6. D.S. Richeson, L.M. Tonge, J. Zhao, J. Zhang, H.O. Marcy, T.J. Marks, B.W. Wessels and C.R. Kannewurf, Appl. Phys. Lett. 54 (1989) 2154
7. W. Gosling in Superconductivity: The Opportunities, Implications and Applications, edited by A. Conway (IBC Technical Services Ltd, London 1988), p.25
8. Science and the Citizen, Scientific American 264 (1991) 15
9. Industry News, Semiconductor International, April 1993, p.18
10. News Item, Chem. Eng. P. (Sept. 1989), p.44
11. J. Zhao, C.S. Chern, Y. Q. Li, D.W. Noh, P. Norris, P. Zawadzki, B. Kear and B. Gallois, J. Cryst. Growth 107 (1991) 699
12. M.L. Hitchman, D.D. Gilliland, D.J. Cole-Hamilton and S.C. Thompson, in New Materials and their Applications (Inst. Phys. Conf. Ser. III), (D. Holland, ed.), Institute of Physics, Bristol, 1991 p.305
13. M. Leskela, H. Molsa and L. Niinisto, Supercond. Sci. Technol., 6 (1993) 627
14. S.C. Thompson, D.J. Cole-Hamilton, D.D. Gilliland, M.L. Hitchman and J.C. Barnes, Adv. Mat. Opt. Electron., 1 (1992) 81
15. S.H. Shamlian, M.L. Hitchman, S.L. Cook and B. C. Richards, J. Mater. Chem., 4 (1994) 81
16. M.L. Hitchman, S.H. Shamlian, D.D. Gilliland, D.J. Cole-Hamilton, J.A.P. Nash, S.C. Thompson and S.L. Cook, J. Mater. Chem., 5 (1995) 47
17. K. Timmer, K.I.M.A. Spee, A. Mackor and H.A. Meinema, Inorg. Chim. Acta, 190 (1992) 109
18. A.F. Bykov, P.P. Semyannikov and K. Igumenov, J. Thermal Anal., 38 (1992) 1463
19. A.F. Bykov, P.P. Semyannikov and K. Igumenov, J. Thermal Anal., 38 (1992) 1477
20. B.C. Richards, S.L. Cook, D.L. Pinch, G.W. Andrews, G. Lengeling, B. Schulte, H. Juergenson, Y.Q. Shen, P. Vase, T. Freltoft, C.I.M.A. Spee, J.L. Linden, M.L. Hitchman, S.H. Shamlian and A. Brown, Physica C, 252 (1995) 229
21. A.Y. Kovalgin, F. Chabert-Rocabois, M.L. Hitchman, S.H. Shamlian and S.E. Alexandrov, J. Phys. IV, Colloque C5 (1995) 357
22. M.L. Hitchman, S.H. Shamlian, G.G. Condorelli and F. Chabert-Rocabois, J. Alloys Compd., 251 (1997) 297
23. G.G. Condorelli, M.L. Hitchman, A.Y. Kovalgin and S.H. Shamlian, in Chemical Vapor Deposition (M.D. Allendorf and C. Bernard, eds.), The Electrochemical Society, Pennington, N.J., 1997, p.901
24. K. Nakamoto, Infrared Spectra of Inorganic and Co-ordination Compounds, Wiley, N.Y., 1963, p.216
25. P. Wiseman, An Introduction to Industrial Organic Chemistry, Applied Science Publishers, London, 1979, p.69
26. O. Levenspiel, Chemical Reaction Engineering, 2nd Ed., Wiley, N.Y., 1962, p.177

ELECTRODEPOSITION OF HTSC FILMS

M.Á. ALARIO-FRANCO, M.S. MARTÍN-GONZÁLEZ, J. GARCÍA-JACA, E. MORÁN.
Departamento Química Inorgánica, Facultad de Ciencias Químicas, Universidad Complutense, 28040-Madrid Spain (UE).

1. Abstract

We have deposited in different substrates the metallic components of several HTSC materials by cathodic electrodeposition from the different nitrates in DMSO. In a second step, the thermal treatment of the deposit, under adequate conditions, allows one to obtain the superconducting films.

We have followed the chemical reactions that lead from the metallic deposit to the HTSC film by means of IR spectroscopy, X-ray diffraction, thermal analysis and Scanning Electron Microscopy.

Among the materials that we have deposited we will briefly describe the case of films of $Bi_2Sr_2CaCu_2O_{8+x}$, $YBa_2Cu_3O_{7-x}$, $BaBiO_3$ and $Ba_{1-x}K_xBiO_3$. In particular, this seems to be the first preparation of films of this last materials by this procedure and the measured T_c is of the order of 35 K.

2. Introduction

As the commercial applications of HTSC materials are slowly developing, the preparation of films in a not too expensive way becomes of interest. Electrodeposition, a well established chemical process, is both economical and often used at the industrial scale, so that if a procedure is found for electrodepositing well formed films of HTSC, it will be relatively easy to scale up the process to the industrial technology.

Since HTSC materials are constituted by brittle ceramics, it is clear that their eventual applications in the form of wires, tapes or other surfaces requires the adapting the metal oxide powders to those shapes. In the case of the wires, the so-called OPIT (from Oxide Powder in Tube) seems to be the most common procedure [1]. In the case of detectors or tapes, a wide range of deposition process are being used, among which one can quote sputtering [2], laser ablation [3], MOCVD [4,5], etc. Electrodeposition of metals followed by an adequate oxidation process has also been successfully employed in making HTSC films [6-13], the advantages of this procedure

G. Van Tendeloo et al. (eds.), High-Temperature Superconductors and Novel Inorganic Materials, 51–56.

are well known since it has been used for a long time in industrial processes such as electroplating, were virtually any shape can be covered. It is indeed a relatively cheap procedure that can be handled without complicated equipment, a simple electrolytic cell is almost all that is required. Obviously, to obtain the oxides, a second, oxidation-step is needed.

Cathodic electrodeposition has already been used [6-13], in the preparation of HTSC films of Ybco, the bismuth, thallium and mercury superconductors; yet, there are some aspects relating the way in which the deposition process takes place which are not yet well known, and also it is of interest to know the intermediate species that intervene in the formation of the final phase.

In the present paper, we will describe our attempts to prepare by electrodeposition/oxidation HTSC films of some of the well known superconducting oxides including both "cuprates" and copper free materials.

3. Experimental

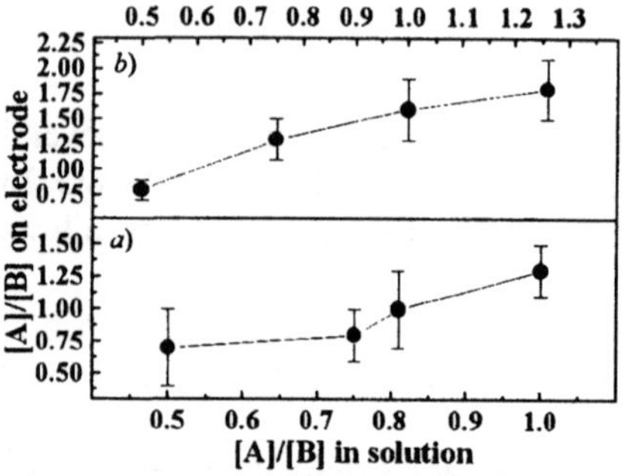

Figure 1. **Solution *versus* electrode cationic proportions of a) Sr:Bi and b) Cu:Bi.**

Films were deposited from a solution of the reagent grade metal nitrates or hydrated nitrates in dimethylsulfoxide (DMSO). In order to attain the adequate superconducting compositions in the films, a first part of the study consisted in establishing a correlation between the concentration of the different metals on the films and that of the ions present in the solution, which in fact seldom coincides. Figure 1 shows these data in the case of the a) Sr:Cu and b) Cu:Bi pairs.

Chronoamperometry was performed at a sufficiently low potential so as to be able to deposit the highly electropositive alkali-earth metals. For these reasons, a potential of V = -3.5 V versus Ag/AgCl was employed.

Table 1 shows the particular conditions used for each of the superconducting materials prepared.

TABLE 1 Synthesis Conditions

Superconducting Material	Synthesis Conditions (at room temperature) Concentration of different species (mM/l)
$Bi_2Sr_2CaCu_2O_{8-x}$	Bi: 20, Sr: 20, Ca: 13, Cu: 15 E= -3,5 V $^{(*)}$/ 30s + 0 V$^{(*)}$/ 30s
$YBa_2Cu_3O_{7-\delta}$	Y: 20, Ba: 36, Cu: 44 E= -3,5 V $^{(*)}$ 30 min.
$Ba_{1-x}K_xBiO_3$	Ba: 9.5, Bi: 20, K: 3 E= -3,5 V $^{(*)}$ 30 min.

(*) *versus* Ag/AgCl

Silver was used as substrate in the form of rectangular plates (0.5 x 1.0 cm^2) and platinum as counter electrode. Once the metals were deposited, oxidation so as to obtain the SC oxides was performed in air by heating at increasing temperatures for different periods of time.

The formation of the HTSC films was followed by means of IR spectroscopy, TG analysis, X-ray diffraction and Scanning Electron Microscopy (SEM) and Microanalysis (EDS).

Superconducting temperatures were measured through magnetic susceptibility measurements performed in a Lake Shore 7225 susceptometer measuring in AC mode at 125 Hz with different magnetic films in the temperature range 5-120 K.

4. Results and discussion

4.1 $Bi_2Sr_2CaCu_2O_{8+x}$

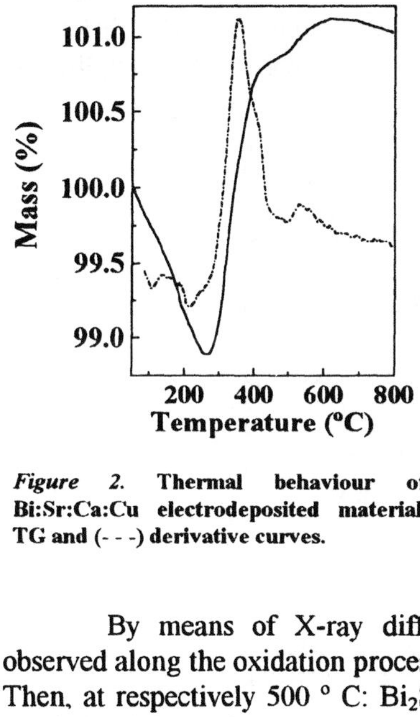

Figure 2. **Thermal behaviour of the Bi:Sr:Ca:Cu electrodeposited material: (—) TG and (- - -) derivative curves.**

Figure 2 shows the TGA in air for the adequate Bi/Sr/Ca/Cu film. Two, apparently very simple, processes can be observed: one, at relatively low temperature, between room temperature and 250 ° C, that can be attributed to the desorption of the solvent and some water and a second one, characterised by a weight gain, that follows immediately afterwards and that ends up around 500 ° C; this can be assigned to an oxidation process. The situation is however somewhat more complicated, as already suggested by the derivative plot, were more than one peak, implying more than one oxidation reaction, are observed. A detailed analysis of this will be published elsewhere.

By means of X-ray diffraction, the following species were successively observed along the oxidation process. Up to 200 ° C a low crystalline deposit is present. Then, at respectively 500 ° C: $Bi_2Sr_2CuO_x$, Bi_2CuO_4; 600 ° C: $Bi_2Sr_2CuO_x$, Bi_2CuO_4; 700 ° C: $Bi_2Sr_2CuO_x$, $Sr_3Bi_2O_6$; 800 ° C: $Bi_2Sr_2CaCu_2O_8$, and a small amount (>5 %) of $Bi_2Sr_2CuO_x$.

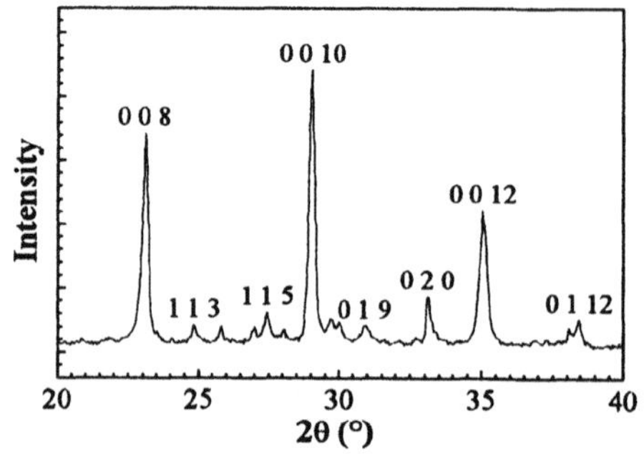

Figure 3. **X-ray diffraction pattern of Bi-2201 deposit where the relative intensity of the 00*l* lines. can be apreciated**

Heating at 800 ° C seems to be the adequate treatment to obtain $Bi_2Sr_2CaCu_2O_8$; a portion of the X-ray diffraction pattern appears on Figure 3; the relative intensity of the 00*l* lines suggests that there is a certain degree of preferential orientation of the Bisco film parallel to the substrate.

Heating it up to 900 ° C originates the decomposition of the superconductor with $Ca_4Bi_6O_{13}$, $Sr_3Bi_2O_6$ and $SrCu_2O_3$ as the main decomposition products.

Magnetic susceptibility measurements Figure 4.a, shows a Tc of the order of 95 K, noticeably higher than that obtained from a powder of a Bisco prepared by the ceramic procedure in the same conditions, Figure 4.b. This difference, can be attributed to differences in the degree of oxygen doping between the two samples.

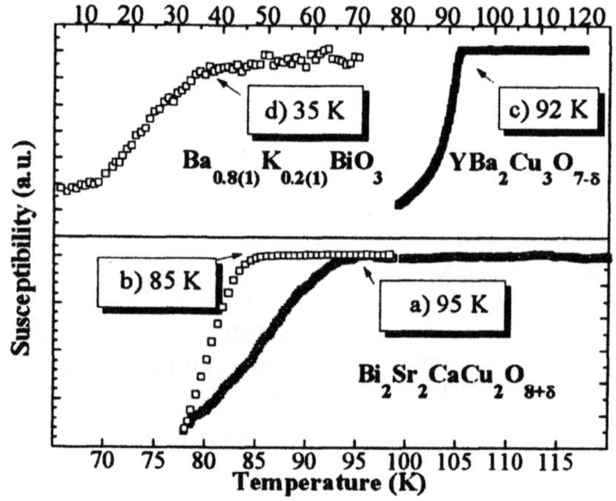

Figure 4. **AC susceptibilty of a) $Bi_2Sr_2CaCu_2O_{8+x}$/Ag films, b) ceramic $Bi_2Sr_2CaCu_2O_{8+x}$, c) $YBa_2Cu_3O_{7-\delta}$/Ag and d) $Ba_{1-x}K_xBiO_3$/Ag**

4.2 $YBa_2Cu_3O_{7-\delta}$

A similar study for preparing Ybco has shown that the conditions to obtain superconducting films of this material are 900 ° C in air. A subsequent treatment in 1 atm of oxygen at 450 ° C induces superconducting properties, Figure 4.c.

However, the films obtained in a first deposition are, after oxidation, not continuous. Figure 5.a shows that a series of wide cracks are separating the particles. A vertical cut of the film, Figure 5.b shows a kind of columnar growth pattern. In order to improve the quality of the film, a second metal deposition was done followed by a similar oxidation process. Figure 5.c shows that the area covered by the film is now higher, and that the film height has also increased, Figure 5.d, *i.e.* the film is now thicker. Yet some cracks still remain in the deposit. It then seems that successive depositions do improve the quality of the films and work is in progress along this line trying to make them fully continuous.

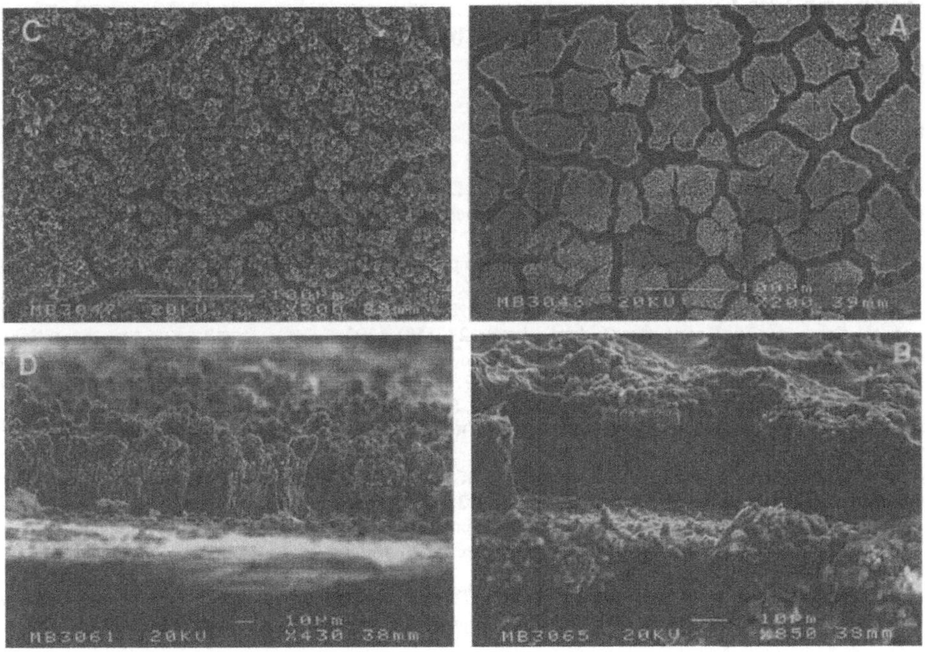

Figure 5. **Scanning electron micrographs corresponding to $YBa_2Cu_3O_{7-\delta}$:a) & b) first deposition, c) & d) second deposition, respectly vertical and horizontal views**

4.3 $Ba_{1-x}K_xBiO_3$

In the case of this copper-free material, the presence of potassium makes things a little more complicated [14] and it is only after leaving the metal deposit in air, for at least three days when potassium bicarbonate is formed, that one can proceed with the decomposition step. Otherwise, potassium does not appear in the final film, probably because of volatilization. Also here the deposited material obtained after oxidation is very much cracked and, again, re-deposition should be used to improve the quality of the film. Nevertheless, susceptibility measurements confirmed that the films are superconducting, Figure 4.d.

5. Conclusions

We have been able to grown thin films of several HTSC materials on silver, by electrodeposition of the corresponding metals followed by an adequate thermal treatment.

$Bi_2Sr_2CaCu_2O_{8-x}$ can be grown parallel to the substrate by heating up to 800 ° C; at higher temperatures if decomposes.

Redeposition of the metals is helpful in improving quality of the films as shown with $YBa_2Cu_3O_{7-\delta}$.

We have grown, for the first time by electrodeposition, superconducting films of $Ba_{1-x}K_xBiO_3$.

6. References

1. Sarma, M.S., Syamaprasad, U. and Damodaran, A.D. (1997) Powder compaction characteristics and tube dimensions in PIT fabrication of Ag/BPSCCO superconducting tapes. *Superconductor science & technology.* **10**, 337-341
2. Semenov, A. P. (1993) Deposition of Thin High-Temperature Superconductor Films by Ion-Beam Sputtering (Review). *Instruments and experimental techniques.* **36**, 173-183.
3. Jackson, T.J. and Palmer, S.B. (1994) Oxide superconductor and magnetic metal thin film deposition by pulsed laser ablation: a review. *J. of physics D: Appl. Physics.* **27**, 1581-1593.
4. Schulz, D.L. and Marks, T.J. (1994) MOCVD Routes to Thin Metal Oxide Films for Superconducting Electronics, *Advanced materials.* **6**, 719-779.
5. Weiss, F., Schmatz, U., Pisch, A., Felten, F., Pignard, S., Senateur, J., Abrutis, A., Frohlich, K., Selbmann, D. and Klippe, L. (1997) HTS thin films by innovative MOCVD processes, *J. of alloys and compounds*, **251**, 264-269.
6. R. N. Bhattacharya, R. Noufi, L. L. Roybal, and R. K. Ahrenkiel, (1991) YBaCuO superconductor thin films via an electrodeposition process, *J. Electrochem. Soc.*, **138**, 1643-1645.
7. Ondoño-Castillo, S. Fuentes, A. Pérez, F. Gómez-Romero P. and Casañ-Pastor, N. (1995) Superconducting $YBa_2Cu_3O_{7-x}$ coatings by simultaneous electrodeposition of Y, Ba and Cu in the presence of cyanide. *Chem. Mater.* **7**, 771-779.
8. S. Ondoño-Castillo y N. Casañ-Pastor. (1997) Thermal and mechanical effects on the properties of $YBa_2Cu_3O_{7-d}$ deposits obtained by electrodeposition of precursors in presence of cyanide. *Physica C*, **276**, 251-258.
9. Maxfield, M. Eckhardt, H. Iqbal, Z. Reidinger F. and Baughman, R. H. (1989) Bi-Sr-Ca-Cu-O and Pb-Bi-Sr-Ca-Cu-O superconductor films via an electrodeposition process, *Appl. Phys. Lett.*, **54**, 1932-1933.
10. Richardson, K. A., Arrigan, D. M. W., de Groot, P. A. J., Lanchester, P. C. and Bartlett, P. N. (1996) *Electrochimica Acta*, **41**, 1629-1632.
11. Bhattacharya, R. N., Duda, A., Ginley, D. S., DeLuca, J. A., Ren, Z. F., Wang, C. A., and Wang, J. H., (1994) Superconducting thallium oxide films via an electrodeposition precursor, *Physica C*, **229**, 145-151.
12. Bhattacharya, R. N. Parilla P. A. and Blaugher, R. D. (1993) Processing of thick-films electrodeposited thallium oxide superconductors, *Physica C*; **211**; 475-485.
13. Pawar, S.H., Ubale, M.J. and Kulkarni, S.B. (1994) Electrochemical synthesis of Hg-Ba-CuO superconducting thin films from a aqueous solution, *Materials Letters*, **20**, 279-282.
14. Martín-González, M.S. García-Jaca, J. Morán, E. Alario-Franco, M.Á. (1998) *Physica C*, **297**, 185-191.

CRYOCHEMICAL TECHNOLOGY OF SUPERCONDUCTORS AND RELATED MATERIALS

YU.D.TRETYAKOV, O.A.SHLYAKHTIN
Department of Chemistry, Moscow State University,
119899 Moscow, Russia

Along with various chemical methods of synthesis such as sol-gel technique, spray drying and aerosol spray pyrolysis, the cryochemical technology is one of the most efficient methods for preparation of multicomponent and multiphase materials. Cryochemical processing is based on appropriate balance between cold and heat, with a cold being a key element of technology and being used to prevent uncontrolled changes in intermediate and final products. Since the first publication of Landsberg and Campbell in 1965 [1], the scope of this technology has extended and now includes various innovative processes like cryocrystallization, freeze drying, cryoprecipitation, cryoextraction, cryoimpregnation and cryogrinding (Fig.1). Fundamental and applied aspects of these processes are generalized in recently published monograph [2].

Specific features of cryochemical method applications are closely related to specific nature of materials. Cryochemical technology has many advantages, since it allows to mix components with high precision at the stage of solutions preparation and stabilizes physico-chemical prehistory of materials, because the nucleation of crystal phases in microvolumes of substance occurs at identical conditions (cryocrystallization

G. Van Tendeloo et al. (eds.), High-Temperature Superconductors and Novel Inorganic Materials, 57–62.

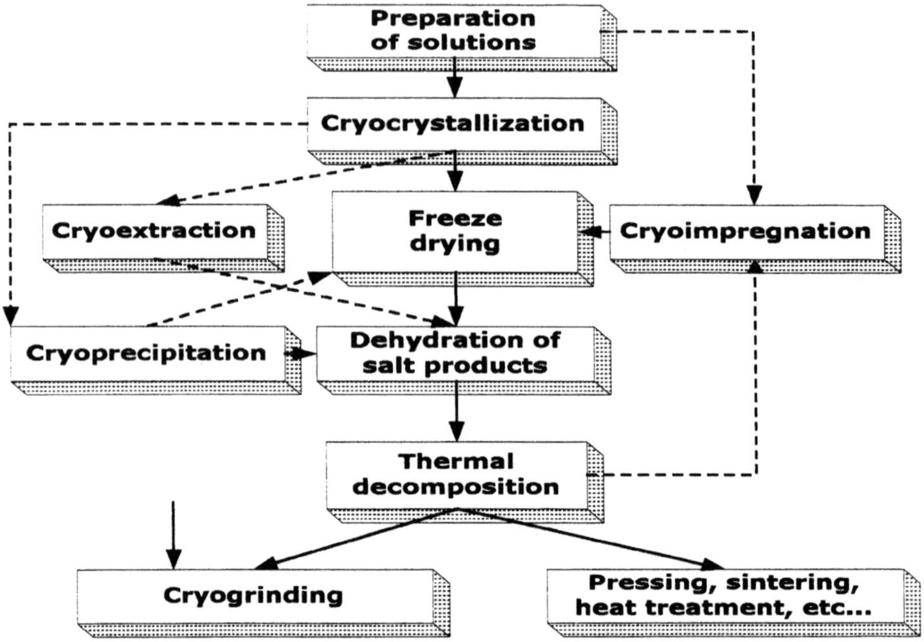

Figure 1. Main processes of cryochemical technology

stage). Multicomponent cryochemical powders are very homogeneous at the deeply submicron level, which is also essential for most applications.

Meanwhile, there are some features common for all cryochemical powders irrespectively of their composition. First of them is small size of crystallites of oxide powders obtained at the stage of thermal decomposition. Their typical size (20-70 nm) is larger than obtained by sol-gel method but still small enough for most applications. Thermal treatment of powders and application of various methods of powder engineering allow to enlarge crystallite size to 500 - 2000 nm if necessary. A small size of crystallites provides high sintering activity of these powders, especially important in preparing refractory and other materials requiring high temperatures of sintering.

Another feature belongs to the powders obtained by freeze-drying technique (approx. > 90 % of applications). Sublimation of water from frozen droplets results in snow-like powders with very porous salt framework composed of strong aggregates of salt crystallites. The proper choice of thermal treatment conditions and effect of topochemical memory allow to reproduce this kind of structure in the oxide powders. This type of microstructure is rather desirable and suitable for applications where stable and extended surface area is needed (i.e catalysts, sorbents, foams etc.). At the same time this specific aggregate structure makes obstacles in preparing high density (> 95 %) ceramics by sintering at moderate temperatures (900 - 1300 C). The problem can be solved by using hot pressing methods resulted in preparation of high density ceramics with relatively small grain size (1-2 μm).

The mentioned characteristics of cryochemical powders as well as an ability of freeze drying technique to be used as simple and effective laboratory method to prepare high quality powders of multicomponent oxide compounds brought out a large numbers of applications. For 30 year of its history, cryochemical technology has been successfully used for preparation of ferrites and ferroelectrics, pigments and refractory ceramics, solid state electrolytes and alloys [2], while many other potential applications still remain underestimated.

The discovery of high-temperature superconductors gave rise to a new wave of interest to cryochemical techniques. This kind of materials has been the most popular object of cryochemical synthesis for last 8 years - over 100 publications since 1988 [2]. The efficiency of application of freeze-drying technique to these complex chemical objects may be illustrated by sputtered neutral mass spectrometry (SNMS) of $YBa_2Cu_3O_x$ ceramics obtained by different routes [3]. Standard deviations of analytical signal intensities during layer-by-layer composition analysis of bulk samples are twice (Ba) to 5 times (Y) lower for the freeze-dried sample compared to spray-dried. This difference indicates clear distinction of chemical homogeneity even between these chemically prepared ceramics.

Most evidently, the advantages of cryochemical precursors was demonstrated during preparation of Ag-sheathed (Bi,Pb)-2223 tapes and wires. Powder-in-tube technology of their production involves processing of substantially multiphase oxide precursors - intermediate products of the synthesis of 2223 phase. Homogeneous distribution of these intermediates within powder particles and small size of the last ones not only increased the target phase formation but also enhanced uniformity of reaction zone development through the core, leading to continuous and uniform microstructure formation. The influence of aggregation in this case is not essential due to application of thermomechanical treatment of composite tape promoting better orientation of crystallites and elimination of micropores [4]. Advancement of technology of these composite tapes allowed to achieve Jc > 10^5 A/cm^2 at T = 77 K which is the best value attained for (Bi,Pb)-2223 based materials until now [5].

The porous structure of cryochemical precursors can give additional advantages when the mechanism of the synthesis of HTSC phase includes transport via gas phase. Thus, better permeability of freeze-dried precursors for Hg-vapors during ampoule synthesis of Hg-HTSC phases substantially accelerates the synthesis and enhances the yield of higher homologues of $HgBa_2Ca_{n-1}Cu_nO_x$ series [6].

One of the perspective direction for HTSC applications of cryochemical technology is synthesis of HTSC microcomposites with nonsuperconducting phases. Trapping of microparticles of alien phases by HTSC grains allows to generate new pinning centers and enhance intragrain Jc values which is especially essential for HTSC of Bi-series due to poor field stability of their Jc. Chemical homogeneity of cryochemical precursors ensures smaller size and better distribution of alien phase inclusions leading to substantial Jc enhancement for Bi-2212 based ceramics [7].

Further progress of superconducting materials is connected with development of artificial structures and composites containing micro- and nanosized HTSC elements. A large variety of synthesis techniques and appropriate comprehension of synthesis processes makes possible and necessary the next turn of freeze-drying synthesis - from powder preparation to powder engineering and constructing new materials at the stage of precursor processing.

More attention should also be paid to the further advancement of cryochemical processing techniques. The growing role of nanostructures and nanoceramics makes reasonable wider application of nonaqueous solutions of inorganic salts in various cryochemical techniques, including freeze-drying. More intensive drop of solubility at low temperatures allows to minimize recrystallization processes and to keep unchanged nanosized salt crystals formed at the freezing stage.

Another way of improvement of efficiency of cryochemical processes is based at new, nontraditional concepts and methodologies. One of the efficient ways of the modelling and prediction of cryochemical experiment results is based on consideration of the synthesis as a system evolution from stochastic to ordered state. In this case application of fractal concepts [8] allows to explain many of observed phenomena.

Considerable progress can be achieved on the way to better reproducibility of synthesis results through deterministic chaos approach [9] when all structure-sensitive and physico-chemical properties of material are considered as fundamentally irreproducible due to the nature of processes taking place during preparation of material. In terms of this theory, we cannot ensure complete reproducibility of properties but can improve it by prevention of chaos appearance in different ways, acceleration of synthesis in order to minimize time-dependent chaos or suppression of chaotic behaviour by counteraction based on feedback relations. Further studies in this field may lead to serious reconsideration of many traditional techniques and expediencies in materials technology, especially when it concerns multicomponent systems.

Concerning the perspectives of further development of cryochemical technology it should be noted that the best results were achieved when the proper combination of high temperature and low temperature processes was used. Cold, for the most part, supplements traditional technological procedures and makes them more controllable and efficient. This concerns the situations wherein instead of chemical precipitation at room temperature the cryoprecipitation is used, drying processes can take place at low temperatures under residual gas pressures (freeze drying) and so on. Another way, which is perspective from technological point of view but remained unused until now, is combination of cooling with other extreme actions. The close interdependence of physical, chemical and biological phenomena gives reasons to affirm that investigation of cryoeffects in any particular branch can lead to fundamental discoveries, which in turn will be necessary for the development of new advanced technologies.

Acknowledgments. Authors are grateful to A.L.Vinokurov for his kind help during manuscript preparation. The work is supported by the National Program «Actual directions in the physics of condensed matter» and State Program «Integration».

References.

1. Landsberg, A. and Campbell, T.T. (1965) Freeze-dry technique for making ultra-fine metal powder, *J. of Metals*, 856
2. Tretyakov, Yu.D., Oleinikov, N.N. and Shlyakhtin, O.A. (1998) *Cryochemical Technology of Advanced Materials*, Chapman&Hall, London
3. Strelkov, A.V., Zhilin, A.B., Gaskov, A.M. and Tretyakov Yu.D. (1993) Characterization of high-Tc superconductors by mass soectral methods of surface analysis (LIMS, SNMS), *J.Alloys and Compounds* **195**, 145-148.
4. Nikulin, A.D., Shikov, A.K. et al. (1994) Ag-2223 (Bi) tapes structure and superconducting properties in dependence of initial powder type, *IEEE Transactions on Magnetics* **30**, 4(2), 2368-2371
5. Lelovic, M., Krishnaraj, P., Eror, N.G., Iyer, A.N., Balachandran, U., (1996)Transport critical current density above 10^5 A cm^{-2} at 77 K in $Bi_{1.8}Pb_{0.4}Sr_{2.0}Ca_{2.2}Cu_{3.0}O_y$ superconducting tapes made by the Ag wire-in-tube method, *Supercond.Sci.Technol.* **9**, 3, 201-204
6. Lee, S.R., Shlyakhtin, O.A., Mun, Mi-Ock, Bae, Myong-Kwang, Lee, Sung-Ik (1995) A freeze-drying approach to the preparation of $HgBa_2Ca_2Cu_3O_{8+x}$ superconductor, *Supercond.Sci.Technol.* **8**, 60-64 .
7. Pupysheva, O.V., Shlyakhtin, O.A., Lennikov, V.V., Putlyaev, V.I. and Tretyakov, Yu.D. (1997) Preparation of $Bi_2Sr_2CaCu_2O_x$ - $SrZrO_3$ Composites from Freeze-dried Mixtures by Melt Processing, *Inorg. Materials* **33**, 9, 951-955
8. Peitgen, H.-O. and Richter, P.H. (1986) *The Beauty of Fractals*, Springer-Verlag, Berlin
9. Markworth, A.J., Stringer, J. and Rollins, R.W. (1995) Deterministic Chaos Theory and Its Applications to Materials Science. *MRS Bulletin*, 7, 20-28

PHASE EQUILIBRIA OF THE (Bi,Pb)-Sr-Ca-Cu-O SYSTEM PERTAINING TO THE 2212 AND 2223 PHASES

W.Wong-Ng, L.P. Cook, A. Kearsley[*], G. Lawrence[*], and W. Greenwood[**], Ceramics Division, [*]Applied Mathematics Division, National Institute of Standards and Technology (NIST), Gaithersburg, MD 20899; [**]Geology Department, University of Maryland, College Park, MD 20742.

1. Introduction

The presence of melt has been reported to improve the grain alignment and associated properties of high T_c superconductors in the (Bi,Pb)-Sr-Ca-Cu-O (BSCCO) system [1-2]. In this regard, the powder-in-tube (PIT) processing technique, which involves melt processing, has been found to be a feasible technique for large scale production of wires and tapes [3-4]. Therefore a detailed understanding of melt formation and the related equilibria is important for the development of high T_c superconductors. Currently, industrial emphasis in the BSCCO system is on the 80K Pb-free 2212 and the 110K Pb-doped 2223 compounds.

Phase diagrams (both subsolidus and liquidus) provide fundamental processing information for the complex multicomponent BSCCO system. The primary objective of the NIST high T_c phase diagram program is to develop the portions of diagrams that are most relevant to the processing of Pb-free 2212 and Pb-doped 2223 compounds. This paper discusses the determination of the subsolidus phase equilibria and the primary phase field (crystallization field) of the Pb-free 2212 and the Pb-2223 phases. An example of the application of this data is given. The PIT processing method utilizes Ag-tubing, and so the effect of Ag on the phase equilibria will also be discussed.

2. Experimental Methods

Reagent grade Bi_2O_3, PbO, $SrCO_3$, $CaCO_3$ and CuO were used for sample preparation. All samples were prepared by solid state calcining. To capture liquid during the experiments, a small piece of MgO wick was mixed in with the samples. The samples were placed in MgO crucibles and annealed in the appropriate atmosphere (air, or a mixture of 7.5 vol. % O_2 + 92.5 vol. % Ar), then quenched in liquid nitrogen-cooled helium. X-ray diffraction was used to study the quenched samples. Quantitative energy dispersive x-ray spectrometry (EDS) was applied to obtain the composition of the melt captured in the wicks as well as the composition of crystalline phases. The EDS data was collected and reduced according to conventional methods [5] via the DTSA software package [6].

G. Van Tendeloo et al. (eds.), High-Temperature Superconductors and Novel Inorganic Materials, 63–68.

3. Results and discussion

3.1 THE 2212 PHASE

A total of 10 phases was found to be in equilibrium with the 2212 phase [7]. For convenience, symbols (in the order Bi:Sr:Ca:Cu) are used to represent the ternary and quaternary oxide phases. The equilibrium phases were 0x21 [$(Ca,Sr)_2CuO_3$], 11 9 x 5 [$(Bi,Pb)_{2.2}Sr_{1.8-x}Ca_xCuO_z$], 2110 [$Bi_{16}(Sr,Ca)_{14}O_z$], 0 14 x 24 [$(Sr,Ca)_{14}Cu_{24}O_{41}$], 2310 [$Bi_2(Sr,Ca)_4O_z$], 4805 [$Bi_4Sr_8Cu_5O_z$), 2201 [$(Bi,Pb)_2Sr_{2-x}Ca_xCuO_z$], (Ca,Sr)O, CuO, and 0x11 [$(Sr_{1-x}Ca_x)CuO_2$, Ca-rich]. Sixteen 4-phase equilibrium volumes containing 2212 as a member were determined and are listed in Table 1 together with the initial melting temperatures and the corresponding melt compositions. These volumes are self-consistent in that there is no overlapping or unfilled phase space. The melt compositions were used to construct a mathematical model for the liquidus surface (primary phase field) as shown in Figure 1. In this diagram the volume is expressed in Cartesian coordinates and the cotectic faces that are visible in the orientation shown are labelled. The initial melting temperatures are indicated at the corners of the polygon. Inside the volume,the 2212 phase is in equilibrium with liquid (L). On each face 2212 is in equilibrium with L and the labelled phase. Along the edges, 2212 is in equilibrium with 2 phases plus L, and at the corners, 2212 is in equilibrium with 3 phases with L.

TABLE 1. Initial melts of 4-phase volumes containing the 2212 phase.

Sample ID	4-phase equilibrium	DTA T(°)	Melt Composition (Mole fraction %) Bi	Sr	Ca	Cu
1	2212-2110-119x5-CuO	825	41.7	16.1	22.3	19.9
2	2212-0x21-2110-CuO	830	30.4	7.4	26.6	35.6
3	2212-0x21-CaO-2110	838	35.2	20.8	24.6	19.4
4	2212-014x24-CuO-119x5	856	27.6	25.1	8.8	38.5
5	2212-2110-119x5-CaO	856	37.8	18.0	19.6	24.6
6	2212-014x24-0x11-CuO	861	26.2	24.2	22.3	27.3
7	2212-014x24-0x11-0x21	863	25.3	23.6	18.9	32.2
8	2212-119x5-2310-CaO	873	29.8	23.2	21.9	25.1
9	2212-4805-0x21-CuO	875	25.9	21.8	14.2	38.1
10	2212-2201-2310-014x24	877	23.9	32.8	11.7	31.6
11	2212-4805-0x11-CuO	877	26.4	26.3	12.7	34.6
12	2212-119x5-2201-2310	877	32.8	30.7	4.5	32.0
13	2212-119x5-014x24-2201	878	27.7	26.9	2.5	42.9
14	2212-0x21-014x24-2310	885	28.7	21.4	20.8	29.1
15	2212-4805-0x11-0x21	887	29.9	24.7	14.9	30.5
16	2212-2310-0x21-CaO	889	26.9	24.7	17.6	30.8

In Figure 1, a point representing a bulk composition inside the volume (Bi:Sr:Ca:Cu = 26.9:24.7:17.6:30.8) is also shown [8]. When this composition was completely melted and slowly cooled in air large 2212 crystals formed initially along the crystallization path within the 2212 primary phase volume (Figure2). During the crystallization process, the liquid moved away from 2212 along the line joining 2212 and the bulkcomposition. With falling temperature, the liquid continued to move along this line, crystallizingonly 2212 until it intersected the outer surface of the volume, as defined by the 119x5 cotectic face.

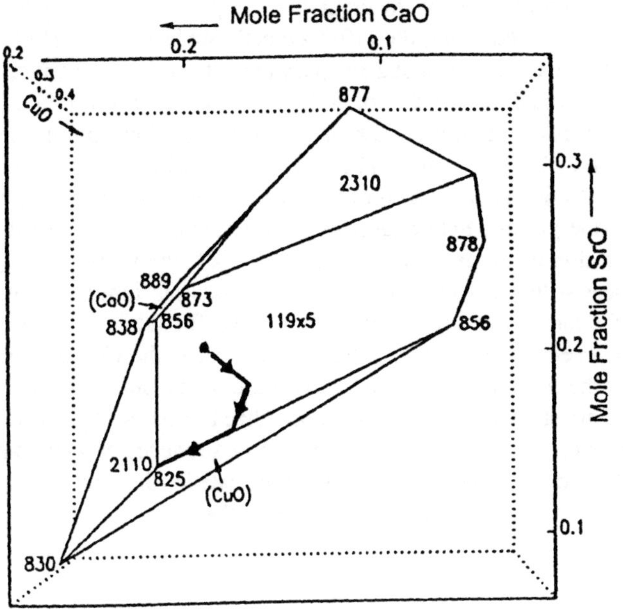

Fig. 1. Primary phase field of the 2212 phase in the BSCCO system. The proposed crystallization path of composition Bi:Sr:Ca:Cu = 26.9:24.7:17.6:30.8 is also shown.

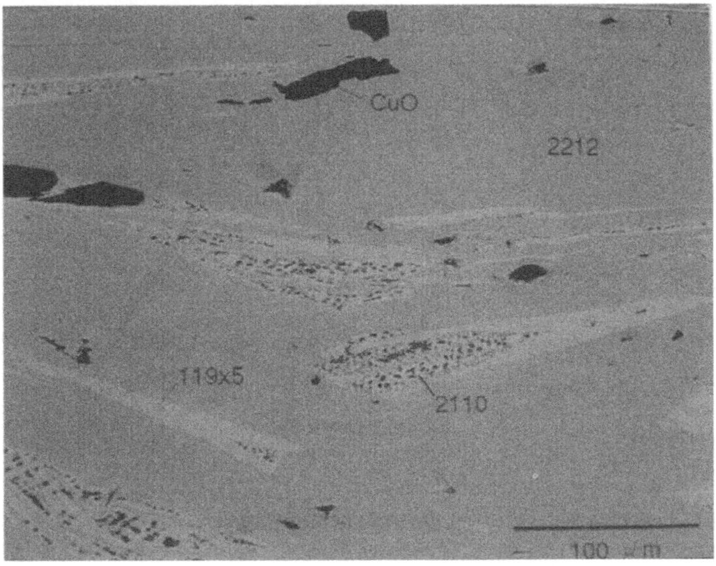

Fig. 2. Backscattered electron micrograph of slow cooling experiment on bulk composition Bi:Sr:Ca:Cu = 26.9:24.7:17.6:30.8

At this point, 119x5 also began to crystallize (Figure 2, light intergrowths), and the crystallization path moved along the 119x5 cotectic surface. The 119x5 and 2212 phases continued to crystallize together along this path until it reached the edge, where CuO also began to crystallize (medium-sized dark grains in Figure 2). The liquid continued to move along the edge. Finally at the corner of the crystallization volume, as defined by the intersection of the CuO, 119x5 and 2110 cotectic crystallization surfaces, the 2110 phase also appeared. Eutectic crystallization of the four phases continued until all the liquid was consumed (the fine-grained heterogeneous microstructure in the interstices of Figure 2).

With the addition of excess silver (≈ 30 weight %), the initial melting temperature of the 2212 4-phase volumes were all lowered, by amounts ranging from 4 °C to 22 °C. Silver entered the melt, with a mole fraction of 6 to 8 %. While in most cases, the amount of Bi, Sr and Ca remained similar to the Ag-free melts, the Cu concentration decreased substantially for the more Cu-rich liquids. In Figure 3, the outline of the Ag-free and Ag-containing melts are shown as projections (normalized in terms of the CaO, CuO and Bi_2O_3 contents). In Figure 3, Ag addition is manifested in a reduced copper content of the melts. Since Cu and Ag mostly exist in the 1+ oxidation state in the melt, the compositional shift may be interpreted as substitution of one for the other

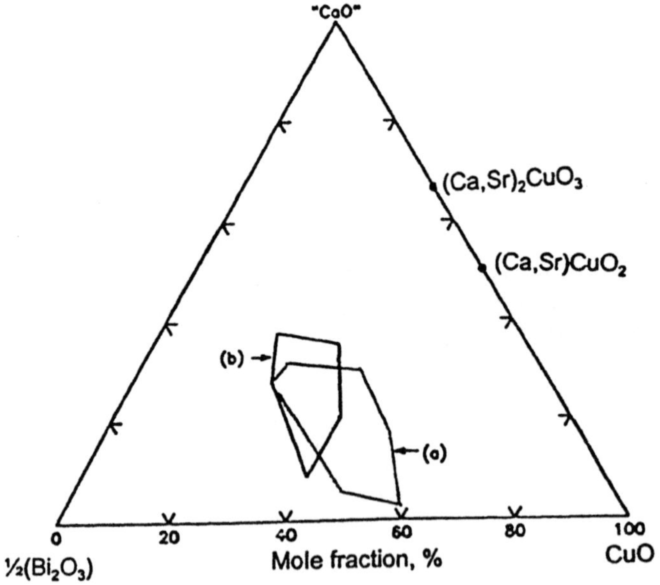

Fig. 3. Projection of the 2212 primary phase field onto the ternary plane $BiO_{1.5}$-CaO-CuO. (a) without the addition of Ag, and (b) with Ag addition.

in the liquid phase. This is reasonable since they have similar ionic radii (1.26 and 0.96 Å for Ag^{1+} and Cu^{1+}, respectively [9]).

3.2. THE Pb-2223 PHASE

Procedures similar to those used for the 2212 phase were employed for the determination of the 2223 primary phase field in the 5-component BSCCO system. First, the 11 phases that are in equilibrium with the Pb-2223 phase were determined : 1x20 [$(Ca,Sr)_2PbO_4$], 0x21, CaO, CuO, 3221 [$(Pb,Bi)_3(Sr,Ca)_5O_x$]), 0x11 [Ca-rich], 0x11' [Ca-poor], 2310, 119x5 and 2212. The total of 330 potential 5-phase combinations including the Pb-2223 phase was narrowed by first treating only those combinations which include 2212 and 2223. The number of initial possibilities is thereby reduced by a factor of nearly 3. The first 16 volumes in Table 2 define the Pb 2223 + 2212 portion of the Pb-2223 subsolidus and have been demonstrated to fit well together without overlapping[10]. The twenty-nine 5-phase assemblages that involve Pb-2223 as a member were subsequently determined and are shown in Table 2.

TABLE 2. Set of 29 Pb-2223 five-phase volumes (7.5 vol. % O_2 / 92.5 vol. % Ar)

2223-2212-1x20-119x5-2310	2223-2212-014x24-1x20-119x5	2223-2212-0x11-2310-119x5
2223-2212-1x20-2310-CaO	2223-2212-2310-0x11-3221	2223-2212-119x5-0x11-CuO
2223-2212-0x21-1x20-CuO	2223-2212-014x24-119x5-CuO	2223-2212-0x21-014x24-CuO
2223-2212-1x20-0x11-3221	2223-2212-014x24-0x21-1x20	2223-2212-0x11-0x11'-CuO
2223-2212-1x20-CuO-0x11	2223-2212-3221-CaO-1x20	2223-2212-0x11-0x11'-2310
2223-2212-3221-CaO-2310	2223-0x11'-2310-0x11-119x5	2223-0x11'-2310-119x5-014x24
2223-1x20-0x11'-CuO-0x11	2223-0x11'-014x24-119x5-CuO	2223-0x11'-CaO-0x11-3221
2223-1x20-CuO-0x21-014x24	2223-1x20-CuO-0x11'-01424	2223-2310-3221-CaO-0x11
2223-1x20-CaO-0x11'-2310	2223-1x20-2310-014x24-0x11'	2223-1x20-0x11'-3221-0x11
2223-1x20-CaO-0x11'-3221	2223-0x11'-CaO-2310-0x11	

Data on the first-melting liquids of the 29 5-phase volumes was modeled using the convex hull technique [11]. All 29 points were found to lie either on, or close to, the convex hull. The results indicate a well-formed, closed volume in 5-dimensional space. The convex hull has 162 "faces" defined by the interconnected data points, and is defined by the equation $\mathbf{Ax} - \mathbf{b} \leq 0$, which involves a matrix, A, of dimension 162 x 5, and vectors **x** and **b**. The vector **b** describes the position of the faces relative to the origin in 5-component space, and **x** is an arbitrary data point lying on the convex hull.

To graphically represent the primary phase field, an isopleth projection, i.e. a section of constant composition through the phase space, was employed to give a general sense of the "shape". Figure 4 shows an isoplethal section of the primary phase field volume made by holding the $BiO_{1.5}$ and PbO values constant at the median values for the 29 data points, with the remaining oxides plotted relative to the three Cartesian axes. The view in Figure 3 appears to have an oval shape, whereas in a 90° view the volume appears to be thin and elongated (not shown). The equation $\mathbf{Ax} - \mathbf{b} \leq 0$ can be used to determine if a given composition lies inside the Pb-2223 primary phase field. If the difference of the two terms on the left side of the equation is negative, the particular composition lies inside the crystallization volume. If the value is zero, the composition is on the surface, and if it is positive, it is outside the volume. Compositions falling outside this volume should be avoided for the processing of the Pb-2223 phase, as they would be expected to produce large primary crystals of unwanted phases.

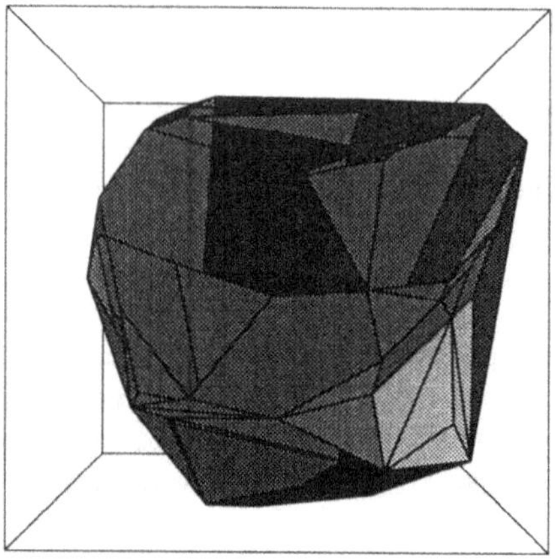

Fig. 4. The primary crystallization field of the Pb-2223 phase. An isoplethal section made by holding the $BiO_{1.5}$ and PbO values constant at the median values for the 29 data points ($BiO_{1.5}$= 20.4, and PbO= 6.9).

As was observed for the 2212 phase, the addition of Ag did not appear to significantly alter the Pb-2223 subsolidus relationships. In general, the melt temperature was lowered by 2 - 25 °C among the multiphase volumes. Our current efforts involve continuation of quantitative melt analysis in order to determine the effect of Ag on the primary phase field of the Pb-2223 phase.

4. References

1. Zhang, W. and Hellstrom, E.E. (1993) The development of an aligned microstructure during melt processing of $Bi_2Sr_2CaCu_2O_x$ (BSCCO-2212) doctor- bladed films, Physica C **218** 141-132.
2. Aksenova,T.D., Bratukhin, P.V., Shavkin, S.V., and Melnikov,V. L., Antipova,E.V., Khlebova, N.E., and Shikov, A.K. (1993) Texture formation in $Bi_2Sr_2CaCu_2O_x$/Ag tapes prepared by partial melt process, *Physica* **C205** 271-279.
3. Balachandran, U., Iyer,A.N., Haldar,P. Hoehn, Jr. J.G., and Motowidlo, L.R. (1994) Progress in development of tapes and magnets made from Bi-2223 superconductors, in K. Kristen and C. Burnham (eds.), *Proceedings of The Fourth International conference and Exhibition: World Congress on Superconductivity*, Vol. II, Orlando, FL. pp. 639-649.
4. Sandhage, H., Riley, Jr., G.N., and Carter, W. (1991) Critical issues in the OPIT processing of high-J_c BSCCO superconductors, *J. Metal*, **43** 21-27.
5. Heinrich, K.F.J. (1981) *Electron Beam X-ray Microanalysis*, Van Nostrand Reinhold Co., New York.
6. Fiori, C.E., Swyt, C. R. and Myklebust, R.L. (1991) *NIST/NIH Desktop Spectrum Analyzer Program and X-ray Database*, NIST Standard Reference Database No. 36.
7 Wong-Ng, W., Cook, L.P., and Jiang, F. (1998) The Primary Crystallization Phase Field of "2212" Phase and the Effect of Ag Addition," *J. Amer. Ceram. Soc.*, in print.
8. Cook, L.P. and Wong-Ng, W. (1998) Primary crystallization volume of BSCCO 2212: Applications to Crystal Growth and Melt Processing, in W. Wong-Ng, U. Balachandran and A.S. Bhalla (eds.), *Impact of Recent Advances in synthesis and Processing of Ceramic Superconductors*, Amer. Ceram. Soc., Westerville,OH, in print.
9. Shannon, R.D. (1976) Revised effective ionic radii and systematic studies of interatomic distances in halides and chalcogenides, *Acta Crystallogr.* **A32** 751-760.
10. Wong-Ng, W., Cook, L.P., Jiang, F., Greenwood, W., Balachrandran, U., and Lanagan, M. (1997) Subsolidus Phase Equilibria of the High T_c Pb-2223 Superconductor in the (Bi,Pb)-Sr-Ca-Cu-O System Under 7.5% O_2, *J. Mater. Res.* **12** [11] 2855-2863.
11. Barber, C.B., Dobkin, D.P. and Huhdanpaa, H. (1996). The quickhull alogorithm for convex sets, *ACM Transaction on Mathematical Software* **22** [4] 469-483.

FORMATION OF STABLE PHASE INCLUSIONS IN Bi-2212 AND (Bi,Pb)-2223 MATERIALS.

P.E.KAZIN, V.V.POLTAVETS, V.V.LENNIKOV,
R.A.SHUBA, E.A.EREMINA, YU.D.TRETYAKOV
Chemistry Department, Moscow State University, 119899 Moscow, Russia
M.JANSEN, B.FREITAG
Institut für Anorganische Chemie, Universität Bonn, 53121 Bonn, Germany
G.F. DE LA FUENTE, A.LARREA
Instituto de Ciencia de Materiales de Aragón, CSIC-Universidad de Zaragoza, 50015 Zaragoza, Spain

1. Introduction.

A promising way to enhance critical current density and irreversibility field of high T_c superconductor materials is to form material with complex microstructure, such as composite consisting of dense superconducting matrix with uniformly distributed fine inclusions of non-superconducting phases. These inclusions and related defects can act as effective flux pinning centers.

The second phase can be obtained by shifting chemical composition of the system from a single phase region, or by addition of new component(s) to the system. In the latter case a variety of alien phases based on the additional component can be formed along with the main superconductor phase. In the present work we consider the formation of such alien phases in Bi-2212 and Bi-2223 based materials.

2. Experimental.

The system Bi-(Pb)-Sr-Ca-Cu-A-O, where A is an additional chemical element, was considered. The homogeneous precursors, containing all components mixed at submicron level, were processed to obtain two phase composites {superconductor phase / A-rich phase inclusions}. The precursors were prepared using oxalate co-precipitation, sol-gel and nitrate solution decomposition methods. Phase assemblage was investigated in solid state for

G. Van Tendeloo et al. (eds.), High-Temperature Superconductors and Novel Inorganic Materials, 69–74.

Bi-2212 and (Bi,Pb)-2223 containing systems as well as above the melting point for Bi-2212 containing system. The Bi-2212 based samples were fabricated as sintered ceramics (840-860°C) , melt textured bulk materials (900-840°C) and in particular cases as silver sheathed tapes. The (Bi,Pb)-2223 based samples were prepared as sintered ceramics only (845-860°C). Fraction of the second phase in the composites was up to 50 mass %.

In choosing the additional component the following requirements for the doping were taken into account: i) chemical compatibility of the alien phase with the superconductor at the processing conditions; ii) low solubility of the dopant A in the superconductor phase; iii) ability of the alien phase to form fine and well dispersed inclusions.

3. Results and discussion.

3.1. BI-2212 WITH COMPATIBLE PHASES.

3.1.1. *Phase relations in the doped system.*

Some results of the Bi-2212 system doping with different metal cations are presented in Table 1.

TABLE 1. Bi-2212 with A-containing surpluses.

Doping element A	Solubility of A in Bi-2212 (mol/mol)	T_c (K)	A-rich phase compatible with Bi-2212	Solubility of A in BSCCO melt (mol A / mol 2212), at temp.
Zr	<0.05	93	$SrZrO_3$	<0.05, 1300°C
Hf	<0.05	93	$SrHfO_3$	<0.05, 1300°C
Sn	<0.03	90	$Sr_{1-x}Ca_xSnO_3$ ($x \approx 0.1$)	<0.05, 1300°C
Mg	<0.05	92	$Mg_{1-x}Cu_xO$ ($x \approx 0.1$)	<0.1, 1300°C
Al	<0.03	90	$Sr_{1.7}Ca_{1.3}Al_2O_6$ $BiSr_{1.5}Ca_{0.5}Al_2O_z$	ca. 0.05, 900°C ca. 1, 1300°C
Ga	0.15	80	$Sr_{1.3}Ca_{1.7}Ga_2O_6$ $Bi_{0.2}SrCa_{0.8}Ga_2O_z$ Bi-Sr-Ca-Cu-Ga-O*	0.6-0.8, 900°C
In	<0.05	82	$Sr_{1-x}Ca_xIn_2O_4$ ($x \approx 0.55$)	<0.1, 900°C

*approx. comp. $Bi_{1.7}Sr_{2.3}Ca_{0.6}Cu_{1.6}Ga_{0.4}O_z$ by EDX

The solid state solubility of the dopants in Bi-2212 phase proved to be limited to low values, as was determined by EDX analysis. As a consequence the T_c of the superconductor has not been changed significantly by the doping. Presence of corresponding surpluses results in formation, along with Bi-2212, some compatible phases enriched with component A. These phases equilibrate with Bi-2212 at processing conditions.

The second phase compositions can be considered in terms of Lewis acid-base properties. The BSCCO system is quite basic. Most of the binary oxides AO_x are relatively acid, so they consume preferably SrO and CaO as most basic oxides and reduce their acidity to be adjusted to that for Bi-2212 phase. Magnesium oxide is basic enough itself so that it retains its structure, although small substitution of Mg by Cu takes place.

Majority of the alien phases co-exist with BSCCO melt as well. The solubility of Zr-, Hf-, Sn-, and Mg-based phases is very low even at 1300°C. This feature can be of advantage to produce crucibles and substrates stable against BSCCO melt. The aluminates proved to be slightly soluble at 900°C, but considerably - at 1300°C. Ga is found in the melt in large amount even at 900°C, In is not observed in the melt at this temperature.

The presence of Zr-, Hf-, Sn-, Mg-, and In-containing surpluses do not change phase assemblage in BSCCO around the melting point of Bi-2212 while Al- and Ga-based surpluses influence strongly on the mechanism of Bi-2212 melting. The following reactions, taking place on the system melting in air, can be outlined.

$$\text{Undoped Bi-2212} \xrightarrow{885^\circ C} \text{Liquid} + (Sr,Ca)CuO_2 + \text{Cu-free} \qquad [1]$$

$$\text{Bi-2212} + BiSr_{1.5}Ca_{0.5}Al_2O_z \xrightarrow{870^\circ C} \text{Liquid} + (Sr,Ca)_3Al_2O_6 \qquad [2]$$

$$\text{Bi-2212} + Bi_{0.2}SrCa_{0.8}Ga_2O_z \xrightarrow{860^\circ C} Bi_{1.7}Sr_{2.3}Ca_{0.6}Cu_{1.6}Ga_{0.4}O_z + L_1 \rightarrow$$

$$\xrightarrow{<900^\circ C} L_2 + (Sr,Ca)_3Ga_2O_6$$

Bi-2212 with $BiSr_{1.5}Ca_{0.5}Al_2O_z$ melts at 870°C giving liquid and another Al-based solid phase. No cuprates and Cu-free appear as it is in the case of undoped Bi-2212. The similar situation takes place for the Ga-based phase surplus, but in addition between 860 and 900°C a new complex phase forms which includes all cations of the doped system.

3.1.2. *Melt processed composites {Bi-2212 / A-rich phase inclusions}.*

In the doped systems the crystallization of Bi-2212 phase from the melt takes place when alien A-rich phase presents as solid particles. The particles can be trapped by growing Bi-2212 grains or displaced forming agglomerations. The size of the particles depends significantly on the preliminary treatment of the precursors and nature of the alien phase.

Characteristics of the melt-crystallized composites {Bi-2212 / alien phase} are presented in Table 2. The average size of alien particles varies from submicron to tens of microns for different alien phases. It correlates qualitatively with solubility of the dopants in the BSCCO melt. Different affinity to aggregation is observed. The most uniform distribution of the alien precipitates in the superconductor matrix is found for $Sr_{1.7}Ca_{1.3}Al_2O_6$

and $Mg_{1-x}Cu_xO$. Zr-, Sn-, and In-rich grains are partially imbedded in large Bi-2212 lamellas, partially aggregated on the superconductor grain boundaries. It is interesting that two Al-based phases behave differently in the composite: $Sr_{1.7}Ca_{1.3}Al_2O_6$ grains are found imbedded in Bi-2212 lamellas while $BiSr_{1.5}Ca_{0.5}Al_2O_z$ aggregated in-between. The Ga-containing phases form big grains that is apparently connected with high solubility of Ga in the melt. The grains are surrounded by the matrix consisted of thin intergrown lamellas of Bi-2212 and Bi-2201 phases. In contrast to other alien phases $(Sr,Ca)In_2O_4$ forms anisotropic rod-like grains which can be more effective centers for flux pinning.

TABLE 2. Characteristics of melt-crystallized composites {Bi-2212 / A-rich phase}.

A-rich phase	Precursor (method)	Average grain size	State in the Bi-2212 matrix	Pinning effect: R^* value
$SrZrO_3$	oxalates	0.2 μm	in 2212, agglom.	0.092 (tapes)
$Sr_{1-x}Ca_xSnO_3$	sol-gel	0.2 μm	in 2212, agglom.	0.074 (bulk)
$Mg_{1-x}Cu_xO$	nitrates	1 μm	in 2212	0.021 (bulk)
$Sr_{1.7}Ca_{1.3}Al_2O_6$	nitrates	2-5 μm	in 2212	0.026 (bulk)
$BiSr_{1.5}Ca_{0.5}Al_2O_z$	nitrates	2-5 μm	agglom.	0.016 (bulk)
$Sr_{1.3}Ca_{1.7}Ga_2O_6$	nitrates	20 μm	between 2212	-
$Bi_{0.2}Sr_{0.9}Ca_{0.9}Ga_2O_z$		10 μm	lamellas	
$Sr_{1-x}Ca_xIn_2O_4$	nitrates	0.5x5 μm rods	in 2212, agglom.	-
undoped	oxalates			0.01 (bulk) 0.03 (tape)

$^*R=\Delta M(60K,\ 10mT)/\ \Delta M(5K,\ 1T)$

The interface between Bi-2212 and alien phase inclusion is usually sharp without any intermediate layers which supports compatibility of the alien phases with the superconductor phase. The inclusions of $Mg_{1-x}Cu_xO$ and $Sr_{1.7}Ca_{1.3}Al_2O_6$ exhibit especially sharp and smooth interface. One can speculate that most uniform distribution of such inclusions is connected with lower interface energy in comparison with the other alien phases. Although there is no crystallographic orientation of the inclusions in reference to the matrix even in the latter cases.

In the last column of Table 2 *R* values are shown which reflect the effect of the second phase on the pinning. The *R* value is defined as a ratio between hysteresis magnetization at 60K and 10mT to that at 5K and 1T. The increase in this value expresses the increase in number of pinning centers with high energy which are still active at high temperatures. Substantial increase of *R* is observed for $SrZrO_3$ and $(Sr,Ca)SnO_3$ precipitates which have the smallest average grain size of 0.2 μm. For the other second phases the increase of this value is not pronounced.

Earlier we have reported on pinning effect of the second phase of $SrZrO_3$ [3], $(Sr,Ca)SnO_3$ [4], $Mg_{1-x}Cu_xO$ [5], and $Sr_{1.7}Ca_{1.3}Al_2O_6$ [2]. The presence of the precipitates was usually exhibited in increase of J_c and its field stability at high temperatures while at low temperature the J_c adopted close values for the doped and undoped samples.

3.2. (BI,PB)-2223 WITH COMPATIBLE PHASES.

It is well known that (Bi,Pb)-2223 phase has substantially narrower compositional and temperature stability range and slower formation kinetics than Bi-2212. So the more delicate approach is required in search for compatible phases. In our work the compounds were used which had been found to be compatible with Bi-2212 as well as some other dopants which were known not to be bad for (Bi,Pb)-2223 phase formation and properties.

The data on the (Bi,Pb)-2223 based systems are shown in Table 3.

TABLE 3. (Bi,Pb)-2223 with A-containing surpluses.

Doping element A	Solubility of A in (Bi,Pb)-2223 (mol/mol)	T_c (K)	A-rich phase compatible with (Bi,Pb)-2223	Dopant effect on (Bi,Pb)-2223 phase formation
Zr	<0.05	110	$Sr_{1-x}Ca_xZrO_3$, $x \approx 0.1$	suppresses
Sn	<0.05	109	$Sr_{1-x}Ca_xSnO_3$	suppresses
Al	<0.03	108	$BiSr_{1.5}Ca_{0.5}Al_2O_z$	improves
V	ca. 0.05	106	$Sr_{2.9}Ca_{0.2}Cu_xV_2O_z$	improves
B	-	109	$(Ca,Sr)_2B_2O_5$	improves
S	<0.1	109	$SrSO_4$	improves

Solid state solubility of the additional elements in (Bi,Pb)-2223 has been determined to be very low, and T_c has remained nearly the same as for undoped (Bi,Pb)-2223. The compatible complex oxides, in addition to element A, include mostly Sr and Ca cations as in the case of Bi-2212 system. It is interesting that only one Al-rich phase was found to equilibrate with Bi-2223 in contrast to two phases for Bi-2212. The other phase $(Sr,Ca)_3Al_2O_6$ destroys (Bi,Pb)-2223 yielding (Bi,Pb)-2212 phase and $BiSr_{1.5}Ca_{0.5}Al_2O_z$ among the products.

The alien phase surplus in the initial precursor influences significantly on (Bi,Pb)-2223 formation process. Zr- and Sn-based surpluses decrease the higher homologue formation rate. For $SrZrO_3$ surplus it was observed that agglomerates of submicron $Sr_{1-x}Ca_xZrO_3$ particles attract large amount of Pb-rich liquid thus hindering the Bi-2212 to Bi-2223 transformation due to lack of the liquid [6].

The surpluses of Al-, V-, B-, and S-based phases accelerate the Bi-2223 phase formation process. XRD patterns of partially reacted samples reveals the presence of Bi-2201 phase. This phase can be formed from liquid on cooling of the samples. Additional amount of the liquid during the processing can be the reason for the improved Bi-2223 phase formation.

The alien phase particles are usually of order of several microns and are distributed between Bi-2223 lamellas. The presence of low fraction of the alien phase does not affect hysteresis magnetization for the composites at 5 and 77K High fraction of V-containing phase and $SrSO_4$ causes decrease in hysteresis magnetization at 77K.

4. Conclusion.

Therefore the range of alien phases was established to be compatible with Bi-2212 and (Bi,Pb)-2223, without significant change of the superconductor phase composition and T_c. These phases formed stable precipitates in Bi-2212 melt-crystallized composites and had different ability to segregation. The effect of alien phase precipitates for Bi-2212 was expressed in the increase of J_c and its field stability at high temperatures. The strongest positive effect on flux pinning was caused by the smallest precipitates of $SrZrO_3$ and $(Sr,Ca)SnO_3$. The presence of alien phase surplus in the BPSCCO precursor either favored or hindered the (Bi,Pb)-2223 formation.

5. Acknowledgments.

The work was supported by Russian Ministry of Science (project "Composite"), Russian Foundation of Basic Research (project 97-03-33249a), and BMBF (Germany, project 13N6761).

6. References.

1. Hellstrom, E.E. and Zhang, W. (1997) Important aspects of the melt related to processing $Bi_2Sr_2CaCu_2O_x$ conductors, in Y.Abe (ed.), *Superconducting Glass-Ceramics in Bi-Sr-Ca-Cu-O: Fabrication and its Application*, World Scientific Publishing Co., Singapore, pp.19-43.
2. Kazin, P.E., Poltavets, V.V., Tretyakov, Yu.D., Jansen, M., Freitag, B., and Mader, W. (1997) Study on the superconducting composite material formation in the system $Bi_2Sr_2CaCu_2O_{8+x}$/Al-containing phases, *Physica C* **280**, 253-266.
3. Kazin, P.E., Jansen, M., Larrea, A., de la Fuente, G.F., and Tretyakov, Yu.D. (1995) Flux pinning improvement in Bi-2212 silver sheathed tapes with submicron $SrZrO_3$ inclusions, *Physica C* **253**, 391-400.
4. Kazin, P.E., Shuba, R.A., Poltavets, V.V., Tretyakov, Yu.D., Jansen, M., and Wagner, N. (1996) Doping of Bi-2212 with Sn and Al containing phases for flux pinning improvement, *Abstr. IX Trilateral Seminar on HTSC*, September 23-26, Gabelbach, Germany.
5. Lennikov, V.V., Kazin, P.E., Putlayev, V.I., Tretyakov, Yu.D., and Jansen, M. (1996) Influence of magnesium oxide on the properties of $Bi_2Sr_2CaCu_2O_{8+x}$ high T_c superconductor synthesized by melt techniques, *Zhurn. Neorgan. Khimii (Russ.)* **41**, 911-915.
6. Kazin, P.E., Makarova, M.V., Jansen, M., Adelsberger, Th., and Tretyakov, Yu.D. (1997) Interaction of Bi(Pb)-2223/2212 ceramics with $Sr_{1-x}Ca_xZrO_3$, *Supercond. Sci. Technol.* **10**, 616-620.

ATOMIC LAYER MOCVD OF OXIDE SUPERCONDUCTORS AND DIELECTRICS

S. Oda, S. Yamamoto, Z. Wang, H. Tobisaka and K. Nagata
Research Center for Quantum Effect Electronics
Tokyo Institute of Technology
O-okayama, Meguro-ku, Tokyo 152-8552, Japan

1. Introduction

Although metalorganic chemical vapor deposition (MOCVD) is regarded as a manufacturable technology and widely used in semiconductor industries, particularly for the feature of high throughput and large area uniformity, it has not been a common method for the preparation of thinfilms of oxide superconductors [1-3] because of several problems, such as degradation of precursors, poor reproducibility, and outgrowth of boulders. Recently, a growth technology of YBCO thinfilms has been improved significantly by newly developed third generation precursors, ultrasonic sensors, and layer-by-layer deposition method based on understanding of growth mechanism.

2. Precursors

Beta-diketonate, such as dipivaloylmethanate (DPM), compounds of metals are used as precursors for MOCVD of oxide superconductors. Among three metallic sources of YBCO, barium precursors often cause a problem of degradation. Since the size of barium atom is large compared to ligands, barium precursors tend to form an oligomer by associating neighboring molecules resulting in the formation of less volatile complex. The second generation barium precursors [4] employ adducts, such as phenanthroline, to prevent oligomerization by the effect of steric hindrance and partially solved the problem of degradation by reducing operating temperature from 220°C to 180°C.

Recently, the third generation precursors with a low melting point and can be used in the liquid form have been developed [5,6]. They are sufficiently volatile as a source gas for MOCVD even at 100-130°C and free of the problem of sintering as is encountered in the case of solid precursors. Although fluorinated sources are more volatile, one needs to incorporate water vapor in the reactor tube to eliminate fluorine compounds in the films [5,7,8]. Superconductivity of thinfilms is degraded by water vapor and residual fluorine compounds.

Ba(DPM)2-tetraene, originally synthesized for Sr-precursor in the Tri Chemical Laboratory Inc. [6], is a nonfluorinated source and the most commonly used today for MOCVD of YBCO and related HTS films [9-13]. High volatility of Ba(DPM)2-tetraene

G. Van Tendeloo et al. (eds.), High-Temperature Superconductors and Novel Inorganic Materials, 75–78.

made possible high deposition rate of YBCO films at a lower vaporizing temperature. Compositional deviation of the films was within 4% [9]. A YBCO film of good surface morphology over 60μm x 60μm and good crystalinity with a minimum channeling yield of 2.6% was obtained [10]. Constant supply rate of Ba was achieved for 130h when the vaporizing temperature was kept at 135°C[11]. A diffusion layer between MgO substrate and the YBCO film, found from samples prepared by fluorinated precursors, disappeared from samples prepared by nonfluorinated source [12]. The stability of precursors allowed us a systematic study of layer-by-layer growth of YBCO films [13].

3. Process monitoring

In MOCVD the source gas supply rate is determined by the saturated vapor pressure of precursors (defined by the vaporizer temperature) and the flow rate of carrier gas, which are measured and controlled precisely by using PID temperature controller and the mass flow controller, respectively, as long as the vapor pressure of the precursor and the efficiency of mass transport are constant as is the case of MOCVD of common semiconductors. However, this is not the case of MOCVD of HTS. Therefore, in situ monitoring of the precursor flow rate is desirable for a reproducible growth of HTS thinfilms. Process diagnostic methods for MOCVD precursors of HTS include ultrasonic sensors [14,15], ultraviolet spectroscopy [16] and infrared absorption spectroscopy [17].

The ultrasonic sensors, applied for the concentration monitor of binary gas mixture, are based on the principle that the velocity of sound is determined by the molecular mass and the ratio of the specific heats (Cp/Cv) of the species in the gas phase. Researchers in STI used the ultrasonic sensors to control the composition of metal precursors and successfully prepared high quality YBCO films for microwave applications [14,15].

We also have studied the effect of the ultrasonic sensor for the process monitoring of MOCVD of YBCO. We found that the use of helium gas, instead of argon, allowed higher accuracy measurements of concentration of precursors due to a large difference of molecular mass. To our surprise, precursor supply rates fluctuated even when we used Cu(DPM)2, which was believed much more stable than Ba precursors. We found out that the fluctuation was due to improper layout of temperature controller of the vaporizer. The validity of realtime monitoring of precursor was proved. We have successfully stabilized the precursor supply rate of Ba by employing feedback systems.

Ultraviolet absorption of precursors was employed for in-line process monitoring of MOCVD of YBCO [16]. Infrared absorption spectroscopy was used to study decomposition process of precursors of MOCVD [17].

Optical reflectance measurements also provided a method for *in situ* diagnostics of crystal growth process in MOCVD of HTS thinfilms [18]. The thickness of films and the mode of crystal growth are deduced from the reflectance measurements. Recently, we investigated spectroscopic ellipsometry for the realtime monitoring of the crystal growth process of YBCO. The crystal growth mechanism in the initial stage of the growth is distinguishable from the measurement of time-dependent ellipsometry signal. Phase transition due to oxyden doping, which is closely related to the critical temperature is also observable.

4. Boulders

The problem of boulders, or second phase precipitates, formation was solved using atomic layer-by-layer supply of precursors in MOCVD. A very smooth surface film of c-axis oriented YBCO with roughness of less than monomolecular layer over 10μm x 10μm, free of boulders, has been obtained [4].

We have clarified that the mechanism of the boulder formation is due to the preferential nucleation of CuO at defects of substrate (mostly SrTiO3) surface [19]. CuO, once formed, grows much faster than YBCO layers. The important thing to eliminate the boulder formation is to block the contact between Cu atoms and the defects. Atomic layer-by-layer method is very effective for that purpose. We have successfully prepared YBCO films free of boulders by starting with BaO layer, which is not so stable as CuO and presumably changes to YBCO structure after the first cycle of gas source supply is completed. We have also found that a block-by-block deposition method is effective to avoid boulder formation.

A very large monomolecular terrace with length of 0.33μm for YBCO films on SrTiO3 substrates and 0.66μm on NdGaO3 substrates may be due to the enhanced migration of molecular species on the surface [13]. Very high critical current densities of $3x10^7 A/cm^2$ at 4.2K and $3x10^6 A/cm^2$ at 77K have been obtained. These films are promising for the application of superconductor-insulator-superconductor trilayer junctions.

5. Dielectrics

Dielectric materials for tri-layer superconducting junctions were investigated. The problem of surface roughness of CeO2 films deposited on YBCO was solved by two step growth; the first 2nm-thick film was deposited at 570°C, while the rest of the film was grown at 650°C. The resistivity of PrBaCuO films was increased by Ga doping.

We also found that atomic layer MOCVD was effective for the epitaxial growth of SrTiO3 films with a very smooth surface and high crystalline quality on MgO substrates. The FWHM of X-ray rocking curve of 0.18° is much narrower than ever reported for this system [20].

6. Device applications

Superconducting devices were prepared using YBCO ultrathin films derived by atomic layer MOCVD. The crystal structure of HTS is consisted of layers of superconductors and insulators, which is considered as natural Josephson junction. We have investigated the intrinsic Josephson effect in YBCO thinfilms patteren by AFM lithography. We have found voltage steps and hysterisis effect unique to the intrinsic Josephson effects [21].

We have proposed a novel structure of superconducting devices, planar type field effect devices. Contrary to the conventional stacked-type field effect devices, the planar structure is free from the problem of degradation of interface between superconducting layer and dielectric layer. We have prepared a narrow YBCO strip line with 2mm-wide and side gate 2mm-apart. We have found that current is modulated by gate bias of 20V [22,23].

Conclusion

Improved significantly, MOCVD has become a suitable choice for thinfilm growth of HTS.

Acknowledgements

Financial support by a Grant-in-Aid for Scientific Research from the Ministry of Education, Science, Culture and Sports, by NEDO under the management of FED, and by the Yazaki Memorial Foundation is greatly acknowledged.

References

[1] M. Leskela, H. Molsa and L. Niinisto, Supercond. Sci. Technol. **6**, 627 (1993).
[2] D.L. Schulz and T.J. Marks, in CVD of Nonmetals, Ed. W.S. Rees, (VCH, Weinheim, 1996)
[3] I.M. Watson, Chem. Vap. Deposition, **3**, 9 (1997).
[4] S. Oda, H. Zama and S. Yamamoto: J. Cryst. Growth, **145**, 232 (1994).
[5] B.C. Richards, S.L. Cook, D.L. Pinch and G.W. Andrews, J. Phys. (Paris) IV, **5**, C5-407 (1995).
[6] T. Kimura, H. Yamauchi, H. Machida, H. Kokubun and M. Yamada, Jpn, J. Appl. Phys. **33**, 5119 (1994).
[7] Y. Ito, Y. Yoshida, Y. Mizushima, I. Hirabayashi, H. Nagai and Y. Takai, Jpn. J. Appl. Phys. **35**, L825 (1996).
[8] Y. Toshida, Y. Ito, I. Hirabayashi, H. Nagai and Y. Takai, Appl. Phys. Lett. **69**, 845 (1996)
[9] H. Zama and T. Morishita, Jpn. J. Appl. Phys. **35**, L770 (1996).
[10] H. Zama and T. Morishita, Ext. Abst. Int. Workshop on Superconductivity, Iwate, 179 (1996).
[11] H. Zama, N. Tanaka, T. Morishita, Mater. Sci. Eng. B, to be published (1998).
[12] H. Nagai, Y. Yoshida, Y. Ito, S. Taniguchi, I. Hirabayashi, N. Matsunami and Y. Takai, Supercond. Sci. Technol. **10**, 213 (1997).
[13] S. Yamamoto, A. Kawaguchi, K. Nagata, T. Hattori and S. Oda, Appl. Surf. Sci. **112**, 30 (1997).
[14] E.J. Smith, Ext. Abst. Int. Workshop on Superconductivity, Maui, 117 (1995).
[15] J. Musolf, J. Alloys Compounds, **251**, 292 (1997).
[16] W.J. DeSisto, B.J. Rappoli and J.C. Pazik, Ext. Abst. Int. Workshop on Superconductivity, Maui, 473 (1995).
[17] A.Y. Kovalgin, F. Chabert-Rocabois, M.L. Hitchman, S.H. Shamlian and S.E. Alexandrov, J. Phys. (Paris) IV, **5**, C5-357 (1995).
[18] S. Oda, K. Sakai and H. Zama, Appl. Surf. Sci. **75**, 259 (1994).
[19] S. Oda, S. Yamamoto and A. Kawaguchi, J. Phys. (Paris) IV, **5**, C5-379 (1995).
[20] Z. Wang and S. Oda, Jpn. J. Appl. Phys. **37**, in press (1998).
[21] S. Yamamoto, A. Kawaguchi and S. Oda, Physica C, **293**, 244-248 (1997).
[22] S. Suzuki and S.Oda, Physica C, **282-287**, 2495-2496 (1997).
[23] S.Suzuki, H.Tobisaka and S.Oda, Jpn. J. Appl. Phys. **37**, 492-495 (1998).

Single source MOCVD of HTSC films onto movable substrates

G Wahl, O Stadel, L Klippe
Institut für Oberflächentechnik und Plasmatechnische Werkstoffenentwickling (IOPW), Technische Universität Braunschweig, Bienroder Weg 53, 38108 Braunschweig, Germany

A R Kaul, O Y Gorbenko, S V Samoylenkov
Chemistry Department, Moscow State University, 119899 Moscow, Russia

Abstract. Superconducting $YBa_2Cu_3O_7$ films on various substrates were grown by a liquid single source MOCVD technique on moving substrates. A CVD reactor was constructed for a continuous deposition on tapes using band evaporator. $YBa_2Cu_3O_7$ films were deposited on IBAD-buffered YSZ tapes, polycrystalline YSZ tapes and single crystal $SrTiO_3$, $LaAlO_3$, MgO and YSZ substrates. Strong preferential c-axis oriented growth and in-plane a-b texture of $YBa_2Cu_3O_7$ films on these substrates were observed by XRD. The critical current density of $YBa_2Cu_3O_7$ / $SrTiO_3$ films at 77K was larger than 10^6 A/cm^2. On IBAD tapes we obtained $j_c \geq 4.7\ 10^4$ A/cm^2.
Another reactor for the simultaneous deposition of oxide films on both sides of the substrates over the area as large as 20 × 20 mm^2 was designed. The (00*l*) epitaxial thin films of (Ce, Gd)O_2 on (1-102)Al_2O_3, as well as $HoBa_2Cu_3O_7$/ (Ce,Gd)O_2 / Al_2O_3 heterostructures were prepared by the technique.

1. Introduction

The production of high quality HTSC-films on long tapes promises industrial applications like superconducting fault current limiters, transformers and magnets.
For high critical current densities of YBCO it is necessary to use substrates, which induce in plane textured films. Mechanically treated, rolling-assisted biaxially textured metal substrates (RABiTS) [1] or oriented buffer layers on polycrystalline tapes deposited by Ion Beam Assisted Deposition or by inclined deposition seem to be possible technical substrates.
Microwave devices, such as resonators are possible applications for HTSC-films, too. For resonators double sided deposited HTSC-films are of large practicable interest.
Generally MOCVD allows high growing rates and industrial large scale applications. In conventional multi source MOCVD systems degradiation of the Ba-precursor

G. Van Tendeloo et al. (eds.), High-Temperature Superconductors and Novel Inorganic Materials, 79–84.

complicates the time stability of film stoichiometry. This problem can be solved by single source techniques [2, 3].

2. Experimental

2.1. EVAPORATING SYSTEMS

The precursors used in the following evaporating systems are $Y(thd)_3$, $Ba(thd)_2$ and $Cu(thd)_2$ with various adducts (triglyme, tetraglyme and ortho-phenantroline).
The band evaporator is a liquid single source system. The solution is feeded continuously onto a band and in two different evaporating zones (50-70°C and 270-300°C) the solvent is separated from the precursors. With N_2 carrier gas the precursors are transported into the reactor and the solvent into a cooling trap, where it condense. So no solvent vapours are present in the reaction zone and the solvent can be recycled. The system can be continuously used during more than 20 hours for one deposition process. The system is reported in detail elsewhere [4].

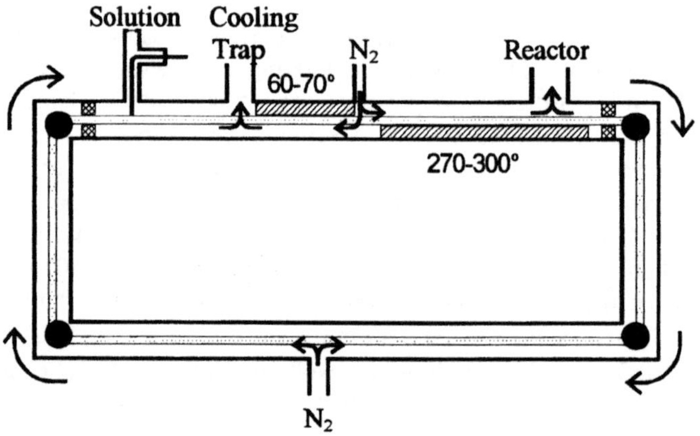

Figure 1: Band evaporator.

The aerosol system is another liquid single source technique [5]. The aerosol, droplets of micronic size, are generated by an ultrasonic excitation. The aerosol is transported into a heated zone, where the precursors and the solvent evaporate. This gas phase react with oxygen in the CVD reactor.
A solid single source system is the powder flash evaporator. The solid precursors are mixted and then filled in a horizontal tube. The powder is moved by a vibration device and falls into the evaporation zone. The evaporated precursors are transported into the CVD reaction chamber by a nitrogen or argon gas flow[6].

2.2. MOCVD REACTORS

The CVD reactor constructed for the deposition on long tapes is a cylindrical hot wall reactor (inner diameter 4 cm, height 5 cm) with two slits on each side. The tape is continuously moved through these slits and the deposition zone. The deposition rate can be more than 1,5 µm/min. As it was demonstrated by deposition of high textured YSZ on a unmoved polycrystalline Al_2O_3 substrate (thickness of 50 µm in 30 minutes) [7].

TABLE 1. Experimental conditions for YBCO deposition on tapes

Deposition temperature	800-850°C
Heated pipes	270-300°C
Diglyme evaporation zone	60-70°C
Precursor evaporation zone	270-300°C
Total pressure	500 Pa
Oxygen partial pressure	100-450 Pa
Deposition rate	3-15 cm/h (thickness 0.5-1 µm)
Total gas flow	22.8 l_N/h
Precursor mole fraction	1-4 10^{-3}

For the deposition on both sides of a substrate during one deposition process a substrate holder in a hot wall reactor was constructed. By rotation of the substrate holder the position of the surface in relation to the gas flow was continuously changed. The metal frame was asymmetrical connected with rotation axes. In this way the rotation around the axis normal of the substrate surface resulted in the turning over of the frame around the perpendicular axis (Fig. 2). The construction allowed the simultaneous deposition on both sides of a square substrate with a size 20 mm x 20 mm. The substrate rotation frequency was 5-10 Hz.

Figure 2: Substrate holder for the double side deposition.

3. Results

3.1. TAPE REACTOR

To show the capability of the system YBCO-films were deposited on single crystalline substrates (YSZ, MgO, $SrTiO_3$, $LaAlO_3$). The size of the substrates were 1 cm x 1 cm. These substrates were fixed on a substrate holder and moved continuously through the tape reactor during deposition. We obtained on all of these substrate strong c-axis oriented YBCO growth (Fig. 3) and in plane orientation with FWHM(103 or 102) = 0.6-1.8°. On (100)YSZ we optimised the deposition conditions and measured only 90°-in plane orientation of the YBCO(103)-peak without buffer layer (Fig. 4).

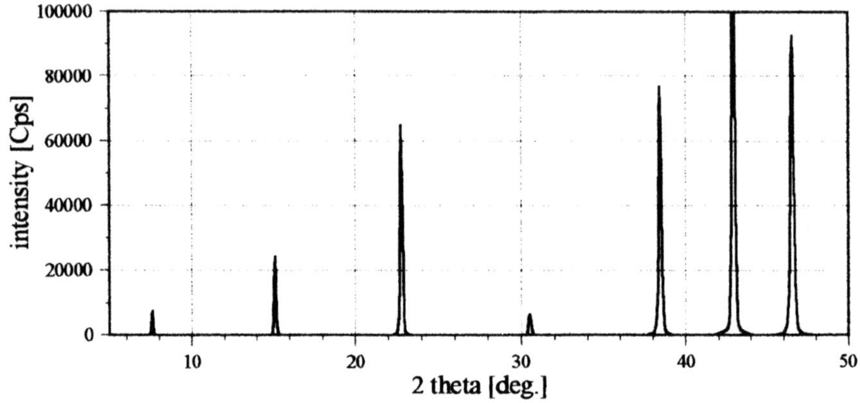

Figure 3: 2 theta scan of YBCO on (100)MgO.

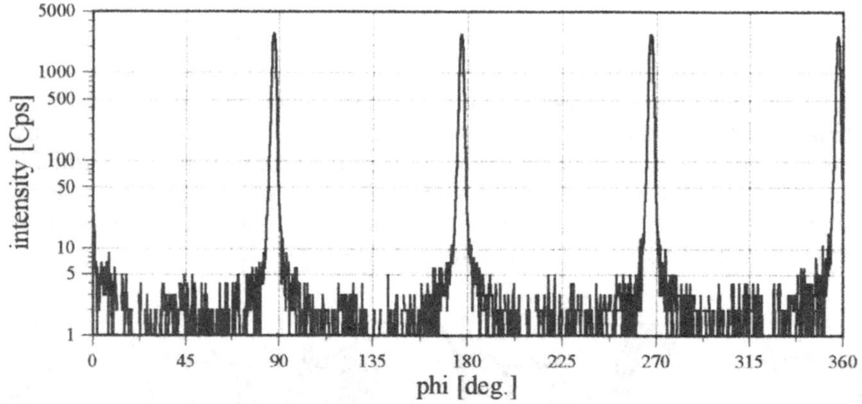

Figure 4: Phi-scan of (103) of YBCO on (100)YSZ; FWHM = 1.8°.

The superconducting transition temperature of the YBCO film on (100)YSZ was T_c = 90 K, measured by ac-magnetisation measurement.

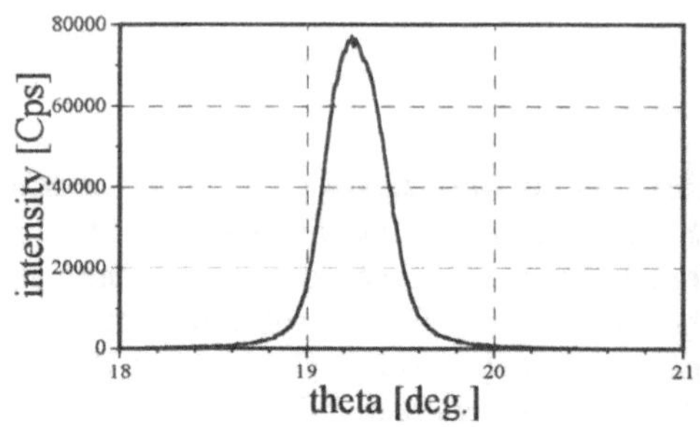

Figure 5: Rocking curve of YBCO(005)-peak on (100)MgO; FWHM = 0,37°.

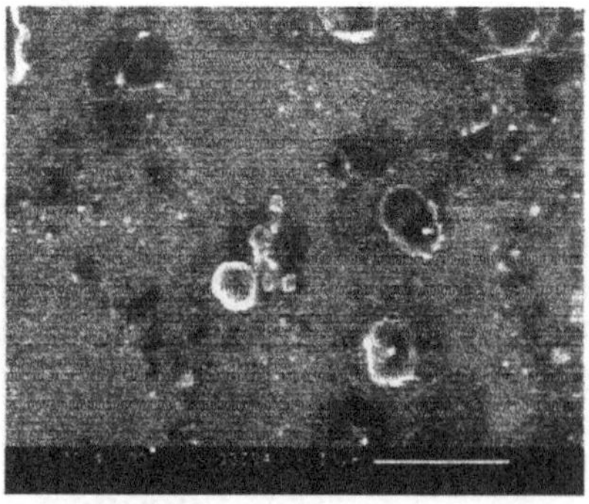

Figure 6: SEM photo of YBCO single crystalline $LaAlO_3$; thickness: 1μm.

On (100)MgO we produced YBCO-films (T_c = 88 K) with a high quality in plane (FWHM(103) = 0,95°) and out of plane orientation (rocking curve of FWHM(005) = 0,37°, s. Fig. 5).
YBCO-films on single crystalline $LaAlO_3$ (SEM-photo s. Fig. 6) and $SrTiO_3$ exhibit as well strong c-axis and in plane orientation (FWHM(102) = 0,6°) and the transition temperature were 90 K. Only on $SrTiO_3$ the critical current density was measured (j_c (77 K) = 2 10^6A/cm^2).
To obtain high quality YBCO on tapes we started deposition on IBAD buffered polycrystalline YSZ tapes. The in plane orientation of the YBCO was FWHM(103) = 10° (Fig. 7), which depended on the YSZ buffer layer (FWHM(220) = 16°). The critical current density was determined by dc-I_c-measurement (I_c (77 K) ≥ 4.7A, j_c (77 K) ≥ 4.7 10^4A/cm^2; at I = 4.7 A the sample broke, therefore I_c could not be detected). With a theta-2theta scan some a-axis orientation was measured. So on IBAD buffered tapes the deposition conditions must be further optimised.

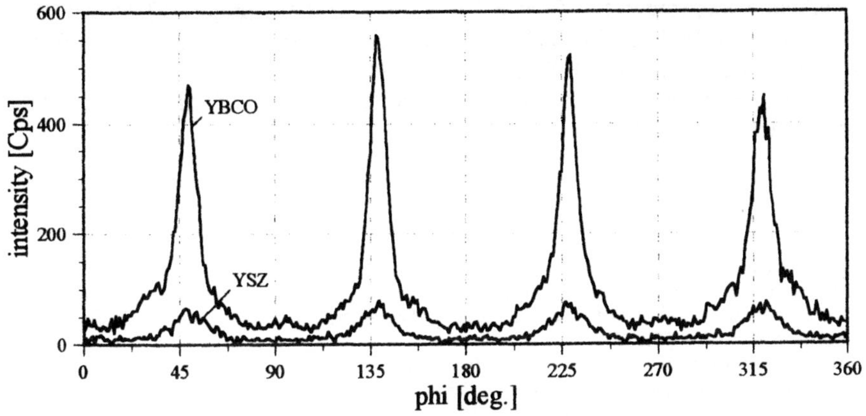

Figure 7: Phi-scan of (103) of YBCO (FWHM = 10°) and (220) of YSZ IBAD buffer (FWHM = 16°).

3.2. DOUBLE SIDE DEPOSITION

In this MOCVD-system were prepared HoBCO / (001)$CeO_2(Gd_2O_3)$ / (1-102)Al_2O_3 heterostructures and HoBCO films on single crystalline (100)YSZ.
The uniformity of the Gd concentration for $CeO_2(Gd_2O_3)$films, measured by EDX over the main diagonal of the sample (22 mm), was achieved with 15±1%. The (100) $CeO_2(Gd_2O_3)$ films were grown epitaxial on the Al_2O_3 substrate with a variation of the thickness between 13-33 nm.
The deposition of HoBCO films on (001)$CeO_2(Gd_2O_3)$ buffered Al_2O_3 and on (100)YSZ resulted in c-oriented superconducting films. The following uniformity over the main diagonal of the sample (22 mm) was obtained: 14.9±0.4% Ho; 31.0±1.3% Ba; 53.1±0.7% Cu on $CeO_2(Gd_2O_3)$ buffer layer and 17.0±1.0% Ho; 32.2±0.8% Ba; 50.8±1.2% Cu on YSZ. The HoBCO films on YSZ were superconducting with T_c = 90 K and j_c = 5 10^5 A/cm^2.

4. Conclusion

In two different MOCVD-systems superconducting films were successfully prepared on moving substrates in a continuous process. The YBCO films on travelling single crystalline substrates ($SrTiO_3$, $LaAlO_3$, MgO and YSZ) and IBAD buffered tapes exhibit high superconducting T_c and j_c values. On IBAD buffered tapes the deposition will be further optimised.
For industrial applications the deposition zone of the demonstrated MOCVD-reactor (diameter: 1.5 cm) must be enlarged. So on long tapes the YBCO films will be deposited in a appropriate time. The main task for the future is to scale up this MOCVD-system and to increase the deposition rate.

Acknowledgements

The authors acknowledge the financial support of BMBF (project 13 N 6947/5), EU (Brite-EuRam II Project No.7887), Volkswagen Foundation (project I/73 628) and Russian National Program on HTSC (grant No. 93-458).
Authors are also thankful to Dr. Neumüller and Dr. Schmidt (Siemens AG, Erlangen) for the collaboration.

References

1. Norton, D. P., Goyal, A., Budai, J. D., Christen, D. K., Kroeger, Specht, E. D., He, Q., Saffian, B., Paranthaman, M., Klabunde, C. E., Lee, D. F., Sales, B. C., List, F. A. (1996) Epitaxial YBCO on Biaxially Textured Nickel(001): An Approach to Superconducting Tapes with High Critical Current Density, *Science 274, 755-757*
2. Onabe, K., Nagaya, S., Shimonosono, T., Iijima, Y., Sadakata, N., Saito, T. and Kohno, O. (1996) Fabrication of YBCO Superconducting Tape by Continuous MOCVD Technique, *Proc. of ICEC16/ICMC: PSI-e252, Kitakyushu, Japan.*
3. Feng-Chi, Y., Yi-Yuan, X., Ji-Ping, C., Guo-Wen, Y. and Bin-Ji, C. (1996) Preparation of YBCO/Ag superconducting tapes with a high current-carrying capacity by MOCVD using a single solution source from a supersonic atomizer', *Supercond. Sci. Technol. 9, 991-993.*
4. Klippe, L., Wahl, G. (1995) Preparation of superconducting YBCO films on polycrystalline YSZ-substrates by single-source MOCVD, *Inst. Phys. Conf. Ser. No. 148, 611-614.*
5. Weiss, F., Fröhlich, K., Haase, R., Labeau, M., Selbmann, D., Senateur, J. P., Thomas, O. (1993) Preparation of YBCO films by low pressure MOCVD using liqued solution sources, *J. de Physique IV, C3, 321-328.*
6. Samoylenkov, S. V., Gorbenko, O. Y., Graboy, I. E., Kaul, A. R., Tretyakov, Y. D. (1996) LuBCO thin films prepared using MOCVD, *J. Mater Chem. 6 (4), 623-627.*
7. Wahl, G., Pulver, M., Decker, W., Klippe, L. (1997) CVD processes for coatings and surface modifications *Proc. E-MRS Spring Meeting Strasbourg, to be published.*

THE LIQUID SOURCE INJECTION TECHNIQUE : PRINCIPLE AND APPLICATION TO THE DEPOSITION OF OXIDE LAYERS AND MULTILAYERS

V. GALINDO, J.P. SÉNATEUR, F. WEISS, A. ABRUTIS*

LMGP- ENSPG - BP 46 - 38402 - SAINT MARTIN D'HERES CEDEX - FRANCE
* VILNIUS UNIVERSITY - Dept. of General and Inorganic Chemistry - 24 Naugarduko -2600 VILNIUS - LITHUANIA

1. INTRODUCTION

The possibility to growth by Metal-Organic Chemical Vapor Deposition (MOCVD) high temperature superconductors (HTS) thin films and other complex compounds containing heavy metallic elements, like barium, strontium or rare earth, has been demonstrated by several authors (1, 2). The potential applications of these layers cover a large field, from the micro electronic (filters, antennas...) to the high power applications (cables, fault current limiters, transformers...).
The performances of CVD systems depends on precise regulation of the distribution of the vapor phase at the vicinity of the substrate. The main difficulty for the MOCVD of complex compounds using conventional sources is the precise control of the gas phase composition. Even with precursors thermally stable at their volatilization temperature, the stabilization of the vapor pressure is not a trivial problem : controlling vaporization is a delicate matter for a liquid, and a real challenge for a solid. A strong effort to alleviate these problems is made through the recent development of improved solid or liquid source system. Most of them use liquid precursors, all are based on the same principle : the main part of the precursor is maintained at room temperature in a close vessel, few amounts are continuously introduced in an evaporator held at high temperature where a flash evaporation occurs. The difference with conventional MOCVD sources is that the partial vapor pressure of the precursors depends only on the flow rate of the solution inside the evaporator. Different systems have been proposed for the control of this flow rate :

- Aerosol (3-6)
- Micro-pump or syringe (7)
- Liquid mass flow controller (8)

If the precursors are solids, they are dissolved in a convenient solvent, which is selected so that it doesn't change the chemistry of the precursors and it is easily vaporized
In all these systems, the liquid is slowly and continuously injected inside the evaporator held under low pressure at high temperature : the liquid being in contact with the low pressure inside the reactor, selective volatilization of the solvent may occurs, leading to precipitation of solid which may affect the flow rate.

G. Van Tendeloo et al. (eds.), High-Temperature Superconductors and Novel Inorganic Materials, 85–90.

2. PRINCIPLE OF THE INJECTION - CVD

In LMGP, we have recently tested and developed, in collaboration with Vilnius University, a new liquid delivery source for CVD (9), based on the injectors used for the injection of fuel in thermal motors. The principle is a sequential computer-controlled injection of micro amounts of solution of precursors into the evaporator, where a flash volatilization occurs. The solution of precursors (or the liquid precursors) is maintained at room temperature, in a close vessel pressurized under 2 Bar of pure inert gas. The solution is pushed by this gas pressure toward the injector, which is a high-speed micro electro valve driven by a computer or a pulse generator.
A rough representation of a direct injection reactor is presented in Fig. 1, different geometries have been tested, but they all are based on the same principle.

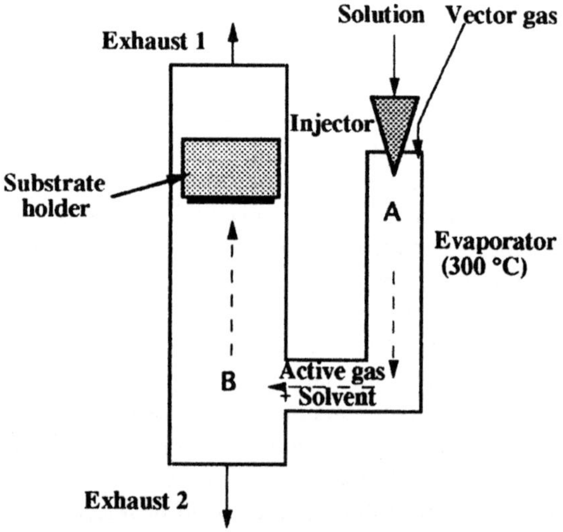

Fig 1 - Schematic representation of the direct injection reactor. The droplets of solution are injected into the evaporator (A), where they are "flash evaporated". The vapors are carried by the vector gas toward the reactor (B). During deposition, the vector gases are introduced both from the input A and from the exhaust 2, the gases are ejected toward the pump by the exhaust 1. Before and after the deposition, the vector gases are introduced both from A and from the exhaust 1 and are ejected by the exhaust 2.

This system is quite different from the other liquid delivery systems : the solution having no contact with the low pressure inside the evaporator, even highly volatile solvents can be used. The flow rate of the precursor, which defines the vapor pressure of the active species, is controlled by three parameters :

a) The volume of each injection, controlled by the duration of the aperture of the valve (typically 2 milliseconds., corresponding to about 4 mm^3 at each injection).

b) The frequency of injection (between 0,5 and 8 Hz in our experiments, but which can reach 100 Hz !). Typically, the flow rate of the solution of the precursors has been controlled between 2 and 40 cc/hour in our experiments. The reproducibility of the volume delivery is better than 5%.

c) The concentration of the solution (from 100 % for liquid precursors, to 1 weight % in organic solvent. THF, hexane or monoglyme have been used)

3. EXPERIMENTAL

The deposition of a wide variety of different layers has been realized, both in LMGP and in Vilnius University :

- Metallic layers like Cu and Ti(N,C)(10)
- Simple oxides like Y_2O_3, CeO_2 ,Ta_2O_5, SiO_2 (11)TiO_2
- Binary or ternary oxides like $ZrO_2(Y)$ (12), $YBa_2Cu_3O_7$ (13), $La_{(1-x)}$ ($)_xMnO_3$ (14), $Sr_{(1-x)}Ba_xTiO_3$.
- Multilayered structures : YBCO/CeO_2 (15) and Ta_2O_5/SiO_2 (16) using two injectors working separately.

The quality of the layers already obtained is at the level of the state-of-the-art, showing that the presence of the solvent in the reaction zone and that the pulsed growth regime induced by the sequential injection did not affect the quality of the films. A peculiarity of our injection source is that multilayers are easily realised with two (or more) injectors working alternatively. We will present here the first results obtained on the synthesis of PrBCO/ $YBa_2Cu_3O_7$ superlattices by injection MOCVD.

3.1. DEPOSITION OF PrBCO/ $YBa_2Cu_3O_7$ SUPERLATTICES

3.1.1 Deposition of YBCO layers

The conditions for the deposition of epitaxial, c textured, $YBa_2Cu_3O_7$ superconducting layers by injection-CVD is extensively studied, both in Vilnius University and in LMGP. The layers are obtained in hot wall reactors with different geometries, using only one source containing a mixture of tetramethyl heptane dionates of Y, Ba and Cu in solution in monoglyme. The best layers (Tc>90 K, Jc (77 K)>>10^6 A/cm^2) are obtained on single crystal substrates for deposition temperature between 825 and 860 °C. The deposition conditions are : evaporator temperature = 280 to 300 °C, $Y(thd)_3$ concentration in monoglyme = .02 Mol./l., volume injected = 2 to 4 mm^3/injection (which corresponds to a mean thickness of .5 to 1 nm. deposited at each injection), reactor pressure = 4.6 hPa, gas speed = 3m/sec., oxygen/argon concentration = .5, gas flow perpendicular to the substrate.

The main conclusions are :

a) High performances epitaxial superconducting layers are obtained over a surprisingly wide range of composition on SrTiO3, NdGaO3 and LaAlO3 substrates, this composition range is more limited for layers obtained on MgO substrate.

b) The composition of the layer is far from the composition of the solution, showing that the different precursors have very different reaction yields (Fig 2). Assuming that the deposition yield for the copper is one, the observed shift in composition between the layer and the solution can be explained by a reaction yield around .7 for the deposition of Y, but only .5 for Ba.

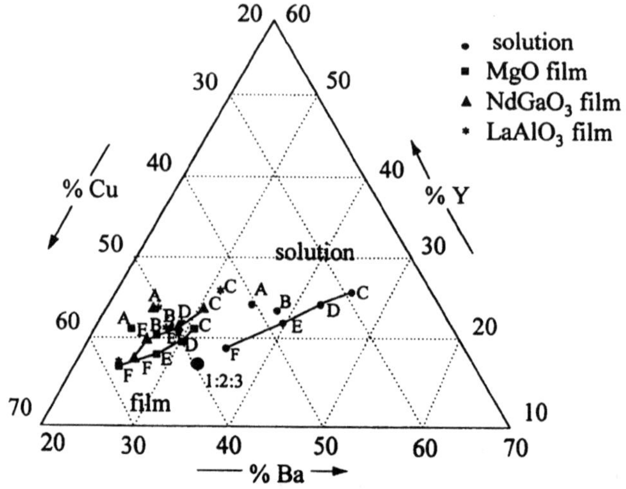

Fig 2 - Composition of the source solutions and of the corresponding layers obtained on NdGaO3, LaAlO3 and MgO. A large shift is observed : the layers have a lower Ba and Y concentration than the solution.

3.1.2. Deposition of PrBCO layers.

$Pr_{(1-x)}Ba_{(2+x)}Cu_3O_7$ layers have been obtained using the same deposition conditions as for $YBa_2Cu_3O_7$ .Highly c-textured samples are obtained over a wide range of solutions compositions, the shift between the compositions of the solution and of the layer is very similar to that obtained for $YBa_2Cu_3O_7$. The resistivity of the layers obtained with different source composition varies between .5 and 15 Ohms.cm. at 100 K, in relation with a partial substitution Pr-Ba which may occurs in this compound. The concentration of the solution used for the synthesis of superlattices corresponds to a resistivity at 100 K = 2 Ohms.

3.1.3. Deposition of PrBCO/ $YBa_2Cu_3O_7$ superlattices.

The deposition of layered samples is achieved by the use of two injection sources, connected on the same evaporator and working alternatively. The films are composed by 20 (PrBCO/ $YBa_2Cu_3O_7$) double layers on $SrTiO_3$ or $LaAlO_3$ single crystals. The thickness of each layer is fixed by the number of droplets injected by each injector, which is easily controlled by the computer. The X-ray diffraction patterns and pole figures experiments show that the samples are « epitaxied » on the substrates, with the orientations already observed for $YBa_2Cu_3O_7$ or PrBCO layers obtained by PVD or by conventional MOCVD process. The difference with conventional MOCVD is a considerably higher reproducibility. The superlattice structure is evidenced by the presence of satellite lines close to the 001 x-ray lines, which allows the determination of the modulation length of the superlattice (Fig 4.). The increase of thickness at each injection is determined by a linear fit of the variation of the modulation length versus the number of drops injected for growing the $YBa_2Cu_3O_7$ layer (Fig 4) The results are reported in table 1. The concentration of the solution of precursors in monoglyme being very low (.003 Mol/l for Y), the increase of thickness at each injection is very weak (respectively .073 nm for $YBa_2Cu_3O_7$ and .048 nm for PrBCO). The shift between the composition of the source and of the layers is quite the same as observed with higher concentrations of precursors : the effect of dilution is only to decrease the mean thickness deposited at each injection.

TABLE 1. Parameters used for the synthesis of PrBCO/$YBa_2Cu_3O_7$ superlattices, critical temperatures and thicknesses of individual layers deduced from the fit of the superlattice modulation versus the number of injections on YBCO layer.

Number of Y/Pr injections	Superlattice modulation (nm.)	Tc onset (K)	Tc width (K)	Y/Pr layer's thicknesses (nm.)
40 / 100	6.8	40	5	29 / 46
100 / 100	11.8	83	1.6	73 / 46
160 / 100	17.3	84	.8	117 / 46
250 / 100	23.3	89	.3	182 / 46
400 / 100	33.3	90	.4	292 / 46

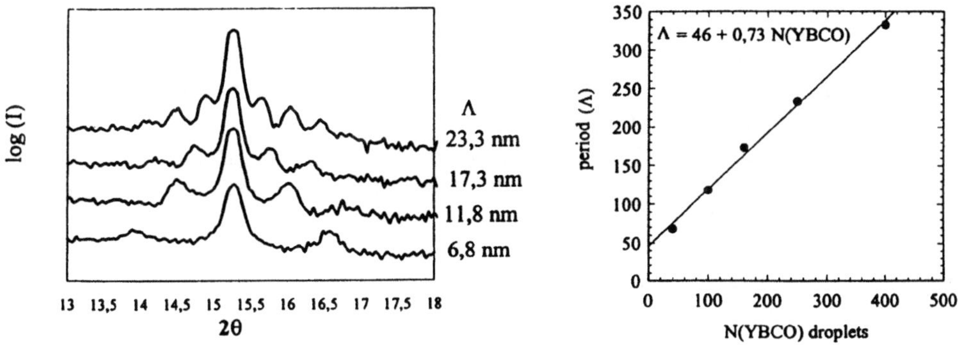

Fig 3 a. Evolution of the (002) X ray diffraction and of the satellites lines on superlattices obtained with different superlattice wavelengh, PrBCO thicknesses is constant. (See Table 1) - 3-b. Linear fit of the wavelengh modulation versus the number of injection during the growth of the YBCO layers.

4. CONCLUSIONS

The injection - MOCVD technique, using computer-driven fast micro electrovalves for the control of the flow rate of liquid precursors in a high temperature evaporator is very efficient for the synthesis of a large variety of thin layers or multilayers. The system is very versatile : the growth rate can be varied over a wide range, only by changing the injection frequency, the pulse width or the dilution of the solution. The pulsed growth regime did not affect the properties or the morphology of the layers : superlattices of complex compounds are obtained in a very simple way.

References

1. Leskelä M. , Mölsa H., Niinistro L. (1993), Chemical Vapor Deposition of High-Tc superconducting thin films (Review article). *Supercond. Sci. And Technol.* 6 627
2. Watson Ian M., (1997) Metal-Organic CVD of the high Tc Superconductor YBCO (Review article). *Chem. Vap. Deposition* 3 n°1
3. Jergel M. (1995) Synthesis of high-Tc superconducting films by deposition from an aerosol (Review article). *Supercond. Sci. And Technol.* 8 67
4. Weiss F. ,Schmatz U. , Pisch A., Felten F., Sénateur J.P., Abrutis A., Frohlich K., Selbmann D., Klippe L. (1997) - HTS thin films by innovative MOCVD processes- *J. Of Alloys and Compounds* 251 - 264
5. Klippe L., Stolle R., Decker W., Nurnberg A., Wahl G., Gorbenko Yu., Erokhin Yu, Graboy I., Kaul A.R., Pozigun S.A. (1993)- CVD of High-Tc superconductors, *"Applied Superconductivity 93"*, H.C. Freihardt Ed.- Vol 1p 407.
6. Gorbenko Yu., Kaul A. R., Babushkina N.A., Belova L.M., (1997), Giant magnetoresistive thin films (La,Pr)0.7(Ca,Sr)0.3MnO3 prepared by aerosol MOCVD, *J. Mat. Chem.* 7 (5) 747
7. Matsuno S., Unemura T., Uchikawa F., Ikeda. B. (1995) - YBCO Thin films with Yttria stabilized Zirconia buffer layer on metal substrate by liquid source CVD , *Jpn. J. Appl. Phys.* 34 2293.
8. Studebaker D.B., Zhang J., Marks T.J., Wang Y.Y., Dravid V.P., Schindler J.L., Kannewurf C.R. (1998) Liquid source metal-organic chemical-vapor deposition of high-quality YBCO films on polycrystalline silver substrates, *Appl. Phys.* Letters - 72 -10 - 1253.
9. Patent n°93/08838 PCT N° FR94/00858. Licensed : Amitis Technology - 9 Parc d'Activité Pré Millet - 38330 Montbonnot - France
10. Gilles S., Bourhila N., Senateur J.P., Madar R., Levy R.A., Blanquet E., (1997) Deposition of Ti(C,N) thick films by ILPCVD starting from TDMAT, *Proceedings of MRS Falls Meetings Boston.*
11. Bourhila N., Felten F., Sénateur J.P., Schuster F., Madar R., Abrutis A. (1997) - deposition and characterisation of ZrO2 and yttria-stabilized ZrO2 films using injection-LPCVD- Proceedings of EUROCVD 11 - Electrochemical Society Proceedings Vol. 97-25, p. 417
12. Sénateur J.P., Felten F., Pignard S., Weiss F., Abrutis A., Bigelyte V., Teiserskis A., Saltyte Z., Vengalis B. (1997) - Synthesis and characterization of YBCO thin films grown by injection MOCVD. *J. of Alloys and Compounds* 251 288
13. Pignard S., Vincent H., Sénateur J.P., Pierre J., Abrutis A. (1997) - Annealing effect on magnetic and electrical properties of epitaxial LaMnO3 thin films grown by chemical vapor deposition, *J. Appl. Phys.* 82 9 4445
14. Abrutis A., Kubilius V., Bigelyte B., Teiserskis A., Saltyte Z., Sénateur J.P., Weiss F. (1997) - In-Situ heteroepitaxial growth of CeO2/YBCO films on sapphire by injection-MOCVD, *Mat. Letters* 31 201
15. Felten F., Sénateur J.P., Labeau M., Yu-Zhang K., Abrutis A. - Deposition of Ta2O5/SiO2 multilayers films by a new process "injection MOCVD", *Thin Solid Films (1997) 296 79*

OXIDE-FLUORIDES: NEW PARENT STRUCTURES FOR SUPERCONDUCTING MATERIALS

C. GREAVES, J.L.KISSICK AND L.D. AIKENS
University of Birmingham
School of Chemistry, University of Birmingham, Birmingham B15 2TT, UK

Abstract. The low temperature fluorination of low dimensional mixed metal oxides containing copper can be achieved in a variety of ways to achieve new structures which have the appropriate characteristics to support superconductivity. The reaction mechanisms can be quite complex, and this paper describes the synthesis of two new oxide-fluorides - $La_{1.9}Sr_{1.1}Cu_2O_5F_2$ and $PbSr_2YCu_3O_6F_4$ - via processes which involve anion insertion, anion substitution and cation extraction.

1. Introduction

The insertion of fluorine into mixed copper oxides to control carrier density and hence induce superconductivity, *e.g.* the fluorination of La_2CuO_4 to form $La_2CuO_4F_\delta$ [1], was established in the early days of high temperature superconductor research. Only more recently has the potential of fluorination for the synthesis of totally new superconducting structures been established. Although this synthetic strategy depends on the ability of fluorine to insert into fairly open structural frameworks, it is the variety of possible subsequent processes which renders the method so useful for the design of new materials. Initial observations related to the fluorination of Sr_2CuO_3 to form superconducting $Sr_2CuO_2F_{2+\delta}$ [2], a process which involves not only insertion of fluorine and the substitution of fluorine for oxygen, but also - and more importantly - a structural rearrangement. This transformation produces the CuO_2 layers which are essential for superconductivity but are not present in the Sr_2CuO_3 precursor oxide.

Several other oxide-halide superconducting phases have since been reported, with high pressure syntheses producing materials with T_c as high as 111K [3]. The variety of structural effects is demonstrated by the insertion of fluorine into reduced $YBa_2Cu_3O_{6+\delta}$ [4]. Whereas the insertion process appears related to that occurring in Sr_2CuO_3, in this case no structural transformation or F/O exchange occurs, and a superconducting phase (T_c=94K) with approximate composition $YBa_2Cu_3O_6F_2$ is formed.

G. Van Tendeloo et al. (eds.), High-Temperature Superconductors and Novel Inorganic Materials, 91–96.

In this paper, two new oxide-fluoride materials are reported, both of which are derived from oxides related to superconductors: $La_{1.9}Sr_{1.1}Cu_2O_6$ and $Pb_2Sr_2YCu_3O_8$. Although both phases have the structural requirements to support superconductivity, appropriate electronic control has not yet been achieved.

2. Experimental

$La_{1.9}Sr_{1.1}Cu_2O_6$ was prepared conventionally from high purity La_2O_3, $SrCO_3$ and CuO by heating in air at 1075°C for three periods of 15 h, with a thorough grinding between heatings. Although a variety of fluorination procedures may be employed for materials of this type [5], this study used an intimate mixture of $La_{1.9}Sr_{1.1}Cu_2O_6$ and excess NH_4F (molar ratio 1:5, 350°C for 20 h). The process was repeated four times to give an overall molar ratio of 1:20.

The sample of $Pb_2Sr_2YCu_3O_8$ was synthesised from a precursor obtained by heating a mixture of $SrCO_3$, Y_2O_3 and CuO at 900°C for 5 h in flowing O_2. The stoichiometric amount of PbO was mixed with the precursor and heated at 900°C in N_2 for three periods of 12 h. Fluorination was achieved by mixing with NH_4F (molar ratio 1:6) and heating in air or flowing N_2 at 350°C for 8 h. Similar, but not identical, products have been obtained using F_2 gas or CuF_2 as fluorinating agents.

Stuctural characterisation was achieved using Rietveld refinement of X-ray powder diffraction (XRD) data (Siemens D5000, transmission mode, primary beam monochromator, 8° PSD) and neutron powder diffraction data (POLARIS, ISIS, Rutherford Appleton Laboratory). The programs FULLPROF [6] and GSAS [7] were used for refinements using X-ray and neutron diffraction data, respectively.

3. Results and Discussion

3.1 FLUORINATION OF $La_{2-x}Sr_{1+x}Cu_2O_6$

As previously described [5], fluorination of the cuprates $Ln_{2-x}A_{1+x}Cu_2O_6$ (Ln=La,Nd; A=Ca,Sr) results in significant expansion along *c*. In the present study, the unit cell of $La_{1.9}Sr_{1.1}Cu_2O_6$ (tetragonal; *a*=3.851(1) Å, *c*=20.048(6) Å) expanded to an orthorhombic cell with *a*=3.8540(2) Å, *b*=3.9200(2) Å *c*=21.425(1) Å. Chemical analysis has indicated that under the fluorination conditions used, the Cu oxidation state increases slightly (to approximately 2.1) and approximately 2F per formula unit are incorporated into the structure. Three phases were present in the sample: the major fluorinated product, and two decomposition products, one related to fluorite (probably La-doped SrF_2) and CuO. Structural analysis was performed using X-ray diffraction data, and the fitted profile is shown in Figure 1. Structural data for the principal phase are given in Table 1. Since X-ray diffraction cannot differentiate between O and F ions, the site distribution shown in Table 1 has been derived from bond distance and Madelung energy arguments which will be presented elsewhere [8].

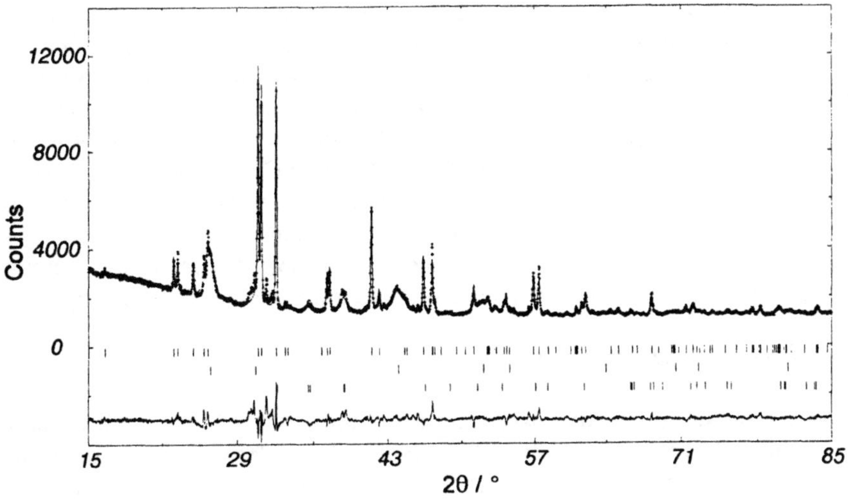

Figure 1. Observed and calculated XRD profiles, fluorinated $La_{1.9}Sr_{1.1}Cu_2O_6$.

TABLE 1. Refined parameters, $La_{1.9}Sr_{1.1}Cu_2O_5F_2$

Atom	*x/a*	*y/b*	*z/c*	B/Å^2	Cell occupancy
La	0	0	0	2.4(1)	2
La/Sr	0	0	0.1775(2)	2.4(1)	1.8/2.1
Cu	0	0	0.5873(5)	2.0(2)	4
O1	0	0.5	0.093(2)	3.0(5)	4
O2	0.5	0	0.084(2)	3.0(5)	4
O3	0	0	0.5	3.0(5)	2
F1	0.5	0	0.25	3.0(5)	2.5(2)
F2	0	0	0.677(2)	3.0(5)	1.8(2)
Immm; *a*=3.8540(2) Å, *b*=3.9200(2) Å, *c*=21.425 Å; R_{wp}=5.54%, R_{exp}=2.29%					

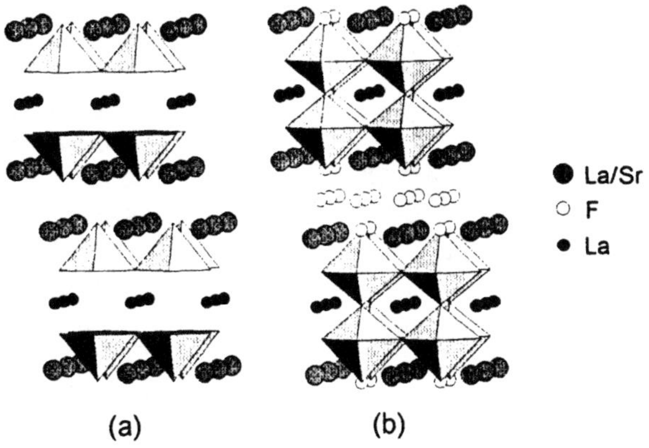

Figure 2. Strucutres of (a) $La_{1.9}Sr_{1.1}Cu_2O_6$; (b) $La_{1.9}Sr_{1.1}Cu_2O_5F_2$

Figure 2 shows the structures of $La_{1.9}Sr_{1.1}Cu_2O_6$ and the fluorinated product with approximate composition $La_{1.9}Sr_{1.1}Cu_2O_5F_2$. It can be seen that O^{2-} ions completely occupy the anion positions between the Cu positions which are vacant in $La_{1.9}Sr_{1.1}Cu_2O_6$, and a disordered array of F^- ions provides partial filling of the apical Cu coordination sites and the interstitial (fluorite-block) sites. The Cu-O-Cu links (Figure 2) are unusual and probably form due to the large Cu-Cu separation (3.712(4) Å [9]) which is present in the precursor oxide. Interestingly, only a small expansion occurs (to 3.74(2) Å) when O^{2-} ions form Cu-O-Cu links. Although very small Meissner signals have been observed in some samples [5], in general the materials appear to be underdoped and further attention to electronic control is necessary.

3.2 FLUORINATION OF $Pb_2Sr_2YCu_3O_8$

The products obtained by fluorinating $Pb_2Sr_2YCu_3O_8$ depend on the fluorination method employed; here, fluorination by NH_4F is considered. X-ray diffraction revealed substantial structural changes to the primary phase, and the production of a fluorite related phase with *a*=5.779(1) Å (GSAS refinement, see below). This phase has been assumed to be PbF_2, despite the unit cell being significantly smaller than that reported (5.925(4) Å [10]), and structure refinement was consistent with the stoichiometry PbF_2. Neutron powder diffraction suggested the primary phase to have composition $PbSr_2YCu_3O_6F_4$ with the structure shown in Figure 3; this is consistent with the substantial expansion observed along [001]. The fitted profile is shown in Figure 4 and refined structural parameters are given in Table 2. The assignment of anion sites to O^{2-} and F^- ions was based on Madelung energy calculations [8].

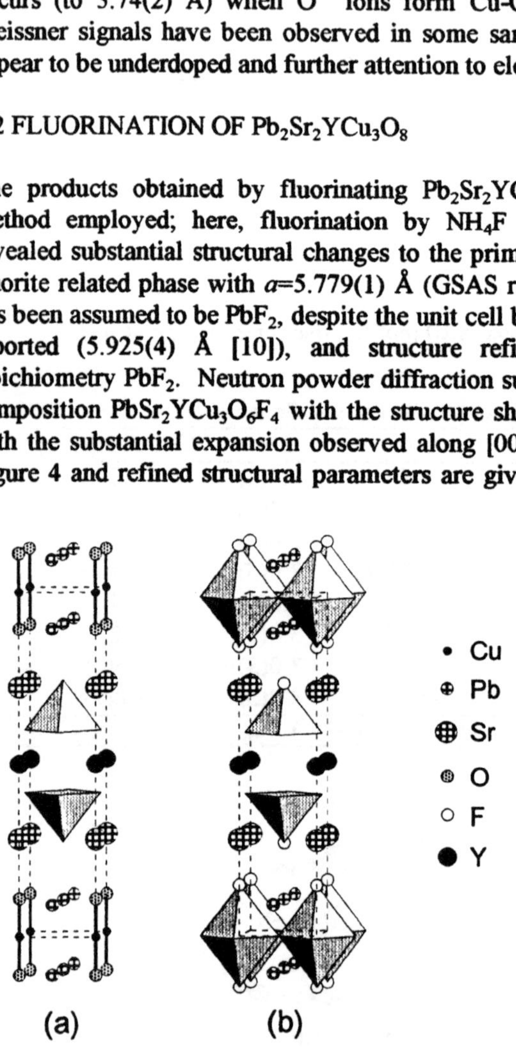

Figure 3. Structures of (a)$Pb_2Sr_2YCu_3O_8$; (b)$PbSr_2YCu_3O_6F_4$

Fluorine enters the $Pb_2Sr_2YCu_3O_8$ lattice in the vicinity of the PbO-Cu-PbO regions (Figure 3) and causes structural rearrangement to provide octahedrally coordinated Cu with a new CuO_2 plane and two apical fluorine ions. The structure therefore appears to contain two separate regions which are capable of supporting superconductivity: a single layer of corner linked CuO_4F_2 octahedra and a double layer of CuO_4F square pyramids. In this fluorination, several interesting chemical processes seem to be involved:

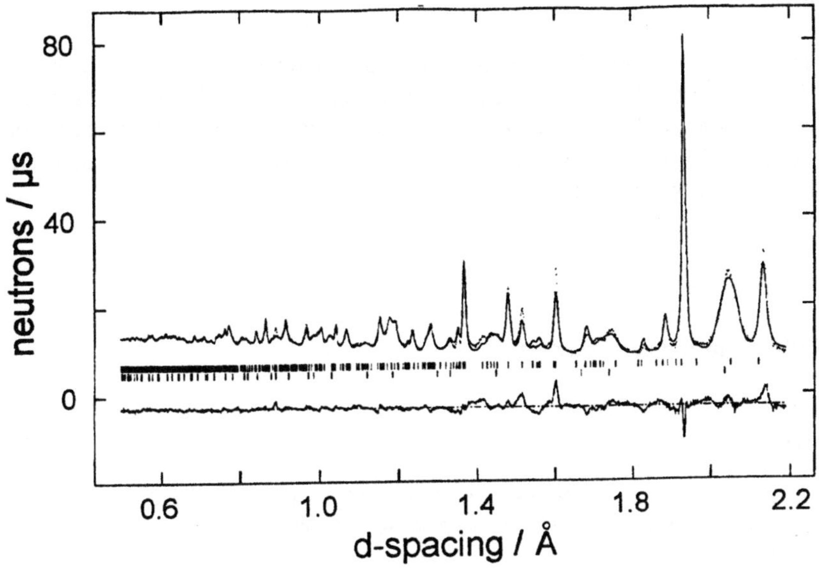

Figure 4. Neutron diffraction profiles for fluorinated $Pb_2Sr_2YCu_3O_8$

i F⁻ for O^{2-} substitution within the PbO layers
ii rearrangement of the O^{2-} ions to create new CuO_2 layers
iii overall oxidation of the Cu-O framework caused primarily by abstraction of 50% of the Pb^{2+} cations.

TABLE 2. Refined structural parameters, $PbSr_2YCu_3O_6F_4$

Atom	*x/a*	*y/b*	*z/c*	B/Å^2	cell occupancy
Pb	0.5	0.5	0.0957(5)	0.47(5)	1
Sr	0	0	0.2875(7)	3.4(2)	2
Y	0	0	0.5	0.47(5)	1
Cu1	0	0	0	0.47(5)	1
Cu2	0.5	0.5	0.3977(4)	0.47(5)	2
O1	0.5	0	0.4122(3)	1.00(6)	4
O2	0.5	0.0	0.0138(6)	1.00(6)	2
F1	0.408(2)	0.408(2)	0.2702(7)	1.00(6)	2
F2	0.063(1)	0.063(1)	0.1507(9)	1.00(6)	2
P4/mmm; *a*=3.8601(2) Å, *c*=17.062(3) Å; R_{wp}=2.84%, R_{exp}=0.78%					

Certain structural features indicated in Table 2, *viz* the thermal parameter for Sr and the displacements of F1, are attributed to structural variations associated with the Pb vacancies. It is believed that the overall energetics of the reaction may be influenced by the high lattice energy of the PbF_2 phase formed [8].

4. Conclusions

Low temperature fluorination of oxides containing fairly open structural frameworks can result in a variety of structural effects involving fluorine insertion, fluorine substitution and structural rearrangement. New structural elements capable of supporting superconductivity can be created, which offers potential for the design and synthesis of new superconducting materials.

5. Acknowledgements

We thank EPSRC for financial support.

6. References

1. Chevalier, B., Tressaud, A., Lepine, B., Amine, K., Dance, J.M., Lozano, L., Hickey, E., and Etourneau, J. (1990) Stabilisation of a new superconducting phase by low temperature fluorination of La_2CuO_4, *Physica C* **167**, 97-101.
2. Al-Mamouri, M., Greaves, C., Edwards, P.P., and Slaski, M. (1994) Synthesis and superconducting properties of strontium copper oxy-fluoride, $Sr_2CuO_2F_{2+\delta}$, *Nature (London)* **369**, 382-384.
3. Kawashima, T., Matsui, E., and Takayama-Muromachi, E. (1996) New Oxyfluoride superconductors $Sr_2Ca_{n-1}Cu_nO_{2n+\delta}F_{2\pm y}$ (n=2; T_c=99K, n=3; T_c=111K) prepared at high pressure, *Physica C* **257**, 313-320.
4. Shpanchenko, R.V., Rozova, M.G., Abakumov, A.M., Ardashnikova, E.I., Kovba, M.L., Putilin, S.N., Antipov, E.V., Lebedev, O.I., and Van Tendeloo, G. (1997) Inducing superconductivity and structural transformations by fluorination of reduced YBCO, *Physica C* **280**, 272-280.
5. Slater, P.R., Hodges, J.P., Francesconi, M.G., Greaves, C., and Slaski, M. (1997) Fluorination of the Ruddlesden-Popper type cuprates, $Ln_{2-x}A_{1+x}Cu_2O_{6-y}$ (Ln=La,Nd; A=Ca,Sr), *J. Mater. Chem.* **7**, 2077-2083.
6. Rodriguez-Carajal, J. (1997) FULLPROF, version 3.2, based on the original code by Wiles, D.B., and Young R.A., *J. Appl. Crystallogr.* **14**, 149-151.
7. Larson, A.C., and Von Dreele, R.B. (1994) General Structure Analysis System, University of California.
8. Greaves, C, Kissick, J.L., Francesconi M.G., Gillie, L., and Aikens L.D. (1998) Synthetic strategies for new inorganic oxide-fluorides and oxide-sulphates, *J. Mater. Chem.* submitted.
9. Lobo, R.C. (1990) Structural studies of non-stoichiometric mixed metal copper oxides, Ph.D. thesis, University of Birmingham.
10. Ito, Y. and Koto, K. (1983) Thermal hysteresis of anion disorder in beta PbF_2, *Solid State Ionics* **9-10**, 527-530.

HIGH-PRESSURE SYNCHROTRON DIFFRACTION STUDY OF THE SUPERCONDUCTING SPIN-LADDER COMPOUND $(Sr,M)_{14}Cu_{24}O_{41}$, WITH M=Ca, Ba, Nd

P. BORDET[*], *C. CHAILLOUT*[*], *C. DARIE*[*], *S. PACHOT*[*], *R. J. CAVA*[**], *M. HANFLAND*[***], *M. MAREZIO*[****], *H. TAKAGI*[*****]
[*] *Laboratoire de Cristallographie CNRS, BP166, 38042 Grenoble Cedex 9, FRANCE*
[**] *Department of Chemistry and Materials Institute, Princeton University Princeton NJ 08540, USA*
[***] *ESRF, BP220, 38043 Grenoble, FRANCE*
[****] *MASPEC-CNR, via Chiavari 18/A, 43100 Parma, ITALY*
[*****] *Institute for Solid State Physics, University of Tokyo, Roppongi Minato-ku, Tokyo 106, JAPAN*

1. Introduction

Following the discovery of high-Tc superconducting cuprates, the theoretical investigation of the magnetic and transport properties in low-dimensional copper oxide systems has become increasingly active. Recently, the so-called "spin-ladder" systems have drawn much of the attention, due to their predicted striking properties. For example, Dagotto and Rice [1] predicted the existence of a spin gap, when the antiferromagnetic coupling along rungs is stronger than along legs, and Rice et al. [2] showed later that this property would only exist for even leg ladders. They also predicted the appearance of superconductivity when holes are introduced in this system. Superconductivity was detected for the first time in a compound containing spin ladders by Uehara et al.[3], who measured the resistivity as a function of pressure and temperature in the $(Sr_{14-x},Ca_x)Cu_{24}O_{41}$ compounds. These authors reported the appearance of a superconducting transition in the 3-5GPa range, with a maximum transition temperature of about 12K, in samples with a large calcium concentration (x=13.6). The structure of $(Sr_{14-x},Ca_x)Cu_{24}O_{41}$ can be described as the alternate stacking of layers having the two-leg-ladder arrangement and Cu_2O_3 composition, and layers made of isolated chains of edge-sharing CuO_4 squares, separated by planes containing the alkaline earth cations. The ladder planes are formed by the corner sharing connection of double chains of edge-sharing CuO_4 squares. Within the double chains, the Cu cations are linked through 90° Cu-O-Cu bonds, leading to a very weak magnetic coupling, and the ladders may be considered as magnetically quasi-isolated objects. The CuO_4 squares belonging to the chains above and below are oriented at 45° with respect to the CuO_4 squares belonging to the ladders. The incommensurate $\sqrt{2}$ ratio between the

G. Van Tendeloo et al. (eds.), High-Temperature Superconductors and Novel Inorganic Materials, 97–102.

Cu-Cu separation in both types of layers resulting from this arrangement, leads to a composite type of structure, which in first approximation can be described by using two orthorhombic unit cells with parameters : $a_l \approx 11.4$Å, $b_l \approx 13.0$Å and $c_l \approx 2.75$Å for the $Sr(Cu_2O_3)Sr$ ladder-containing slabs, and $a_c=a_l$, $b_c=b_l$ and $c_c \approx 3.92$Å $\approx 10/7 \times c_l$ for the chain layers. The a direction is in the plane of, and perpendicular to the chains and ladders ; the b direction is perpendicular to the stacking and the c direction is along the chains and ladders. However, this schematic description of the structure may be oversimplified, and thorough crystallographic studies [4,5] pointed out the presence of buckling in both ladder and chain layers, and of interlayer Cu-O bonds which may play an important role in the doping mechanism.

Optical measurements have shown [6] that the intrinsic hole doping due to the 2.25+ average copper valence is gradually transferred from the chains to the ladders on increasing calcium concentration. Since this substitution is isovalent, the doping modification is induced by altering the average size of the (Sr,Ca) mixed site, which leads to changes of the Cu-O bond lengths in both chains and ladders. Due to the smaller ionic size of Ca cations, the effect of Ca substituting for Sr can be viewed as a chemical pressure effect. Since the superconducting properties of the $(Sr_{14-x},Ca_x)Cu_{24}O_{41}$ compound appear only at high pressure and for specific calcium concentrations, it is of the highest relevance to investigate the evolution of the structure with pressure and composition. Therefore, we have carried out a high pressure synchrotron diffraction experiment at room temperature in the 0-10 GPa range for a set of $(Sr_{14-x},M_x)Cu_{24}O_{41}$ samples with M=Ca, Ba, Nd and x between 0 and 13.6 in order to investigate the combined effects of applied pressure, and modification of the average ionic size and/or valence of the (Sr,M) mixed site.

2. Experimental

Five different samples of $(Sr_{14-x},M_x)Cu_{24}O_{41}$ general formula were used for this experiment, hereafter denoted as : Sr14 : x=0 ; Sr8Ca6 : M=Ca, x=6 ; SrCa13 : M=Ca, x=13.6 (superconducting composition) ; Sr10Ba4 : M=Ba, x=4 ; Sr8Nd6 : M=Nd, x=6. The first four samples were used to investigate the effect of pressure as a function of the average size of the mixed site, while the data comparison between Sr8Ca6 and Sr8Nd6 allows us to compare the effect of changing only the average valence of this site (Nd^{3+} and Ca^{2+} cations have very close ionic sizes). They were previously characterized for cationic stoichiometry and phase purity by EDS analysis and laboratory powder x-ray diffraction.

The high pressure synchrotron diffraction experiment was carried out at the ID9 beamline of the European Synchrotron Radiation Facility (ESRF) using angle dispersive powder diffraction with image plates as detectors. Samples were loaded into a 130 μm diameter and 35 μm high gasket hole of a membrane diamond anvil cell (MDAC). Pressures were measured using the ruby fluorescence method [7]. Silicon oil was used as pressure transmitting medium. Though it is known to be a rather poorly hydrostatic medium above 6 GPa, it was chosen because of its inert character with respect to the samples. In order to check the effect of pressure gradients induced by this medium at higher pressures, two additional experiments were carried out on samples Sr8Nd6 and

SrCa13, by using a 4:1 methanol/ethanol mixture, known to remain a better pressure transmitter to 10 GPa. The results obtained were similar for the Sr8Nd6 sample, but the SrCa13 sample started to decompose above 3GPa in the methanol/ethanol mixture, leading to the appearance of calcium and copper containing phases. Nevertheless, the similar results obtained for the Sr8Nd6 sample with both transmitting media indicate that the observed effects described below are not due to pressure gradients brought about by silicon oil at high pressures.

The monochromatic beam (λ=0.458Å) was selected by a horizontally focusing asymmetrically cut bent Si(111) Laue monochromator [8]. It was vertically focused by a curved Pt-coated Si mirror. The beam size on the sample was 30x30 μm. Diffraction rings from the powder samples were recorded on an A3 size Fuji image plate located at 441 mm from the sample. The plates were scanned on a Molecular Dynamics image plate reader, and processed by using the Fit2D software developed at the ESRF [9]. The images were corrected for spatial distortion effects. Corrections for the image plate tilt with respect to the direct beam were applied by using images from a standard silicon powder, which were also used to calibrate the wavelength and the sample to detector distance. The corrected images were averaged over 360° about the direct beam position, yielding spectra similar to those obtained by classical diffraction techniques.

Due to the complexity of the structure, we concentrated our analysis on the evolution of the cell parameters with pressure. For this purpose, we selected for each sample a set of diffraction lines (generally at low 2θ angles) which were well isolated in the whole pressure range, and obtained their angular positions by least square refinement using a pseudo-Voigt line shape function. The cell parameters were obtained by transforming the 2θ positions into d values using the Bragg's law. Note that in most cases, it was not possible to identify isolated peaks from the "chain" unit cell, so only the parameters $a_c=a_l$, $b_c=b_l$ and c_l from the "ladder" unit cell could be obtained.

3. Evolution of the cell parameters with composition at room pressure

The relative evolution of the cell parameters at P≈0.1GPa is shown in figure 1, as a function of the average ionic radius of the mixed site. The normalisation was made with respect to the SrCa13 sample. The cell parameter increase with increasing average ionic radius of the mixed site is strongly anisotropic, with a much larger effect in the b direction, and the smallest effect along c. This is most probably due to the weakness of ionic bonds along the stacking direction with respect to the strong in-plane Cu-O bonds inside the ladder and chain layers. A similar anisotropy is to be expected for the effect of pressure. It is interesting to point out that the a and b cell parameters of the Sr8Nd6 sample follow the regular evolution due to size effects, while the c parameter value is markedly higher. This indicates that the charge effect brought about by the replacement of Ca^{2+} by Nd^{3+} cations leads to modifications of the Cu-O bonding scheme mainly along the chain and ladder directions.

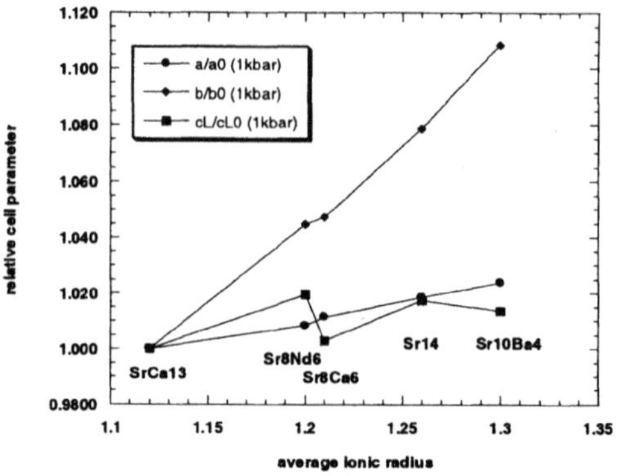

Figure 1 : Relative evolution of the cell parameters as a function of average ionic radius of the mixed cation site (Å) at≈ 0.1GPa..

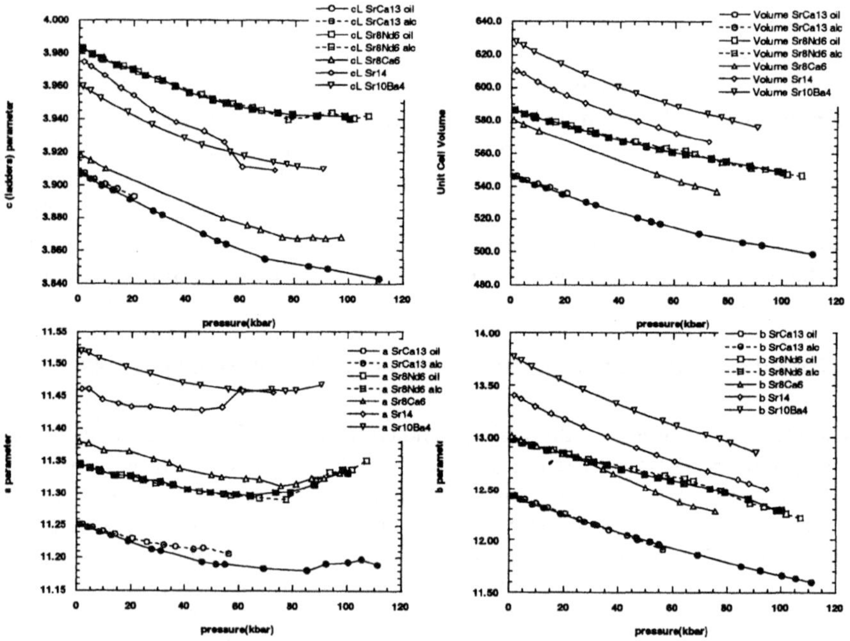

Figure 2 : Evolution of the cell parameters (Å) and volume (Å^3) up to ≈ 1GP

4. Evolution of the cell parameters as a function of composition and pressure

The evolution of the cell parameters and volume as a function of pressure is shown for each sample in figure 2. At first glance, all samples, except Sr8Nd6, exhibit a similar behavior, with roughly parallel decreases of the cell parameters and volume with increasing applied pressure. Axis compressibilities are shown in figure 3.
The compressibility is the smallest in the a axis direction, with values ranging from 1.3 to 1.6×10^{-3} GPa^{-1}. The compressibility along c is about twice as large as that along a, with values ranging from 2.1 to 2.9×10^{-3} GPa^{-1}. These values are close to those reported for the in-plane compressibilities of superconducting cuprates[10]. The b axis

compressibility is much larger than the in-plane ones, with values ranging between 6.3 and $11x10^{-3}$ GPa^{-1}. These values are twice as large as the largest c axis compressibilities reported for the superconducting cuprates. Moreover, the b axis compressibility markedly increases with increasing average ionic radius of the mixed cation site. This could indicate an increase of interlayer interactions when the size of the mixed site is decreased, i.e. when the ladder and chain layers come closer to each other.

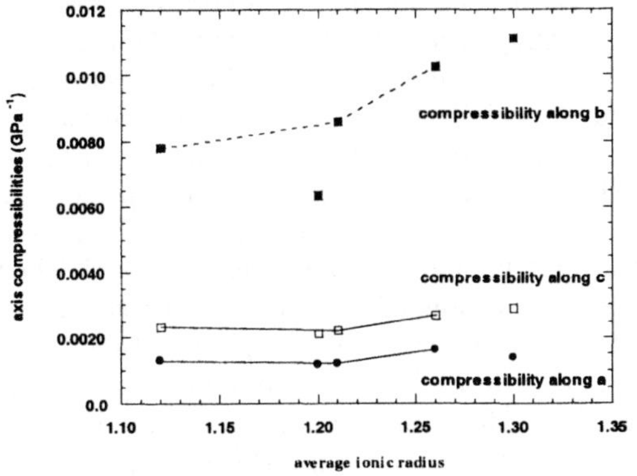

Figure 3 : Axis compressibilities as a function of average ionic radius of the mixed cation site (Å).

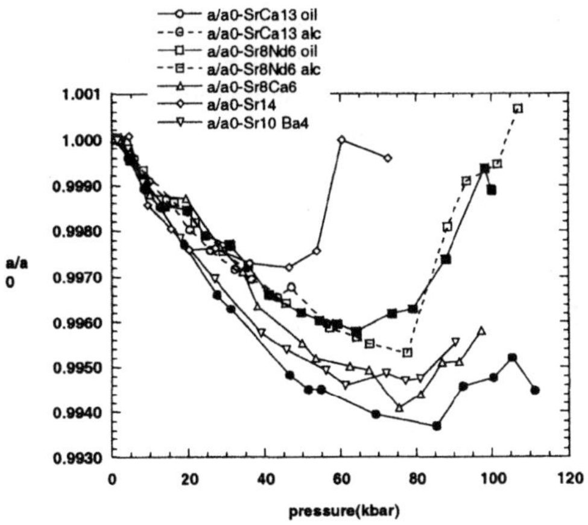

Figure 4 : Relative evolution of the a cell parameter as a function of pressure.

In figure 4, we present the relative evolution of the a cell parameters as a function of pressure. The most striking feature observed, is the presence of a lattice anomaly consisting in a strong increase of the a cell parameter above 6-8GPa, depending on sample composition. The pressure value at which the anomaly takes place seems to increase with increasing Ca content, going from ≈5GPa for Sr14, to ≈7.5GPa for Sr8Ca6, and to ≈8.5GPa for SrCa13. For Sr10Ba4 and Sr8Nd6, the anomaly appears at ≈7.5GPa. For the latter sample, above the a axis anomaly pressure value, the c parameter seems to become stable, and the b axis compressibility starts to increase again. The former effect seems to exist also for the other samples, even though less pronounced. The diffraction spectra recordered at pressures above the lattice anomaly do not present noticeable differences, such as peak splitting or superlattice reflections, from those recordered below, indicating that no major structural rearrangement is occuring at the transition.

Although the present data do not allow us to draw definitive conclusions about the nature of this lattice anomaly, a possible origin could be the abrupt decrease of the ladder and chain layers buckling induced by the applied pressure. Such an effect could indeed lead to an increase or stabilization of the in-plane cell parameters a and c, and allow to restore the compressibility in the b axis stacking direction . The effect of applied pressure and lattice anomaly on the physical and superconducting properties of the $(Sr_{14-x},M_x)Cu_{24}O_{41}$ compounds may be discussed on the basis of the present results and already reported structural and physical measurements [4-6]. The main effect of the substitution of Ca for Sr and of applied pressure is a strong decrease of the b cell parameter, which corresponds to an increased coupling between the ladder and chain layers, and to a charge transfer from the chains to the ladders. The similarity between the substitution and pressure effects on the cell parameter indicates that the cation substitution acts as an internal chemical pressure. However, the substitution effect alone is not sufficient to induce superconductivity at low temperature, and an additional external pressure has to be applied, with appearance of the superconducting state at ≈3GPa for the $(Sr_{0.4},Ca_{13.6})Cu_{24}O_{41}$ compound, which corresponds to a cell parameter b≈12.15Å. The disappearance of superconductivity at ≈6GPa might be related to the lattice anomaly, even though it seems to appear at higher pressures (≈8.5GPa for the SrCa13 sample) at room temperature. Low temperature and high pressure diffraction experiments are needed to confirm this model.

5. References

1 E. Dagotto and T.M. Rice, Science 271, 618 (1995)

2 T.M. Rice, S. Gopalan and M. Sigrist, Europhys. Lett. 23, 445 (1993)

3 M. Uehara, T. Nagata, J. Akimitsu, H. Takahashi, N. Mori and K. Kinoshita, J. Phys. Soc. Jpn. 65, 2764 (1996)

4 A. Frost-Jensen, B. B. Iversen, V. Petricek, T.M. Schultz and Y. Gao, Acta Cryst. B53, 113 (1997)

5 T. Ohta, F. Izumi, M. Onoda, M. Isobe, E. Takayama-Muromachi and A.W. Hewat, J. Phys. Soc. Jpn. 66, 3107 (1997)

6 T. Osafune, N. Motoyama, H. Eisaki and S. Uchida, Phys. Rev. Lett. 78, 1980 (1997)

7 H. K. Mao, J. Xu, and P. M. Bell, *J. Geophys. Res.*, **91**, 4673 (1986)

8 C. Schulze, U. Lienert, M. Hanfland, M. Lorenzen, and F. Zontone, Journal of Synchrotron Radiation, to be published

9 A. P. Hammersley, S. O. Svensson, M. Hanfland, A. N. Fitch, and D. Häusermann, High Pressure Research **14**, 235 (1996)

10 H. Takahashi and N. Mori, in "Studies of High Temperature Superconductors", Ed. A. Narlikar, Nova Science Publishers, Inc., vol. 16, p. 1 (1996)

STRUCTURAL STUDIES OF CUPRATES AND BISMUTHATES BY ELECTRON MICROSCOPY

C. CHAILLOUT*, S. KAZAKOV*,**, J. KARPINSKI***,
H. SCHWER***, E.M. KOPNIN***, P. BORDET*, S. PACHOT*,
E.V. ANTIPOV**, J.J. CAPPONI***
* *Laboratoire de Cristallographie, CNRS, BP166,
38042 Grenoble cedex 9, France*
** *Department of chemistry, Moscow State University,
Moscow 119899, Russia*
*** *Laboratorium für Festkörperphysik, ETH Honggerberg,
CH-8093 Zürich, Switzerland*

1. Introduction

After ten years of very intensive studies on cuprates superconductors, our research effort has been devoted recently to new systems showing interesting magnetic and transport properties, including possibly superconductivity. Two different class of materials have been investigated : first of all, we have been looking for new families of bismuth-based superconductors. We have been able to synthesize $SrBiO_3$ and to induce superconductivity by K or Rb doping. The second direction has involved low-dimensional copper oxide systems, such as $Sr_{n-1}Cu_{n+1}O_{2n}$, $Sr_{14-x}A_xCu_{24}O_{41}$ (A=Ca, Nd, Ba,...), $Sr_{1-x}CuO_2$. For these two class of materials, part of the structural studies has been done by electron microscopy using a Philips CM microscope operated at 300kV and equipped with a Kevex system for Energy Dispersive X-ray Spectroscopy analysis (EDX). The results are reported here.

2. Bismuth-related superconductors

The first bismuth-oxide-based superconductor ($T_c \approx 12K$), $BaPb_{1-x}Bi_xO_3$, was discovered in 1975 by Sleight et al. [1]. In the superconducting region ($0.05 \leq x \leq 0.35$), this compound has a tetragonal symmetry. Its structure is related to the perovskite one and is described in the I 4/mcm space group which allows only one position for the (Bi/Pb) cations.
Later on, following the prediction of Mattheiss [2], that doping on the Ba site rather than on the Bi one would induce superconductivity with a higher T_c, $Ba_{1-x}K_xBiO_3$ was obtained. In the superconducting region ($T_c \approx 40K$), ($0.35 \leq x \leq 0.5$), this compound has a cubic symmetry [3].
Recently, we have been able to synthesize $SrBiO_3$ and to induce superconductivity by K or Rb doping [4]. The doping was made possible by the use of high pressure.

G. Van Tendeloo et al. (eds.), High-Temperature Superconductors and Novel Inorganic Materials, 103–108.

2.1 $SrBiO_3$

All electron diffraction patterns of $SrBiO_3$ can be indexed on the basis of a perovskite cell with a primitive lattice of parameters $a \approx b \approx a_p\sqrt{2}$ and $c \approx 2a_p$, where a_p refers to the ideal cubic cell ($a_p \approx 4$Å). Those taken along the [001] zone axis, as in figure 1a, show a splitting of all reflections except those along the [110]* direction. This indicates that the crystallites are twinned, the twinning plane being (110). This is very similar to what is observed for $YBa_2Cu_3O_7$ for example [5]. The twinning angle ε is about 1.3°. This value is in good agreement with the one calculated from the lattice parameters values obtained by synchrotron powder x-ray diffraction (a = 5.9480 (2) Å, b = 6.0951 (2) Å, c = 8.4854 (3) Å, β = 90.06(1)°), the formula $\varepsilon = \arcsin [(b^2-a^2)/(b^2+a^2)]$, giving ε = 1.4°.
It should be mentioned that from electron diffraction it has not been possible to decide between the orthorhombic Pbnm and the monoclinic $P2_1/n$ space groups. In principle, the 0kl reflections with k odd and l different from 0 should exist only in the $P2_1/n$ space group. However, due to the combined effects of twinning and double diffraction which may generate otherwise forbidden reflections, the conclusion is not so trivial. On the other hand, $P2_1/n$ is the only space group in which the Bi cations occupy two different crystallographic sites and permits therefore to account for the insulating properties of the compound.

EDX analyses have revealed that the Sr/Bi ratio was always larger than one, up to 1.5. Single phase samples were indeed only obtained when this ratio was larger than 1.2. The structure refinements have confirmed that the largest of the two Bi sites was partially occupied by Sr cations.

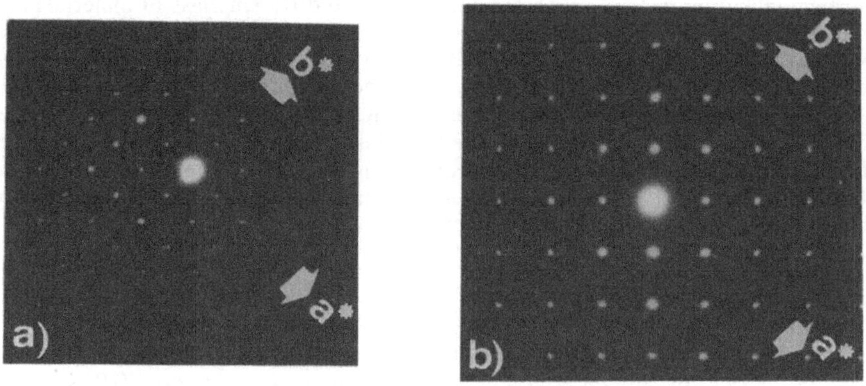

Figure 1. Electron diffraction pattern of $SrBiO_3$ (a) and $Sr_{0.4}K_{0.6}BiO_3$ (b) taken along the [001 zone axis

2.2 DOPED SAMPLES

As a function of K or Rb doping, the symmetry changes. Samples with a low K content (≤45%) are usually not single phase. For K substitutions larger than ≈45%, which corresponds to the superconducting region, the symmetry becomes tetragonal. The

solubility limit of K is ≈60%. From electron diffraction patterns the lattice is body centered (figure 1b) and the reflections can be indexed using the I4/mcm space group (as for $BaPb_{0.75}Bi_{0.25}O_3$) and the parameters a=b=5.9416 Å, c= 8.4394 Å.
On most of the HREM images (figure 2a) the presence of three types of domains is observed, corresponding to the three perpendicular possible orientations for the c-axis. This effect is a common feature in compounds involving a superstructure of the basic perovskite. The domains are about 200Å wide. The corresponding diffraction patterns are built up by the superimposition of the three variants (figure 2b, for a $Sr_{0.73}K_{0.27}BiO_3$ crystallite).

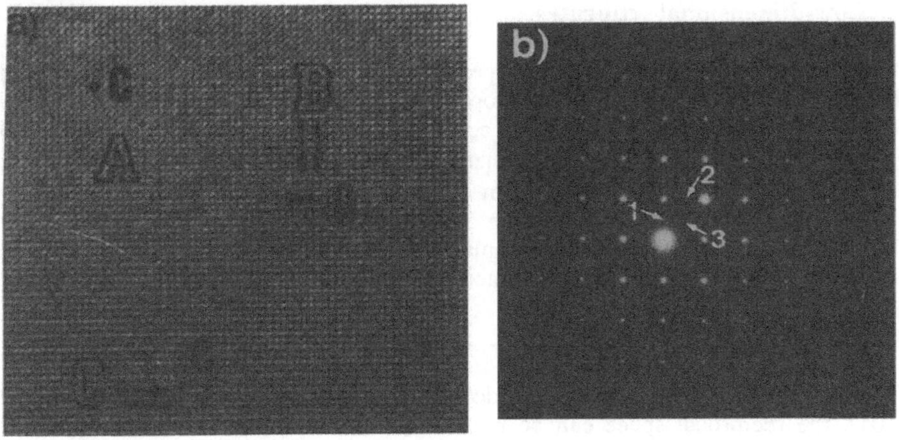

Figure 2. HREM image (a) and corresponding diffraction pattern (b) showing the presence of three domains

Figure 3 is an image taken along the [001] zone axis for a $K_{0.45}Sr_{0.55}BiO_3$ crystallite. The contrast consists in bright dots associated with Sr(K) or Bi cations and forming a square lattice of ≈ 4.2Å. However, the contrast is not uniform and clusters of a few greyish dots appear randomly. This could be attributed to a local change in the oxygen environnement of the cations.

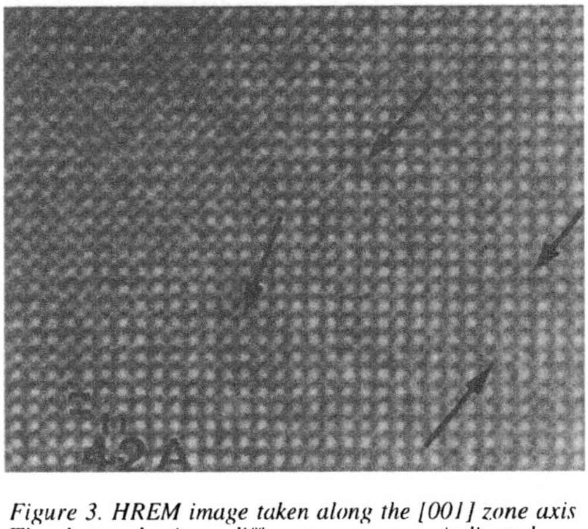

Figure 3. HREM image taken along the [001] zone axis The clusters having a different contrast are indicated.

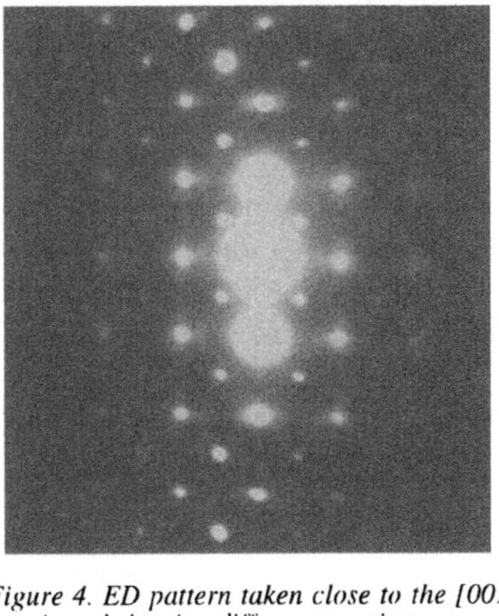

Figure 4. ED pattern taken close to the [001] zone axis and showing diffuse scattering

Finally, for some crystallites we have observed diffuse scattering on the diffraction patterns, in the form of crosses around the Bragg reflections (figure 4). The same geometry has been observed in the $Ba(Pb,Bi)O_3$ system [6]. This is very likely related to a departure of oxygen under the electron beam.

The symmetry changes also as a function of Rb doping. In the superconducting region, close to $Sr_{0.5}Rb_{0.5}BiO_3$ it looks like cubic from powder x-ray diffraction.

3. Low-dimensional cuprates

In the last years, low-dimensional copper oxide systems have drawn a lot of attention due to their interesting magnetic properties, including even superconductivity in $Sr_{14-x}Ca_xCu_{24}O_{41}$ [7]. Recently, a compound containing monodimensional CuO_4 chains, $Sr_{1-x}CuO_2$, has been synthesized [8], and we have investigated some of its structural aspects by electron microscopy.

From EDX analysis, the powder sample was found homogeneous, and the nominal composition, $Sr_{0.73}CuO_2$, was confirmed.

3.1. ELECTRON DIFFRACTION

From electron diffraction patterns recorded along different zone axes, [001], [010], [110], [101], the reciprocal space can be regarded as consisting of two sublattices. The diffraction spots of the two sublattices satisfy the extinction rules of F and C-centered space groups, respectively. From several diffraction patterns, we have been able to get values of the lattice parameters of the two sublattices:

C-centered sublattice	F-centered sublattice
a_C=3.77 (3) Å,	a_F=2.73 (1) Å
b_C=6.811(3) Å,	b_F=6.811 (3) Å
c_C=5.525 (2) Å	c_F=11.05 (3) Å

These values are in good agreement with those found from x-ray diffraction. The C lattice as been associated with the Sr arrangement and the F one with the CuO one. Such a description has already been reported for $Sr_{14-x}Ca_xCu_{24}O_{41}$ [9] and for $Ca_{0.85}CuO_2$ [10].

Figure 5a shows the [001] zone axis in which both sublattices are simultaneously revealed. Besides the main spots corresponding to the two sublattices, weaker reflections are present. They can be interpreted as double diffraction effect of one sublattice towards the other: if one takes a strong reflection, such as 200, of the C-lattice as the origin, then, some weak spots describe the reciprocal pattern of the F-lattice, and vice versa. Triple diffraction spots can even be observed (small arrows). The rows of satellites are indicated by large arrows. Another way to consider the weaker spots is to take the C-centered lattice as the basic lattice. Then, weaker spots appear as satellite reflections on each side of the main spots. The periodicity associated with these satellites is in fact related to the difference between a_F and a_C, namely $\Delta g=1/(2a_F^*-2a_C^*) \approx 4.95$Å. This period was found to be about the same for all different crystallites.

It should be noticed that for all orientations, the spots of both sublattices appear sharp and well defined. This is different from the case of $Sr_{14}Cu_{24}O_{41}$ samples for which the spots associated with the Cu_2O_3/Sr sublattice on one hand and the CuO_2 chains on the other hand have a different aspect: those related to the CuO_2 chains are heavily streaked in some orientations.
In case of figure 5b, both sublattices are not perfectly aligned. This effect can be observed when looking at the two satellite reflections shown by the small arrows. These two satellites are due to double diffraction effect when 200_F or -200_F are taken as the origin of the double diffraction for the C-sublatice, respectively. It is clear that the two satellites do not lie on a line parallel to either the a_F* or the a_C* directions.

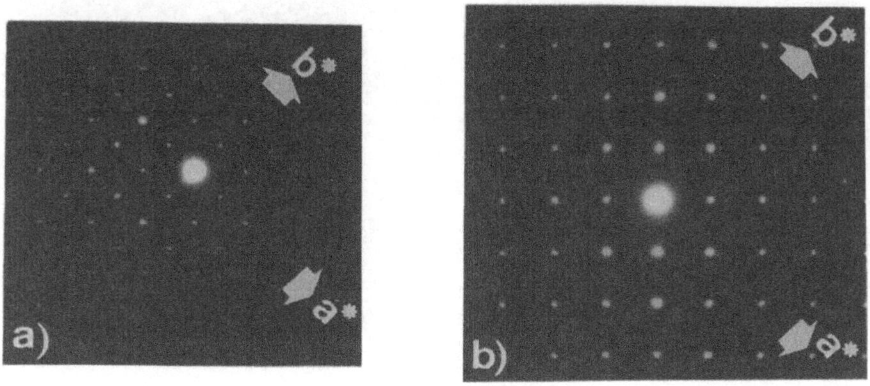

Figure 5. ED patterns of $Sr_{0.73}CuO_2$ taken along the [001] zone axis . For (b) the sublattices are not well aligned

3.2 HIGH RESOLUTION IMAGES

Figure 6 is a high resolution image taken along the [001] zone axis and corresponding to the diffraction pattern of fig. 5a. Depending on the areas of the image, different contrasts are observed. In areas like A, a regular arrangement of bright dots forming centered rectangles can be seen. The dimension of the rectangular array is ≈3.8 x 6.8 Å, which is in good agreement with the Sr (-C) sublattice in this projection. In other parts of the image, such as B, one can notice, along the a-axis, a succession of three brighter dots with a spacing of ≈8 Å and then less intense dots with a smaller spacing. The periodicity of this sequence is ≈28Å. The brighter dots are not aligned in a direction parallel to the b-axis but give rise to a periodic arrangement of dark and bright lines as indicated by the large arrows. The periodicity is ≈6.8Å and the direction of these lines has an angle of ≈85° with the a-axis. These periodicity and orientation are related in the reciprocal space to the satellite indicated by the small white star in figure 5a.

3.3 DARK FIELD IMAGES

Dark field images have been obtained along the [010] zone axis by selecting the following reflections: 200_C, 200_F, 201_C, $20\text{-}1_C$, 202_F, $20\text{-}2_F$. An example is given in figure 7. The contrast of the image consists of bright dots located on parallel rows separated by ≈ 5.55Å, that is the c-parameter. The rows can be interpreted as produced by the interference between spots of the same sublattice. The dot spacing along the row is related to the difference between the a parameters of the two sublattices (interference between 200_C and 200_F). Along a row, it is not constant ; furthermore, the positions of the dots may vary from one row to the next one. Consequently, the dots do not form a straight but a wavy line along the c-axis (indicated in figure 7). Depending on the crystallites, the waving aspect is more or less pronounced. According to Milat et al. [9], this would imply that the relative positions of the two sublattices may be different in adjacent rows.

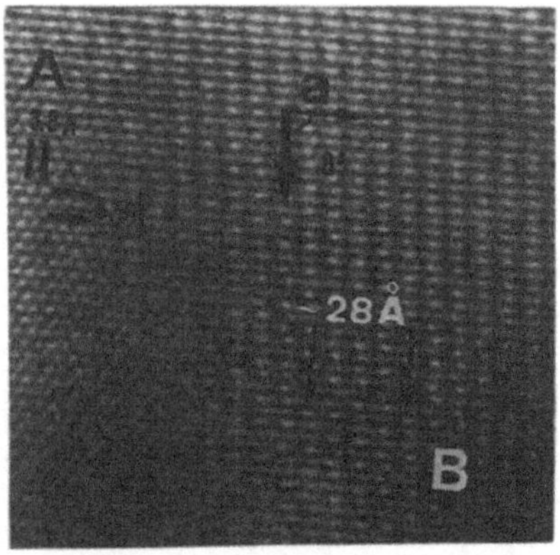

Figure 6. HREM image taken along the [001] zone axis.

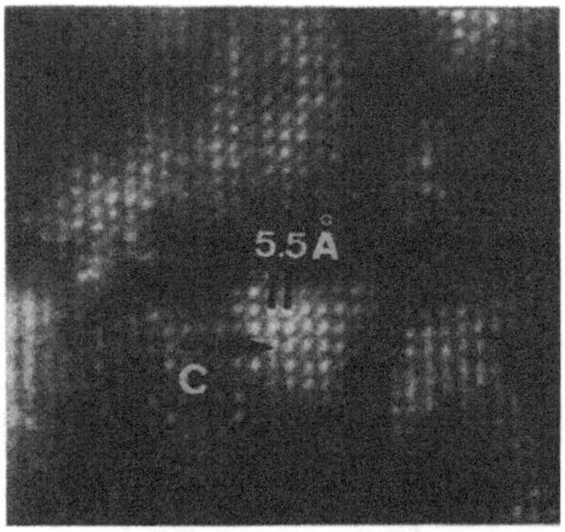

Figure 7. Dark field image taken along the [010] zone axis.

References

1. Sleight, A.W., Gillson, J.L., Bierstedt, P.E., (1975) High-temperature superconductivity in the $BaPb_{1-x}Bi_xO_3$ system, *Solid State Comm.* **17,** 27-28
2. Mattheiss, L.R., Gyorgy, E.M., Johnson, D.W., (1988) Superconductivity above 20K in the Ba-K-Bi-O system, *Phys. Rev. B* **37,** 3745-3746
3. Hinks, D.G., Dabrowski, B., Jorgensen, J.D., Mitchell, A.W., Richards, D.R., Pei, S., Shi, D., (1988) Synthesis, structure and superconductivity in the Ba1-xKxBiO3-y system, *Nature* **333**, 836-838
4. Kazakov, S.M., Chaillout, C., Bordet, P., Capponi, J.J., Nunez-Regueiro, M., Rysak, A., Tholence, J.L., Radaelli, P.G., Putilin, S.N., Antipov, E.V., (1997) Discovery of a second family of bismuth-oxide-based superconductors, *Nature* **390**, 148-150
5. Hodeau, J.L., Chaillout, C., Capponi, J.J., Marezio, M., (1987) Twinning in $Ba_2YCu_3O_{6+x}$ single crystals, *Solid State Comm.* **64**, 1349-1352
6. Chaillout, C., (1986) Thesis, Université J. Fourier, Grenoble
7. Uehara, M., Nagata, T., Akimitsu, J., Takahashi, H., Mori, N., Kinoshita, K., (1996) Superconductivity in the ladder material $Sr_{0.4}Ca_{13.6}Cu_{24}O_{41.84}$, *J. Phys. Soc. Jpn.* **65**, 2764-2768
8. Karpinski, J., Schwer, H., Meijer, G.I., Conder, K., Kopnin, E.M., Rossel, C., (1997) High-oxygen-pressure synthesis, structure and properties of the infinite-chain compound $Sr_{0.73}CuO_2$, *Physica C* **274**, 99-106
9. Milat, O., Van Tendeloo, G., Amelinckx, S., Mehbod, M., Deltour, R., (1992) The incommensurate structure of $(Sr,Ca)_{14}Cu_{24}O_{41}$: a study by electron diffraction and high-resolution microscopy, *Acta Cryst* **A48**, 618-625
10. Milat, O., Van Tendeloo, G., Amelinckx, S., Babu, T.G.N., Greaves, C., (1992) Structural variants of $Ca_{0.85}CuO_2$, *J. Solid State Chem.*. **101**, 92-114

NOVEL STRUCTURAL PHENOMENA AT THE MAXIMUM T_c IN 123 AND $HgBa_2CuO_{4+\delta}$ SUPERCONDUCTORS: EVIDENCE FOR A STRUCTURAL RESPONSE THAT COMPETES WITH SUPERCONDUCTIVITY

J. D. JORGENSEN[a], O. CHMAISSEM[a,b], D. G. HINKS[a], A. KNIZHNIK[c], Y. ECKSTEIN[a,c], H. SHAKED[a,b,d], J. L. WAGNER[a,e], B. DABROWSKI[a,b,f], S. SHORT[a], J. F. MITCHELL[a], J. P. HODGES[a]

[a]*Materials Science Division and* [b]*Science and Technology Center for Superconductivity, Argonne National Laboratory, Argonne, IL 60439;*

[c]*Department of Physics and Crown Center for Superconductivity, Technion, Haifa 3200, Israel;*

[d]*Department of Physics, Ben Gurion University, Beer-Sheva 84105, Israel;*

[e]*Department of Physics, University of North Dakota, P. O. Box 7129, Grand Forks, ND 58202;*

[f]*Department of Physics, Northern Illinois University, DeKalb, IL 60115.*

Abstract

Structural distortions that compete with superconductivity have been investigated in two systems where oxygen content can be used to vary the doping continuously from the under doped state, through the maximum T_c into the over doped state. In the 123 system, $(La_{1-x}Ca_x)(Ba_{1.75-x}La_{0.25+x})Cu_3O_{7+\delta}$, the buckling of the CuO_2 planes goes through a maximum at the maximum T_c. In $HgBa_2CuO_{4+\delta}$, where buckling of the CuO_2 planes is not available as a structural degree of freedom, there is a plateau at the maximum T_c where the unit cell volume expands as oxygen is added while the charge transfer and T_c remain constant. These unusual structural phenomena upon crossing through the maximum T_c are hypothesized to be a response of the crystal structure to the electronic structure, with the structural distortions competing with superconductivity, or lowering the T_c from what it would otherwise be.

G. Van Tendeloo et al. (eds.), High-Temperature Superconductors and Novel Inorganic Materials, 109–116.

1. Introduction

Structural distortions that compete with, or destroy, superconductivity are a common feature of conventional superconductors.[1-5] The general argument for explaining such behavior has been that as chemical or structural variables are changed to increase T_c by raising the electronic density of states, the free energy can be lowered by distorting the structure to open a gap at the Fermi energy, transforming the material to an insulator. No analogous behavior has been reported in the high-T_c superconductors; that is, there is no case where a structural distortion or phase transformation destroys superconductivity upon varying the chemistry to increase T_c and, thereby, places a limit on the maximum T_c. However, there is ample evidence for a close connection between T_c and particular features of the crystal structure. In the $La_{2-x}M_xCuO_4$ system, several experiments have demonstrated a relationship between the buckling angle of the CuO_2 planes and T_c.[6-9] When the buckling angle is varied by the application of pressure or chemical substitution, T_c decreases with increasing buckling angle. The observation of an unusual phase separation phenomena at the maximum T_c in $YBa_2Cu_3O_{6+\delta}$, with the two phases having different buckling angles and different T_c's, leads to the same conclusion for that compound.[10] In the present work, we investigate the crystal structure and superconducting properties of two high-T_c systems where the doping can be varied continuously through the maximum T_c. We interpret the behavior of T_c and the crystal structure as both being responses to the changing electronic structure as the doping is varied. In both systems, there is a structural distortion at the maximum T_c that competes with superconductivity, lowering T_c from what it would otherwise be.

2. CuO_2 Plane Buckling in the 123 Structure

$(La_{1-x}Ca_x)(Ba_{1.75-x}La_{0.25+x})Cu_3O_{7+\delta}$ is a compound with the 123 structure in which oxygen content can be used to continuously vary the doping from the under doped regime, through the maximum T_c, far into the over doped regime.[11] Powder samples with x=0.1, and 0.4 were synthesized by the methods described previously [11]. Oxygen contents were varied by subsequent annealing of parts of these initial samples in oxygen pressures from 10^{-4} to 1100 atm. The samples were confirmed to be single phase by x-ray and neutron diffraction. The oxygen contents of the samples were determined by iodometric titration and T_c's were obtained from magnetic susceptibility and resistivity measurements. Sharp superconducting transitions were observed for all samples. Time-of-flight neutron powder diffraction data were collected at room temperature on the Special Environment Powder Diffractometer (SEPD) at Argonne's Intense Pulsed Neutron Source (IPNS) [12] and analyzed by the Rietveld technique using the GSAS code [13] over a d-spacing range of 0.5 to 4 Å, which included up to 597 Bragg peaks.

The structures are tetragonal, space group P4/mmm, at all compositions as a result of the La substitution on the Ba site disrupting the normal oxygen chain ordering. The same defect chemistry -- partial occupancy of the single oxygen site in the "chain-Cu" plane -- is observed in the under doped and over doped regimes. This means that oxygen content can be used to vary the carrier concentration smoothly from the under doped regime to the over doped regime. The metal-site substitutions and the total oxygen contents of the samples determined from the neutron refinements are in agreement with the starting composition and with titration. Oxygen contents as high as 7.3 were achieved in both the x=0.1 and x=0.4 systems. The lattice parameters vary linearly with oxygen content. The Cu2-O4 apical copper-oxygen bond length also varies linearly over the full range, suggesting a smoothly increasing charge transfer. Thus, the hole concentration in the CuO_2 planes, deduced from bond valence sum calculations, increases linearly with oxygen content, as shown in Fig. 1. These results confirm that the hole concentration scales with oxygen concentration over the entire composition range.

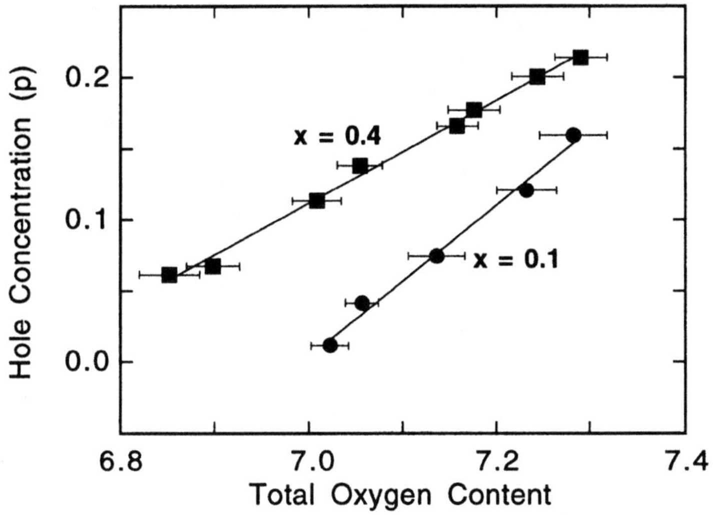

Figure 1. Hole concentration per CuO_2 plane vs. total oxygen for $(La_{1-x}Ca_x)(Ba_{1.75-x}La_{0.25+x})Cu_3O_{7+\delta}$. Round symbols are for x=0.1; square symbols are for x=0.4.

For both the x=0.1 and x=0.4 compounds, a clear maximum in T_c is observed at a total oxygen content of about 7.15, as shown in Fig. 2. The buckling of the CuO_2 planes (defined as the angle by which the plane oxygen atoms are out of the plane of the copper atoms) also goes through a maximum at the same oxygen content, as is shown in Fig. 2. The relationship between T_c and buckling angle is the same in the under doped and over doped regimes, within our experimental uncertainties. This remarkable scaling of the buckling angle and T_c in both the under doped and over doped regimes, while all other structural parameters that are sensitive to charge transfer show only monotonically changing behavior over the whole range, leads us to conclude that the buckling is a response to the same feature of the electronic structure that controls T_c. Thus, we argue that the buckling angle is an indirect probe of the Fermi surface topology.

Drawing on the arguments put forward for conventional superconductors (discussed in the introduction) our results for this 123 system can be understood within the framework of the following assumptions: (1) T_c increases with increasing electronic density of states at the Fermi energy. (2) The free energy of the system increases with increasing electronic density of states at the Fermi energy. (3) If possible, the structure will relax/distort to minimize the free energy. Thus, the maximum relaxation/distortion will occur when the Fermi energy is at a peak in the density of states.

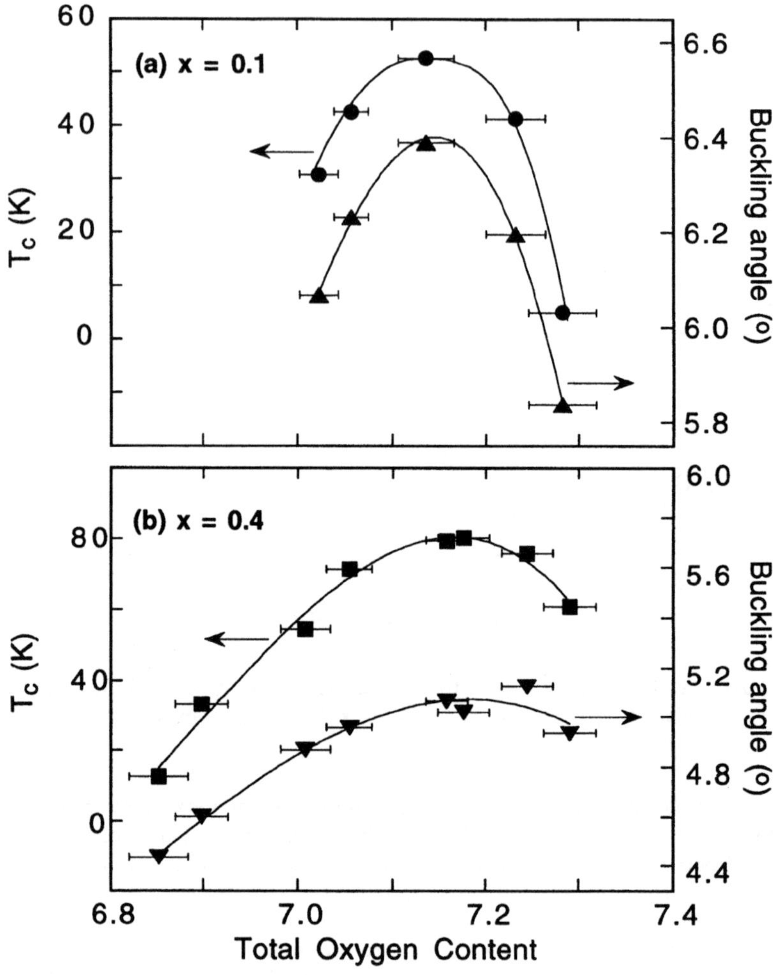

Figure 2. T_c (from resistivity measurements) and buckling angle of the CuO_2 plane vs. total oxygen content for $(La_{1-x}Ca_x)(Ba_{1.75-x}La_{0.25+x})Cu_3O_{7+\delta}$ with x=0.1 (a) and x=0.4 (b).

The buckling of the CuO_2 planes is the structural parameter that is expected to be most sensitive to the relevant features of the electronic structure. Band structure calculations [14] and photoemission studies [15] for the $YBa_2Cu_3O_{6+\delta}$ structure show that the CuO_2 planes and the one-dimensional CuO chains both contribute to the peak in the density of states, but that the chain contribution diminishes as the chain oxygen is depleted and disordered. The 123 compound we have studied is tetragonal and does not contain ordered CuO chains. At most, it can have only disordered chain fragments in the "chain plane," consistent with the average tetragonal symmetry. Its electronic structure has not been calculated, but we assume that the CuO_2 planes are the dominant contribution to peak in the density of states. Buckling would be expected to modify the Fermi surface so as to lower and/or broaden the peak, as has been calculated [16] and shown experimentally [6-9] for the La_2CuO_4 structure. Thus, the observed changes in buckling angle vs. oxygen content indicate that the structure distorts to lower its free energy as the Fermi energy passes though the peak in the density of states.

3. Cell Volume Expansion in the $HgBa_2CuO_{4+\delta}$ Structure

In light of these results, it is important to ask what structural distortion that competes with superconductivity will occur in a high-T_c compound where CuO_2 plane buckling is not available as a structural degree of freedom. The $HgBa_2CuO_{4+\delta}$ system, where the CuO_2 planes are constrained by symmetry to be flat, [17] offers an opportunity to investigate this question.

A single batch of $HgBa_2CuO_{4+\delta}$ powder was divided into smaller samples which were annealed at 300-400°C in oxygen partial pressures from 10^{-7} to 150 atm to produce samples with oxygen contents over the range $0.01 \leq x \leq 0.22$. T_c's were determined from ac susceptibility measurements and oxygen contents were determined by Rietveld refinement using neutron powder diffraction data.

The T_c's as a function of oxygen content are shown in Fig. 3. The samples clearly extend from the under doped regime, through the maximum T_c, into the over doped regime. The maximum T_c is nominally constant for oxygen contents from $x \approx 0.06$ to 0.16, giving a well defined plateau. The presence of the plateau shows that oxygen content alone is not sufficient to uniquely determine the T_c. It is instructive to

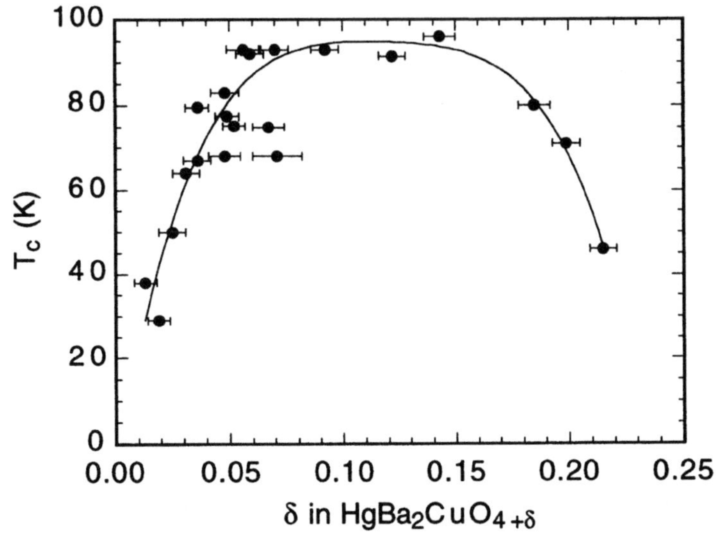

Figure 3. T_c vs. oxygen content for $HgBa_2CuO_{4+\delta}$.

plot T_c as a function of the amount of charge transfer. The structural variables most sensitive to charge transfer are the z coordinates of the Ba and apical O (O2) atoms. These atoms can be viewed as moving up or down (in opposite directions) along the z direction in response to the charges in the CuO_2 plane and the Hg-O_δ layer. Thus, the parameter most sensitive to the charge transfer is [z(Ba)-z(O2)]. Fig. 4a shows T_c plotted as a function of [z(Ba)-z(O2)]. The T_c data map onto a smooth curve when plotted with respect to this variable with the data at maximum T_c clustered around a single point. Thus, these maximum T_c's all correspond to the same charge transfer, in spite of the different oxygen contents.

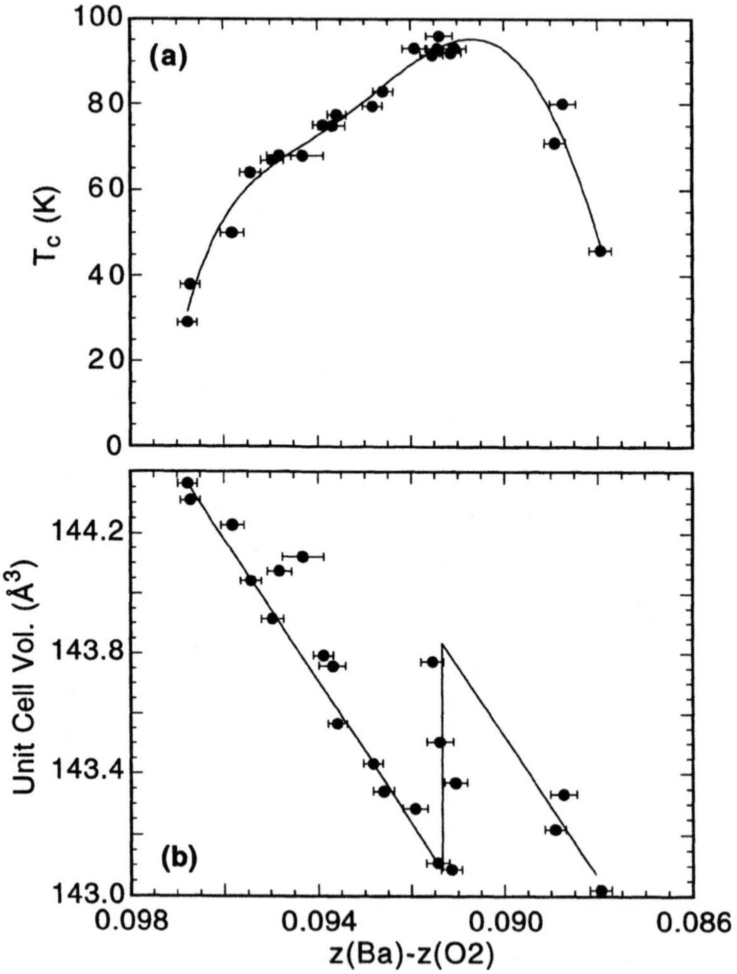

Figure 4. (a) T_c and (b) unit cell volume of $HgBa_2CuO_{4+\delta}$ vs. the structural parameter [z(Ba)-z(O2)], which is a measure of the charge transfer.

As oxygen is added to cross the plateau, the charge transfer remains constant as a result of a structural distortion that occurs over this range of oxygen compositions. This can be clearly seen by plotting structural variables as a function of [z(Ba)-z(O2)]. Fig. 4b shows the behavior of the unit cell volume. In the under doped and over doped regimes, the unit cell volume decreases smoothly with increasing charge transfer, with nominally the same slope. However, in the plateau region, where both charge transfer and T_c are constant, the volume increases. We interpret this unusual increase in cell volume as a response to features of the electronic structure at the maximum T_c. The increasing cell volume results in a constant charge transfer and T_c in spite of the increasing oxygen content. This effect produces the plateau in T_c rather than the expected inverse parabolic behavior.

One important implication of this hypothesis is that the maximum T_c of $HgBa_2CuO_{4+\delta}$ would be considerably higher if the structural distortion could be suppressed. Because the distortion is manifest as an increase in the cell volume, the application of pressure is an obvious way to attempt to suppress the distortion. Ignoring other effects of pressure on T_c, one would predict that pressure would substantially raise T_c for optimally doped samples; but, that the pressure effects would be much smaller for under doped and over doped samples. Qui et al. have reported the pressure dependence of T_c for $HgBa_2CuO_{4+\delta}$ for a range of oxygen contents extending into the over doped regime.[18] Their results are consistent with the prediction. The largest pressure-induced increases in T_c are observed for optimally doped samples. For over doped samples, applied pressure reduces T_c by a small amount.

4. Conclusions

These results show that high-T_c superconductors exhibit structural distortions that compete with superconductivity. Thus, the distortions are largest at the maximum T_c. These distortions are thought to be a response to the same features of the electronic structure that give rise to the high T_c. In the case of the 123 structure, the relevant distortion is buckling of the CuO_2 planes, which is known from other experiments to lower T_c. In the case of the $HgBa_2CuO_{4+\delta}$ structure, the distortion is an unusual expansion of the unit cell volume, which is expected from pressure experiments on this compound to lower T_c. In contrast to the behavior often observed in conventional superconductors, these structural distortions do not give rise to a phase transformation that destroys superconductivity. However, we conclude that the distortion lowers the maximum T_c and that the maximum T_c would be increased if the distortion could be suppressed.

Acknowledgments

This work was funded by the US Department of Energy, Division of Basic Energy Sciences - Materials Science, contract No. W-31-109-ENG-38 (JDJ, DGH, SS, JFM, JPH), the National Science Foundation, Office of Science and Technology Centers, grant No. DMR 91-20000 (OC, HS, BD), and the Israel Science Foundation (AK, YE).

References

1. R. Baillif, A, Dunand, J. Muller, K. Yvon, Phys. Rev. Lett. 47, 672 (1981); B. Lachal, R. Baillif, A. Junod, J. Muller, Solid State Commun. 45, 849 (1983); J. D. Jorgensen, D. G. Hinks, G. P. Felcher, Phys. Rev. B 35, 5365 (1987).
2. S. Pei, J. D. Jorgensen, B. Dabrowski, D. G. Hinks, D. R. Richards, A. W. Mitchell, J. M. Newsam, S. K. Sinha, D. Vaknin, A. J. Jacobson, Phys. Rev. B 41, 4126 (1990).
3. L. R. Testardi, Rev. Mod. Phys. 47, 637 (1975) and the references cited therein.
4. J. A. Wilson, F. J. DiSalvo, S. Mahajan, Adv. Phys. 24, 117 (1975).
5. J. M. Williams, A. J. Schultz, U. Geiser, K. D. Carlson, A. M. Kini, H. H. Wang, W.-K. Kwok, M.-H. Wangbo, J. E. Schirber, Science 252, 1501 (1991); D. Jerome, Science 252, 1509 (1991); J. M. Williams, J. R. Ferraro, R. J. Thorn, K. D. Carlson, U. Geiser, H. H. Wang, A. M. Kini, M.-H. Whangbo, "Organic Superconductors (Including Fullerenes) Synthesis, Structure, Properties, and Theory," (Prentice-Hall Inc., New Jersey, 1992).
6. Y. Yamada and M. Ido, Physica C 203, 240 (1992).
7. H. Takahashi, H. Shaked, B. A. Hunter, P. G. Radaelli, R. L. Hitterman, D. G. Hinks, J. D. Jorgensen, Phys. Rev. B 50, 3221 (1994)
8. B. Dabrowski, Z. Wang, K. Rogacki, J. D. Jorgensen, R. L. Hitterman, J. L. Wagner, B. A. Hunter, P. G. Radaelli, D. G. Hinks, Phys. Rev. Lett. 76, 1348 (1996).
9. M. Braden, O. Hoffels, W. Schnelle, B. Buchner, G. Heger, B. Hennion, I. Tanaka, H. Kojima, Phys. Rev. B 47, 12288 (1993).
10. E. Kaldis, J. Rohler, E. Liarokapis, N. Poulakis, K. Conder, P. W. Poeffen, Phys. Rev. Lett. 79, 4894 (1997); Ch. Kruger, K. Conder, H. Schwer, E. Kaldis, J. Solid State Chem. 134, 356 (1997).
11. D. Goldschmidt, G. M. Reisner, Y. Direktovitch, A. Knizhnik, E. Garstein, G. Kimmel, Y. Eckstein, Phys. Rev. B 48, 532 (1993); Physica C 217, 217 (1993); D. Goldschmidt, A. Knizhnik, Y. Direktovitch, G. M. Reisner, Y. Eckstein, Phys. Rev. B 49, 15928 (1994).
12. J. D. Jorgensen, J. Faber, Jr., J. M. Carpenter, R. K. Crawford, J. R. Haumann, R. L. Hitterman, R. Kleb, G. E. Ostrowski, F. J. Rotella, and T. G. Worlton, J. Appl. Crystallogr. 22, 321 (1989).
13. A. C. Larson and R. B. VonDreele, General Structure Analysis System, LAUR 86-748 (1985-1990).
14. See, for example, W. E. Pickett, Rev. Modern Phys. 61, 433 (1989) and the references cited therein.
15. R. Liu, B. W. Veal, A. P. Paulikas, J. W. Downey, H. Shi, C. G. Olson, C. Gu, A. J. Arko, J. J. Joyce, Phys. Rev. B 45, 5614 (1992); R. Liu, B. W. Veal, A. P. Paulikas, J. W. Downey, P. J. Kostic, S. Fleshler, U. Welp, C. G. Olson, X. Wu, A. J. Arko, J. J. Joyce, Phys. Rev. B 46, 11056 (1992).
16. M. R. Norman, G. J. McMullan, D. L. Novikov, A. J. Freeman, Phys. Rev. B 48, 9935 (1993).
17. J. L. Wagner, P. G. Radaelli, D. G. Hinks, J. D. Jorgensen, J. F. Mitchell, B. Dabrowski, G. S. Knapp, M. Beno, Physica C 210, 447 (1993); O. Chmaissem, Q. Huang, S. N. Putilin, M. Marezio, A. Santoro, Physica C 212, 259 (1993).
18. X. D. Qui, Q. Xiong, L. Gao, Y. Cao, Y. Y. Xue, C. W. Chu, Physica C 282-287, 885 (1997).

BA-CU-O-CL COMPOUNDS WITH NON-EQUIVALENT CU SITES: PHASE EQUILIBRIA AND CRYSTAL STRUCTURE

K. Ruck[1], G. Krabbes[1], M. Ruck[2]
[1]*Institut für Festkörper- und Werkstofforschung Dresden (IFW), Postfach, D-01171 Dresden, Germany*
[2]*Institut für Anorganische Chemie, Universität (TH) Karlsruhe, Engessersstraße, D-76128 Karlsruhe, Germany*

1. Introduction

The interest in studying the alkaline earth copper oxide chlorides started with compounds showing structural similarity to the cuprate superconductors because they behave as model substances with undoped Cu-O planes. Concerning the Ba-Cu-O-Cl system at least five quaternary phases are known [1,2]: $Ba_2Cu_3O_4Cl_2$, $Ba_2CuO_2Cl_2$ (stable at high pressure) both with planar Cu-O planes and $Ba_3Cu_2O_4Cl_2$, $Ba_9Cu_7O_{15}Cl_2$, Ba_2CuO_2Cl showing different structural features. Recently, we reported on the magnetic properties of $Ba_3Cu_2O_4Cl_2$ and $Ba_2Cu_3O_4Cl_2$ [3]. Both compounds order antiferromagnetically at low temperatures. Whereas $Ba_3Cu_2O_4Cl_2$ exhibits a metamagnetic spin flop transition below T_N=20K, the magnetic behaviour of $Ba_2Cu_3O_4Cl_2$ is more complicate. According to its crystal structure the compound exhibits two magnetic subsystems which order at $T_{N1}\approx$340K and $T_{N2}\approx$33K, respectively. The dynamic of holes in the valence bands of $Ba_3Cu_2O_4Cl_2$ and $Ba_2Cu_3O_4Cl_2$, as expressed via the screening of a Cu 2p core hole, was studied by high-resolution XPS measurements [11]. For $Ba_3Cu_2O_4Cl_2$ a locally screened final state was found. In case of $Ba_2Cu_3O_4Cl_2$ an additional nonlocal screening via Zhang-Rice singlet formation and the O 2p band is discussed.

2. Experimental

The compounds were prepared by conventional solid state techniques. In order to prevent the formation of carbonates CO_2-poor air was used especially for the synthesis of samples with high Ba-content. Phase diagram investigations were performed with DTA-/ TG-measurements (DTA/TG92, Setaram) and soaking experiments. For further details see also [4]. The solid phases were characterized using a powder diffractometer

G. Van Tendeloo et al. (eds.), High-Temperature Superconductors and Novel Inorganic Materials, 117–122.

PW 1820 (Phillips). The single crystal structure determination of $Ba_3Cu_2O_4Cl_2$ was performed with a four circle diffractometer AED2 (Siemens / Stoe).

3. Phase Equilibria

The projection of the quaternary system Ba-Cu-O-Cl into the pseudo-ternary phase diagram BaO-$BaCl_2$-CuO_x is drawn in *Figure 1*. It can be devided into a subsolidus part right beside the line Ba_4OCl_6 - $Ba_3Cu_2O_4Cl_2$ - $BaCuO_2$ and the area of the BaO-rich region where melt already occurs at the applied conditions (temperature, oxygen partial pressure).

The phase compatibilities of the subsolidus part determined by X-ray analysis are quite clear: Stable quaternary phases are $Ba_3Cu_2O_4Cl_2$ and $Ba_2Cu_3O_4Cl_2$. Considering ternary phases, Ba_4OCl_6 (the only known compound of the pseudo-binary system $BaCl_2$ - BaO) and $BaCuO_2$ become stable. At least at higher temperatures $BaCu_2O_2$ was found to crystallize from oxygen poor melts. As can be seen in *Figure 1* $BaCuO_2$ exhibits a homogeneity range concerning metal and oxygen content [5]. REM-investigations give additional evidence for a partly substitution of chlorine for oxygen.

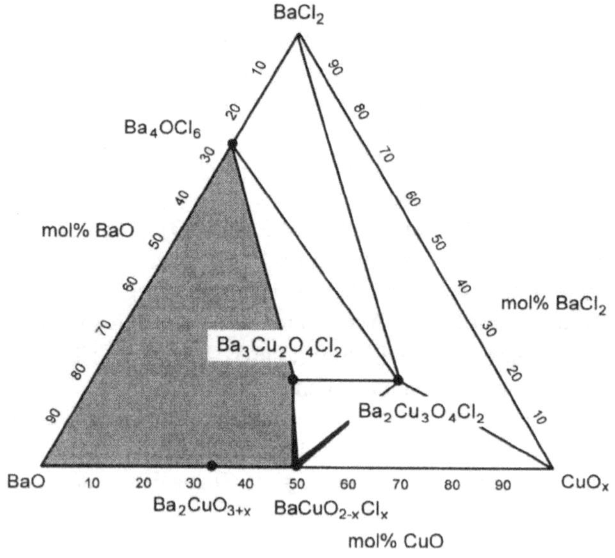

Figure1. Pseudo-ternary phase diagram BaO - $BaCl_2$ - CuO_x at T≈750°C and $p(O_2)$=0.21atm. The white area represents the subsolidus region of which the stable phases and compatibility lines are drawn. The gray part of the diagram is not discussed in this work.

Within this area, the melt occurs above 900°C. There is not only one primary occuring melt which expands over the entire diagram with rising temperature. At least three different kinds of melt have been found to develop independently from each other. Two of them result from univariant reactions, which are discussed later. The third one is formed by the eutectic reaction between $BaCuO_2$ and CuO near 905°C, which has been intensively studied in the literature [6,7].

3.1 UNIVARIANT REACTIONS IN AIR

The knowledge about phase relations and decomposition behaviour of the above mentioned stable compounds is a useful tool for preparation and crystal growth. Except for $BaCuO_2$, which melts at about 993°C [6,7], no literature data concerning ternary and quaternary phases exist. DTA-investigations of Ba_4OCl_6 reveal a single endothermic signal at 1015°C indicating a nearly congruent decomposition behaviour. In contrast to the other compounds Ba_4OCl_6 decomposes without change of the oxygen content.
At a temperature of about 905°C Ba_4OCl_6 reacts with $Ba_3Cu_2O_4Cl_2$ according to (1):

$$\alpha\ Ba_3Cu_2O_4Cl_2 + \beta\ Ba_4OCl_6 \rightleftarrows \chi\ Ba_2Cu_3O_4Cl_2 + \delta\ L + \varepsilon\ O_2 \qquad (1)$$

The resulting phase relations at 910°C and 930°C proofed by soaking experiments are drawn in *Figure 2*. Especially the formation of $Ba_2Cu_3O_4Cl_2$ intersecting the line $Ba_3Cu_2O_4Cl_2$ - Ba_4OCl_6 above 905°C gives evidence for the proposed mechanism. A further slight increase in temperature leads to a reaction resulting in the decomposition of the line $Ba_2Cu_3O_4Cl_2$ - Ba_4OCl_6 at about 932°C.

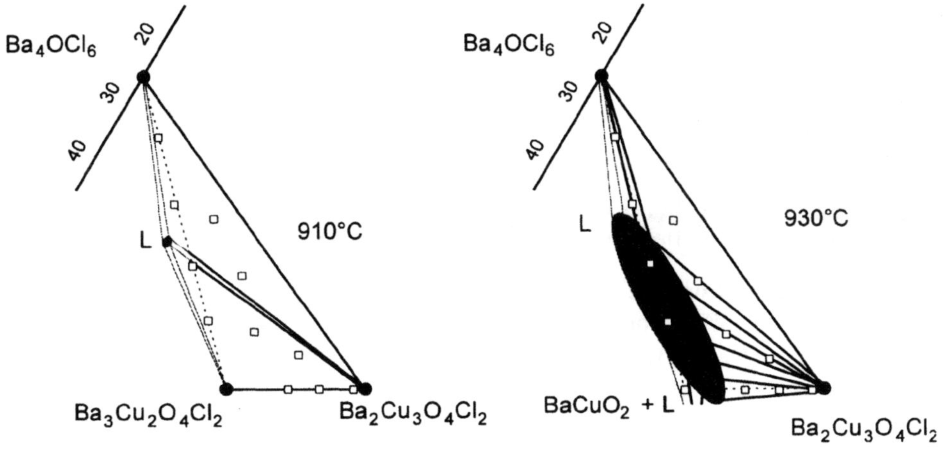

Figure 2. Isothermal sections of the Gibbs triangle $Ba_2Cu_3O_4Cl_2$ - $Ba_3Cu_2O_4Cl_2$ - Ba_4OCL_6 (*Figure 1*) at temperatures above the univariant reaction (1). The squares represent samples studied via soaking experiments and DTA-measurements leading to the phase compatibilities at 910°C and 930°C, respectively.

In the temperature range between the two isothermic sections $Ba_3Cu_2O_4Cl_2$ decomposes peritectically at about 925°C into $BaCuO_2$ and probably $Ba_2Cu_3O_4Cl_2$ [4]. However, the the coexistence line $BaCuO_2$ - $Ba_2Cu_3O_4Cl_2$ is stable up to the same temperature within the error limits (927°C). So, the experimental proof becomes extremely difficult.

A further univariant reaction occurs between $Ba_2Cu_3O_4Cl_2$ and $BaCl_2$ at 901°C following the equation (2):

$$\alpha\, Ba_2Cu_3O_4Cl_2 + \beta\, BaCl_2 \rightleftarrows \chi\, CuO + \delta\, L + \varepsilon\, O_2 \qquad (2)$$

As can be seen in *Figure 3*, the residual liquid has a $BaCl_2$-rich composition, which leads to the expansion of the primary crystallization field of CuO at least over the largest part of the $Ba_2Cu_3O_4Cl_2$ - $BaCl_2$ - CuO triangle and stretches into the triangle $Ba_3Cu_2O_4Cl_2$ - Ba_4OCl_6 - $Ba_2Cu_3O_4Cl_2$.

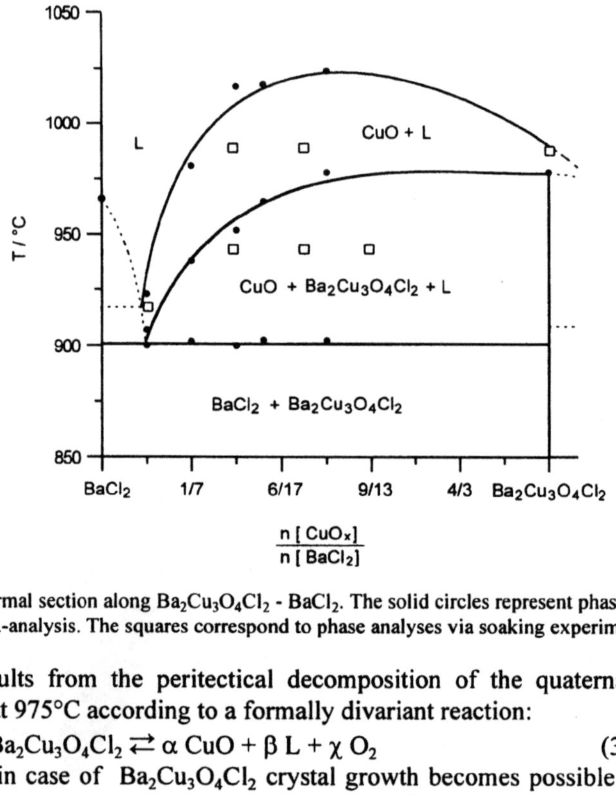

Figure 3. Polythermal section along $Ba_2Cu_3O_4Cl_2$ - $BaCl_2$. The solid circles represent phase transitions measured by DTA-analysis. The squares correspond to phase analyses via soaking experiments.

The latter results from the peritectical decomposition of the quaternary compound $Ba_2Cu_3O_4Cl_2$ at 975°C according to a formally divariant reaction:

$$Ba_2Cu_3O_4Cl_2 \rightleftarrows \alpha\, CuO + \beta\, L + \chi\, O_2 \qquad (3)$$

Nevertheless, in case of $Ba_2Cu_3O_4Cl_2$ crystal growth becomes possible from the pure compound because its primary crystallization field directly adjoints the primary crystallization field of CuO.

Single crystals of $Ba_3Cu_2O_4Cl_2$ were obtained by a flux method using $BaCl_2$-excess.

4. Crystal Structures of the quaternary compounds

$Ba_2Cu_3O_4Cl_2$. With alternate stacking of planar Cu_3O_4 planes and BaCl block layers the crystal structure of $Ba_2Cu_3O_4Cl_2$ has a close relationship to the high temperature superconducting oxides [1,8]. Characteristic structural features are the two

crystallographically different copper sites within the planes, which are responsible especially for the interesting magnetic properties.

$Ba_3Cu_2O_4Cl_2$. $Ba_3Cu_2O_4Cl_2$ contains also two non-equivalent copper sites in the lattice. Both are square planar coordinated by oxygen atoms, an additional chlorine atom completes the coordination sphere of one type of copper atoms to a pyramidal polyhedron. They are linked together forming a folded chain along the a-axis (*Figure 4*). The crystal structure was first reported by Müller-Buschbaum et al. [1]. On account of the rising interest concerning electronic structure and properties of this compound, the anisotropic displacement parameters as well as the structural data at low temperatures were determined (Table 1).

Table 1. Structural parameters of $Ba_3Cu_2O_4Cl_2$ determined by single-crystal X-ray diffraction at 20°C and **-146°C**.

	x	y	z	U_{11}	U_{22}	U_{33}	U_{13}	U_{eq}
Ba1	0.25	0.5	0.44867(5)	58(3)	77(2)	132(3)	0	89(1)
			0.44899(4)	**33(1)**	**47(1)**	**80(2)**		**53(1)**
Ba2	0.25	0	0.21161(5)	80(3)	79(2)	106(3)	0	89(1)
			0.21150(4)	**41(2)**	**44(1)**	**61(2)**		**49(1)**
Ba3	0.75	0.5	0.20398(5)	96(3)	78(2)	117(3)	0	97(1)
			0.20368(4)	**50(2)**	**42(1)**	**72(2)**		**55(1)**
Cu1	0	0	0.5	61(6)	63(4)	99(5)	-10(5)	74(2)
				38(3)	**39(3)**	**55(4)**	**- 2(3)**	**44(1)**
Cu2	0.75	0	0.32970(9)	33(6)	50(4)	111(6)	0	65(2)
			0.32945(7)	**23(3)**	**30(3)**	**81(5)**		**45(2)**
Cl1	0.75	0	0.0867(2)	292(17)	147(9)	120(11)	0	187(6)
			0.0864(2)	**154(7)**	**80(6)**	**78(8)**		**104(3)**
Cl2	0.25	0.5	0.1161(2)	242(17)	107(10)	187(14)	0	179(6)
			0.1151(2)	**99(7)**	**49(6)**	**129(10)**		**92(3)**
O	0.9681(4)	0.2217(3)	0.3619(2)	117(18)	78(11)	119(15)	-12(14)	104(6)
	0.9695(3)	**0.2221(3)**	**0.3617(2)**	**55(10)**	**46(7)**	**90(12)**	**-2(10)**	**64(4)**

The anisotropic displacement parameters are defined for exp $[-2\pi^2 (U_{11}h^2a^{*2}+.....+2U_{23}klb^*c^*)]$. U_{12} and U_{23} are restrained to zero for all atoms of special positions. The refined values for O are U_{12}=-33(12)/**-2(7)** and U_{23}=16(9)/**19(7)**, respectively. U_{eq} are defined as one third of the trace of the orthogonalized U_{ij} tensor. Summary of crystal data and intensity collection: space group: Pmma (No.51); formula units: Z=2; V=415/**412**Å^3; d_{calc}=5.38/**5.44**g/cm^3 $\mu(MoK_\alpha)$=197/**198**cm^{-1}; data collection: four circle diffractometer AED2 (Siemens / Stoe), $2\theta_{min}$=3°, $2\theta_{max}$=50/**55**°; $0 \le h \le 6$, $-7 \le k \le 7$, $-12 \le l \le 12$ / $\mathbf{0 \le h \le 8, -7 \le k \le 7, -13 \le l \le 13}$; ω-scan; number of reflections: 1370/**1981**, independent: 411/**565**; absorption correction: analytical ABSPSI [9], R_{int}=0.034/**0.04**, R_{sigma}=0.024/**0.025**; extinction parameter: 2.3x10^{-3}/**1.9x10^{-3}**, refinement: SHELXL97 [10], R_1(aniso)=0.029/ **0.031**, wR_2(aniso)=0.023/**0.02** weight: $1/\sigma^2(F_o^2)$.

As can be seen in Table 1, no dramatic shift of the atomic parameters occurs down to 127K. The lattice constants at 293K (a=6.548(1), b=5.991(1), c=10.585(1) Å) change at the low temperature to a=6.525(5), b=5.979(5), c=10.558(8) Å. They show the strongest compression along the direction of the CuO chain reflecting the further reduction of the short Cu-Cu distances (*Figure 4*). An opposite shift of Ba1- and Ba2-type atoms can be observed along the c-axis.

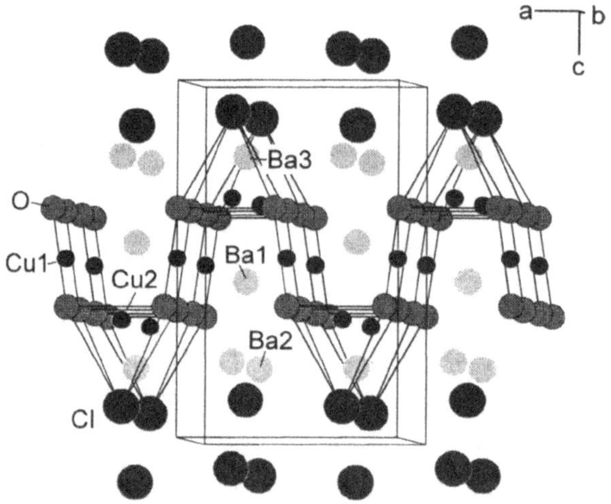

Interatomic distances [Å] at 20°C and **-146°C**

Cu1-O	1.986 (2)	**1.984 (2)**
Cu2-O	1.980 (2)	**1.982 (2)**
Cu2-Cl	2.572 (2)	**2.565 (3)**
Cu1-Cu2	2.4350(7)	**2.429 (1)**

Figure 4. Unit cell of $Ba_3Cu_2O_4Cl_2$ with coordination polyhedra around the copper atoms.

Acknowledgment

Support by the Fond der Chemischen Industrie is gratefully acknowledged.

References:

1. Müller-Buschbaum, Hk. (1991) The Crystal Chemistry of Copper Oxometalates, *Angew. Chem. (Int. Ed. Engl.)* **30 (7)**, 723-744.
2. Tatsuki, T., Tokiwa-Yamamoto, A., Tamura, T., Moriwaki, Y., Wu, X.-J., Adachi, S., Tanabe, K. (1996) A new oxychloride, $Ba_2CuO_2Cl_2$, prepared under high pressure, *Physica C* **265**, 323-327.
3. Eckert, D., Ruck, K., Wolf, M., Krabbes, G., Müller, K.-H. (1998) Magnetic behavior of the low dimensional compounds $Ba_3Cu_2O_4Cl_2$ and $Ba_2Cu_3O_4Cl_2$, J. *Appl. Phys.* **83,** in press.
4. Ruck, K., Eckert, D., Krabbes, G., Wolf, M., Müller, K.-H., Phase diagram investigations in the system Ba-Cu-O-Cl and magnetic properties of $Ba_3Cu_2O_4Cl_2$ and $Ba_2Cu_3O_4Cl_2$, submitted to *J.Solid State Chem.*
5. Paulus, E.F., Miehe, G., Fuess, H., Yehia, I., Loechner, U. (1991) The crystal structure of $BaCuO_2$, *J. Solid State Chem.* **90**, 17-26.
6. Zhang, W., Osamura, K., Ochiati, Sh. (1990) Phase diagram of the BaO-CuO binary system, *J. Am. Ceram. Soc.* **73**, 1958-1964.
7. Krabbes, G., Bieger, W., Schätzle, P. (1995) Phase equilibria- a fundamental for growth and processing of Y-Ba-Cu-O and related high temperature superconductors, *Current topics in crystal growth Res.* **2**, 359-375; and references therein.
8. Pitschke, W., Krabbes, G., Mattern, N. (1995) Powder Diffraction data and Rietfeld refinement of the compound $Ba_2Cl_2Cu_3O_4$, *Powder Diffraction* **10(4)**, 282-287.
9. Adams, K.-G. (1995) *ABSPSI, Program for analytical absorption correction*, modified version of the program by Marks and Alcock, University of Karlsruhe, Germany.
10. Sheldrick G.M. (1997) *SHELXL97, Program for crystal structure refinement*, University of Göttingen, Germany.
11. Böske, T., Maiti, K., Knauff, O., Ruck, K., Golden, M.S., Krabbes, G., Fink, J., Osafune, T., Motoyama, N., Eisake, H., Uchida, S. (1998) Cu-O network-dependent core hole screening in low-dimensional cuprate systems: A high resolution x-ray photoemission study, *Phys. Rev. B* **57(2)**, 1-4; and references therein.

La DOPING EFFECTS ON STRUCTURE AND CONDUCTIVITY OF $(Ba,Sr,Ca)PbO_3$

A. YAMAMOTO and T. ONO
Superconductivity Research Laboratory,
International Superconductivity Technology Center,
1-10-13 Shinonome, Koto-ku, Tokyo 135, Japan

$A_{1-x}La_xPbO_3$ (A=Ba,Sr) and $Ca_{1-x}La_xPbO_3$ samples were synthesized by the solid state reaction method under ambient and high pressure (5GPa), respectively, and La doping effects on their structure and transport properties were studied. Solubility limit of La in the alkaline earth element's site was found to be about 10%. Temperature dependencies of electrical resistivity and Seebeck coefficient were indicated that La doped samples are metallic conductors. Absolute values of resistivity for La doped $SrPbO_3$ and $CaPbO_3$ samples were significantly lower than those of undoped ones. Seebeck coefficients of these samples were negative, and absolute values were reduced by the doping. These results suggest that the La doping introduces electrons into the conduction band Superconductivity was not observed in these materials above 4K.

Key words; $(Ba,Sr,Ca)PbO_3$, La doping effect, transport properties

1. Introduction

The simple perovskite $BaPbO_3$, one of the well-known conducting oxides [1], is the basis for the superconductor $BaPb_{1-x}Bi_xO_3$ (T_c=12K) [2]. The high conductivity of these materials is explained by the formation of a broad band due to mixing of lead 6s and oxygen 2p states [3]. Carrier doping in these lead oxides by chemical methods, for example, A-site substitution [4, 5] and reducing oxygen content [6], is of interest in comparison to the copper-oxide superconductors.

Itoh *et al.* [4] reported the crystal structure and the transport properties of $Ba_{1-x}Sr_xPbO_3$ (x=0-1.0). They found that electrical resistivity of these samples increased and carrier concentration decreased with increasing Sr content. The carrier concentration of $BaPbO_3$ (2.70x 10^{20} per cm^3 at 77K) and $SrPbO_3$ (2.06x10^{18}), indicated that these materials are in a low carrier system compared to high-T_c copper oxides. They also concluded that mobility of electrons is seriously influenced by the randomness in the A-site. M.V.

G. Van Tendeloo et al. (eds.), High-Temperature Superconductors and Novel Inorganic Materials, 123–128.

Lobanov *et al.* [5] reported the resistivity properties of undoped and La doped $SrPbO_3$. They demonstrated that La doping in the Sr site leads to a drastical decrease of the absolute value of the resistivity. Cava *et al.* [6] tried to introduce electrons in the conduction band of $BaPbO_3$ and $Ba_4Pb_3O_{10}$ by the oxygen deficiency, however, the change in Hall coefficient with oxygen content was much less than is expected.

Recently, we have successfully synthesized perovskite-type $CaPbO_3$ under high pressure [7]. The X-ray analysis revealed that the structure is strongly distorted with $GdFeO_3$-type. The electrical resistivity of this compound is considerably high in comparison with those of $BaPbO_3$ and $SrPbO_3$, provably due to less overlap or gap formation between Pb6s-O2p σ and σ^* bands.

In the present work, La doping effects on the structure and transport properties of the $APbO_3$ (A=Ba,Sr,Ca) were investigated. The reason why we selected La for a doping ion is that substitution of trivalent ion for divalent ion may introduce electrons into the conduction band. La doping is also expected to suppress the oxygen deficiency. Samples were carefully synthesized to keep the stoichiometry, and transport properties were studied by both electrical resistivity and thermopower. It was found that La doping introduced electrons in the conduction band, irrespective of ionic size of the A-site.

2. Experimental

$APbO_3$ (A=Ba,Sr) and their La doped samples were prepared from high-purity $BaCO_3$, $SrCO_3$, La_2O_3 and Pb_3O_4 by the solid state reaction. Mixed and palletized powders were calcinated at 800℃ in air, sintered at 900℃ for 48 hours several times with flowing O_2, and finally annealed at 500℃ for 12 hours to suppress oxygen deficiency. $CaPbO_3$ and its La doped samples were prepared at 1000℃ for 1 hour under 5 GPa using a cubic-anvil-type apparatus from mixed powder of high-purity CaO, La_2O_3, PbO_2, PbO and Ca_2PbO_4, which prepared from $CaCO_3$ and PbO_2. Samples were examined by X-ray powder diffraction (XRD) using Cu Kα radiation, and the unit cell dimensions were determined using the XRD data by the least-squares method. Metal composition was confirmed by inductively coupled plasma (ICP) analysis. Oxygen content was determined by the iodometry. Electrical resistivity and thermopower measurements were performed by conventional methods.

3. Results and discussion

To determine solubility limit of La in the A site, we prepared $A_{1-x}La_xPbO_3$ (A=Ba,Sr, Ca, x=0,0.1,0.2) samples and examined by XRD. Impurity phases were detected in the samples with x=0.2 irrespective of element spices. Thus, we studied $APbO_3$ (A=Ba,Sr, Ca) and their 10% La doped samples in the present work. Unit cell dimensions are summarized in the Table 1. No change in the symmetry with La doping was observed, and the degree of the distortion of the doped samples is comparable to that of undoped ones. The unit cell dimensions increased with La doping, although ionic radius of La^{3+}(C.N.=12) is smaller than those of Ba^{2+}(C.N.=12) and Sr^{2+}(C.N.=12). This can be explained by redution of Pb oxidation states. This idea is consistent with results of the averaged Pb valence, which was calculated by charge neutrality, as indicated in the Table 1. Analyzed metal compositions of $BaPbO_{3-\delta}$ and $SrPbO_{3-\delta}$ were found to be Ba:Pb=0.98(2):1.00(2) and Sr:Pb=0.98(2):1.02(2), respectively. This indicated that the stoichiometry of the metals was kept during the sintering within the error bar of this analysis. Oxygen content of these samples were also listed in Table 1. They are almost constant to 3.00, suggesting no substantial oxygen deficiency in these samples. The oxygen content of the La doped samples is always slightly bigger than that of undoped ones, although increase of the oxygen content dose not completely cancel the $Pb^{(4-\delta)+}$ formation by La^{3+} doping.

TABLE 1. Unit cell dimensions, oxygen content and averaged Pb valence of $A_{1-x}La_xPbO_{3-d}$ (A=Ba,Sr,Ca, x=0,0.1)

composition	a / Å	b / Å	c / Å	v / Å^3	oxygen content	$Pb^{\alpha+}$
$BaPbO_{3-\delta}$	6.056(1)	6.009(4)	8.496(3)	309.2	2.98 (±0.01)	3.96
$Ba_{0.9}La_{0.1}PbO_{3-\delta}$	6.058(2)	6.021(4)	8.504(5)	310.2	3.01	3.92
$SrPbO_{3-\delta}$	5.854(1)	5.950(1)	8.318(1)	289.7	2.98	3.96
$Sr_{0.9}La_{0.1}PbO_{3-\delta}$	5.869(3)	5.974(7)	8.302(8)	291.1	3.01	3.92
$CaPbO_{3-\delta}$	5.658(3)	5.871(3)	8.132(5)	270.2	2.97	3.94
$Ca_{0.9}La_{0.1}PbO_{3-\delta}$	5.694(1)	5.906(1)	8.191(2)	275.5	3.00	3.90

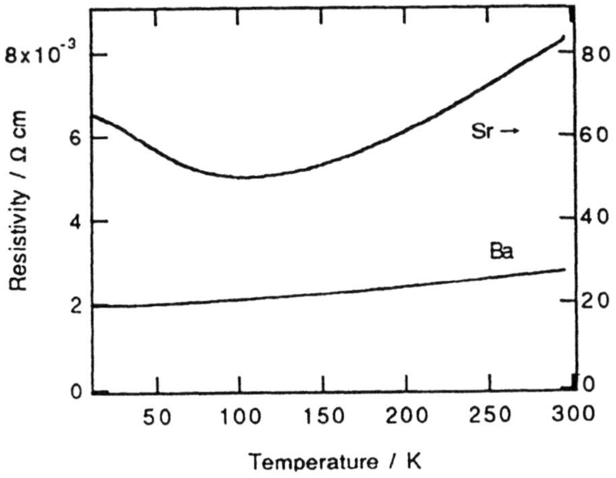

Fig. 1 (a) Temperature dependence of electrical resistivity of $A_{1-x}La_xPbO_{3-\delta}$ (A=Ba,Sr, x=0)

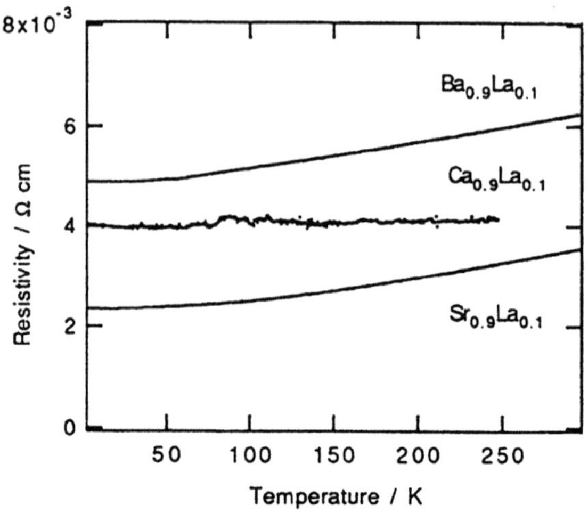

Fig. 1 (b) Temperature dependence of electrical resistivity of $A_{1-x}La_xPbO_{3-\delta}$ (A=Ba,Sr,Ca, x=0.1)

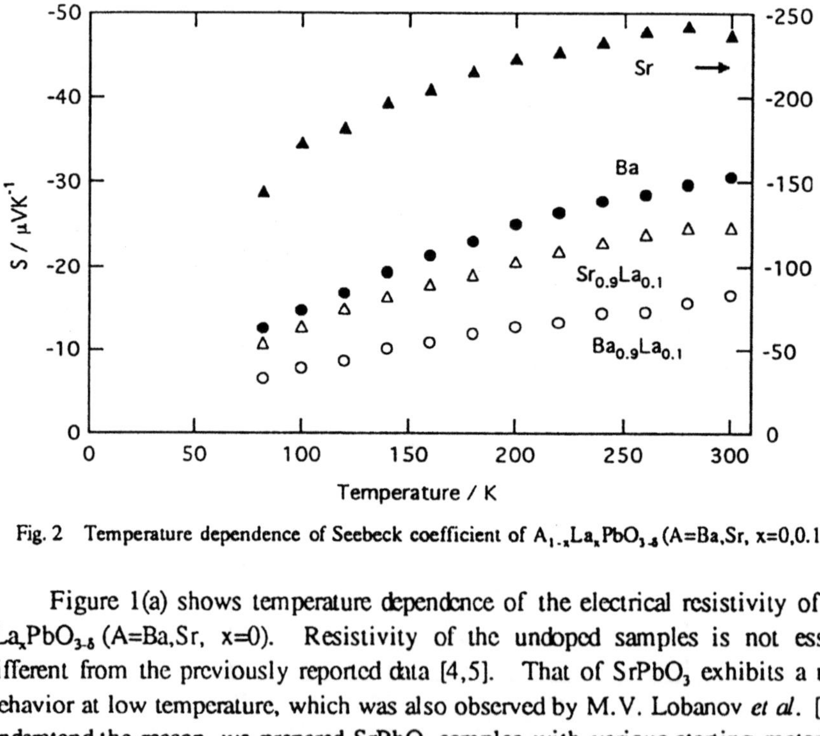

Fig. 2 Temperature dependence of Seebeck coefficient of $A_{1-x}La_xPbO_{3-\delta}$ (A=Ba,Sr, x=0,0.1)

Figure 1(a) shows temperature dependence of the electrical resistivity of the $A_{1-x}La_xPbO_{3-\delta}$ (A=Ba,Sr, x=0). Resistivity of the undoped samples is not essentially different from the previously reported data [4,5]. That of $SrPbO_3$ exhibits a uptumed behavior at low temperature, which was also observed by M.V. Lobanov *et al.* [5]. To understand the reason, we prepared $SrPbO_3$ samples with various starting materials (for example, $Sr(NO_3)_2$) and sintering condition. However, the temperatures at which starts to increasing resistivity have no reproducibility even in the same conditions. This may suggest that the electrical state of this material at low temperature is very delicate because of, for example, localization of carrier. Resistivity of $CaPbO_3$ was considerably high at room temperature, and it could not measured in our experimental apparatus. As shown in Fig. 1(b), absolute values of resistivity for La doped $SrPbO_3$ and $CaPbO_3$ were significantly lower than those of undoped ones, while that of La doped $BaPbO_3$ is higher than that of undoped one. Increase of resistivity in $BaPbO_3$ is speculated that much difference in ionic radius between Ba^{2+} and La^{3+} leads to lower mobility and lower conductivity probably due to randomness, as observed in $Ba_{1-x}Sr_xPbO_3$[4]. It is noted that the La doped $CaPbO_3$ exhibit the high conductivity with keeping strong distortion. This demonstrated that insulator-metal transition was induced by the La doping. Figure 2 shows temperature dependence of Seebeck coefficients of these samples. This shows that the all samples are metallic conductors with electron carrier. Considerably high value of Seebeck coefficient of $SrPbO_3$ may correspond to the low carrier concentration reported by Itoh *et al.* [4]. The absolute values were reduced by the doping in both A=Ba and Sr. This interpreted that the electrons were introduced into the conduction band. Thermopower study of the $CaPbO_3$ and its La doped samples are in progress.

4. Conclusion

$APbO_3$ (A=Ba,Sr,Ca) and their La doped samples with perovskite-type structure were synthesized, and La doping effects on their structure and transport properties were studied. Solubility limit of La in the A site was found to be about 10%. The unit cell dimensions increased with La doping, suggesting reduction of Pb oxidation state. Temperature dependencies of electrical resistivity and Seebeck coefficient of La doped samples revealed that they are metallic conductors. Absolute values of resistivity for La doped $SrPbO_3$ and $CaPbO_3$ were significantly lower than those of undoped ones. These results suggest that La doping introduces electrons in the conduction band. In this study, it was demonstrated that the La doping in the A-site is a effective method of electron doping for the A-Pb-O system. It is of particular interest to introduce carrier in the layered lead oxides series, $A_{n+1}Pb_nO_{3n+1}$, for the search of non-copper superconductors.

Acknowledgment

The authors thank Dr. I. Terasaki of Waseda University for useful discussions and fruitful suggestions. This work was supported by NEDO.

References

[1] Shannon, R.D., and Bierstedt, P.E, Single-crystal growth and electrical properties of $BaPbO_3$, *J. Amer. Ceram Soc.*, **53** (1970) 635-636.

[2] Sleight, A.W., Gilson, J.L., and Bierstedt, P.E., High-temperature superconductivity in the $BaPb_{1-x}Bi_xO_3$ system, *Solid State Commun.*, **17** (1975) 27.

[3] Mattheiss, L.F., and Hamann, D.R., Electronic structure of $BaPb_{1-x}Bi_xO_3$, *Phys Rev. B* **28** (1983) 4227-4241.

[4] Itoh, M. Sawada, T. Kim, I.S., Imaguma, Y. and Nakamura, T., Composition dependence of carrier concentration and conductivity in $(Ba_{1-x}S_x)PbO_{3-\delta}$ system, *Solid State Commun.*, **83** (1992) 33-36.

[5] Lobanov, M.V., Kopnin, E.M., Xenikos, D. Grippa, A.Ju., Antipov, E.V., Capponi, J.J., Marezio, M. Julien J.P., and Tholence., J.L., Synthesis and resistivity properties of $Sr_{1-x}La_xPbO_{3-\delta}$ and $SrPb_{1-x}Bi_xO_{3-\delta}$, *Mater. Res. Bull.*, **32** (1997) 983-992.

[6] Cava, R.J., Takagi, H. Karajewski, J.J., Peck,. Jr. W.R., Hwang, H.Y., Oxygen-deficient barium lead oxide perovskites, *Phys Rev. B* **47** (1993) 11525-11528.

[7] Yamamoto, A. Khasanova, N. and Tanabe, K., Synthesis and characterization of perovskite-type $CaPbO_3$, *Solid State Ionics*, 1998, in press.

EFFECTS OF Sr SUBSTITUTION ON THE T_c AND CRYSTAL STRUCTURE OF $Tl_2(Ba_{1-x}Sr_x)_2Ca_2Cu_3O_y$

Y. Idemoto[1], F. Izumi[2], Q. Huang[3], A. Santoro[3], M. Matsuzawa[1], and N. Koura[1]

[1] *Department of Industrial Chemistry, Faculty of Science and Technology, Science University of Tokyo, Noda, Chiba 278-8510, Japan*
[2] *National Institute for Research in Inorganic Materials, Tsukuba, Ibaraki 305-0044, Japan*
[3] *NIST Center for Neutron Research, National Institute of Standards and Technology, Gaithersburg, Maryland 20899, USA*

1. Introduction

The physical properties of superconducting copper oxides are strongly affected by substituting cations in charge reservoirs and/or changing oxygen contents. Therefore, the dependence of oxygen deficiencies, interatomic distances, and bond angles on chemical compositions has been extensively investigated. The highest superconducting transition temperature, T_c, of $Tl_2Ba_2Ca_2Cu_3O_y$ is 125 K. The T_c of $Tl_2(Ba_{1-x}Sr_x)_2Ca_2Cu_3O_y$ (TBSCCO, Figure 1) prepared by solid-state reactions dropped with increasing Sr content up to $x = 0.5$ [1]. On the other hand, T_c reached the highest value at $x = 0.5$ in TBSCCO prepared by mist microwave-plasma chemical vapor deposition [2]. Replacement of Ba by Sr must change the oxygen content, oxidation state of Cu, and crystal structure, which leads to a change in T_c. In this work, we prepared TBSCCO with different Sr contents and determined their crystal structures by neutron powder diffraction to clarify the relationship between their structures and superconductivity.

2. Experimental

Precursors with chemical compositions of $(Ba_{1-x}Sr_x)_2Ca_2Cu_3O_y$ ($x = 0.15$, 0.20, and 0.25) were prepared by coprecipitation from oxalic acid-ethanol solutions with subsequent decomposition at 400 °C. The resulting solids were heated at 900 °C in air for 24 h. The precursors were mixed with excess amounts of Tl_2O_3 with a nominal Tl composition of 2.4, pressed into pellets, and calcined at 850 °C with a

G. Van Tendeloo et al. (eds.), High-Temperature Superconductors and Novel Inorganic Materials, 129–132.

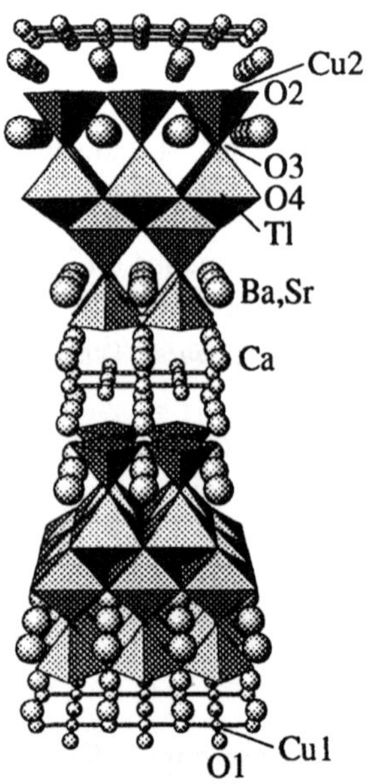

Figure 1. Idealized crystal structure of TBSCCO.

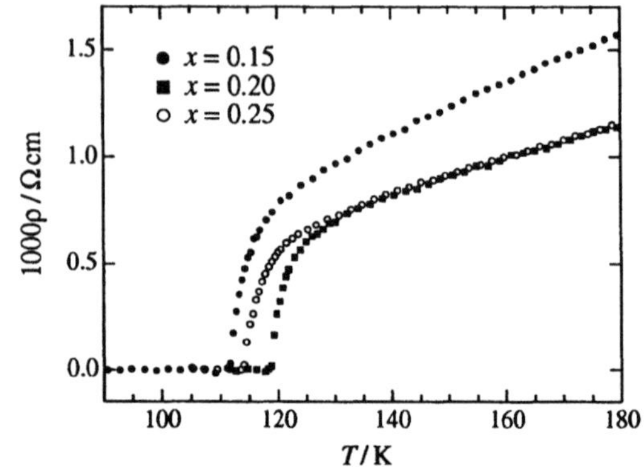

Figure 2. Resistivity *versus* temperature for $Tl_2(Ba_{1-x}Sr_x)_2Ca_2Cu_3O_y$.

powder of the same superconductor to prevent Tl from vaporizing. Resistivities, ρ, of TBSCCO were measured by the four-probe dc method. Metal contents in TBSCCO were determined by ICP emission spectrometry.

Neutron powder diffraction data were measured at room temperature on the BT-1 neutron powder diffractometer at the NIST research reactor. The white beam of neutrons from the reactor was monochromatized to a wavelength of 1.5401 Å with a Cu(311) monochromator. Horizontal divergences were 15′, 20′, and 7′ for in-pile, minochromatic-beam, and diffracted-beam collimators, respectively.

3. Results and Discussion

The three samples of TBSCCO were nearly pure with a constant Tl valence of 3.0. The $T_c(\rho = 0)$ of TBSCCO was 110 K for $x = 0.15$, 119 K for $x = 0.20$, and 114 K for $x = 0.25$ (Figure 2).

The neutron diffraction data were analyzed on the basis of space group I4/mmm with a Rietveld-refinement program RIETAN-98. Preliminary refinements indicated that Tl and O4 atoms at 4e positions on a TlO sheet displayed unreasonably large apparent thermal motion. Tl contents were determined to be smaller than the stoichiometric value of 2: 1.78 for $x = 0.15$, 1.76 for $x = 0.20$, and 1.85 for $x = 0.25$. In addition, the thermal vibration of O4 was anisotropically marked on

TABLE 1. Results of the Rietveld refinements for $Tl_2(Ba_{1-x}Sr_x)_2Ca_2Cu_3O_y$ ($Z=2$). Space group: I4/mmm (No. 139). M: $Ba_{1-x}Sr_x$, g: occupation factor, B: isotropic atomic displacement parameter.

$x = 0.15$: $a = 3.84897(12)$ Å and $c = 35.5352(14)$ Å.
$R_{wp} = 5.44\%$ ($R_e = 4.93\%$), $R_p = 4.43\%$, $R_B = 6.45\%$, and $R_F = 4.00\%$.
$x = 0.20$: $a = 3.84697(13)$ Å and $c = 35.4833(15)$ Å.
$R_{wp} = 4.70\%$ ($R_e = 4.55\%$), $R_p = 3.87\%$, $R_B = 6.10\%$, and $R_F = 3.84\%$.
$x = 0.25$: $a = 3.84615(13)$ Å and $c = 35.4424(15)$ Å.
$R_{wp} = 5.62\%$ ($R_e = 4.78\%$), $R_p = 4.57\%$, $R_B = 6.48\%$, and $R_F = 4.05\%$.

Atom	Site	Symmetry	x	y	z	g	B (Å^2)
$x = 0.15$							
Tl	4e	4mm	1/2	1/2	0.22021(10)	0.889(12)	1.78(8)
M	4e	4mm	0	0	0.14489(14)	1	1.10(7)
Ca	4e	4mm	0	0	0.04589(15)	1	0.51(6)
Cu1	2b	4/mmm	1/2	1/2	0	1	0.52(6)
Cu2	4e	4mm	1/2	1/2	0.09001(9)	1	0.68(5)
O1	4c	mmm.	1/2	0	0	1	0.83(6)
O2	8g	2mm.	1/2	0	0.08835(9)	1	0.79(4)
O3	4e	4mm	1/2	1/2	0.16425(13)	1	1.47(7)
O4	16n	.m.	0.602(2)	1/2	0.2757(2)	0.216(6)	1.8(3)
$x = 0.20$							
Tl	4e	4mm	1/2	1/2	0.22023(11)	0.886(12)	1.91(8)
M	4e	4mm	0	0	0.14515(16)	1	1.26(7)
Ca	4e	4mm	0	0	0.04616(16)	1	0.62(7)
Cu1	2b	4/mmm	1/2	1/2	0	1	0.56(7)
Cu2	4e	4mm	1/2	1/2	0.09054(9)	1	0.67(5)
O1	4c	mmm.	1/2	0	0	1	0.92(7)
O2	8g	2mm.	1/2	0	0.08875(10)	1	0.87(4)
O3	4e	4mm	1/2	1/2	0.16463(15)	1	1.70(8)
O4	16n	.m.	0.601(2)	1/2	0.2758(2)	0.219(7)	1.4(3)
$x = 0.25$							
Tl	4e	4mm	1/2	1/2	0.22014(10)	0.905(12)	1.99(8)
M	4e	4mm	0	0	0.14505(15)	1	1.23(7)
Ca	4e	4mm	0	0	0.04616(16)	1	0.62(7)
Cu1	2b	4/mmm	1/2	1/2	0	1	0.57(7)
Cu2	4e	4mm	1/2	1/2	0.09054(9)	1	0.69(5)
O1	4c	mmm.	1/2	0	0	1	0.90(7)
O2	8g	2mm.	1/2	0	0.08885(9)	1	0.95(4)
O3	4e	4mm	1/2	1/2	0.16437(14)	1	1.65(8)
O4	16n	.m.	0.608(2)	1/2	0.2760(3)	0.223(7)	2.2(3)

TABLE 2. Selected geometric parameters (Å or °) in $Tl_2(Ba_{1-x}Sr_x)_2Ca_2Cu_3O_y$. m: number of equivalent bonds. Symmetry codes: (i) x, $-y$, z.

	m	$x = 0.15$	$x = 0.20$	$x = 0.25$
Cu1–O1	4	1.9245(1)	1.9235(1)	1.9231(1)
Cu2–O2	4	1.9254(2)	1.9243(2)	1.9240(2)
Cu2–O3	1	2.638(5)	2.635(6)	2.617(5)
Cu2–O2–Cu2^{i}		176.5(3)	176.6(3)	176.4(3)

the *ab* plane. Then, we adopted a structural model where O4 is displaced along the *a* and *b* axes to be split at four 16n positions [3], and the Tl site is partially deficient. Final lattice and structure parameters as well as R factors are listed in Table 1, and geometric parameters associated with Cu in Table 2.

The lattice parameters, a and c, and the Cu–O distances decreased with increasing Sr content owing to the smaller ionic radius of the Sr^{2+} ions (1.36 Å) than that of the Ba^{2+} ion (1.52 Å). The Cu2–O2–Cu2 bond angles, which serve as a measure of the 'flatness' of the CuO_2 sheet, were nearly constant regardless of x. Hence, the dependence of T_c on x can be hardly explained in terms of differences in geometrical configurations.

The g(Tl) values of *ca.* 0.9 are consistent with the Tl contents of *ca.* 1.8. However, the Tl site may be not *ca.* 10% deficient but partly substituted by Cu because Cu contents were 3.12 for $x = 0.15$, 3.13 for $x = 0.20$, and 3.04 for $x = 0.25$. Such metal nonstoichiometry must more or less affect the hole concentration on the CuO_2 sheet. On changing oxygen contents, T_c *versus* y curves had maxima in samples with $x = 0.15$ and 0.20, which indicates the presence of under- and over-doping regions in TBSCCO. Table 1 shows a decrease in g(O4) with increasing Sr content. The g(O4) values correspond to y values of 9.73 for $x = 0.15$, 9.75 for $x = 0.20$, and 9.78 for $x = 0.25$. These facts suggest that the oxygen content as well as the Tl/Cu contents may be optimized for superconductivity at $x = 0.20$.

References

1. Soeta, A., Suzuki, T., Takeuchi, S., Kamo, T., Usami, K., and Matsuda, S. (1989) Relationship between crystal structures and solid solution of Tl–Sr–Ca–Cu–O and Tl–Ba–Ca–Cu–O superconductors, *Jpn. J. Appl. Phys.* **28**, L1186–L1189.
2. Takahashi, N., Koukitu, A., and Seki, H. (1995) Influence of Sr content on Tl–Ba–Ca–Cu–O superconducting thin films prepared by the mist microwave-plasma chemical vapor deposition method, *Jpn. J. Appl. Phys.* **34**, L109–L111.
3. Torardi, C.C., Subramanian, M.A., Calabrese J.C., Gopalakrishnan, J., Morrissey, K.J., Askew, T.R., Flippen, R.B., Chowdhry, U., and Sleight, A.W. (1988) Crystal structure of $Tl_2Ba_2Ca_2Cu_3O_{10}$, a 125 K superconductor, *Science* **240**, 631–634.

ANION ORDERING IN FLUORINATED LA_2CUO_4.

J. HADERMANN[1], A.M. ABAKUMOV[1,2], G. VAN TENDELOO[1], R.V. SHPANCHENKO[2], P.N. OLEINIKOV[2], E.V. ANTIPOV[2]

[1] *EMAT, University of Antwerp (RUCA), Groenenborgerlaan 171, B-2020 Antwerp, Belgium*

[2] *Department of Chemistry, Moscow State University, 119899 Moscow, Russia*

1. ABSTRACT

La_2CuO_4 has been fluorinated at 200°C, 250°C, 300°C and 400°C using XeF_2. The resulting products were studied by X-Ray diffraction (XRD), electron diffraction (ED), high resolution electron microscopy (HREM) and microanalysis (EDX). The higher fluorination temperatures produce a new monoclinic phase with cell parameters a=17.36Å, b=5.36Å, c=10.59Å, β=91.5° and spacegroup C2/m. Fluorine places itself in part of the interstitial positions, which is accompagnied by the removal of the apical oxygens in between fluorine atoms. This results in an alternation of 2 NaCl-type slabs with one fluorite-type slab along the [110] direction of the K_2NiF_4-subcell. No superconducting transition was found for these samples.

2. INTRODUCTION

The hole concentration, which is necessary for the appearance of superconductivity in La_2CuO_4 can be achieved either by heterovalent cationic substitution on the A-sublattice or the insertion of additional anions into the interstitial positions between the double LaO layers. For the latter different techniques are available: high oxygen pressure annealing electrochemical oxidation, oxidation by $KMnO_4$ solution and halogen treatment. La_2CuO_4 being treated by diluted or pure F_2 gas as a fluorinating agent reveals a superconducting transition with T_c(35-40K [1]. A superconducting oxyfluoride was observed only for a low fluorination level, which corresponds to a low temperature of the fluorine treatment (150-200^{0}C). It was found that incorporation of a small amount of fluorine into the La_2CuO_4 structure (up to $La_2Cu(O,F)_{4.18}$) is accompanied by the enhancement of the orthorhombic distortion and an increase of the c parameter of the unit cell [2]. Neutron powder structure refinement as well as EXAFS studies revealed that no significant structural transformation occurs during the fluorination; fluorine atoms are located at interstitial

G. Van Tendeloo et al. (eds.), High-Temperature Superconductors and Novel Inorganic Materials, 133–138.

positions, tetrahedrally coordinated by La atoms and placed between two successive LaO layers [3]. This behavior is similar to that previously found for oxygen-doped La_2CuO_4 samples [4].

An increase of the fluorination temperature does lead to a structural rearrangement. The formation of a new, non-superconducting, tetragonal phase with cell parameters *a*=5.7(1)Å and *c*=13.1(1)Å was observed by X-ray diffraction when La_2CuO_4 was treated by F_2 at 230 °C [5]. Electron diffraction revealed the presence of a superstructure indexed on the basis of a monoclinic unit cell [6]. An incommensurate superstructure due to the presence of shear planes was also found in limited regions of the sample. The origin of this ordering was not determined however and the structure of the compound formed at a high level of fluorination is still unclear.

Recently we showed that XeF_2 is a soft and effective fluorinating agent which can be used for the preparation of different superconducting oxyfluorides: $Sr_2CuO_2F_2$ [7], $YBa_2Cu_3O_6F_2$ [8], $Y_2Ba_4Cu_7O_{14}F_\delta$ [9], $HgBa_2CuO_4F_\delta$ [10]. We recently used XeF_2 to fluorinate La_2CuO_4.

3. EXPERIMENTAL

The basic La_2CuO_4 compound was prepared by intimate mixing of stoichiometric amounts of La_2O_3 and CuO. This mixture was then pressed into pellets and heated in air at 950°C for 40h. The lattice parameters were determined as *a*=5.3562(2)Å, *b*=5.4015(2)Å and *c*=13.1531(8)Å, in good agreement with those previously published. 0.4g of this La_2CuO_4 was mixed with XeF_2 in a molar ratio of 1:1 and grinned in agate mortar, then placed in a Ni-crucible and sealed into a copper tube. Annealing took place at 200°C, 250°C, 300°C or 400°C for 40h, afterwards the mixture was furnace cooled to room temperature. Phase composition and lattice parameters were determined by X-Ray diffraction using a focusing Guinier-camera. AC susceptibily measurements were performed in a temperature range 12-100K at an external field amplitude of 1 Oe and at a frequency of 27 Hz. Electron diffraction (ED) and high resolution electron microscopy (HEM) were performed using a JEOL 4000EX operating at 400kV. EDX analysis was performed on a Philips CM20 microscope with a LINK-2000 attachment. The image simulations were made using MacTempas software.

4. RESULTS

A significant structural rearrangement, associated with the formation of a tetragonal K_2NiF_4-type structure (I-centred lattice with cell parameters *a*=4.0383(6)Å, *c*=13.093(3)Å) occurs at 250 °C. Electron diffraction shows weak superstructure reflections which were indexed on a body-centred tetragonal lattice with doubled cell parameters of the K_2NiF_4-subcell (*a*=8.06Å, *c*=26.2Å). The weakness of the superstructure spots and their unstability under electron beam irradiation allows us to connect them to an ordering in the anion sublattice. Details can be found in [11]. In the present contribution we will however concentrate on the structural changes taking place after heat treatment at 300°C or 400°C.

After fluorination at 300°C or 400°C, X-ray diffraction reveals the presence of a tetragonal K_2NiF_4-based structure with spacegroup I4/mmm and lattice parameters a=4.0383(6)Å and c=13.093(3)Å,. However there is a decrease of symmetry and a distortion of the tetragonal unit cell, as evidenced by the broadening and increased splitting of some reflections. Badly cristallized LaOF is also present in small quantities, as revealed by the presence of a number of extremely wide but weak peaks. Measurements of the temperature dependence of the magnetic susceptibility down to 12K showed no superconducting transition.

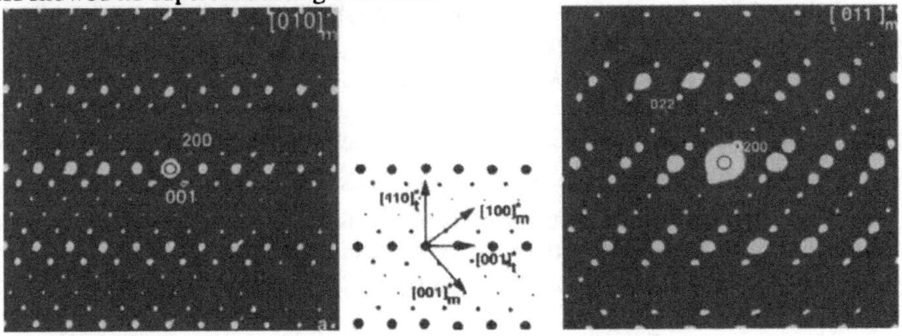

Fig.1 Electron diffraction patterns along (a) the $[010]_m$* and (b) the $[01\ 1]_m$* direction. Next to the diffraction patterns the orientation of the reciprocal axes of the monoclinic and the K_2NiF_4-type subcell is shown.

Electron diffraction patterns (Figs. 1a,b) show a basic structure which can be indexed using the unit cell derived from X-ray diffraction, but also the presence of a superstructure. For this superstructure the shortest reciprocal vectors lie in the$[116]_t$* and the $[\bar{1}\bar{1}\ 4]$ directions. Consequently, the $1/5[116]_t$* and $1/5[\bar{1}\bar{1}\ 4]$ vectors were chosen as the a_m* and c_m* vectors respectively, which makes β equal to 91.5°. The subscript t refers to the K_2NiF_4 sublattice and m to the monoclinic cell. The zone axis of this diffraction pattern is $[010]_m$*.

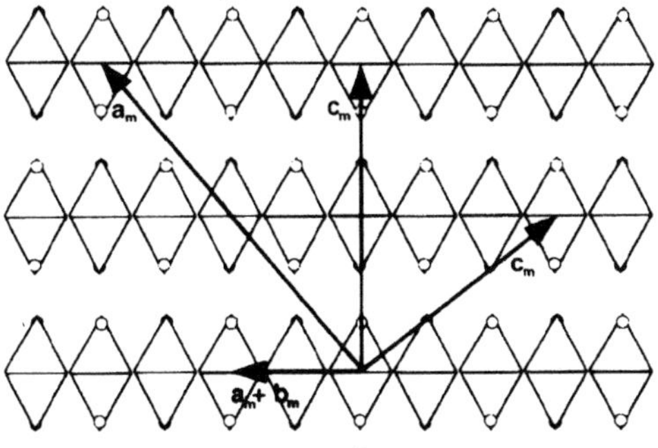

Fig.2 Schematic representation along $[1\bar{1}\ 0]_t$ of the orientation of the axes of the monoclinic cell relative to those of the K_2NiF_4-type subcell.

A schematic representation of the relative orientation of the axes of the tetragonal cell to those of the monoclinic cell is given in fig.1 for reciprocal space and in fig.2 for real space.

The period along the *b*-direction of the monoclinic cell was chosen equal to the one along $[\bar{1}\ 1\ 0]_t$, i.e. 5.4-5.7Å. This assumption is supported by fig.1b, which also shows doubling of the a-parameter and the presence of systematic extinctions. The extinction condition derived from this pattern is $h+k=2n$, which allows the proposition of the spacegroups C2, Cm and C2/m. The cell parameters following from these observations are $a\approx 17.36$Å, $b\approx 5.62$Å, $c\approx 10.59$Å and $\beta\approx 91.5°$.
The relation of the monoclinic cell to the K_2NiF_4-type cell is given by:

$$h^*_m = 2h^*_t + 2k^*_t + l^*_t$$
$$k^*_m = -h^*_t + k^*_t$$
$$l^*_m = -3/2\ h^*_t + 3/2\ k^*_t + 1/2 l^*_t\ .$$

On the HREM images (Fig.3) the bright rows are the projections of the (CuO_2)-planes and the rows in between are the (La_2O_2)-layers, as was deduced from image simulations. An alternation between dark and light is visible along the (110)-direction in these (La_2O_2)-layers. The period of this alternation corresponds to the monoclinic unit cell deduced from the EDP's. In the thicker part of the crystal, there is also a modulation in the (CuO_2)-planes.

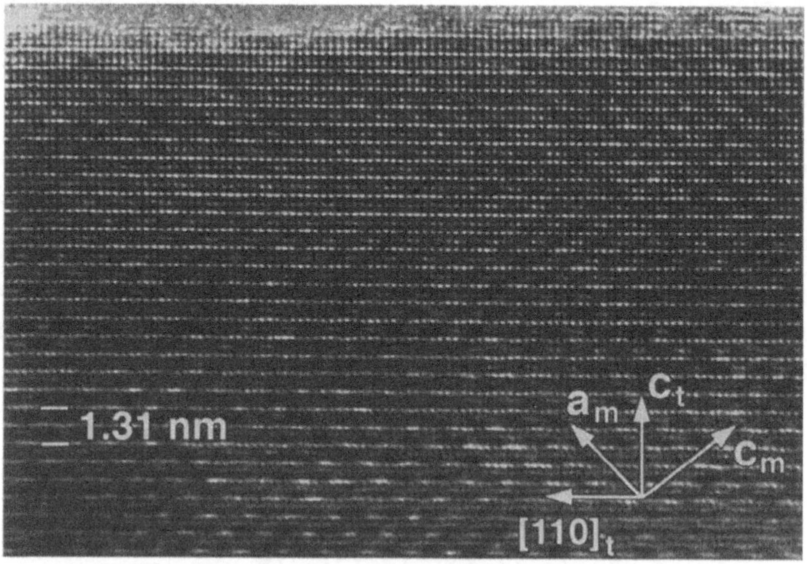

Fig.3 HREM image along $[1\bar{1}\ 0]_t$ corresponding to the ED-pattern of fig.1.

EDX-analysis shows the presence of fluorine. As the basic structure is still that of the K_2NiF_4-type subcell, one can assume that the ordering takes place by a rearrangement between oxygen and fluorine atoms and that the cation sublattice remains unchanged. Extra anions are known to be incorporated in interstitial positions of

La_2CuO_4, coordinated by four La^{3+}-ions and four apical oxygens [3, 4]. Fluorine in some of the interstitial positions could already partly explain the modulation observed in the HREM images such as fig. 3. However, this would result in a too short distance between fluorine and the apical oxygens. This necessitates the removal of the apical oxygens lying in between the fluorines. Secondly, in order to further increase the distance between F and neighbouring O_{ap}, a tilting of the CuO_6 octahedra at both sides of the fluorine atom occurs, which also causes the CuO_5 piramid to be pushed towards the fluorine. This enhances the modulation in the (CuO_2)-planes observed on the HREM image of fig.3. The observed image contrast can further be explained by an alternation of these fluorite-type blocks (T'-structure) with the La_2CuO_4 blocks (T-structure). Anion exchange has thus taken place between apical oxygens and interstitial fluorine. Such an exchange has been observed earlier for fluorination in a closed vessel above 300°C for other superconductors such as YBCO or Hg1201 [8, 9].

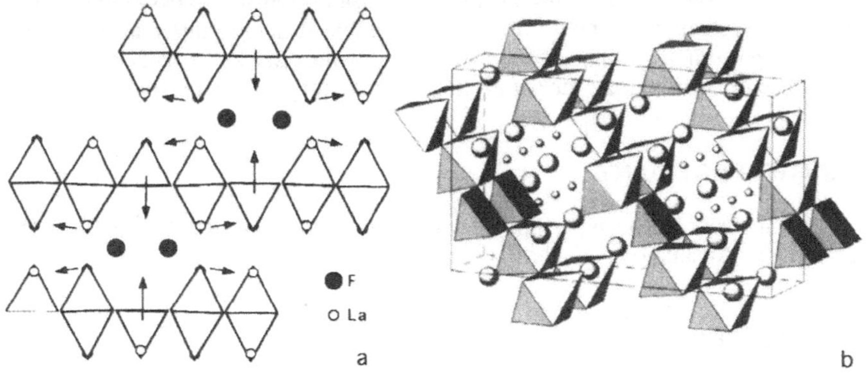

Fig.4 (a) Schematic representation along $[010]_m$ of the displacements induced by the incorporation of interstitial fluorine. (b) Model for the monoclinic supercell.

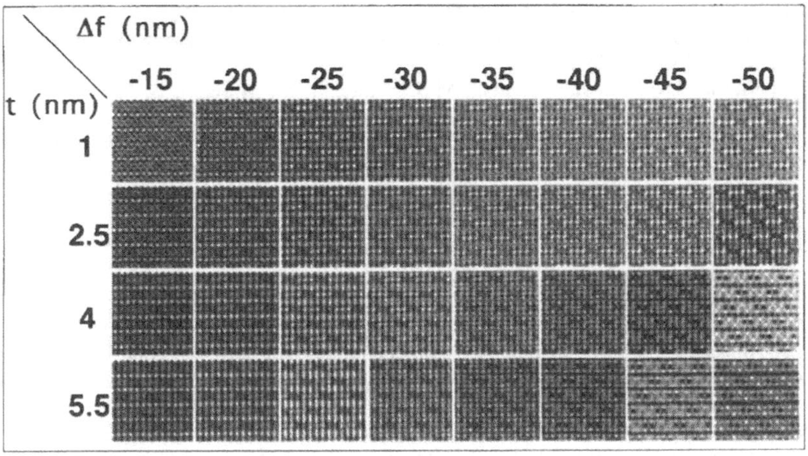

Fig.5 Images calculated along the $[010]_m$ direction of the model for different thicknesses and defocus.

There are several indirect confirmations of this anion exchange. Traces of CuO_2 are formed at the inner surface of the copper tube used for fluorination, the *c*-parameter has decreased to 13.093(3)Å, which agrees with the presence of the fluorite slabs in the structure, since the transformation from T to T' is accompanied by a compression of the unit cell along the c-axis, and the in-plane Cu-O distances are extremely large for Cu-cations with an oxidation state between 2^+ and 3^+. This last argument indicates a possible partial reduction of the (CuO_2)-planes.

A model of the monoclinic unit cell was constructed (Fig.4). The atomic coordinates were calculated using the C2/m space group, as this is the most symmetrical one compatible with the extinction conditions. The images calculated based on this model are shown in fig.5; they are in good agreement with the experimental HREM image of fig.3. The composition of this phase, calculated from the model, is $La_2CuO_{3.6}F_{0.8}$. Both the insufficient copper valence, which is +2, and the distortion of the (CuO_2)-planes, explain the disappearance of superconductivity. Some of the crystals also show an incommensurable structure, different however from the one observed in [6]. This incommensurate structure can be explained by assuming the presence of shear planes; the details however are not yet fully resolved.

ACKNOWLEDGEMENT

This work has been performed within the framework of IUAP 4/10, a DPWB contract of A. Abakumov and a NATO linkage contract.

REFERENCES

1. B.M. Tissue, K.M. Cirillo, J.C. Wright, M. Daeumling, D.C. Larbalestier, Solid State Comm. 65 (1988) 51
2. B. Chevalier, A. Tressaud, B. Lepine, K. Amine, J.M. Dance, L. Lozano, E. Hickey, J. Etourneau, Physica C 167 (1990) 97
3. M.H. Tuilier, B. Chevalier, A. Tressaud, C. Brisson, J.L. Soubeyroux, J. Etourneau, Physica C 200 (1992) 113.
4. C. Chaillout, S.W. Cheong, Z. Fisk, M.S. Lehmann, M. Marezio, B. Morozin, J.E. Schriber, Physica C 158 (1989) 183.
5. B. Chevalier, A. Tressaud, B. Lepine, C. Robin, J. Etourneau, J.Less.Comm.Metals 164-164 (1990) 832
6. F. Weill, B. Chevalier, M. Chambon, A. Tressaud, B. Darriet, J. Etourneau, G. Van Tendeloo, Eur.J.Solid State Inorg.Chem. 30 (1993) 1095
7. E.I. Ardashnikova, S.V. Lubarsky, D.I. Denisenko, R.V. Shpanchenko, E.V. Antipov, G. Van Tendeloo, Physica C 253 (1995) 259
8. R.V. Shpanchenko, M.G. Rozova, A.M. Abakumov, E.I. Ardashnikova, M.L. Kovba, S.N. Putilin, E.V. Antipov, O.I. Lebedev, G. Van Tendeloo, Physica C 280 (1997) 272
9. A.M. Abakumov, M.G. Rozova, R.V. Shpanchenko, M.L. Kovba, S.N. Putilin, E.V. Antipov, O.I. Lebedev, G. Van Tendeloo, E.M. Kopnin, J. Karpinski, Physica C (1998) in press.
10. A.M. Abakumov, V.L. Aksenov, V.A. Alyoshin, E.V. Antipov, A.M. Balagurov, D.A. Mikhailova, S.N. Putilin, M.G. Rozova, Phys.Rev.Lett. 80 (1998) 385
11. A.M. Abakumov, J. Hadermann, G. Van Tendeloo, R.V. Shpanchenko, P.N. Oleinikov, E.V. Antipov, Physica C (in the press)

INTERFACE DEFECTS INDUCED BY EPITAXIAL GROWTH MECHANISM IN YBa2Cu3O7 THIN FILMS ON SrTiO3 SUBSTRATE: APPLICATION TO THIN FILMS BICRYSTALS AND MULTILAYERS

J. AYACHE[1], A. THOREL[2], S.J. KIM[1], U. DAHMEN[3], J.P CONTOUR[4], M.DROUET[4], and J. LESUEUR[1]

1. CSNSM CNRS-IN2P3, Batiment 108, 91405 Orsay, France.
2. Ecole des Mines de Paris, BP87 91003 Evry, France.
3. National Center for Electron Microscopy, Lawrence Berkeley Laboratory, MCSD Division, University of California CA 94720 , Berkeley
4. Unité Mixte de Recherche CNRS / Thomson-Csf, 91404 Orsay, France.

Abstract : We have combined TEM, HREM and AFM (Atomic Force Microscope) techniques to study growth induced defects on YBCO thin film bicrystals and on YBCO based superlattices. AFM images clearly evidence "in plane wavyness" of the Grain Boundary (GB) and a quantitative analysis of HREM images shows unambiguously the oscillation of the grain boundary plane. This GB structure and morphology result from the island growth mechanism of the film. Studying YBa2Cu3O7/SrTiO3 (YBCO/STO) heterostructures , we show that the interface roughness increases across the superlattices. Extended defects like twins (due to the heteroepitaxy of the orthorhombic c-axis YBCO on top of the cubic STO) and conical defects related to a plastic deformation of the superlattices have been found.

1- Introduction

Artificial structures such as YBCO grain boundary junctions and YBCO based heterostructures are promising parts of superconducting electronic devices. To get very high quality thin films, STO single crystals have been widely used as substrates ; however, STO can also be used as insulating barrier in SIS heterostructures. Therefore, we mainly focus this paper on the microstructure of YBCO/STO films and [YBCO/STO]/STO supêrlattices, grown by a Pulsed Laser Deposition (PLD) process. This technique is a reasonable compromise between growth rate, cost and quality of the deposit. Using TEM/HREM analysis, we describe the structure of the interfaces in thin film bicrystrals and in c- and a-axis oriented multilayers.

2- Experimental

The YBCO thin film bicrystals were deposited on melt-grown SrTiO3 bicrystals with a misorientation of 24° around the <100> tilt axis. The growth

G. Van Tendeloo et al. (eds.), High-Temperature Superconductors and Novel Inorganic Materials, 139–144.

procedure has been previously described in details [1,2]. The supperlattices consist in a ten fold stacking of $YBa_2Cu_3O_{7-y}$ /$SrTiO_3$ (YBCO/STO) bilayers with and without a $PrBa_2Cu_{3-x}Ga_xO_{7-y}$ (PBCGO) buffer layer grown on a [100] $SrTiO_3$ substrate. This structure has been grown at 710°C in order to improve the STO crystallinity and to reduce the copper interdiffusion [3].
The morphology and crystallography study of $YBa_2Cu_3O_7$ thin film bicrystals and heterostructures have been carried out from macroscopic to atomic scale by means of analytical electron transmission microscopies SEM/TEM /HREM techniques coupled with quantitative measurements on high resolution TEM images and atomic force microscopy. Mechanical tripod thinning has been used to prepare TEM samples in order to avoid spurious effects (like defect diffusion, amorphization...) due to ion milling bombardment.

3- $YBa_2Cu_3O_7$ thin film bicrystal

Growth islands with well-defined terraces were observed by AFM (fig.1) on both sides of the grain boundary as well as in its immediate vicinity. The size of the islands ranged from 300 to 1200nm in diameter, and their roughness were 10nm in height in the matrix and 3 to 6nm close to the grain boundary. A scheme of the growth islands of YBCO deposited film, resulting from AFM observations, is displaid on figure 2. It consists of an array of terraces domes associated with a corresponding hole array at triple junctions of the grains. Some of them are located at the grain boundary. Figure 1 shows clearly the meandering of the YBCO grain boundary compatible with previously YBCO GB observed structures [1-2, 4-6]

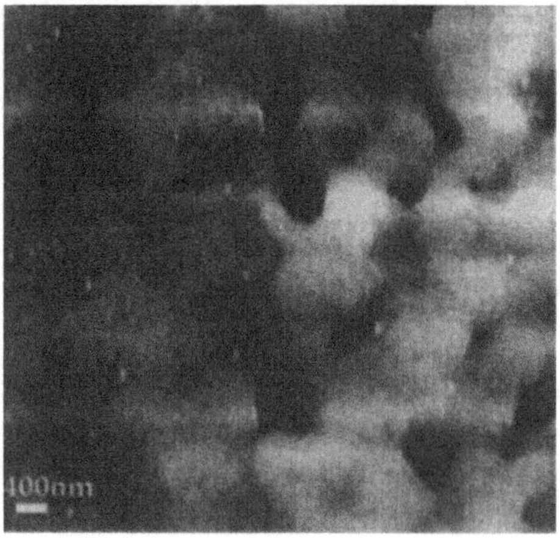

Figure 1
YBCO bicrystal AFM image

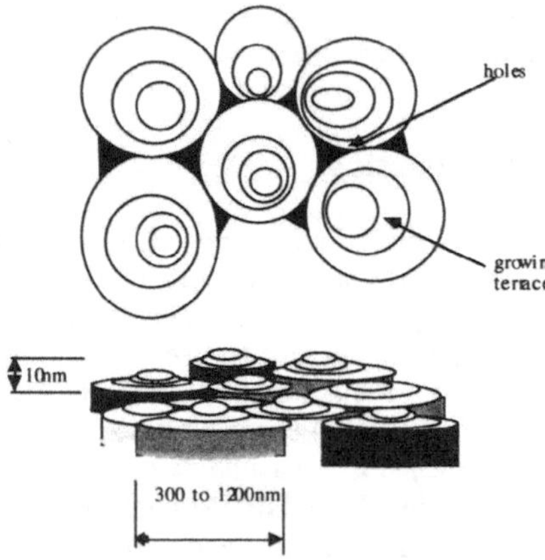

Figure 2
Schematic view of the growth islands compact arrangment inferred from AFM and SEM observations

TEM analysis confirms that the grain boundary in the film did not always follow the substrate grain boundary faithfully, but undulated about the average boundary plane. This undulation is a result of the film growth mechanism. HREM observations (Fig.3) of the apparently complex undulating boundary structures can be explained as a result of an overlap of YBCO grains originated from the two sides of the bicrystal. The wavelength of the undulations was similar to the characteristic lateral film roughness detected by complementary SEM and AFM imaging.
To get a more quantitative description of the GB local structure, we have carried out a Fourier analysis of the HREM images to interpret the observed Moire patterns arising from the underlying crystalline structure at the vicinity of the GB (Fig.3).

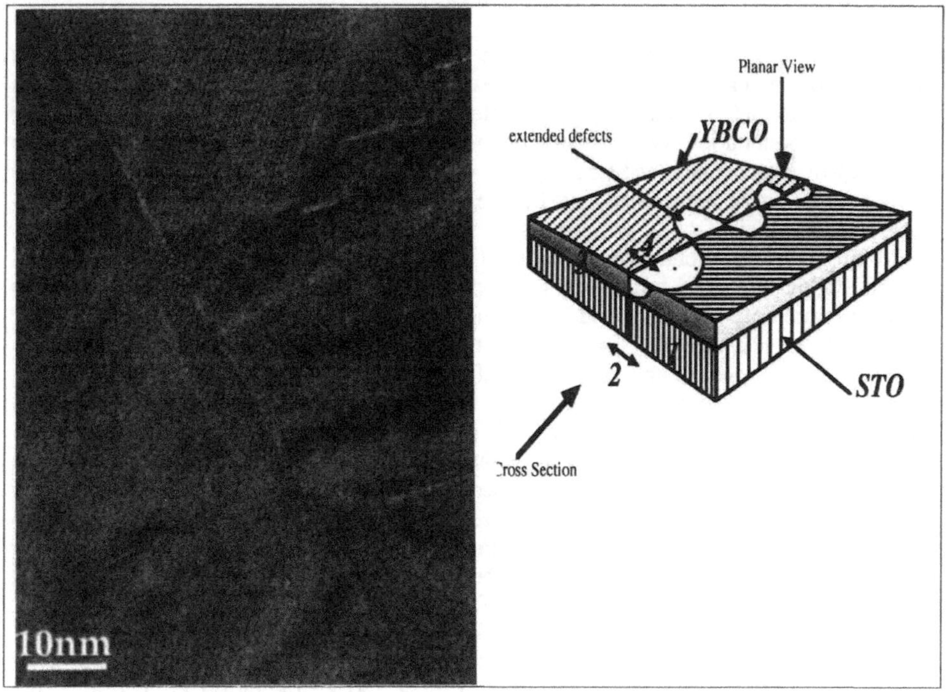

Figure 3
(3a) HREM planar view image of the YBCO grain boundary exhibiting oscillating contrasted analyzed as Moires fringes due to overlap of YBCO grain1 and YBCO grain 2 at the GB.
(3b) schematic view of the undulating grain boundary showing the different parameters to take into account for theoretical calculation of moires

The diffraction diagrams explain all the experimentally observed moirés by an overlap between YBCO crystals only when the lattice parameters were allowed to vary over the range 0.380nm< a(YBCO)<0.385nm and 0.384nm<b(YBCO)<0.392nm. A detailed analysis of the local lattice orientation indicated a ±2° angular variation. Thus the YBCO bicrystal angle was found to vary locally from 22° to 26° It may also occur that locally the YBCO crystal prefers to align its (110) planes to the (110) planes of the cubic STO substrate, leading to a ± 1.8° angular deviation between the (100) planes.

From AFM/SEM/HREM analysis it can be inferred that the YBCO grain boundary plane follows the straight STO interface during the initial growth of the YBCO layer. After that initial stage growth terraces start to alternatively cross over to either side of the bicrystal. As a consequence, the interface follows a sinuous path along the original GB. A wide range of inclinations for the interface plane resulting from the growth mechanism has been observed. As seen in other systems, it appears here that the description of a GB only by the long range cristallographic relationship is far from being sufficient : the local structure -here tailored by growth mechanisms- may have a major influence on the final properties.

4. YBa2Cu3O7/SrTiO3 supperlattices

The heteroepitaxial growth of the YBCO orthorhombic structure on the STO cubic structure leads to intrinsic and extrinsic defects. The first type of defects corresponds to heteorepitaxial growth defects implying formation of twins, stacking faults, grain boundaries and antiphase boundaries. The second type of defect is related to the growth mechanism induced by the deposition process and leads to plastic deformation. These latter later may be due to relaxation defects induced by local non-stoicheometrty [7].

TEM observation emphasizes a regular formation of large conical shaped extended defects regularly spaced which are developped from the bottom to the top of the superstructure both on a -axis and c-axis supperlattices. They are visible on Fig.4a and Fig.4b. These defects play an important part in the local distorsion [7] of the superconducting planes and on the resulting roughness of the superlattices. They correspond to plastic deformation and to the formation of strongly bent layers. The plastic deformation may lead to partial or complete loss of the coherence of the superlattice locally. The relaxation produces cracks formation along these defects, decreasing locally the roughness The special shape of these defects is related to the spiral growth process induced by surface precipitates as described in [1].

c-axis grown multilayers contain large amount of dislocations, stacking faults which lead to the formation of antiphase grain boundaries and twins. These latter propagate across the epitaxial structure through a few stacking of bilayers. They nucleate in YBCO c-axis layer. These defects do not extend in a-axis YBa2Cu3O7 layers. As an example we can see on Figure 5 superlattices with a first YBCO a-axis layer grown on a-axis (PBCGO) buffer layer. The initial YBCO a-axis orientation is lost after the deposition of the first STO layer. The regrowth of YBa2Cu3O7 on the first SrTiO3 layer corresponds to c-axis orientation although the growth temperature is kept the constant. The YBa2Cu3O7 c-axis layers leads to a high concentration of interface growth defects and large interface roughness. Such growth defects have been already found in thin films [8]. They favor the formation of roughness, while a-axis grown layer roughness is minimized eventhough the possible twinning of the film[9]. Quantitative analysis of the interface roughness from HREM images points out that heteroepitaxial growth leads to a range of interface roughness ranging from 2nm to 5nm depending on the orientation of the film. The lower corresponds to a-axis films and the higher to the c-axis orientation. The roughness step hight measured corresponds to the c parameter.

The above results show that the interface roughness of YBCO/STO multilayers varies across the heterostructure. The best roughness results are obtained for a-axis YBCO/STO heterostructures..

5. Conclusion

The present study combines TEM, HREM and AFM techniques : it allows to determine the influence of the growth mechanism on YBCO thin film bicrystals and on YBCO based superlattice structures. In particular, we have shown unambiguously from AFM images the"in plane wavyness" of the Grain Boundary (GB) and from quantitative analysis of HREM images, the oscillation of the grain boundary plane, whose structure and morphology result from an island growth mechanism. Studying YBa2Cu3O7/SrTiO3 (YBCO/STO) heterostructures , we show that the interface roughness increases across the superlattices. Extended defects like twins (due to the heteroepitaxy of the orthorhombic c-axis YBCO on top of the cubic STO) and conical defects related to a plastic deformation of the superlattices have been found.

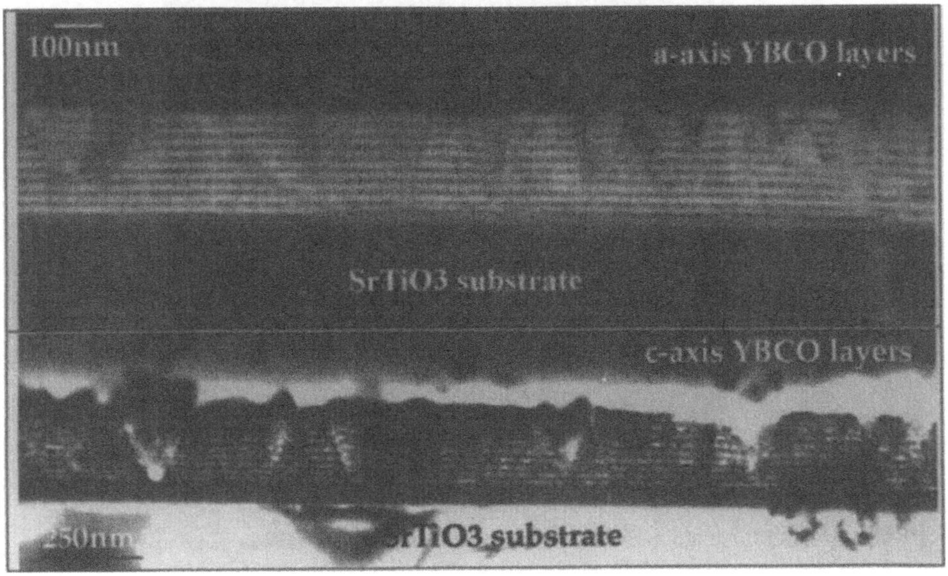

Figure 4: TEM micrograph of a-axis and c-axis multilayer cross sections

6-References

1. Ayache, J. Thorel, A. Le Duigou, J. Bonevich, J. Dahmen, U (1997) Journal of Alloys and Compounds **251** 78-82.
2. Ayache, J. Thorel, A. Lesueur, J. and Dahmen, U. (february 1998). submitted to Journal of Applied Physics.
3. Contour, J.P. Sant, C. Ravelosona, D. FrÈtigny, C. Dolin,C. Rioux, C J. Auvray, P. and Caulet, J. (1994), J. Cryst. Growth,**141**, 141.
4. Traeholt, C. Wen, J.G. Zandbergen, H.W. Shen, Y. Hilgenkamp, J.W.M (1994) Physica **C230** , 425-434.
5. Miller, D.J. Roberts, T.A. Kang, J.H. Talvacchio, J..Bucholz, D.B and.Chang, R.P.H (1995) Appl.Phys.Lett. **66**, 2561.

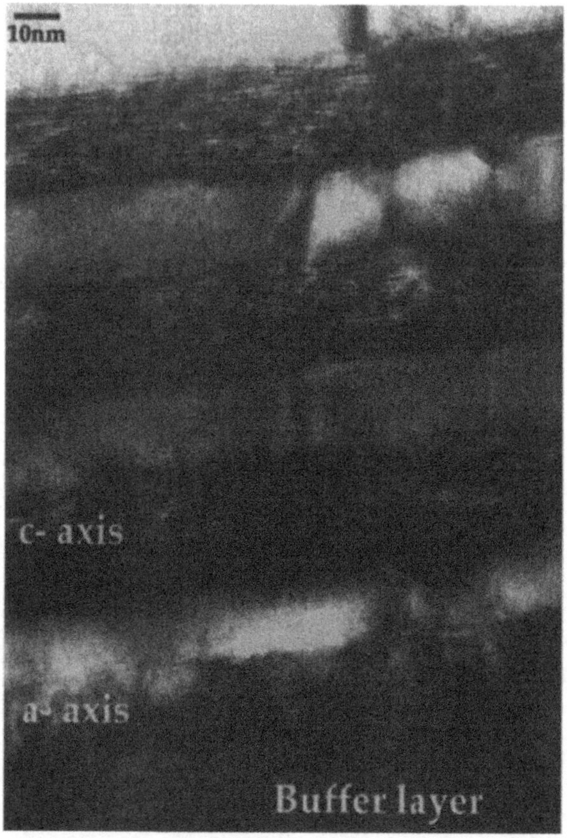

Figure 5
HREM image of the multilayers.where heteroépitaxial interface defects are clarly visible

6. Alarco, J.A.Olsson, E. Ivanov, Z.G. Nilsson, P.A..Winkler, D. Stepantsov, E.A and Tzalenchuk, A.Ya (1993) Ultramicroscopy **51**, 239.
7. Ayache, J. Contour, J.P. Dolin, C. Drouet, M. and Ravelosona, D (1997) Journal of Alloys and Compounds **251**, 185-192.
8. Ramesh, R. Hwang, D.M. Barner, J.B. Nazar, L. Inam, Dutta, A. Wu, X.D and.Venkatesan, T (1990) J. Mater. Res., vol **5**, N°4 , 704.
9. Hamet, J.F. Mercey, B.. Hervieu, M and. Raveau, B (1992) Physica C **193**, 465.

PHASE DIAGRAM AND STRUCTURE - RELATED STUDIES OF THE Nd-RICH $Nd_{1+X}Ba_{2-X}Cu_3O_Z$ SOLID SOLUTION

E.GOODILIN*, N.KHASANOVA*, X.J.WU*, T.KAMIYAMA**, F.IZUMI***, S.TAJIMA*, Y.SHIOHARA*
**Superconductivity Research Laboratory, ISTEC, 1-10-13 Shinonome, Koto-ku, Tokyo 135, Japan, fax 81-3-3536-5714*
***Institute of Materials Science, University of Tsukuba, Ibaraki 305, Japan*
***NIRIM, 1-1 Namiki, Tsukuba, Ibaraki 305, Japan*

Abstract

The Nd213 phase is the most Nd-rich compound among the $Nd_{1+x}Ba_{2-x}Cu_3O_z$ solid solutions. Models of the high-temperature "solid-liquid" phase diagram, polycrystalline material preparation, single crystal growth, and properties of the neodymium-rich solid solutions with compositions close to the Nd213 phase are presented. A tentative diagram for the equilibrium oxygen content is constructed up to 1000°C in air, including a newly investigated x=0.6-0.9 region. Finally, a crystal structure model of $Nd_{1.9}Ba_{1.1}Cu_3O_z$ is suggested and refined on the basis of neutron, electron, X-ray powder diffraction data, HREM, Raman scattering spectra measurements, and the structure-properties correlation is discussed.

1. Introduction

A new high-temperature, $NdBa_2Cu_3O_{7-\delta}$ - based superconductive material has been recently developed for levitation and power applications [1,2]. It was observed that nanoscale phase segregation may result in much better J_c-B characteristics of the melt-processed materials [1] and single crystals [2]. Therefore, many efforts have been done to develop a reliable and well-established scheme of phase relations in the Nd-Ba-Cu-O system. However, cation and oxygen nonstoichiometry of the $Nd(Ba_{1-x/2}Nd_{x/2})_2Cu_3O_z$ solid solution (Nd123ss) and difficulties in applying experimental techniques of the high-temperature analysis, demands several key problems to be carefully investigated, such as
(i) *maximum* peritectic decomposition temperatures T_p and the corresponding *x-values* of the most thermodynamically stable $Nd_{1+x}Ba_{2-x}Cu_3O_z$ at a given pO_2 [3];
(ii) *points of the maximum and minimum* Nd/Ba substitutions, and a possibility of the "negative" (Ba->Nd) substitution in $Nd_{1+x}Ba_{2-x}Cu_3O_z$ [1,3];
(iii) *low-temperature slopes* of the solid solution boundary since neither direct reaction of reagents nor (especially) backward decomposition of the metastable solid solution into the reagents might not be completed because of sluggish kinetics of the processes in a solid state;

G. Van Tendeloo et al. (eds.), High-Temperature Superconductors and Novel Inorganic Materials, 145–150.

(iv) *the neodymium-rich solid solution region* beyond x=0.6-0.7 [4-8] since it is insufficiently studied in literature [1].

To understand an overall picture, we have made in the present work an attempt to investigate phase relations, oxygen nostoichiometry and the structure - properties correlation of the neodymium-rich $Nd_{1+x}Ba_{2-x}Cu_3O_z$ phase. High-temperature phase diagrams, equilibrium solid and liquid compositions should be determined quantitatively to achieve controllable growth of single crystals. This might be helpful from both an academic point of view and for the novel inorganic material engineering, in prediction of processing conditions for materials with a desired composition, microstructure and applicable characteristics.

2. Experimental

Powders and carbon-free ceramic source materials were prepared by the conventional method from dried CuO, $Ba(NO_3)_2$, and Nd_2O_3 powders [5]. The final synthesis was performed by sintering at 1000°C in air for 70 hrs. on single-crystalline MgO plates followed by quenching in liquid nitrogen. After such a treatment, the average composition of 20-30 particles randomly selected in the x=0.9 solid solution sample was determined as $Nd_{1.94(3)}Ba_{1.10(4)}Cu_{2.97(5)}O_z$. Line width of XRD reflexes (particularly, a single *hkl*=113 line) was found to be similar to that of the silicon powder (internal XRD standard). Single crystals of the Nd123ss solid solution were grown by the TSSG - crystal pulling technique [7]. The *ab*-plane orientation of the bottom facet was observed by the Laue XRD. The powder XRD of mechanically crushed crystals did not detect any secondary phase or flux inclusions. EPMA was used to confirm the compositional uniformity of grown crystals (JEOL Superprobe JXA-8621, 15 kV, a $NdBa_2Cu_3O_{7-\delta}$ single crystal as a standard). Oxygenation was performed at 400°C for 100-120 hrs (pO_2=1 atm.).

DTA and TGA were applied with the heating/cooing rates of 2°C/min in air using the calibrated MAC Science thermoanalyzer. In addition to TGA curves, iodometric titration of the samples was performed by a standard technique to calculate the absolute oxygen content in the whole temperature range.

Measurements of neodymium solubility in different Ba-Cu-O solvents were performed in thermally equilibrated MgO crucibles using the ICP analysis (SPS 1200VR Plasma Spectrometer) of directly sampled small melt portions. To study phase relations, 10-20 mg of powdered samples were placed into home-made single crystalline MgO quenching cells, heated up to 970-1060°C in a tubular furnace calibrated in advance and held for 30-40 hrs. Then, the samples were quenched in silicon oil, mounted, polished and quantitatively analyzed by EPMA.

X-ray powder diffraction data ($Cu_{K\alpha}$-radiation) were obtained using the MAC Science MXP3 diffractometer at the room temperature and by means of the Rigaku RINT Ultima diffractometer at high temperatures (1000-1030°C). Measurement of lattice constants was performed with 0.01°/step in a 5-100° 2Θ range at scanning rates of 0.1°/min (25°C), or 4 sec of accumulation time per step (1000°C). In both cases the lattice constants calculation was done by means of the least square method with Si or MgO as internal standards.

Transmission-Electron-Microscopy investigation was carried out by the Jeol JEM 4000EX microscope with accelerated voltage of 400 kV. The Raman scattering spectra (RSS) were measured at the room temperature using a triple spectrometer T64000 Jobin-Ivon equipped with a liquid-nitrogen cooled CCD detector. The neutron powder diffraction experiment was performed in the NIRIM (Tsukuba, Japan) using the time-of-flight scheme.

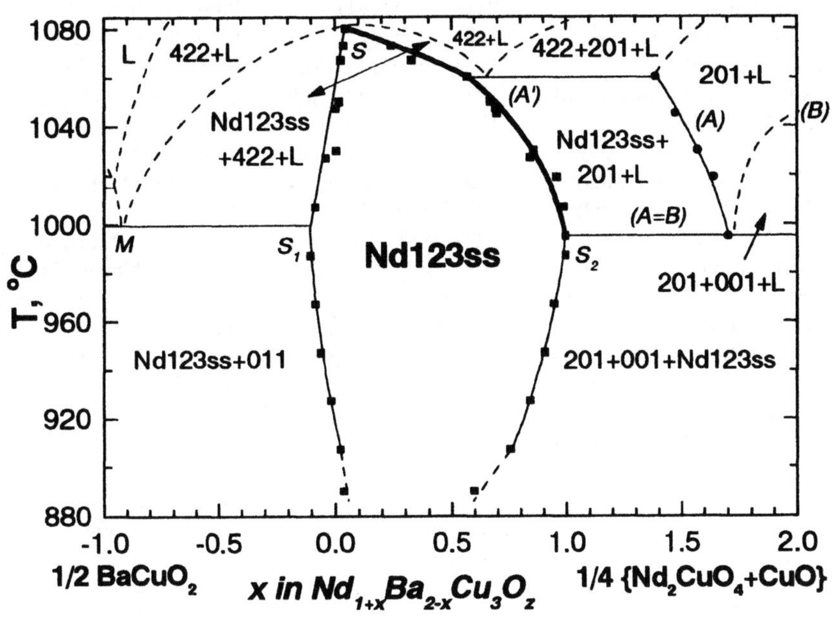

Fig. 1. A scheme of the 1/2 $BaCuO_2$ - 1/4 {Nd_2CuO_4+CuO} quasi-binary section in air

3. Results and discussion

In Fig.1, a scheme of the 1/2 $BaCuO_2$ - 1/4 {Nd_2CuO_4+CuO} quasi-binary section of the equilibrium phase diagram in air is presented with invariant reactions given in Table 1. The $Nd_{2.00(3)}Ba_{1.05(4)}Cu_{2.94(3)}O_z$ composition represents the upper possible level of substitution in the $Nd_{1+x}Ba_{2-x}Cu_3O_z$ solid solution in air near 990-995°C [6], and this temperature has to be accepted as a point of the invariant peritecto-eutectic reaction 213+CuO → 201 + L [8].

A wide temperature range is available for the growth of neodymium-rich Nd123ss single crystals, although a higher viscosity and low Nd solubility in liquid make "eutectic" temperatures below 990°C unattractive for the bulk crystal growth. On the other hand the upper growth temperature is limited by the thermodynamic stability of the solid solution (Fig.1) since it decomposes at lower temperatures as compared to the Nd123 phase (x=0). Above 1060°C, only the less-substituted Nd123ss solid solution is stable as well as the solidus slope changes slightly because the secondary equilibrium solid phase becomes 422 instead of 201 (Fig.1). Thus, a temperature range 990-1060°C is the most suitable for crystal growth of the Nd123ss solid solution with $0.57 < x < 1.0$.

TABLE 1. Invariant reactions in the quasiternary Nd-Ba-Cu-(O) system (pO_2=0.21atm) [2,4,6-8]

Quasibinary system	Reaction (schematically)	Temperature, °C, at%Nd, Ba/Cu ratio in liquid
$BaCuO_2$ - CuO	011+001→L(e1)	906-911, (0%, 3/7)
($3BaCuO_2$-2CuO) - Nd123[#]	123+011→L(e2)	970-976, (≈0.5%, 3/5)
$BaCuO_2$ - Nd123[#]	123+011→L(e3), M(S_1)	≈1000, (≈1%, 3/3)
($3BaCuO_2$-2CuO) - $Nd_4Ba_2Cu_2O_{10}$	123 → 422 + L(p1), S	1086-1090, (3.2%, 3/5)
Nd123ss[##] - CuO	123ss+001→L(e4)	906(e1) -> 990(pe2) *
$Nd_2Ba_1Cu_3O_y$ - "$Ba_3Cu_{16.5}O_z$"	213+001→201+L(pe2),S_2	990 **, (≈4.5%, 1/5.5-1/5.9)
Nd123ss[##]-(Nd_2CuO_4+ CuO)	123ss→201+L(p3), A	990->1060 *, (1/5.5-1/2.8)
Nd123ss[##]-(Nd_2CuO_4+ CuO)	201+L(p4)→422, A'	1060
Nd123ss[##]-(Nd_2CuO_4+ CuO)	123ss→422+L(p5)	1060->1086 *, (1/2.8 - $<$3/5)
Nd_2CuO_4 - CuO	201 + 001→L(e5), B	≈1045, (≈90mol% CuO)

e - eutectic melting, p - peritectic or peritecto-eutectic (pe) reaction;
*The compositional experimental data can be fitted by polynomials above 990°C (EPMA, ICP):
$1+x = -134.54+0.27324T-1.3670\times10^{-4}T^2$ ($x\approx0 \sim 1$ in the solid phase, T in °C)
C_{Nd}^{melt}, at% $= 845413.83-3271.3187T+4.7446T^2-3.057\times10^{-3}T^3+7.3817\times10^{-7}T^4$
$C_{Cu}^{melt}/C_{Ba}^{melt} = -194818.80+746.1754T-1.0713T^2+6.8341\times10^{-4}T^3-1.6345\times10^{-7}T^4$
**985-990°C according to DTA results, (A=B);
[#]assuming no Ba->Nd and/or Nd->Ba substitution, [##]terminal compositions at a given T;
e1-e2-e3 and *e4* correspond to expansion of the $BaCuO_2$-CuO binary eutectic around the Ba-rich, and around the Nd-rich ends of the Nd123ss field, respectively; this liquid approaches the Nd_2CuO_4 - CuO side and assimilates the *e5* binary eutectic at about 1045°C; *p1-p5-p3-pe2* describe the Nd123ss high-temperature peritectic decomposition ($0<x<1$)

A key issue of any melt-preparation technique is a compositional relationship between solid and liquid phases. Consequently, coordinates of the ternary phase equilibrium points of the terminal Nd123ss solid solution, liquid and the 201 (422) phase predict temperatures and initial compositions allowing to form the Nd123ss with maximal substitution (see remarks to Table 1). The Nd-rich (x=1.0) and Nd-poor (x→0) Nd123ss can be in equilibrium with the melts of higher (up to 16:3) and lower (3.5:3 ~ 5:3) Cu:Ba ratio, respectively, i.e. the initial Cu:Ba ratio in the flux allows to control effectively the Nd/Ba ratio in the Nd123ss solid phase at appropriate growth temperatures [2,6-8].

Faceted bulk x=0.85 crystals of about 5~7x6x3-4 mm^3 were grown by the modified TSSG technique, as predicted by the experimental phase relations [6,7]. As-grown $Nd_{1.85}Ba_{1.15}Cu_3O_z$ crystals were found orthorhombic by XRD (a'=2x3.884(1) Å, b'=3.864(1) Å, c'=2x11.585(2) Å). As a consequence, the crystals contain growth twins. The resistivity of the x=0.85 crystals was about 350 kΩxcm at 77K after oxygenation. The relative permittivity varied from $\varepsilon_r \approx$ 1700 (287K, 0kHz) to $\varepsilon_r \approx 60$ (80K, 0kHz) and $\varepsilon_r \approx 20$ (80K, 100 kHz). The crystals demonstrated additional reflexes in electron diffraction patterns (ED), which have to be attributed to a superstructural lattice confirmed by HREM [8].

At the temperature of maximum expansion of the solid solution field, structural phase transitions can be observed in a solid state at x=0.3 and x=0.6

[9]. The first is connected with the oxygen disorder due to the Nd substitution for Ba at random Ba-sites as proved by the apical oxygen mode broadening in Raman spectra [9]. The compositionally-induced phase transition at x≈0.6 originates from the Nd and Ba atoms ordering and subsequent ordering of the oxygen sublattice. Accordingly, the Nd213 phase exhibits properties, which are opposing characteristics of the parent Ba-rich high-temperature superconductor $NdBa_2Cu_3O_{7-\Delta z}$, such as (i) absence of superconductivity, (ii) presence of the copper with an unexpectedly high oxidation number (≈2.05) at high temperatures (1000°C in air), (iii) 2-2.5 times smaller oxygen nonstoichiometry (Δz≈0.33), (iv) orthorhombicity becomes even higher at elevated temperatures (a'=2×3.951(1)Å, b=3.8738(2)Å, c'=2×11.734(1)Å, 1000°C, air, ***in situ*** high-temperature XRD [8]), while the $NdBa_2Cu_3O_{7-\delta}$ phase and other solid solutions (x<0.6) are tetragonal under the same conditions. High- and low-temperature orthorhombic modifications of Nd213 are found for the first time with a transition at 700-720°C (air), and the x-z-T diagram is proposed [9].

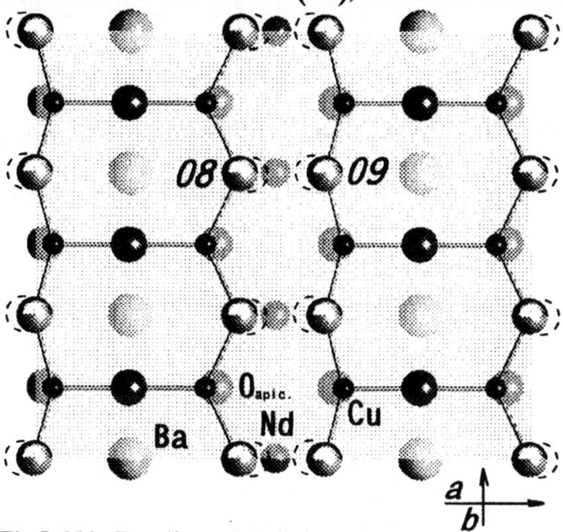

Fig.2. $Nd_{1.9}Ba_{1.1}Cu_3O_z$ basal plane fragment

The Rietveld refinement of the crystal structure for the high-temperature modification of the $Nd_{1.9}Ba_{1.1}Cu_3O_z$ phase was based on the space group B2mm [5] using neutron diffraction data. The structure can be derived from the basic $NdBa_2Cu_3O_7$ structure by doubling ***a*** and ***c*** parameters of the primitive unit cell as primarily caused by the sequential chain-like "in-plane" arrangement of Ba and Nd atoms (Table2). In the high-temperature modification (z≈7.0), the O8 and O9 oxygen positions are partially occupied, whereas the Cu and O atoms form strongly distorted broken chains (Fig. 2). This oxygen disorder is detected also by the Raman spectroscopy [9]. The complete occupancy of this position would lead to the total oxygen content z=7.5 and ordered rows of CuO_5 pyramids in the CuO plane. However, z≈7.33 has been obtained only at pO_2 = 0.21 and 1 atm. When cooling, the ***a*** unit cell parameter stays almost constant, ***b*** increases (giving smaller orthorhombicity) and ***c*** decreases as could be expected from the model. It may produce a variety of modulated superstructures observed by ED. The proposed structure is well-agreeable with other experimental features [9], chemical analysis (EPMA/iodometric titration) and the structure refinement data are consistent well with the expected ideal formula of the compound (Table 2).

The "Nd213" phase study may result in a better understanding of properties of Nd-based cuprate superconductors. From a practical point of view, it can be applied as a substrate material due to its small misfit with an HTSC film, high resistivity, and low dielectric constants [7].

TABLE 2. Structure refinement parameters of the x=0.9 solid solution (quenched, 1000°C, air)

$Nd_{1.9}Ba_{1.1}Cu_3O_z$(ideal),$Nd_{1.94(3)}Ba_{1.10(4)}Cu_{2.97(5)}O_{7.03(2)}$(analyzed#),$Nd_{1.9}Ba_{1.1}Cu_{3.0}O_{6.99}$(refinement)
Space group: B2mm, Z: 4, Unit cell: a = 7.7878*(3)* Å, b=3.8349*(1)* Å, c = 23.1521*(8)* Å

Atom	x	y	z	B	Atom	x	y	z	B
Ba1	0.5000	0.5000	0.0877*(3)*	0.6*(1)*	O2	0.5000	0.0000	0.1810*(3)*	0.46
Nd1	0.5000	0.5000	0.2500	0.06*(5)*	O3	0.2500	0.5000	0.1818*(6)*	0.46
Nd2	0.0000	0.5000	0.0970*(2)*	0.47*(9)*	O4	0.7500	0.5000	0.1858*(6)*	0.46
Cu1	0.2383*(7)*	0.0000	0.0000	1.8*(1)*	O5	0.2059*(5)*	0.0000	0.0800*(2)*	1.17*(9)*
Cu2	0.7617	0.0000	0.0000	1.8	O6	0.7941	0.0000	0.0800	1.17
Cu3	0.2500	0.0000	0.1751*(1)*	0.35*(4)*	O7	0.5000	0.0000	0.0000	1.5*(2)*
Cu4	0.7500	0.0000	0.1751	0.35	O8*	0.880*(4)*	0.5000	0.0000	1.5
O1	0.0000	0.0000	0.1859*(3)*	0.46*(4)*	O9**	0.174*(8)*	0.5000	0.0000	1.5

R_f:1.51%, R_i:2.33%, R_p:4.21%, R_{wp}:5.11%, R_{ex}:4.19%, * 0.77(5), ** 0.22(4) occupancy
$Nd_{2.02(6)}Ba_{1.14(3)}Cu_{2.85(6)}O_z$ as determined by TEM-EDX with the spot size about 0.1 micron

4. Acknowledgment

This work was supported by the New Energy and Industrial Technology Development Organization for R&D of Industrial Science and Technology Frontier Program. The authors thank M.Kambara (Univ. of Tokyo), A.Panfilov, M.Limonov, F.Saba, A.Oka, X.Yao (ISTEC) for help and fruitful discussions.

5. References

1. Murakami, M., Sakai, N., Higuchi, T., Yoo, S.I. (1996) Melt-processed light rare earth element-Ba-Cu-O, *Supercond.Sci.Technol.* **9**, 1015-1032.
2. Shiohara, Y., Endo, A. (1997) Crystal growth of bulk high-Tc superconducting oxide materials, *Materials Science and Engineering* **R19**, 1-86.
3. Wu, H., Kramer, M.J., Dennis, K.W., McCallum, R.W. (1997) Effect of oxygen partial pressure on lower solubility limit of $Nd_{1+x}Ba_{2-x}Cu_3O_7$, *Physica C* **290**, 252-264.
4. Osamura, K., Zhang, W., (1993) Phase diagrams of Ln-Ba-Cu-O systems (Ln=lantanide), *Z.Metallkd.* **84**, 522-528.
5. Goodilin, E., Oleynikov, N., Popov, G., Shpanchenko, V., Antipov, E., Balakirev, G., Tretyakov, Yu., (1996) On the stability region and structure of $Nd_{1+x}Ba_{2-x}Cu_3O_y$ solid solution, *Physica C* **272**, 65-78.
6. Goodilin, E., Kambara, M., Umeda, T., Shiohara, Y. (1997) Construction of a quasi-ternary phase diagram in the $NdO_{1.5}$-BaO-CuO system in the air atmosphere. Part II. Phase equilibria of the neodymium-rich $Nd_{1+x}Ba_{2-x}Cu_3O_z$ solid solution, *Physica C* **289**, 251-264.
7. Goodilin, E., Kambara, M., Umeda, T., Shiohara, Y. (1997) Solubility of neodymium in copper-rich oxide melts in air and growth of $Nd_{1+x}Ba_{2-x}Cu_3O_z$ solid solution single crystals, *Physica C* **289**, 37-50.
8. Goodilin, E., Oka, A., Wen, J., Shiohara, Y., Kambara, M., Umeda, T. (1998) Twins and related morphology of as-grown neodymium-rich $Nd_{1+x}Ba_{2-x}Cu_3O_z$ crystals, *Physica C*, in print
9. Goodilin, E., Limonov, M., Panfilov, A., Khasanova, N., Oka, A., Tajima, S., Shiohara, Y. (1998) Oxygen nonstoichiometry and phase transitions of the neodymium-rich $Nd_{1+x}Ba_{2-x}Cu_3O_z$ solid solution, *Physica C*, accepted
10. Khasanova, N., Goodilin, E., Wu, X., Kamiyama, T., Izumi, F., Tajima, S., Shiohara, Y., in preparation

PHASE STRUCTURE IN THE ASYNNNI MODEL FOR THE SELF-DIFFUSION OF OXYGEN IN 123YBCO

A.S. ILCHEV, O.L. DE LANGE AND S. GUMEDE
Department of Physics, University of Natal, Private Bag X01, Scottsville, Pietermaritzburg 3209, South Africa

Abstract. The ASYNNNI model developed for simulating the self-diffusion of oxygen in 123YBCO is shown to have a non-trivial phase structure. The free energy of the model exhibits different local minima when Monte Carlo simulations are performed for different initial configurations. The barriers separating these minima depend on values of the temperature, oxygen concentration and interatomic potentials.

1. Introduction

Recently there has been interest in the use of Monte-Carlo simulations in the antisymmetric next-to-nearest neighbour Ising (ASYNNNI) model to study oxygen diffusion in $YBa_2Cu_3O_{6+x}$ [1, 2, 3]. Equilibium configurations of the oxygen ions in the canonical ensemble have been obtained using a random start, and these configurations used to study oxygen diffusion in $YBa_2Cu_3O_{6+x}$ [1, 2, 3].

In this paper we consider the following question: Are the equilibrium configurations obtained in the above references the only ones possessed by the ASYNNNI model, or does the free energy of this model possess multi-minima? In particular, we perform Monte Carlo simulations starting with an ordered ("block") start, rather than the random start used in previous work, and we study the evolution of the microstructures as functions of the temperature, oxygen concentration $c = x/2$, and the interatomic potentials.

2. The Model and the Simulation Method

The model considered here is similar to that in [1]: a two–dimensional rectangular lattice of size L_1 by L_2 with periodic boundary conditions. The oxygen ions can occupy the sites of the lattice while the copper ions are

G. Van Tendeloo et al. (eds.), High-Temperature Superconductors and Novel Inorganic Materials, 151–156.

placed between the sites [1]. For a given x, the lattice is populated by N = $0.5xL_1L_2$ oxygen ions.

The aspect of the model which we investigate is whether the free energy has more than one minimum and whether, for certain values of the parameters (temperature, oxygen concentration and interatomic potentials) these minima are separated by impenetrable barriers. If this turns out to be the case, the Markovian random walks which begin from a particular type of initial configuration could be trapped by one of the minima (not necessarily the one corresponding to the lowest value of the free energy) while simulations starting from a different type of initial configuration may end up trapped by another minimum of the free energy, thus leading to different values of the observables. Previously the Monte Carlo studies of the ASYNNNI model have been performed only from a random and uniform initial configuration. After thermalizing for several thousand Monte Carlo steps this leads to a sample of "equilibrium" configurations characterized by the typical chevron-like arrangement of the ions in oxygen-copper chains. The equilibrium density of the occupied sites in rows of these lattices does not show any variation above the statistical noise.

Acting on the assumption that the ASYNNNI model described above may have a nontrivial phase structure due to a more complex form of the free energy we performed Monte Carlo simulations using the Metropolis algorithm and starting from initial configurations in which all the oxygen atoms were ordered with maximum density at one side, leaving the rest of the lattice unoccupied. In what follows we shall refer to this choice of the initial configuration as the block start. It was anticipated that the block start is far from equilibrium, therefore we let the system thermalize through hundreds of thousands of Monte Carlo steps. Since the distribution of the oxygen ions corresponding to a block start is highly inhomogeneous it was expected that, at least for lower temperatures, some relic of this inhomogeneity may survive thermalization over many hundreds of thousands and even millions of Monte Carlo steps, thus signalling a distinct phase in the model.

3. Results

At first we performed simulations with the same values of interatomic potentials as in Ref. [1]. The oxygen density profile was monitored by plotting the ratio of the occupied sites to all sites in each row of the lattice. Simulations performed from a random start reproduced results in [1], see Fig. 1. These microstructures do not show any significant deviation from a homogeneous distribution of oxygen. The oxygen-copper chains appear with equal probability along the O(1) and O(5) sites. This feature of configurations

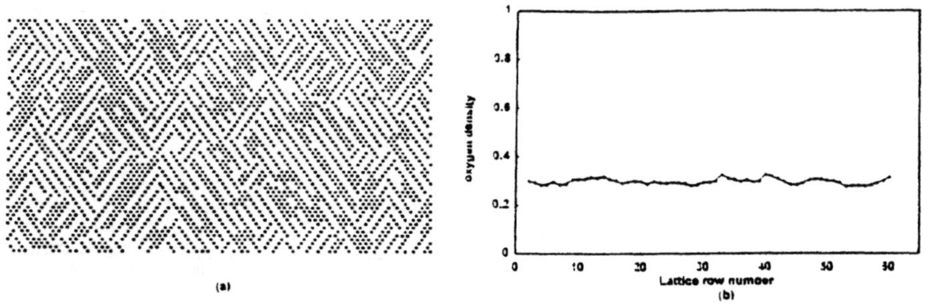

Figure 1. **(a) Configuration for $x = 0.6$ and $T = 0.02$eV after 4kMC steps with a random start. (b) Density profiles in the rows versus the number of the row. The lattice is 150 × 62.**

produced from a random start persists for the whole range of temperatures studied (0.02eV – 0.12eV). The corresponding phase can be denoted H_{10} where the first subscript indicates the presence of chains and the second subscript indicates that these are not preferentially oriented along either the O(1) or O(5) directions. (The orthorhombic structure of the low temperature configurations for $x = 1$ is a particular case of the H_{10} phase.) When T is increased, and for $x < 1$, the oxygen-copper chains appear less frequently and eventually disappear completely: the high temperature phase is characterised by a homogeneous and disordered distribution of oxygen (denoted by H_0 where the subscript 0 indicates the absence of oxygen chains), which is essentially the same as the random initial configuration used in the simulations. These are all the phases of the ASYNNNI model which can be reached by Monte Carlo simulations from a random start.

When the model is simulated from the inhomogeneous block start, a richer phase structure is revealed. For the lowest temperature studied (0.02eV) the system stabilizes at a markedly inhomogeneous oxygen distribution. The oxygen ions tend to stay together and this effect is most pronounced for x between 0.4 and 0.6. Examples of lattice configurations and oxygen density profiles are shown in Figs. 2 and 3; these can be classified as belonging to the phase I_{11} of the model (the first subscript indicates there are chains and the second that they tend to be preferentially ordered in either the O(1) or the O(5) directions).

Three features of the above micro-structures and density profiles are worth noting: 1. Chains of vacancies seem to appear together with the typical oxygen chains (Fig. 3). Other irregular clusters of vacancies are also present. These features are easier to identify at higher oxygen concentrations than that in Fig. 3. A chain of vacancies immersed in a locally

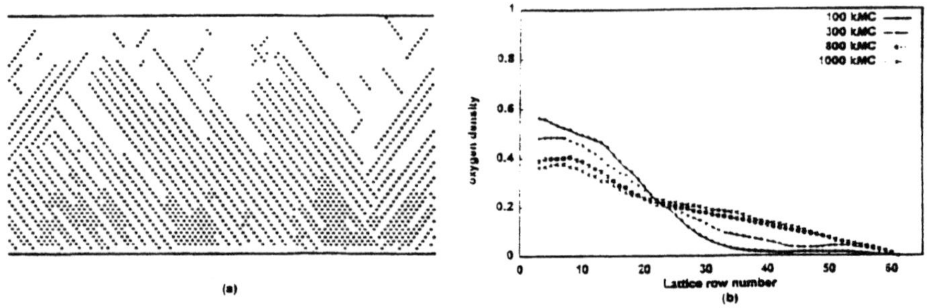

Figure 2. (a) Configuration obtained from a block start with $T = 0.02\text{eV}$ and $x = 0.4$ after 10^6 MC steps. (b) Density profiles for various numbers of MC steps. The lattice is 150×62.

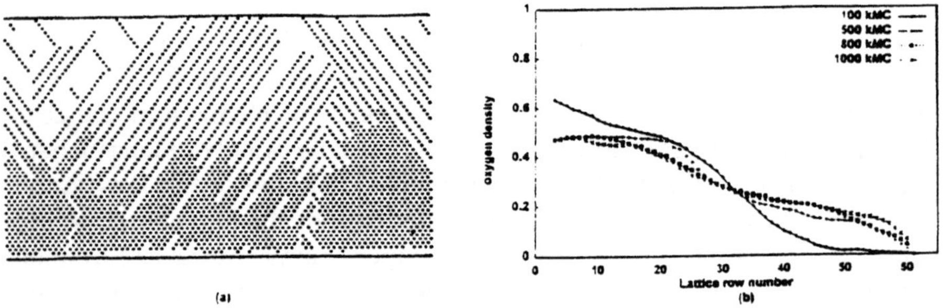

Figure 3. As in Fig. 2 but for $x = 0.6$.

homogeneous environment could move, in principle, without dissipation of energy, and hence without "friction". The formation of vacancy-chains requires both a particular temperature interval and an oxygen concentration which is neither too low nor too high. Could this be an artifact of the model or perhaps it has some bearing on the transport phenomena in YBCO? 2. The oxygen density profiles for different numbers of MC steps seem to pass through the same point. This could correspond to a characteristic length scale associated with the inhomogeneous phases. The dependence of this length on the concentration and temperature is under investigation. 3. The micro-structure of the configurations reached from a block start at low temperatures are anisotropic in the sense that one type of oxygen chain (say O(5)) clearly dominates over the other. The system can end up in either of the two ordered states with equal probability, which is indicative of a spontaneous symmetry breaking. This long-range order is typical for systems of the same universality class as the Ising model for a 2-dimensional ferromagnet. One can expect, therefore, a critical transition between this kind of order (H_{11}) and the phase H_{10}.

For low concentrations the inhomogeneity seems to melt away with MC

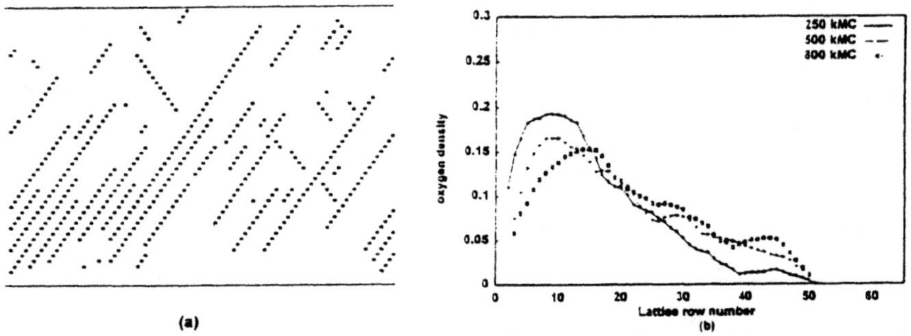

Figure 4. As in Figure 2 but for $x = 0.2$ and after 800 kMC steps. The lattice is 100 × 52.

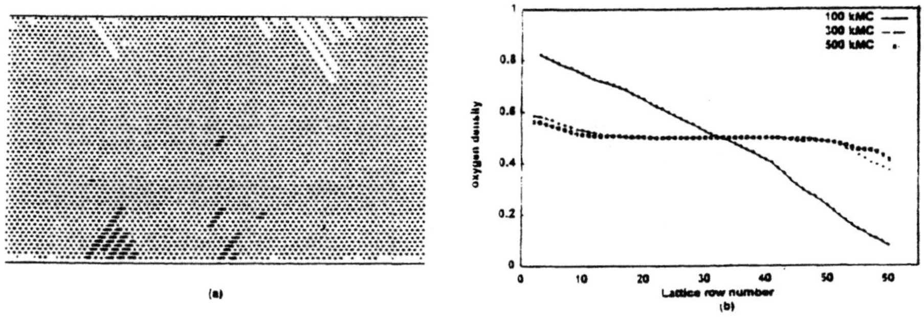

Figure 5. (a) As in Figure 2 but for $x = 1$ and after 200 kMC steps. (b) Density profiles for various numbers of MC steps.

"time" (Fig. 4). For concentrations above $x = 0.8$ the configurations tend to the ordered orthorhombic phase (Fig. 5).

When the oxygen concentration is kept fixed and the temperature is increased, the inhomogeneity of the block start disappears (Fig. 6): initially the anisotropy in the orientation of the oxygen chains persists (phase H_{11}), but at higher temperatures this anisotropy is diminished (phase H_{10}), becoming H_0 (homogeneous without chains, as with the random start) at even higher temperatures.

Our results for the ASYNNNI model can be summarized in a schematic phase diagram (Fig. 7) where OR denotes the orthorhombic phase which is found for $x \gtrsim 1$. This diagram depends strongly on the values of the interatomic potentials. In particular, a 30% increase of the potentials moves the boundary between the I_{11} and H_{11} phases significantly towards higher temperatures. An analysis of this, and other aspects, of the model will be presented elsewhere.

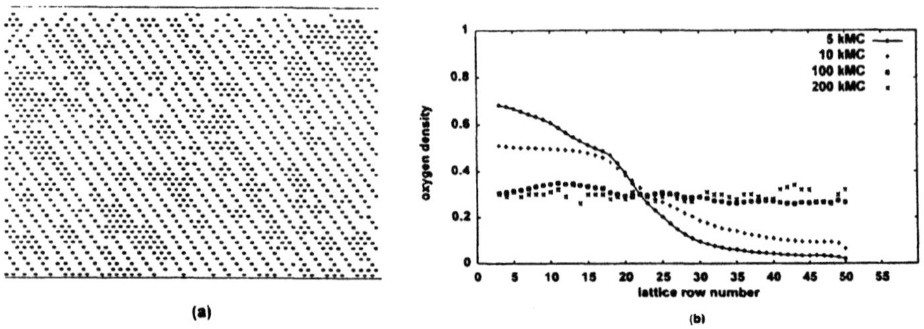

Figure 6. (a) As in Figure 3 but for $T = 0.04$eV and after 700 kMC steps. (b) Density profiles for various numbers of MC steps.

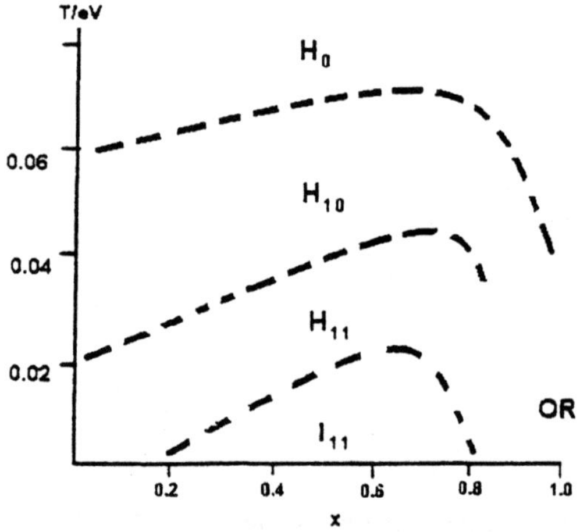

Figure 7. Schematic phase diagram for the ASYNNNI model.

References

1. Solomons, E. and de Fontaine, D. (1990) Monte Carlo study of tracer and chemical diffusion of oxygen in $YBa_2Cu_3O_{6+2c}$, *Phys. Rev. B* **41**, 11159-11167.
2. Pekalski, A. and Ausloos, M. (1994) Monte Carlo simulations of oxygen diffusion in planar model of 123YBCO, *Physica C* **226**, 188-198.
3. Ausloos, M. and Pekalski, A. (1995) Model of oxygen diffusion in $YBa_2Cu_3O_{6+x}$ with next-nearest-neighbour hopping: A Monte Carlo study, *Phys. Rev. B* **52**, 4577-4582.

ANION DOPING AND SUPERCONDUCTING PROPERTIES OF HG-BASED SUPERCONDUCTING CU MIXED OXIDES

E.V. ANTIPOV[1], A.M. ABAKUMOV[1], V.A. ALYOSHIN[1], K.A. LOKSHIN[1], D.A. PAVLOV, D.A. MIKHAILOVA[1], S.N. PUTILIN[1], M.G. ROZOVA[1], A.M. BALAGUROV[2], I.G. KUZEMSKAYA[3], E.S. ITSKEVICH[3]

[1]*Department of Chemistry, Moscow State University, Moscow 119899, Russia*
[2]*Frank Laboratory of Neutron Physics, JINR, Dubna 141980 Russia*
[3]*Institute of High Pressure Physics RAS, Troitsk 142092 Russia.*

1. Abstract

Different members of the $HgBa_2Ca_{n-1}Cu_nO_{2n+2+\delta}$ series with various oxygen concentration were prepared and characterized. Strongly overdoped high members of the series with $n \geq 3$ were prepared only using high pressure high temperature technique. It was found that all members of the series exhibited similar cupola shaped dependencies of T_c vs. δ.

Neutron powder diffraction experiments were carried out for monophase fluorinated $HgBa_2CuO_4F_\delta$ (T_c = 97 and 80 K) samples with different extra fluorine content. Fluorinated series also exhibits the cupola shaped behavior for the T_c vs. δ dependence, but this curve is shifted away from that one for the oxygenated series along the δ axis. NPD showed twice the amount of extra fluorine in comparison with those for the oxygenated Hg-1201 phases with close T_c's. An exchange of the extra oxygen by a double amount of fluorine causes an extended shortening of the apical Cu-O bond distances, while the in-plane distances, as well as T_c do not vary.

2. Introduction

There are two main parameters influencing T_c in $HgBa_2Ca_{n-1}Cu_nO_{2n+2+\delta}$ (n = 1 – 6) family: a concentration (δ) of the extra oxygen located in the Hg layers and a width of a perovskite slab (n). In the former case the dependence of T_c vs. extra oxygen content (δ) exhibits a cupola-shaped behavior for the members with n=1-3 [1-3]. The origin of this phenomenon is a change of a carrier concentration in conducting band from underdoped to overdoped states, followed by a T_c variation with a maximum corresponding to an optimal hole doping. The larger width of the perovskite slab in the structure the more difficult to obtain the overdoped Hg-based superconductor. For instance, a heat treatment

G. Van Tendeloo et al. (eds.), High-Temperature Superconductors and Novel Inorganic Materials, 157–162.

under 100 bar oxygen pressure decreases T_c for the first two members, while no significant changes of T_c was found for the third member. The highest reported value of T_c for each member increases ongoing from the first ($HgBa_2CuO_{4+\delta}$, T_c=97K) till the third member ($HgBa_2Ca_2Cu_3O_{8+\delta}$, T_c=135K) and then decreases to 107K for $HgBa_2Ca_5Cu_6O_{14+\delta}$ [4,5]. Thus a dependence of T_c vs. n exhibits a cupola-shaped behavior also. The probable explanation of this fact might be an insufficient carrier concentration for the higher members of the series with $n \geq 4$ or specific structural features of these phases. Therefore, it would be interesting to prepare Hg-1234 and Hg-1245 samples in overdoped states to accept or reject the first speculation.

Several powder neutron diffraction investigations of $HgBa_2CuO_{4+\delta}$ [6,7] show, in general, a good agreement between the refined structural parameters. However there are severe discrepancies concerning the occupancies of the extra oxygen site. For instance, the high extra oxygen content (up to δ=0.18 for the optimally doped Hg-1201 [7]) determined in several studies is in obvious contradiction with the optimal hole number, p_{opt}, per conduction CuO_2 layer, which is supposed to be $p_{opt}\approx0.16$ [8] and with $\delta_{opt}\sim0.08$ determined by iodometric titration [9]. As a result, the most natural ionic model of Hg-1201 doping (two holes per inserted extra oxygen atom) is presently under discussion and the reduced doping efficiency of oxygen in Hg-1201 has been assumed [10,11].

To verify the different hypotheses, we performed neutron powder diffraction studies of Hg-1201 samples with different extra oxygen [12] and fluorine content. Oxygen and fluorine anions have close crystallochemical behaviors, while their formal charges differ significantly: –2 and –1, respectively. Assuming a simple charge transfer model, we would expect the extra fluorine concentration to be twice as high for the fluorinated phase for the same doping level. It would be also interesting to determine the dependencies of T_c and structural parameters vs. the extra oxygen and fluorine content.

3. Synthesis and samples characterization

HgO, BaO, BaO_2, CaO and CuO were used for preparation of "$HgBa_2Ca_3Cu_4O_x$" and "$HgBa_2Ca_4Cu_5O_x$" samples with a different oxygen concentration (x). The variation of oxygen content was made by changing the BaO/BaO_2 ratio. All preliminary operations were carried out in a glove box filled by nitrogen. The syntheses were performed under high pressure 2-4 GPa at the temperature range 900-1100°C for 2h. Subsequent heat treatment of the obtained samples were performed at 300°C in oxygen and nitrogen flows or under high oxygen pressure 100 and 2000 bar at 300°C.

$HgBa_2CuO_{4+\delta}$ was synthesized according to the procedure described in [13]. The sample was reduced in a dynamic vacuum to the composition $HgBa_2CuO_{4.01}$ (T_c=61K). The reduced Hg-1201 samples were fluorinated by XeF_2 at 150-200°C for 15 h. All operations were made in a glove box in a dried N_2 atmosphere excluding the presence of O_2. Syntheses were carried out in Ni-crucibles placed in N_2-filled and sealed copper tubes. Two fluorinated samples $HgBa_2CuO_4F_\delta$ exhibit T_c = 97K and ~80K, respectively.

X-ray powder diffraction study was performed using a Guinier focusing camera FR-552 (Cu $K\alpha_1$ radiation). The temperature dependence of AC susceptibility was measured with an external field amplitude of 1 Oe at a frequency of 27 Hz. Neutron diffraction experiments were performed with the high resolution Fourier diffractometer

(HRFD) at the IBR-2 pulsed reactor in Dubna. Data processing was carried out by the Rietveld method on the d_{hkl} interval from 0.79 to 2.12Å.

4. Results and discussion

4.1. HIGH PRESSURE SYNTHESIS OF OVERDOPED HG-1234 AND HG-1245

For the preparation of overdoped Hg-1234 and Hg-1245 superconductors we used a variation of oxygen content in the starting mixture "$HgBa_2Ca_4Cu_5O_x$" annealed under fixed conditions (2GPa, 950°C, 2h). Earlier we have prepared overdoped Hg-1223 from the mixture with a starting composition "$HgBa_2Ca_3Cu_4O_x$" where x was equal to 10.4 – 10.6. Now, we decided to use the starting mixture "$HgBa_2Ca_4Cu_5O_x$" with x lying in the range 12.6-13.0. The samples with x=12.6 and x=13.0 contain predominantly either Hg-1234 or Hg-1245, respectively, and their fractions in the samples are not smaller than 80%, while the amount of superconducting impurities (Hg-1245 and Hg-1234, respectively) is not higher than 10%. The as-prepared Hg-1234 and Hg-1245 phases exhibit a relatively low T_c's (116K and 100K, respectively) in comparison with those reported earlier for these optimally doped compounds. Taking into consideration the small values of their a-parameter (and, consequently, the in-plane Cu-O bond distances) we suggested that these superconductors were overdoped. To confirm this assumption the subsequent heat treatments were performed:
1)300°C, oxygen flow, 50 h; 2)300°C, nitrogen flow, 20 h; 3)300°C, oxygen flow, 50h; 4)300°C, $P(O_2)$ = 100 bar, 30 h; 5)300°C, $P(O_2)$ = 2000 bar, 10 h.

The lattice constants of the Hg-1234 and Hg-1245 phases and their T_c's are presented in the Table 1. Both compounds exhibit a similar behavior (T_c and a-parameter variations) caused by heat treatments. The first annealing in oxygen flow did not change the characteristics of the compounds in comparison with those of the as-prepared samples. However, a reduction in nitrogen flow led to a noticeable increase of the a-parameters and a change of T_c's. The subsequent oxidation in oxygen flow resulted in a-parameter decreasing and T_c's increasing (up to 119 and 106 K for Hg-1234 and Hg-1245, respectively). In contrast to the lower members of the series ($n \leq 3$), annealing in oxygen flow does not allow to obtain the optimal T_c's for Hg-1234 and Hg-1245. Moreover, annealing under 100 bar oxygen pressure did not change lattice constants and T_c's. Therefore, a heat treatment under the higher $P(O_2)$ was necessary to achieve an optimal hole concentration. The oxidation under 2000 bar oxygen pressure resulted in a significant increase of T_c's for both phases (125 and 110K, respectively), while the a-parameters decreased. These values

Table 1.
Lattice constants and T_c of Hg-1234 (first line) and Hg-1245 (second line) after subsequent heat treatments.

Treatment	T_c^{onset}, K	a (Å)	c (Å)
as-prepared*	116 100	3.8485(6) 3.8482(5)	19.001(7) 22.149(5)
nitrogen flow, 300°C, 20 h	95 85	3.8532(6) 3.8531(6)	19.003(4) 22.181(4)
oxygen flow, 300°C, 50 h	119 106	3.8522(7) 3.8518(5)	19.009(8) 22.161(6)
$P(O_2)$=2000 bar, 300°C, 10 h	125 111	3.8506(5) 3.8503(5)	19.011(5) 22.157(4)

are close to those reported earlier for these optimally-doped compounds. Summarizing these experiments we can conclude, that the heat treatments changed a carrier concentration from the overdoped state for the initial as-prepared compounds to underdoped phases (after treatments in nitrogen and then in oxygen flows) and, finally, to optimally-doped phases (after a treatment under 2000 bar oxygen pressure). The dependencies of T_c vs. *a*-parameter for the $HgBa_2Ca_{n-1}Cu_nO_{2n+2+\delta}$ superconductors with n = 2-5 demonstrate a cupola-shaped behavior (Fig. 1). On the left side of each dotted curves the superconductors are in overdoped state. Therefore, we should reject the idea of decreasing T_c with increasing width of perovskite-like slab due to underdoping. We may suggest that the origin of this phenomenon has structural reasons, which can be found by a further structural study of these materials.

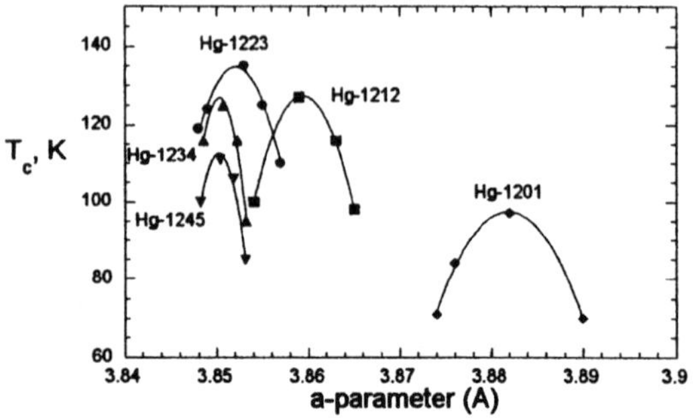

Fig. 1. The dependences of T_c's vs. *a*-parameter for the Hg-1212, Hg-1223, Hg-1234 and Hg-1245. The dotted lines are guides to the eye.

4.2. FLUORINATED HG-1201 PHASES

The results of the neutron data refinement of the fluorinated Hg-1201 samples were described in [14]. The structure refinements revealed an extra anion only in the middle of the Hg mesh. The values of n(F)=0.24(2) and n(F)=0.32(2) were obtained for samples with T_c = 97K and ~80K, respectively. The n(F) values are significantly larger than the values of 0.124(9) and 0.19(1) obtained for the oxygenated Hg-1201 samples with close T_c values. This comparison is quite correct since oxygenated and fluorinated materials were characterized by the same NPD facility and initial Hg-1201 samples were prepared by the same synthesis technique. Fig. 2 shows the dependencies of the T_c values vs. extra oxygen or fluorine concentration. It can obviously be seen that in both cases these dependencies exhibit a parabolic-shape behaviour and the curves are shifted away from each other along the δ axis. One can conclude that fluorine indeed oxidizes the (CuO_2) layers in the Hg-1201 structure, followed by a change in T_c, but the amount of inserted fluorine is approximately twice that of oxygen to achieve the same T_c-values and, consequently, the same doping level. Taking into account the different formal valences of these anions (−1 and −2,

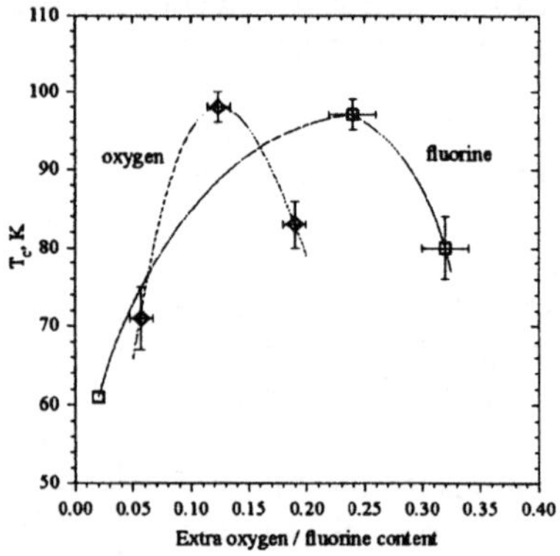

Fig.2 The behavior of T_c vs. extra oxygen or fluorine content. Lines through points are guides to the eye.

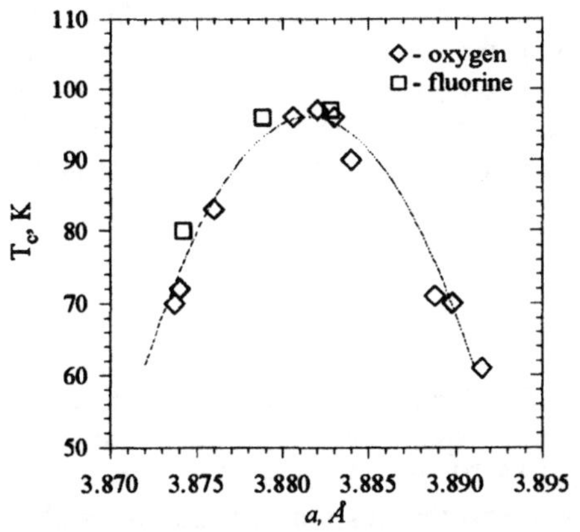

Fig.3 The behavior of T_c vs. the *a*-parameter of the unit cell.

respectively), we conclude that inserted extra oxygen creates the twice as many holes as fluorine. The amount of inserted fluorine (as well as oxygen for the oxygenated samples), however, was found to be significantly larger than could be expected, assuming an optimal hole number p_{opt}=0.16. Therefore, we can conclude that the doping mechanism in the Hg-1201 superconductor is more complex than a simple oxidation of the (CuO_2) layers by an inserted fluorine or oxygen and can include the oxidation of the HgO_2 "dumbbell" as well.

Another important conclusion can be made from the dependence of T_c vs. the *a*-parameter for the oxygenated and fluorinated samples (Fig.3). All data points can be fitted by one parabolic-like function with a maximum around 97 K at *a*=3.882Å. This fact supports the conclusion that this distance and V_{Cu} are crucial parameters determining the magnitude of T_c. Variation of the apical Cu-O2 bond distance caused by the extra anion exchange when $\delta_F \approx 2\delta_O$ is well-pronounced in contrast to the in-plane ones. Fig. 4 shows practically a linear dependence of the apical Cu-O2 distance vs. δ (oxygen or fluorine). An increase in the amount of the extra anion in the Hg-layer results in a compression of the apical Cu-O2 distance. These distances differ significantly between the fluorinated and oxygenated Hg-1201 phases with close T_c and in-plane Cu-O1 bond lengths. For instance, the difference between these distances in the phases with T_c=97 K is about 0.04Å, which is much larger than the standard deviations. At the same time, the apical Hg-O2 bond length in the fluorinated Hg-1201 phases is significantly larger in comparison with that in the oxygenated compounds (Fig.4). This bond is a very strong covalent bond and, even under high pressure, this distance does not decrease significantly: 1.980(4)Å and 1.973(19)Å at ambient conditions and under 5.07 GPa, respectively [15]. Therefore, the more probable origin of the elongation of the Hg-O2 distance is an interaction between Hg and the extra anions. An increase in the coordination number of Hg by the inserted extra anions is accompanied by a shift of the O2 atoms away from the Hg cations towards the Cu atoms, thus elongating the Hg-O2 distance and shortening the Cu-O2 one. We can conclude that the exchange of extra oxygen for double the amount of fluorine causes a variation of the apical Cu-O2 distances, predominantly, while the in-plane distance and T_c remain the same. This transformation can be formally

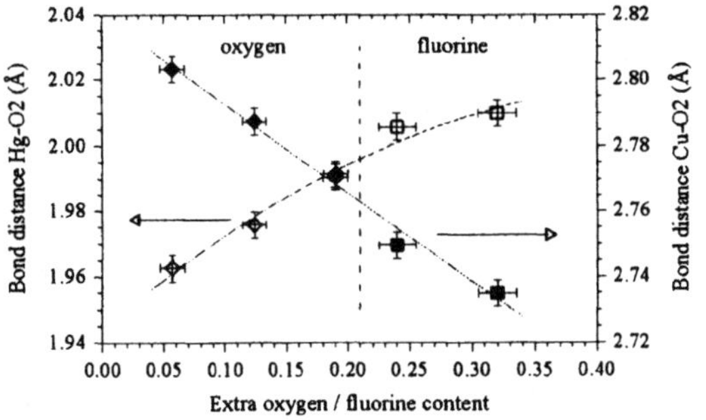

Fig.4 Bond distances Hg - O2 (left scale, open symbols) and Cu-O2 (right scale, full symbols) as a function of extra oxygen or fluorine content. Lines through points are guides to the eye.

considered as anisotropic compression of the CuO_6 octahedron under an application of approximately 2 GPa of uniaxial pressure along the *c* axis [15]. However, our study showed no difference in T_c for the fluorinated or oxygenated optimally-doped Hg-1201 compounds. This result supports the conclusion that compression of the in-plane Cu-O1 distances is the most probable origin of an enhancement of T_c under pressure in the Hg-based superconductors.

5. Acknowledgments

The authors are grateful to P.E.Kazin (MSU) for the magnetic measurements and V.G.Simkin (FLNP) for help with the neutron experiment. This work was partly supported by INTAS, Grant 93-2483ext, INTAS-RFBR, Grant 95-639, the Russian Scientific Council on Superconductivity (Poisk), and the Russian Foundation for Basic Research, Grant 97-02-17103.

6. References

1. Antipov, E.V., Putilin, S.N., Kopnin, E.M., Capponi, J.J., Chaillout, C., Loureiro, S.M., Marezio, M. and Santoro, A. (1994) *Physica C* **21**, 235.
2. Fukuoka, A., Tokiwa-Yamamoto, A., Itoh, M., Usami, R., Adachi, S. and Tanabe, K. (1997) *Phys. Rev. B* **55**, 6612.
3. Lokshin, K.A., Kuzemskaya, I.G., Kulikova, L.F., Antipov, E.V. and Itskevich, E.S. (1997) *Physica C* **279**, 11.
4. Capponi, J.J., Kopnin, E.M., Loureiro, S.M., Antipov, E.V., Gautier, E., Chaillout, C., Souletie, B., Brunner, M., Tholence, J.L. and Marezio, M. (1996) *Physica C* **256**, 1.
5. Scott, B.A., Suard, E.Y., Tsuei, C.C., Mitzi, D.B., McGuire, T.R., Chen, B.-H. and Walker D. (1994) *Physica C* **230**, 239.
6. Chmaissem, O., Huang, Q., Putilin, S.N., Marezio, M. and Santoro, A. (1993) *Physica C* **212**, 259.
7. Huang, Q., Lynn, J.W., Xiong, Q. and Chu, C.W. (1995) *Phys. Rev. B* **52**, 462.
8. Jansen, L. and Block, R. (1996) *Physica A* **230**, 467.
9. Fukuoka, A., Tokiwa-Yamamoto, A., Itoh, M., Usami, R., Adachi, S. and Tanabe, K. (1997) *Phys. Rev. B* **55**, 6612.
10. Singh, D.J. and Pickett, W.E. (1994) *Phys. Rev. Lett.* **73**, 476.
11. Hamed, A., Ortiz, R., Feng, H.H., Li, Z.G., Hor, P.H., Xue, Y.Y., Sun, Y.Y., Xiong, Q., Cao, Y. and Chu, C.W. (1996) *Phys. Rev. B* **54**, 682.
12. Aksenov, V.L., Balagurov, A.M., Sikolenko, V.V., Simkin, V.G., Alyoshin, V.A., Antipov, E.V., Gippius, A.A., Mikhailova, D.A., Putilin, S.N. and Bouree, F. (1997) *Phys. Rev. B* **55**, 3966.
13. Alyoshin, V.A., Mikhailova, D.A. and Antipov, E.V. (1996) *Physica C* **271**, 197.
14. Abakumov, A.M., Aksenov, V.L., Alyoshin, V.A., Antipov, E.V., Balagurov, A.M., Mikhailova, D.A., Putilin, S.N. and Rozova, M.G. (1998) *Phys. Rev. Lett.* **80**, 385.
15. Aksenov, V.L., Balagurov, A.M., Savenko, B.N., Sheptyakov, D.V., Glazkov, V.P., Somenkov, V.A., Shilshtein, S.Sh., Antipov, E.V. and Putilin, S.N. (1997) *Physica C* **275**, 87.

PHASE EQUILIBRIA AND SINGLE CRYSTAL GROWTH OF $Ln_{1-x}Ca_xSr_2Cu_2GaO_7$ (Ln=Y, Ho)

Donggeun Ko, Claude L. Mertzenich, Sophie Boudin, Charlotte L. Stern and Kenneth R. Poeppelmeier
Department of Chemistry and the Science and Technology Center for Superconductivity
Northwestern University, 2145 Sheridan Road, Evanston, IL 60208

Undoped and calcium doped $Ln_{1-x}Ca_xSr_2Cu_2GaO_7$ (Ln=Y, Ho) crystals were grown for the first time using a 0.2SrO/0.8CuO flux by slow cooling through the crystallization region determined from the phase diagram. DTA and X-ray powder diffraction were used to construct the phase diagram, which revealed a solubility of about 10 mol% in the flux without the appearance of peritectic decomposition products. Square plate crystals with ≤ 2 x 2 x 0.5 mm^3 have been obtained by a combination of slow cooling and thermal cycling. EDS x-ray analysis, x-ray powder diffraction and single crystal x-ray diffraction were used to confirm their stoichiometry and structure. Single crystals of $Y_{1-x}Ca_xSr_2Cu_2GaO_7$ with a nominal composition of x=0.2 are superconducting with T_c^{onset} ~ 100 K.

1. Introduction

The anisotropic and two-dimensional CuO_2 layer is an essential structural unit in all known copper oxide superconductors. The CuO_2 layers are bounded by ionic AO metal-oxygen layers (A=Ba, Sr and large lanthanides), thus forming AO-CuO_2-AO structural blocks. For most superconducting compounds, there are multiple CuO_2 layers separated by ionic metal layers A′(A′=Ca, small lanthanides, and Y) within the block and an additional layer of mixed oxidation state cations covalently bonded to oxygen between these blocks [1]. In contrast, some layered copper oxides have ionic, fixed oxidation state, cations in this additional layer [2-4]. The latter compounds are interesting from the point of view of applications and fundamental standpoint because their oxygen stoichiometries are stable. Extensive synthesis and doping efforts have been reported on polycrystalline $LnSr_2Cu_2GaO_7$ (Ln= Y, Lanthanide) [2,5-7]. $LnSr_2Cu_2GaO_7$ crystallizes in the orthorhombic (noncentrosymmetric) space group Ima2 (No.46) as determined from neutron powder diffraction and x-ray single crystal diffraction [2,8]. The structure is similar to $LnBa_2Cu_3O_7$ except that the chains of square planar copper are replaced by chains of gallium tetrahedrally coordinated by oxygen. High oxygen pressure annealing of polycrystalline samples has been used successfully to prevent reduction and incorporate

G. Van Tendeloo et al. (eds.), High-Temperature Superconductors and Novel Inorganic Materials, 163–168.

calcium [5,6]. High oxygen pressure annealed polycrystalline samples exhibit a small Meissner fraction.

The growth of large and aluminum-free gallate single crystals of high quality is essential to understand the relationship between the composition, structure and superconductivity. The crystal should be grown in yttria stabilized zirconia (YSZ) crucibles to prevent incorporation of aluminum from alumina crucibles. Various flux systems have been examined [9]. Crystals of reasonable size and quality can be obtained from the self-flux, SrO-CuO.

The phase diagram, one of the most important parameters for crystal growth, shows the relationship between the chemical composition, the phases, and the temperature under the state of thermodynamic equilibrium. From the phase diagram, crystal growth conditions such as flux composition, starting composition and temperature program can be determined. In our investigation, the phase diagrams of $Ln_{1-x}Ca_xSr_2Cu_2GaO_7$ in the eutectic flux (0.2SrO+0.8CuO) showed up to about 10 mol% solubility and a relatively narrow temperature range of crystallization of ≤100 K for crystal growth as the flux composition was varied from the eutectic composition. DTA was used to construct these phase diagrams. EDAX, SEM and x-ray powder diffraction confirmed the composition, morphology and structure of the crystals and showed that on doping calcium indeed substitutes into the compounds. Thermopower and SQUID magnetic measurements demonstrated that the calcium doped crystals were p-type superconductors.

2. Experimental

2.1. CRYSTAL GROWTH

The phase equilibria of $Ln_{1-x}Ca_xSr_2Cu_2GaO_7$ (Ln=Y, Ho) in the self-flux (0.2SrO+0.8CuO) were investigated for crystal growth using differential thermal analysis (DTA) and x-ray powder diffraction. DTA was performed on a TA Instruments 1600 DTA system and powder diffraction on a Rigaku diffractometer with Cu Kα radiation. $Ln_{1-x}Ca_xSr_2Cu_2GaO_7$ (Ln=Ho or Y) was synthesized from Ln_2O_3, $CaCO_3$, $SrCO_3$, CuO, and Ga_2O_3 of 99.99% purity by solid state reaction. The eutectic flux of composition 0.2SrO/0.8CuO was prepared from the starting materials of $SrCO_3$ and CuO. Mixtures of the self-flux and the compound with varying proportions were heated and cooled at a rate of 5°C/min in air for DTA. The solubility limit and the liquidus temperature were determined and used to build the phase diagrams. The crystals were grown in the liquidus range by slow cooling at a rate of 0.03°C/min or by a combination of slow cooling and thermal cycling at least 4 times between 1050°C and 1060°C. Yttria stabilized zirconia (YSZ) crucibles were used to avoid aluminum contamination.

2.2. CRYSTAL CHARACTERIZATION

Morphological and elemental analyses were performed using a Hitachi Scanning Electron Microscope (SEM) equipped with an Energy Dispersive X-ray Analyzer (EDAX). Secondary electron imaging technique was used for morphological investigation. EDAX analyses were carried out with the accelerating voltage of 25 keV and the working distance of ~22 mm. All crystals showed a plate-like morphology with the square plane developed perpendicular to the a-axis. SQUID magnetic measurement was made on a $Y_{1-x}Ca_xSr_2Cu_2GaO_7$ crystal of approximately $1x1x0.5\ mm^3$ size with a nominal composition of x=0.2 using a Quantum Design MPMS system. An applied field of 100 Gauss and a scan length of 4 cm were used. Thermopower measurements were also carried out on both the doped and undoped $Y_{1-x}Ca_xSr_2Cu_2GaO_7$ crystals as a function of temperature using a slow-ac technique [10].

2.3. CRYSTALLOGRAPHIC DATA COLLECTION

Single crystal data were collected on a $YSr_2Cu_2GaO_7$ crystal. Data collection was carried out on an Enraf-Nonius CAD4 diffractometer with graphite monochromated Mo Kα radiation (λ = 0.71069Å) at a temperature of −120°C. Maximum 2Θ value for the data collection was 60°. An analytical absorption correction was applied using the linear absorption coefficient, which resulted in the transmission factors ranging from 0.058 – 0.694. Corrections for Lorentz and polarization effects as well as secondary extinction were applied. The structure was refined using the Texsan software package [11]. Refinement of the atomic coordinates and the thermal parameters was performed with the full-matrix least-square method. Anomalous dispersion was included in F_{calc}.

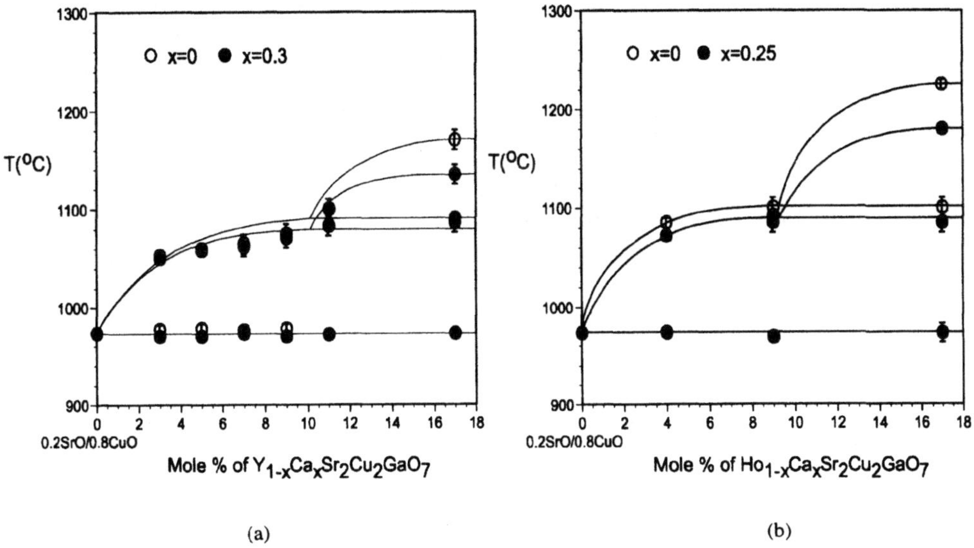

Figure 1. Phase diagrams of $Ln_{1-x}Ca_xSr_2Cu_2GaO_7$, (a) Ln=Y with x=0 and 0.3, (b) Ln=Ho with x=0 and 0.25.

3. Results and Discussion

To construct the phase diagrams, $Ln_{1-x}Ca_xSr_2Cu_2GaO_7$ was mixed with the 0.2SrO/0.8CuO eutectic flux [9] in varying proportions in order to determine the optimal growth conditions. These diagrams are shown in Fig. 1 plotted with respect to the molar percentage of $Ln_{1-x}Ca_xSr_2Cu_2GaO_7$ in the flux. The diagrams for $Y_{1-x}Ca_xSr_2Cu_2GaO_7$ with x=0 and 0.3 and $Ho_{1-x}Ca_x Sr_2Cu_2GaO_7$ with x=0 and 0.25 are shown in Fig. 1(a) and Fig. 1(b), respectively. The eutectic melting temperature is 970°C and the maximum liquidus temperatures are 990°C and 980°C for Ln = Y/Ho and $Y_{0.7}Ca_{0.3}$/$Ho_{0.75}Ca_{0.25}$, respectively. The maximum solubility limit in all the systems was approximately 10 mol% before peritectic decomposition products were observed. Crystals could be grown between 4 and 10 mol% in the appropriate liquidus temperature range. 8 mol% was selected for optimal crystal growth and the melting temperature of the flux chosen to be immediately above the liquidus point of the relevant system. The flux was slow cooled at a rate of 0.03°C/min down to 980°C and quenched to room temperature. This one-step slow cooling procedure produced generally small thin crystals with shiny black square surfaces with the dimension of 1x1x0.01mm^3 or less. Typical thin square plate-like crystals for Ln = Y and $Y_{0.8}Ca_{0.2}$ are shown in the SEM images (see Fig. 2). However these thin crystals are too

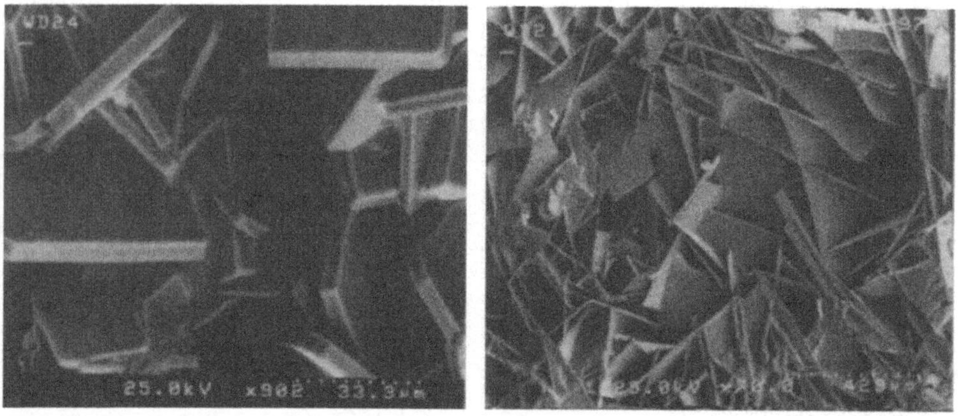

(a) (b)

Figure 2. The SEM images of $Y_{1-x}Ca_xSr_2Cu_2GaO_7$ single crystals as grown in the flux, (a) x=0 and (b) x=0.2.

fragile and small to be used for physical measurements. Thermal cycling with at least 4 cycles in the middle of the liquidus range combined with slow cooling at a rate of 0.03°C/min was employed to improve crystal size and morphology. Indeed crystals larger than 1x1 mm^2 x 0.5 mm(thickness) were obtained. These crystals were large enough to be used for SQUID and thermopower measurements. The compositions of the crystals were monitored through x-ray EDS analysis.

An x-ray single crystal structural determination was performed on an undoped Ln=Y crystal. It was refined in Ima2 (No.46) with improved R (R_w) and B_{eq}/B_{iso} values compared to the previous refinements of a $HoSr_2Cu_2(Ga_{0.72}Al_{0.28})O_7$ crystal [2] or a 10 μm $YSr_2Cu_2GaO_7$ crystallite [8] presumably because of the good crystal quality and absence of aluminum contamination. The refinement results are listed in Table 1. The corresponding atomic positions are shown in Table 2 with the temperature factors.

Table 1. Summary of Crystallographic Data

Formula	:	$YSr_2Cu_2GaO_7$	μ, cm^{-1}	:	34.537
Formula wt	:	572.95	Trans. Factor	:	0.058-0.694
Crystal size	:	0.2 x 0.1 x 0.01mm^3	Temp, °C	:	-120
Space Group	:	Ima2(No.46)	Scan type	:	2Θ/Θ
Lattice Param, Å			2 Θ range, degree	:	2-60
a	:	22.714(1)	Indexes collected	:	+ h, − k, ± l
b	:	5.4785(4)	R (R_w)*	:	0.0545(0.0445)
c	:	5.3861(2)	No. of unique data	:	753
Vol, $Å^3$	:	670.22	No. of unique data with I > 3 σ(I)	:	395
Z	:	4	No. of variables	:	45
Calc. Density, g/cm^3	:	5.678	Secondary extinction coeff	:	4.7(4) x 10^{-7}

*$R = [(\Sigma F_o - F_{calc})]/(\Sigma F_o)$; $R_w = [(\omega(|F_o| - |F_{calc}|)^2/(\Sigma\omega|F_o|^2)]$, $\omega = 1/\sigma^2(F)$

Table 2. Atomic positions for $YSr_2Cu_2GaO_7$

Atoms	Wyckoff Symmetry	x	y	z	B_{eq}*/B_{iso}, $Å^2$	Occ.
Y	4a	0.0	0.0	0.0	0.39(5)*	1.0
Sr	8c	0.84946(9)	-0.0148(6)	-0.004(1)	0.48(4)*	1.0
Ga	4b	0.25	0.5706(7)	0.955(1)	0.50(8)*	1.0
Cu	8c	0.9269(1)	0.4983(8)	-0.004(1)	0.35(4)*	1.0
O1	8c	0.9375(8)	0.747(6)	0.722(6)	0.6(3)	1.0
O2	8c	0.9344(7)	0.251(6)	0.246(6)	0.3(3)	1.0
O3	8c	0.8252(7)	0.548(2)	0.019(4)	0.7(3)	1.0
O4	4b	0.25	0.867(5)	0.111(5)	0.8(4)	1.0

$$*B_{eq} = \frac{8}{3}\pi^2(U_{11}(aa^*)^2 + U_{22}(bb^*)^2 + U_{33}(cc^*)^2 + 2U_{12}aa^*bb^*\cos\gamma + 2U_{13}aa^*cc^*\cos\beta + 2U_{23}bb^*cc^*\cos\alpha)$$

Ln=Y and $Y_{0.8}Ca_{0.2}$ crystals were characterized with thermopower and SQUID measurements. Prior to making the measurements, the crystals were annealed at 700 °C and subsequently slow cooled at a rate of 0.5 °C/min under flowing O_2. Thermopower measurements showed the p-type metallic behavior in Ln=$Y_{0.8}Ca_{0.2}$ crystals with S near 10 μV at room temperature which is typical of other superconducting cuprates. The corresponding value of the undoped crystal was ~ 100 μV. Superconductivity with T_c^{onset} ~ 100 K was observed in a Ln=$Y_{0.8}Ca_{0.2}$ crystal (dimension of ~1x1x0.5 mm^3) in a SQUID measurement.

4. Conclusion

$Ln_{1-x}Ca_xSr_2Cu_2GaO_7$ (Ln=Y, Ho) single crystals were grown as a result of a detailed phase equilibrium study for the first time. Aluminum-free crystals were obtained using YSZ crucibles. The removal of aluminum from the crystal improved the overall refinement including R(R_w) values and temperature factors. The larger and thicker crystals were produced by a combination of slow cooling and thermal cycling in the liquidus range. $Y_{0.8}Ca_{0.2}Sr_2Cu_2GaO_7$ crystals were found to be metallic with S~10 μV at room temperature and superconducting with T_c^{onset} ~ 100 K.

Acknowledgments: The authors would like to thank Dr. C. R. Kannewurf and P. Brazis for conducting the thermopower measurements. This work was supported by the National Science Foundation (Award No. DMR-91-20000) through the Science and Technology Center for Superconductivity and made use of MRL Central Facilities supported by the National Science Foundation, at the Materials Research Center of Northwestern University (Award No. DMR-9632472).

5. References

1. Schuller, I.K., and Jorgensen, J.D. (1989) Structure of high T_c oxide superconductors, *MRS Bulletin* **XIV**, 27-30.
2. Vaughey, J.T., Thiel, J.P., Hasty, E.F., Groenke, D.A., Stern, C.L., Poeppelmeier, K.R., Dabrowski, B., Hinks, D.G., and Mitchell, A.W. (1991) Synthesis and structure of a new family of cuprate superconductors: $LnSr_2Cu_2GaO_7$, *Chem. Mater.* **3**, 935-940.
3. Anderson, M.T., Poeppelmeier, K.R., Zhang, J.P., Fan, H.J., Marks, L.D. (1992) $Ba_2La_2Cu_2Sn_2O_{11}$: A new layered perovskite with a modulated structure, *Chem. Mater.* **4**, 1305-1313.
4. Salvador, P.A., Mason, T.O., Otzschi, K., Greenwood, K.B., Poeppelmeier, K.R., and Dabrowski, B. (1997) Internal chemistry of the pure and chemically substituted quadruple perovskites $Ln'Ln''Ba_2Cu_2Ti_2O_{11}$ (Ln'Ln''=LaY or NdDy), *J. Am. Chem. Soc.* **119**, 3756-3764.
5. Dabrowski, B., Radaelli, P., Hinks, D.G., Mitchell, A.W., Vaughey, J.T., Groenke, D.A., and Poeppelmeier, K.R. (1992) New family of superconducting copper oxides—$GaSr_2Ln_{1-x}Ca_xCu_2O_7$, *Physica C* **193**, 63-67.
6. Zhang, J.P., Groenke, D.A., Dabrowski, B., Poeppelmeier, K.R., Marks, L.D. (1994) Ca induced line defects and superconductivity in $Y_{1-x}Ca_xSr_2Cu_2GaO_7$, *Physica C* **227**, 259-266.
7. Zhang, J.P., Groenke, D.A., Zhang, H., DeLoach, D.I., Dabrowski, B., Poeppelmeier, K.R., Dravid, V.P., and Marks, L.D. (1992) Local structure of $Y_{1-x}Ca_xSr_2Cu_2GaO_7$ superconductors, *Physica C* **202**, 51-60.
8. Roth, G., Adelmann, P., Heger, G., Knitter, R., and Wolf, Th. (1991) The crystal structure of $ReSr_2GaCu_2O_7$, *J. Phys. I* **1**, 721-741.
9. Hijar, C.A. (1996) *Crystal Growth of Layered Cuprates*, Ph.D. Dissertation, Northwestern University.
10. Marcy, H.O., Marks, T.J. and Kannewurf, C.R. (1990) *IEEE Trans. Instrum. Meas.* **39**, 756-760.
11. TEXSAN-TEXRAY (1985) *Structure Analysis Package*, Molecular Structure Corporation.

SUPERCONDUCTIVITY IN $K_xBa_{1-x}BiO_3$ SINGLE CRYSTALS

L. BARIL, W. HARNEIT, T. KLEIN, J. MARCUS AND
C. ESCRIBE-FILIPPINI,
*Laboratoire d'Etudes des Propriétés Electroniques des Solides, CNRS,
BP 166, 38042 Grenoble cedex 9, France*

1. Introduction

$K_xBa_{1-x}BiO_3$ (KBB) as well as $BaBi_{1-x}Pb_xO_3$ is a family of superconductors based on the perovskite $BaBiO_3$. Superconductivity in this compound was discovered by Mattheis et al in 1988 [1] and since KBB has attracted a great deal of attention because of the combination of high Tc ($T_c \approx 30$K for $x \approx 0.4$) with relatively simple isotropic structure (cubic perovskite in the superconducting phase [2]) and lack of Cu. There are several reasons to place KBB in the high T_c superconductors. (i) The value of T_c is exceptionally high in relation to a very low density of state at the Fermi level. Then the application of the BCS mechanism with electron-phonon coupling is questionable for KBB in the same way as for the cuprates. (ii) The maximun T_c is observed near the semiconductor-metal transition. But there are also some differences. The main ones are that (i) KBB is not a layered compound, it has a simple cubic perovskite structure, (ii) the undoped compound $BaBiO_3$ is diamagnetic, (iii) the charge carriers are electrons.

The characteristic linear temperature dependence of the resistivity in cuprates which has received theoritical attention is not observed in this compound. Single crystals exhibit a metallic temperature dependence of the resistivity [3] with saturation at low temperature confirmed by low temperature measurements in high magnetic field [4]. This behaviour is consistent with an electron phonon coupling. At the same time the Hall effect is found to be *n*-type [3]. This result is quite puzzling, but if the substitution of Ba for K suggests that this material should rather be a *p*-doped system, electronic band calculations are in agreement with our results.

Moreover this compound which offers new opportunities to probe the nature of high temperature superconductors without the complication of the strong anisotropy of the cuprates presents the great advantage of having an $H_{c2}(0) \approx 30$T value which can still be reached experimentally. Consequently the vortex dynamics of KBB can be studied in a large part of the H-T diagram contrary to the cuprates.

Here we report magneto-transport and magnetization measurements on KBB single crystals. Our results permitted then to demonstrate that KBB presents an H-T phase diagram in the mixed state very similar to those of high Tc cuprates.
The two main experimental results concerning the mixed state that have been obtained in the cuprates are (i) the remarkable broadening of the resistive transition in the presence of a magnetic field and (ii) the existence of a melting line in H-T phase diagram. We will show that the transport properties of KBB single crystals can be

G. Van Tendeloo et al. (eds.), High-Temperature Superconductors and Novel Inorganic Materials, 169–172.

analysed within the frame work of the vortex glass melting mechanism [5] even if the broadening is not so remarkable. By these measurements we show that a vortex liquid phase can also exist in a 3D superconductor.
From the magnetic hysteresis loop of the high Tc an irreversible magnetic moment, much higher than the reversible one, is deduced. It presents a broad maximum in the intermediate field which is known as "fishtail effect". To explain this effect three main strategies have been proposed : (i) The Jc minimum below the peak is due to relaxation effect, (ii) the maximum field of the peak corresponds to a crossover from 3D to 2D behaviour or to a matching between the vortex lattice and the defect structure, and (iii) two different mechanisms account for the rise and the fall of Jc. Our results show that the first strategies can be rejected.

2. Experimental

The single crystals of KBB studied in this work were grown electrochemically using a technique described elsewhere [6]. This method produces high quality single crystals with T_C values up to 31.5 K and transition width $\Delta T_C \approx 0.2$K.
We used three magnetization techniques involving very different time scales, giving access to a large time window ($t = 10^{-4}$-10^{4}s) : pulsed field measurements, vibrating sample method and SQUID.
The magnetoresistance measurement were performed up to 30 T using four electrical contacts in the standard Van der Pauw configuration. The typical size of the crystals was 0.3x0.3x0.1 mm^3 which minimizes any possible spatial inhomogeneities.

3. Results and discussion

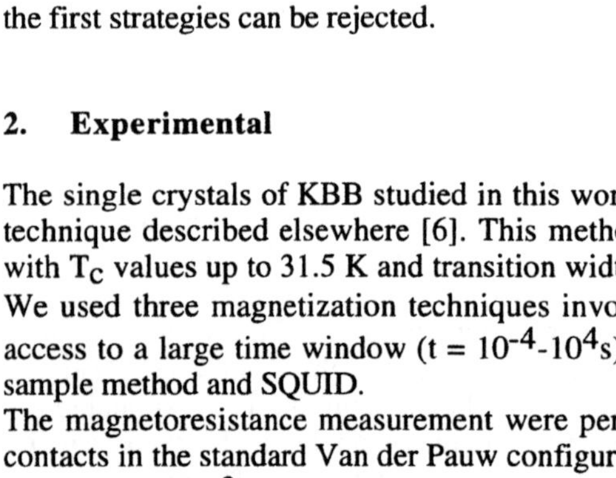

Figure 1. Magnetoresistance of KBB single crystal for various temperature showing the shift in the transition curves as the temperature is decreased.

Contrary to the cuprates, the resistive transition curves are shifted towards lower temperature as the magnetic field is increased without any significant broadening (Fig.1). This is the reason why the upper critical field was first deduced directly from the transition [4].

But analysing the foot of the resistive transition curves through the vortex glass transition theory and showing that the E-J curves present the characteristic power law behavior at Tg we have shown the existence of a vortex glass transition in a 3D superconductor [7].

Another unexpected feature is the "fishtail" effect which has been also observed in the magnetization curves of most of the high Tc oxides [8].As shown in figure 2 such a

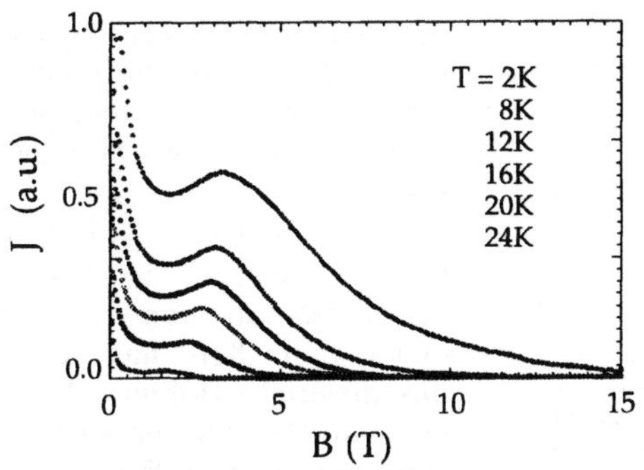

Figure 2. Field dependence of the critical current deduced from the width of the hysteresis loops in KBB single crystal at different temperatures.

"fishtail" is also observed in the KBB compound from 2K to 24K [9]. The rapid decreasing of J at high field can be explained in the frame work of weak collective pinning theory. Comparing the temperature dependence of the collective pinning length L_c of the upper critical field deduced from transport measurements, we could show that pinning is due to spatial fluctuations of the critical temperature δT_c [10]. Indeed the role of oxygen deficiency δ is similar in $K_xBa_{1-x}BiO_{3-\delta}$ and in the copper based high Tc. The critical temperature T_c depends on δ (and x) and on the fluctuations of δ (or x) then lead to fluctuations of T_c.

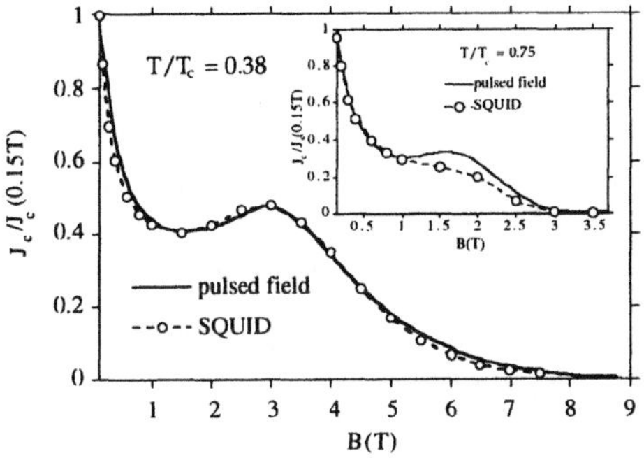

Figure 3. Comparison at different time scales of the fishtail peak measured in the same KBB single crystal at different temperatures.

Fig. 3 shows that despite the very different time scales involved in our experimental techniques, the J(B) curves perfectly superpose at low temperature. The "fishtail" then disappears for long waiting times (SQUID experiment) above 24K (T/Tc ≈ 0.75). These results suggest that, far enough from the vortex glass transition the shape of J(B) curves is not affected by the creep effects which is not the case close to this transition. The variation of the fishtail peak eliminates the matching of the vortex lattice with a static defect structure.

Moreover the lack of anisotropy in KBB leads us to exclude all kinds of 2D-3D crossover models for the explanation of the "fishtail" effect in high Tc superconductors.

Conclusion

As a conclusion we comment the H-T phase diagram (fig. 4) deduced from the experiments described above. Close to T_c the second maximun peak Hp in the magnetization curves varies roughly linearly with H_g. Very similar dependence has been observed too, close to T_c in YBaCuO [11]. At low temperature however , H_p and H_g present a very different behaviour. If Hp is almost temperature independent H_g increases sharply as T tends to zero.

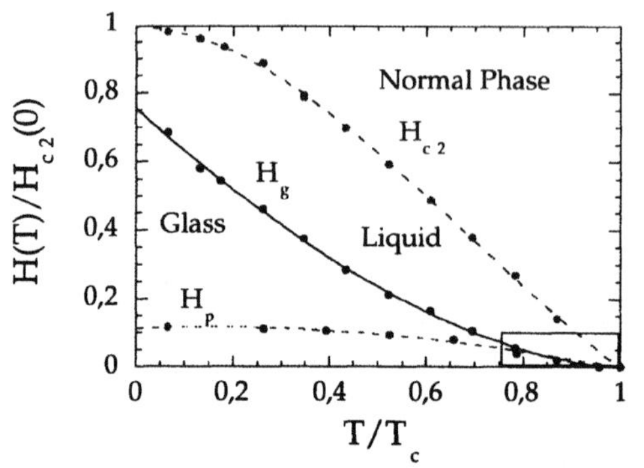

Figure 4. Phase diagram of KKB deduced from our measurements. For comparison we have delimited the domain which can be experimentally explored in cuprates.

We demonstrate clearly by the fact that we can explore the whole H-T diagram that the second peak of magnetization cannot be related to H_g.

This thorough study of the magnetoresistance and irreversibility magnetization on single crystals of the cubic KBB has shown that this material exhibits a behaviour qualitatively very similar to that of cuprrates despite the absence of anisotropy and much smaller thermal fluctuations. This discovery opens a new large range of physical parameters for the study of these fascinating properties and leads to extend or even reconsider some ideas concerning the role of thermal fluctuations and of material disorder.

4. References

1. Mattheis, L.F., Gyorgy, E.M. and Johnson, D.W. (1988) Superconductivity above 20K in the Ba-K-Bi-O system, *Phys Rev B* **37**, 3745-3746

2. Pei, S. et al (1990), Structural phase diagram of the $Ba_{1-x}K_xBiO_3$ system, *Phys Rev B* **41**, 4126-4141

3. Affronte, M., Marcus, J. and Escribe Filippini, C. (1993) Transport properties of $K_xBa_{1-x}BiO_3$ single crystals, *Solid State Commun.* **85**, 501-506.

4. Affronte, M., Marcus, J. and Escribe Filippini, C. et al.(1994), Upper critical field of $K_xBa_{1-x}BiO_3$ single crystals, *Phys Rev B* **49**, 3502-3510.

5. Fisher, D., Fisher, M.P.A., Huse, D.A. (1991) Thermal fluctuations,quenched disorder, phase transitions and transport in type-II superconductors, *Phys Rev B* **43**, 130- 159.

6. Marcus, J., Escribe-Filippini, C., Agarwal, S.K., et al (1991), Electrochemical synthesis and characterization of superconducting $Ba_{1-x}K_xBiO_3$ single crystals, *Solid State Commun.* **78**, 967-969.

7. Klein, T., Baril, L., Escribe Filippini, C., Marcus, J. and Jansen, A.G.M. (1996) Flux motion and phase transition in superconducting $(K,Ba)BiO_3$ single crystals, *Phys Rev B* **53**, 9337-9340.

8 Yeshurun, Y., Malozemoff, A.P. and Shaulov, A. (1996) Magnetic relaxation in high-temperature superconductors, *Rev. of Mod. Phys.* **68**, 911- 949.

9. Harneit, W. et al (1996) Short time study of the fishtail effect in $K_xBa_{1-x}BiO_3$ single crystals, *Physica C* **267**, 270-278.

10. Harneit, W., Klein, T., Baril, L. and Escribe-Filippini, C. (1996), Evidence for collective dTc pinning in superconducting $(K,Ba)BiO_3$ single crystals, *Europhysics Letters* **36**, 141-146.

11. Abulafia, Y. et al (1996), Plastic Vortex Creep in $YBa_2Cu_3O_{7-x}$ Crystals, *Phys. Rev. Lett.* **77**, 1596-1599.

STUDY OF HOLES DISTRIBUTION IN THE $Ln_{1+x}Ba_{2-x}Cu_3O_z$ SOLID SOLUTIONS (Ln = La AND Nd) BY RAMAN SPECTROSCOPY

V.PETRYKIN[1,2], M.KAKIHANA[1], YURII TRETYAKOV[2]
[1]*Materials and Structures Laboratory, Tokyo Institute of Technology, Nagatsuta, Midori, Yokohama 226, Japan*
[2]*Department of Chemistry, Moscow State University, Moscow 119899, Russia*

1. Introduction

From the practical viewpoint besides high value of T_c superconductors must posses also high irreversibility field – maximum field at which critical current is non-zero. This can be reached by stronger coupling between conducting layers due either to decrease of CuO_2-CuO_2 distance [1] or to decrease of electronic anisotropy [2]. The effect of two factors was previously discussed by comparison of different classes of high-T_c superconductors when one of these factors is dominating. The $Ln_{1+x}Ba_{2-x}Cu_3O_z$ solid solutions provide good opportunity to change thickness of blocking layer and electron distribution within one class of superconductors. Despite numerous studies of the crystal structure of solid solutions [3-5] and analysis of lattice dynamics [6,7] it is impossible to predict effect of substitution on H_{irr} *a priori*.
In the present paper the structure of the $Nd_{1+x}Ba_{2-x}Cu_3O_z$ solid solutions (0<x<0.5) was refined using X-ray data and holes distribution was discussed based on Raman spectroscopic data and results of X-ray structure refinement. It was found that holes are redistributed between $d_{x^2-y^2}$ and d_{z^2} orbitals of Cu(2) rather than between Cu(2) and Cu(1). The effect of the increased electronic anisotropy on the H_{irr} was demonstrated.

2. Experimental

Samples of $Nd_{1+x}Ba_{2-x}Cu_3O_z$ (x=0.1,...,0.7) were prepared by solid state reaction from Nd_2O_3, $Ba(NO_3)_2$, and CuO. After decomposition of the nitrate in initial mixtures at 800°C, powders were ground and calcinated at 900°C for 12h in the air, ground again, pressed into pellets and annealed at 940°C in 24h with intermediate grinding. Finally, pellets were sintered at 970°C in oxygen for 24h and quenched into the air. Samples of $Nd_{1+x}Ba_{2-x}Cu_3O_z$ with x=0.0 and 0.9 and $La_{1+x}Ba_{2-x}Cu_3O_z$ were prepared by polymerized complex method [8] using stoichiometric amounts of Nd_2O_3 (La_2O_3), $BaCO_3$, $Cu_2(OH)_2CO_3$, citric acid (CA) and ethylene glycol (EG) (Cation/CA/EG=1/5/20). Solid solutions with x=0.0 and 0.1 were prepared in Ar-O_2 mixture ($Po_2 \approx 10^2$Pa). Oxidation of all samples was performed at 300°C in flowing

G. Van Tendeloo et al. (eds.), High-Temperature Superconductors and Novel Inorganic Materials, 173–176.

oxygen (200 ml/min) for 60h then samples were furnace-cooled. The XRD data were collected by MacScience Diffractometer using CuK_α radiation in scanning step mode - 0.02 deg and acquisition time 5 s per step. Si was used as internal standard for lattice constants evaluation. Superconducting properties of the samples were studied using SQUID MPMS-5. Oxygen content of Nd-123ss was determined by iodometric titration. Raman scattering data were taken in the backscattering geometry by Jobin Yvon T64000 Raman triple spectrometer equipped by Atago Busan microscope. For excitation, wavelength 514.5 nm of Ar^+ laser with laser power 20mW was used.

3. Results and Discussion

All prepared samples were examined by XRD and macro-Raman spectroscopy and no impurity phases were detected. Lattice parameters of solid solutions were calculated by least square method and then refined by Rietveld method together with the other structural parameters. Preferred orientation and surface roughness correction were allowed. Lattice constants and some interatomic distances are presented in Table 1. Relatively high R_{wp} values for solid solutions are due to the presence of Si. We did not observe strong deviation from the linear dependence of **c** axes previously reported in [5] and it can be described by the relation **c** ( ) = 11.7546-0.296**x** for the whole range of solid solutions. The structure of Nd-213 was refined using XRD and TOF data and it will be discussed elsewhere. Oxygen content of solid solutions after oxidation increases with the increase of Nd concentration and average copper valence remains almost unchanged close to the optimal value +2.25 (0.25 hole per Cu atom) which must provide optimal T_c. However, temperature of transition into superconducting state as it is shown in Table 1 decreases gradually and disappearance of superconductivity can not be associated with the orthorhombic to tetragonal phase transformation which occurs at $x \approx 0.2$. Investigation of structure of the solid solutions, refined from XRD data points out that the thickness of insulating block decreases and apical oxygen moves closer to the CuO_2 plane, which means that extra charge must be transferred from CuO chains into the conducting CuO_2 planes. This fact is in the contradiction with the previously discussed simple model [3,9], which implies charge localization on the Cu atoms in the chain site. On the other hand, Nd-Ba(Nd) distance decreases and holes density on the in-

TABLE 1. Lattice parameters and selected interatomic distances in for $Nd_{1+x}Ba_{2-x}Cu_3O_z$. solid solutions refined by Rietveld method. Cu(1) (0, 0, 0), Cu(2) (0, 0, z), Nd (0.5, 0.5, 0.5), Ba(Nd) (0.5,0.5, z) O_{ap} (0, 0, z)

x =	0.0	0.1	0.2	0.3	0.4	0.5
a,	3.9191(2)	3.9182(2)	3.8911(1)	3.8912(1)	3.8872(1)	3.8840(1)
b,	3.8683(1)	3.8764(1)	3.8956(1)	3.8912(1)	3.8872(1)	3.8840(1)
c,	11.7509(5)	11.7366(4)	11.6929(3)	11.6616(3)	11.6313(5)	11.6071(4)
Cu(1)-Cu(2)	4.118(4)	4.118(5)	4.096(5)	4.075(6)	4.085(7)	4.081(5)
Cu(2)-O_{ap}	2.274(12)	2.212(18)	2.146(18)	2.156(18)	2.166(22)	2.194(25)
Nd-Ba(Nd)	3.736(1)	3.739(2)	3.733(2)	3.726(2)	3.699(2)	3.700(2)
R_{wp}, %	5.00	6.62	6.53	6.03	6.40	6.56
R_p, %	3.86	5.12	5.06	4.80	4.82	4.88
χ^2	2.035	2.021	1.844	1.793	1.871	1.805
T_c, K	96	83.7	56.8	38	-	-

plain oxygen atoms must decreases. To take into account both factors bond-valence calculation was performed. However BVC gives quite ridiculous result and predicts that $p(CuO_2)$ stays almost constant till the value at $x \approx 0.3$ and then drops rapidly for higher Nd concentration. Obviously such estimation can not explain experimentally observed T_c and measurements of holes concentration [9].

In the present study to estimate the trend of $p(CuO_2)$ vs. Nd concentration we analyzed asymmetry of in-plain oxygen B_{1g}-like mode in terms of Fano formalism which assumes that quantum interference of bare phonon with free carriers in CuO_2 plain results in the peak shape which can be described by simple relation [10]:

$$I(\omega) = A\frac{(q+(\omega-\omega_0)/\Gamma)^2}{1+((\omega-\omega_0)/\Gamma)^2}+B \quad (1)$$

Where ω_0 is the frequency of bare phonon, q asymmetry parameter, Γ linewidth contribution to the pure lorenzian and B constant background. The density of the continuum states ρ is proportional to the $(\Gamma q^2)^{-1}$. One can see from fig. 1 that estimated density of the continuum states (in other words carrier concentration) decreases by the factor of 2.5 for the region of solid solutions 0<x<0.5.

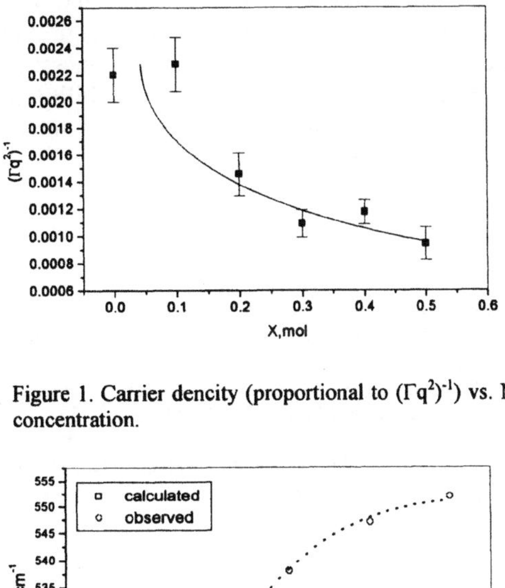

Figure 1. Carrier dencity (proportional to $(\Gamma q^2)^{-1}$) vs. Nd concentration.

In addition we analyzed frequency of apical oxygen vibration since it can be direct probe for the charge redistribution in the vicinity of this atom in high-T_c superconductors [11]. Raman shift of this mode was calculated in the approximation of inharmonic ionic crystal and ω was estimated to vary with $(1/r)^{3/2}$. Calculated and observed frequencies are presented on the fig. 2. Note, calculated frequency for the Nd concentration 0.1-0.3 mole is higher than observed values if we assume bond constants to be equal to those for stoichiometric compound. This is the indication that real value decrease with the Nd substitution and most probable explanation is that holes are localized on $Cu(2)\text{-}O_{ap}$ bond which results in the decrease of electron density in d_{z^2} orbital and weakening of the particular bond constant. One may object that non-isovalent substitution simply effects the frequency of

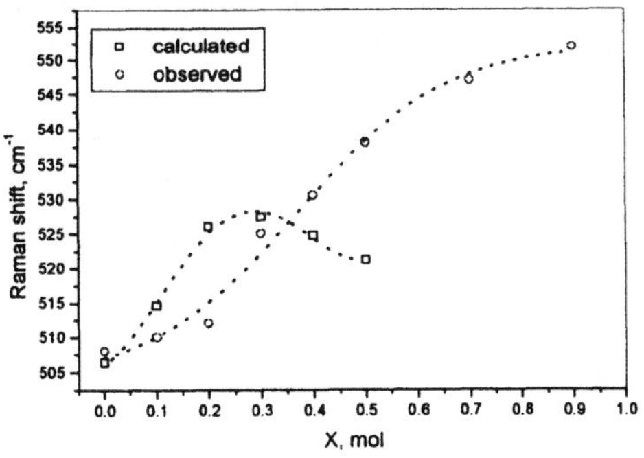

Figure 2. Observerd and calculated Raman shift of apical oxygen mode vs. Nd concentration.

the apical oxygen vibrations. This is partially true since increase of interaction between three valent cation and oxygen atom will increase the frequency of vibration and even decrease the difference of the observed and calculated values which is probably the case for higher Nd concentration.

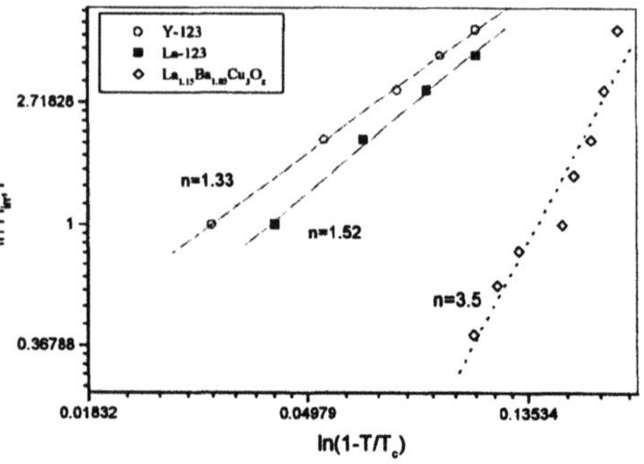

Figure 3. Irreversibility lines for Y-123, La-123 and $La_{1.15}Ba_{1.85}Cu_3O_z$, normalized on T_c.

Such holes redistribution was predicted in the five-band Hubbard model electronic structure calculation [12]. The shortening of $Cu(2)-O_{ap}$ distance can minimize the energy of holes repulsion through the redistribution of holes between $d_{x^2-y^2}$ and d_{z^2} orbitals. Note that from structural point of view substitution of Nd for Ba decreases anisotropy and must provide better coupling between CuO_2 plains, nevertheless electronic anisotropy increases, which must suppress H_{irr}. Fig. 3 demonstrates dramatic change in the irreversibility line for La-123 solid solution when the slope changes from 3/2 for La-123 to 3.5 for La-123ss with x=0.15 which became closer to the anisotropic Bi-HTSC.

4. References

1. Kim, D.H., Gray, K.E., Kampwirth, R.T., Smith,J.C., Richeson, D.S., Marks, T.J., Kang, J.H., Talvacchio, J., Eddy, M. (1991) Effect of Cu-O layer spacing on the magnetic field induced resistive broadening of high-temperature superconductors, *Physica C* **177**, 431-437.

2. Kishio, K., Shimoyama, J., Kimura, T., Kotaka, Y., Kitazawa, K., Yamafuji, K., Li, Q., Suenaga, M., (1994) Carrier doping and interlayer coupling in HTSC single crystals, *Physica C* **235-240**, 2775-2776.

3. Fomichev, D.V., D'yachenko, O.G., Mironov, A.V., Antipov, E.V. (1994) The structures of $Nd_{1+x}Ba_{2-x}Cu_3O_{7-d}$ (x=0.05, 0.22, 0.42) phases refined from X-ray single-crystal data, *Physica C* **225**, 25-33.

4. Takita, K., Katoh, H., Akinaga, H., Nishino, M., Ishigaki, T., Asano H. (1988) X-ray diffraction study on the crystal structure of $Nd_{1+x}Ba_{2-x}Cu_3O_{7-d}$, *Jpn.J.Appl.Phys.* **27**, L57-L60.

5. Goodilin, E., Popov, G., Oleynikov, N., Shpanchenko, V., Antipov, E., Balakirev, G., Tretyakov, Yu. (1996) On the stability region and structure of $Nd_{1+x}Ba_{2-x}Cu_3O_y$ solid solution , *Physica C* **272**, 65-78.

6. Nishitani, R., Yoshida, N., Sasaki, Y., Nishina, Y. (1989) Raman scattering study of the effect of Nd substitution into Ba sites on the phonon spectrum of $Nd_{1+x}Ba_{2-x}Cu_3O_y$, *Jpn.J.Appl.Phys.* **28**, L569-L572.

7. Wegerer, R., Thomsen, C., Cardona, M., Bornemann, H.J., Morris D.E. (1996) Raman investigation of $YBa_{2-x}La_xCu_3O_7$ ceramics, *Phys.Rev.B* **53**, .3561-3565.

8. Kakihana, M. (1996) "Sol-Gel" Preparation of High Temperature Superconducting Oxides *J.Sol-Gel Sci. Tech.* **6**, 7-55.

9. Takita, K., Akinaga, H., Ohshima, T., Takeda, Y., Takano M. (1992) Charge transfer and superconductivity in $Nd(Ba_{1-x}Nd_x)_2Cu_3O_{7+x}$ prepared in high pressure oxygen, *Physica C* **191**, .509-514.

10. Thomsen, C., edited by Cardona, M., Grüntherodt,G. (1991) *Light Scattering in Solids VI*, Springer-Verlag, Berlin., Heidelberg.

11. Kakihana, M., Eriksson, S.G., Börjesson, L., Johansson, L.G., Ström, C., Käll, M. (1993) Charge-transfer and compression effects of isomorphous substitution $YBa_2Cu_3O_7$ *Phys.Rev.B* **47**, 5359-5366.

12. Feiner, L.F., Grilli, M., Castro, C.Di. (1992) Apical oxygen ions and the electronic structure of the high-T_c cuprates *Phys.Rev.B* **6**,10647-10669.

NUCLEAR RESONANCE AND SUSCEPTIBILIY INVESTIGATIONS OF HG-BASED HIGH-T_C SUPERCONDUCTORS

K. LÜDERS, W. HOFFMANN, M. BAENITZ
Fachbereich Physik, Freie Universität Berlin,
Arnimallee 14, D-14195 Berlin, Germany

1. Introduction

The Hg-based high T_C superconductors, discovered 1993 by Putilin et al. [1], comprise systems with the highest transition temperatures [2] ever found for superconducting materials. Therefore, they are of high interest for potential applications. The transition temperature, however, is not the only quantity relevant for applications. Further important properties for instance are the critical current density and the so-called irreversibility field.

In this contribution some results concerning these aspects are summarized. The systems Hg-1201, (Hg,Pb)-1212, and (Hg,Pb)-1223 were investigated magnetically supplemented by some basic investigations by means of NMR.

2. Preparation

Within the frame of cooperation projects the samples investigated (Table 1) were prepared either at the Department of Chemistry, Moscow State University or at the FEM, Schwäbisch Gmünd. The Hg-1201 and Hg-1223 samples were prepared from stoichiometric mixtures of HgO and a precursor, containing Ba, Ca, and Cu, as starting materials [3]. In the case of Hg-1201 $Ba_2CuO_{3+\delta}$ was used [4]. The Hg-1212, (Hg,Pb)-1212, and (Hg,Pb)-1223 samples were synthesized directly from the mixtures of the metal oxides in proper compositions which then were pressed to pellets and heat treated [5, 6].

The substitution of Pb on Hg-sites is motivated by the difficulties which occur in the systhesis of phase pure samples of the higher numbers of the Hg-cuprates. There is a tendency that Pb doping leads to a more facile and reproducible synthesis with relatively large superconducting volume fractions [5, 7]. Even an enhancement of T_C up to 143 K was reported [8].

G. Van Tendeloo et al. (eds.), High-Temperature Superconductors and Novel Inorganic Materials, 177–182.

3. Magnetic investigations

3.1 SUPERCONDUCTING TRANSITIONS

For all samples T_C depends strongly on the oxygen content δ where the function $T_C(\delta)$ shows a bell-like shape. In the case of Hg-1201 for instance optimal doping is reached for δ = 0.08 corresponding to the maximum T_C-value of 96 K and a formal copper valence of 2.16 [9]. Analysing magnetization curves in the low field Meissner regime yields an estimation of the lower critical field B_{C1} (Table 1). Also, B_{C1} is strongly affected by the excess oxygen. In the case of Hg-1201 it is reduced from its optimal value of 27 mT to about 10 mT for δ = 0.01 [9].

Sample	T_C	$B_{C1}(0)$	$\lambda_{a,b}(0)$	j_C (at 5 K)
$HgBa_2CuO_{4+\delta}$ [Hg-1201] (powder)	96 K	27 mT	1540 Å	$\approx 5\times10^9$ A/m^2
$HgBa_2CaCu_2O_{6+\delta}$ [Hg-1212]	125 K	50 mT		$\approx 4\times10^7$ A/m^2
$Hg_{0.7}Pb_{0.3}Ba_2CaCu_2O_{6+\delta}$ [(Hg,Pb)-1212]	130 K	28 mT		$\approx 4\times10^8$ A/m^2
$HgBa_2Ca_2Cu_3O_{8+\delta}$ [Hg-1223] (powder)	132 K			$\approx 10^9$ A/m^2
$Hg_{0.7}Pb_{0.3}Ba_2Ca_2Cu_3O_{8+\delta}$ [(Hg,Pb)-1223]	132 K		1960 Å	$\approx 10^8$ A/m^2

Table 1. Superconducting transition temperatures, lower critical fields, magnetic penetration depths, and critical current densities of some Hg-based Superconductors

Fig. 1 shows an ac susceptibility signal for (Hg,Pb)-1223 [6]. A sharp transition occurs at T_C = 132 K and only one ac loss peak is observed indicating the good quality of this sample. The superconducting volume fraction, however, is approximately 60 % which might be due to the presence of non-superconducting phases.

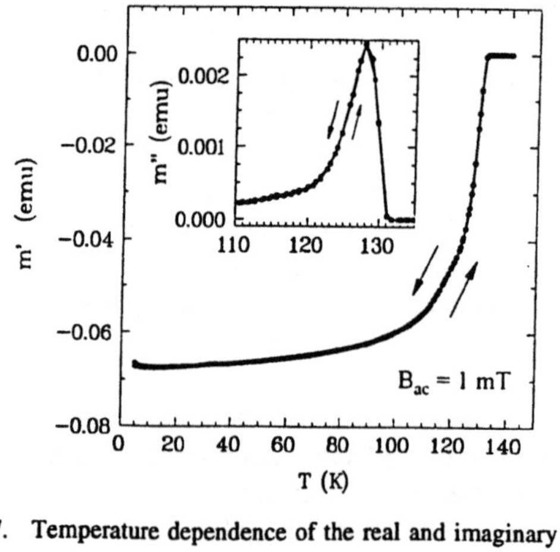

Figure 1. Temperature dependence of the real and imaginary part of the ac susceptibility for (Hg,Pb)-1223

3.2 CRITICAL CURRENT DENSITIES

Critical current densities j_C can be estimated from the width of magnetization loops [6, 10]. With the assumption that the Bean model is applicable, j_C is proportional to $\Delta M/R$ with ΔM = magnetization difference of the two branches of the loop and R = radius of the sample. Fig. 2 shows an example [6]. It should be mentioned that a comparison with respective results obtained from powder samples is difficult. Using average grain diameters (e.g. R = 10 μm) leads to much higher values for j_C [10, 11]. The applicability of the Bean model in these cases, however, might be problematic.

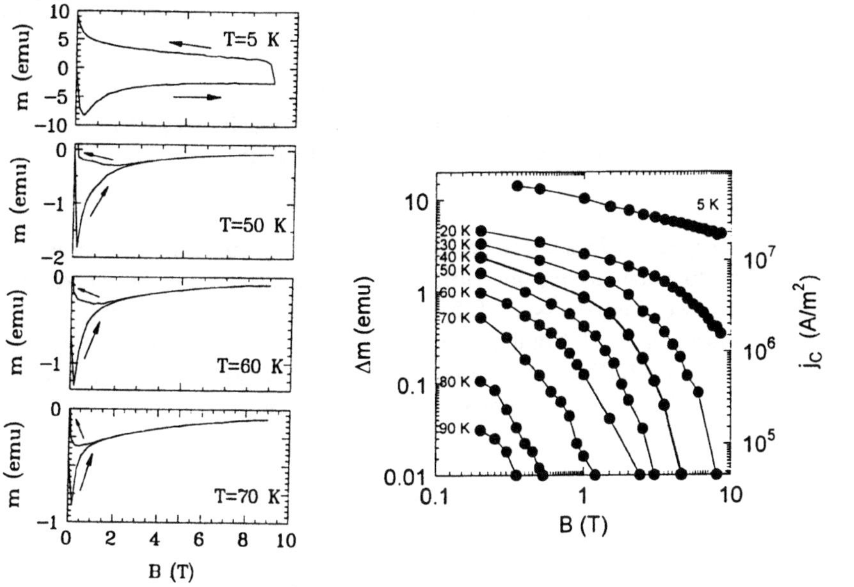

Figure 2. Field dependence of critical current densities (right) obtained from the width ΔM of magnetization loops (left) for a cylindrical bulk (Hg,Pb)-1223 sample with a radius of R = 3 mm

3.3 IRREVERSIBILITY FIELDS

One of the characteristic features of high-T_C superconductors is the existence of an irreversibility line in the B(T) phase diagram. It separates the region with irreversible flux penetration from that where reversible flux motion occurs and was first evident by the merging points of field cooled and zero field cooled transition curves. Further criteria for the determination of this line are the points where magnetization loops close and the peak maxima of the imaginary part of the ac susceptibility. Fig. 3 shows results for Hg-1201 where the ac peak maximum is used [12]. Measurements by using the merging points of fc and zfc curves reveal nearly the same results.

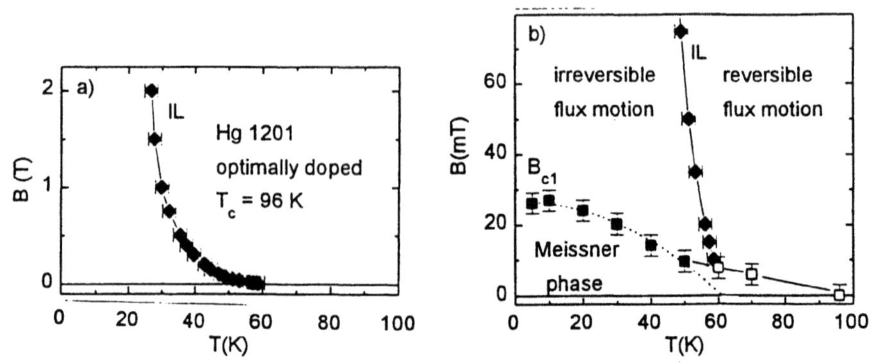

Figure 3. Irreversibility line of an optimally doped Hg-1201 sample (a) and phase diagram on an enlarged scale (b)

Results for a (Hg,Pb)-1223 sample are plotted in Fig. 4 [6]. All three criteria were applied leading to small systematic differences. Nevertheless, the behaviour can be described by a power law $B_{irr} \propto (1-T/T_C)^n$ with n near to 3.5. A similar result was reported by Isawa et al. [11]. Comparison of this exponent with those of other high-T_C systems (n ≈ 1.5 for YBaCuO, n ≈ 5.5 for Bi-cuprates) shows that from the viewpoint of applications the Hg-system is more advantageous than the Bi-system.

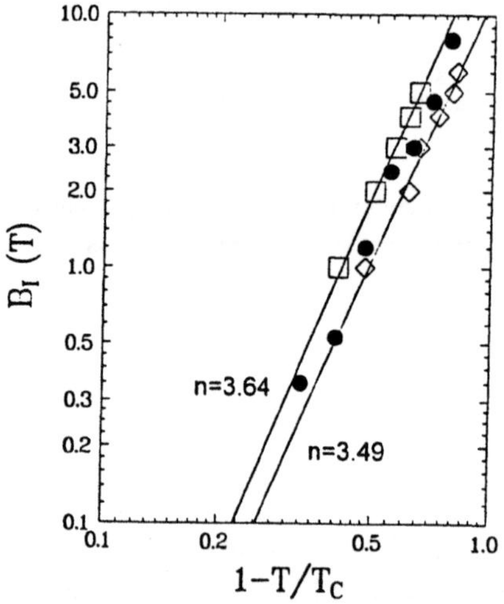

Figure 4. Irreversibility fields of (Hg,Pb)-1223 as a function of $1-T/T_C$ evaluated by dc measurements (●: from magnetization loops, ◊: from zfc and fc curves) and ac measurements (□)

4. NMR Investigations

To get information about the electronic properties of superconductors nuclear resonance experiments (NMR and NQR) are a suitable method. By application of NQR to the Hg-1201 system for instance it could be shown that the NQR frequency increases monotonously with increasing oxygen content δ [13, 14]. Analysing such behaviour [14] leads to an increase of the hole concentration $n_{x^2-y^2}$ in the Cu 3d shell reflecting the importance of the hole concentration between the Cu and O sites in the CuO_2 layers for the superconducting transition.

Concerning the pairing mechanism of high-T_C superconductors the question of s- or d-wave pairing is not convincingly answered. With the aim to contribute to answers of this question, detailed NMR investigations, especially spin-lattice relaxation time measurements, were applied to the Hg-1201 system [15, 16]. ^{63}Cu as well as ^{199}Hg nuclei were used to characterize the spin-lattice behaviour in the temperature range of 4.5 K to 300 K. Fig. 5 shows results of spin-lattice relaxation rate measurements for an optimally doped Hg-1201 sample below T_C. For comparison, numerical calculations were performed for different types of order parameter symmetry: isotopic s-wave, anisotropic s-wave, and d-wave using different values of the gap parameter $2\Delta/kT_C$. The experimental results are in best agreement with the $d(x^2-y^2)$-wave symmetry with a gap value of $2\Delta/kT_C = 7.02$.

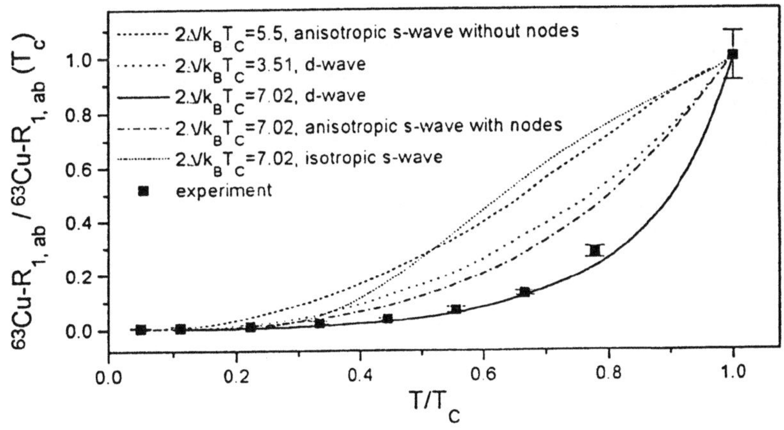

Figure 5. Temperature dependence below T_C of the ^{63}Cu spin-lattice relaxation rate for an optimally doped Hg-1201 sample. For comparison, several numerical calculations for different symmetries are also plotted.

Acknowledgements

The authors would like to thank J. Schmalian and K.H. Bennemann for helpful discussions and the numerical calculations of the spin-lattice relaxation rates.

References

1. Putilin, S.N., Antipov, E.V., Chmaissem, O., and Marezio, M. (1993), *Nature* **362**, 226.
2. Schilling, A., Cantoni, M., Guo, J.D., and Ott, H.R (1993), *Nature* **363**, 56.
3. Marezio, M., Antipov, E.V., Capponi, J.J., Chaillout, C., Loureiro, S., Putilin, S.N., Santoro, A., and Tholence, J.L. (1994) *Physica B* **197**, 570.
4. Alyoshin, V.A., Mikhailova, D.A., and Antipov, E.V. (1995), *Physica C* **255**, 173.
5. Loebich, D. and Khan, H.R. (1996), *Czechoslovak Journal of Physics* **46**, 1183.
6. Baenitz, M. Ecker, C., Lüders, K. Otabe, E.S., Matsushita, T., and Khan, H.R., *Advances in Superconductivity X*, in print.
7. Iqbal, Z., Datta, T., Kirven, D., Lungu, A., Barry, J.C., Owens, F.J., Rinzler, A.G. Yang, D., and Reidinger, F. (1994), *Phys. Rev. B* **49**, 12322.
8. Wu, X.S., Shao, H.M., Jiang, S.S., Gou, C., Chen, D.F., Wang, D.W., and Wu, C.H. (1996), *Physica C* **261**, 189.
9. Baenitz, M., Heinze, M., Hoffmann, W., Lüders, K., and Antipov, E.V. (1997), *Advances in Superconductivity IX*, p. 107.
10. Otabe, E.S., Matsushita, T., Baenitz, M., Ecker, C., Lüders, K., and Antipov, E.V., *Advances in Superconductivity X*, in print.
11. Isawa, K., Higuchi, T., Machi, T., Tokiwa-Yamamoto, H. (1994), *Appl. Phys. Lett.* **64**, 1301.
12. Hoffmann, W., Baenitz, M., Heinze, M, Lüders, K. Gippius, A.A., and Antipov, E.V. (1997), *Physica C* **282-287**, 1381.
13. Gippius, A.A., Antipov, E.V., Hoffmann, W., and Lüders, K. (1996) *Czechoslovak Journal of Physics* **46**, 1149.
14. Gippius, A.A., Antipov, E.V., Hoffmann, W., and Lüders, K. (1997), *Physica C* **276**, 57.
15. Hoffmann, W., Breitzke, H., Baenitz, M., Heinze, M., Lüders, K., Gippius, A.A., Antipov, E.V., Jess, P., Hubler, U., Lang, H.P., and Güntherodt, H.J. (1997), *Proc. 16th Int. Cryogenic Eng. Conf. (Haruyama, T. et al., Eds.)*, p. 1627.
16. Hoffmann, W., Lüders, K., Gippius, A.A., Antipov, E.V., Buntkowsky, G., and Limbach, H.H., to be published.

VORTEX LOCK-IN PHENOMENA IN $YBa_2Cu_3O_y$ SINGLE CRYSTALS

A.A.Zhukov, M.G.Mikheev
Physics Department, Moscow State University, Moscow 117234, Russia
G.K.Perkins, A.D.Caplin
Centre for HTS, Blackett Laboratory, Imperial College, London SW7 2BZ, United Kingdom
H.Küpfer, T.Wolf
Forschungszentrum Karlsruhe, Institut für Technische Physik, D-76021 Karlsruhe, Germany

The vortex lock-in phenomena were studied in $YBa_2Cu_3O_y$ single crystals using vector magnetometer. Several locking structures such as twin boundaries, columnar defects and CuO-planes have been analysed. The regions dominated by locking and trapping of vortices were identified. The temperature and magnetic field dependencies of the characteristic critical locking and trapping angles were studied.

1. INTRODUCTION

The theory [1,2] predicts the appearance of distinct phases, with the vortex direction non-collinear to the applied magnetic field $\boldsymbol{H}$ when $\boldsymbol{H}$ is tilted away from the direction of the correlated planar or linear pinning centre. The first direct experimental observation of such non-collinear vortex phases was reported in reference [3]. The phenomena manifest new type of the interaction between vortices and extended correlated defects. In difference from the vortex pinning that prohibits the translational vortex movement the vortex lock-in blocks the rotation of vortices. This paper reviews our studies [3-5] and compares vortex behaviour induced by different locking structures.

2. EXPERIMENTAL DETAILS

Seven $YBa_2Cu_3O_y$ single crystals grown from the flux [6], differing in twin structure and in concentration of point-like and another defects, were studied here. Two of these samples (MK and WZ) were de-twinned [7]. One crystal (AM) was twinned and had both directions of twin boundaries (TB), but the other (OZ) was cut from a larger crystal and had TB of predominantly one direction. Fifth crystal (VG) was essentially twin-free in as-grown state. We have also studied two $YBa_2Cu_3O_y$ single crystals (AH and I2) with

G. Van Tendeloo et al. (eds.), High-Temperature Superconductors and Novel Inorganic Materials, 183–188.

columnar defects created by high energy ion Au and U irradiation. The fluences correspond to the matching field B_ψ=3T and 0.04T for AH and I2 samples respectively.

A vibrating sample magnetometer (Oxford Instruments 5[H]) was used for these studies; it is equipped with two independent coil sets that measure simultaneously the magnetic moments parallel m_{std} and perpendicular m_{ort} to $\boldsymbol{H}$. This allows the vector magnetic moment and its projection on different directions to be determined. Particularly the magnetic moment parallel to the crystal axis can be calculated from the relations: $m_{ab} = m_{ort}\cos\varphi - m_{std}\sin\varphi$ and $m_c = m_{std}\cos\varphi + m_{ort}\sin\varphi$ where φ is the angle between magnetic field $\boldsymbol{H}$ and *c*-axis. The sample can be rotated *in situ* about the vertical axis, with an angular resolution of 0.01° and reproducibility better than 0.1°.

3. RESULTS AND DISCUSSION

The unique opportunity that presents the use of a vector magnetometer, is the possibility to determine the space direction φ_B of the magnetic induction $\boldsymbol{B}$ and therefore of vortices. This is crucially important for the study of the non-collinearity of $\boldsymbol{B}$ and $\boldsymbol{H}$ induced by the vortex lock-in phenomena. For fields significantly above the self field ($H >> H_{self\text{-}field}$) the components of the magnetic induction parallel to the crystal axis can be calculated from the relations [3]: $B_{ab} = \mu_0\,[\,H_{ab} + (1 - N)\,M_{ab}\,] \approx \mu_0\,[\,H_{ab} + M_{ab}\,]$ and $B_c \approx \mu_0 H_c$. Then the direction of vortices is determined by $\tan\varphi_B = B_{ab}/B_c \approx (\tan\varphi - M_{ab}/M_c)$. This means that for fields close to the *c*-axis the misalignment between vortices

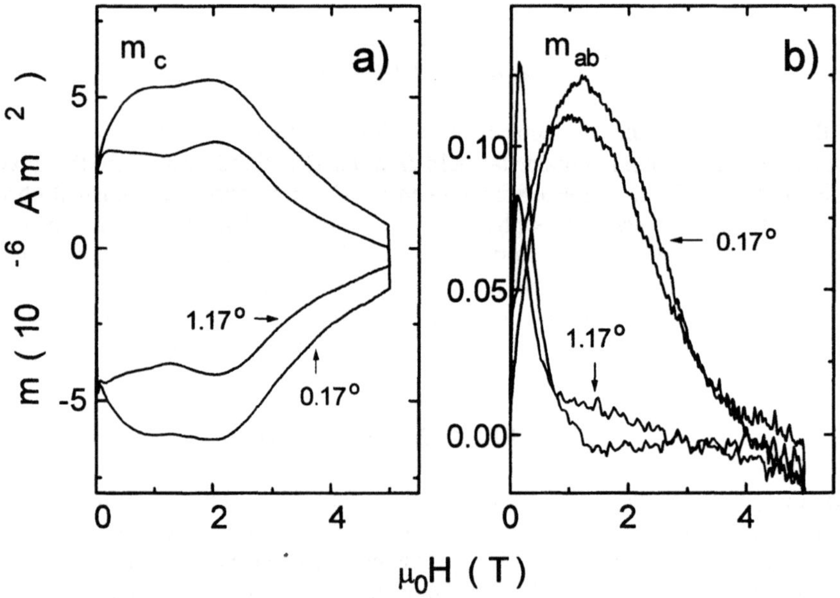

Fig. 1. Magnetic field dependences of the magnetic moments m_c (a) and m_{ab} (b) at 82 K of the twinned crystal OZ for two small angles between $\boldsymbol{H}$ and the TB plane.

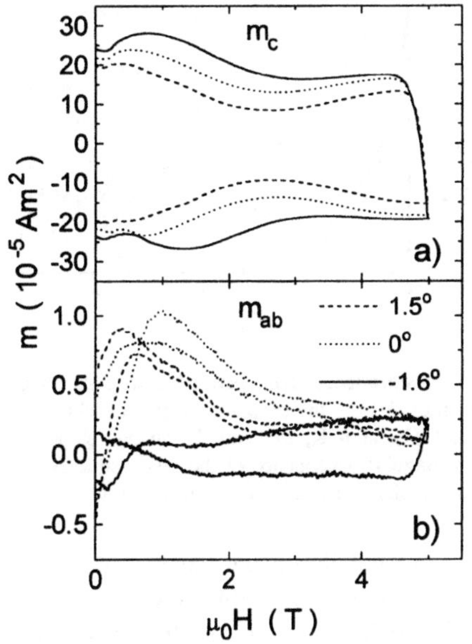

Fig. 2. The field dependencies of the m_c (a) and m_{ab} (b) magnetic moments in the sample AH with irradiation columns for different directions of the magnetic field with respect to the c-axis.

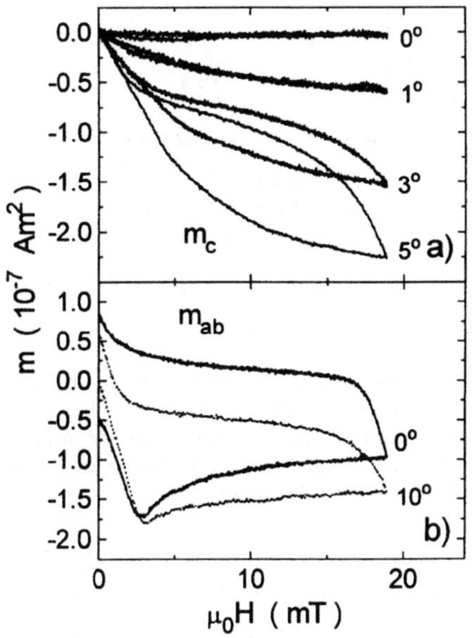

Fig. 3. The field dependencies of the m_c (a) and m_{ab} (b) magnetic moments for the sample VG. Here the angles are measured with respect to *CuO*-plane direction.

and magnetic field $\varphi - \varphi_B \sim M_{ab}$. On the other hand for the fields close to *ab*-plane, this misalignment is proportional to M_c magnetisation. Consequently the vortex lock-in phenomena near c-axis (TB, columns) should be analysed from the magnetic moment m_{ab} whereas vortex localisation by CuO-planes is described by m_c moment.

We have found that for all correlated pinning centres studied (TB (Fig.1), columns (Fig.2) and *CuO*-planes (Fig.3)), the magnetic moment parallel to the centre is highly irreversible whereas transverse magnetic moment is dominated by the reversible component. Such behaviour originate from the different origin of these moments. The transverse magnetisation results from $\boldsymbol{B}$ and $\boldsymbol{H}$ non-collinearity in the locked and kinked vortex phases. It should be essentially reversible because this states are thermodynamically equilibrium. On the contrary the parallel magnetic moment is generated the shielding currents that have opposite direction for the input and output of the magnetic field.

The irreversibility in higher fields seen for localisation near *CuO*-planes, can be related to the geometrical origin from the thin plate-like shape of the sample. Quite generally, the critical state set up in a superconducting plate has its (irreversible) magnetic moment $\boldsymbol{m}$ closely parallel to the plate normal $\boldsymbol{n}$ (in our case the crystalline *c*-axis), even when the applied field $\boldsymbol{H}$ is at large angles to $\boldsymbol{n}$ [8]. However when $\boldsymbol{H}$ is nearly parallel to the plane, within an angle ~ $\theta_c = t/w$ where t is the plate thickness and w its width, the irreversible magnetic moment generated by shielding currents starts to rotate very fast.

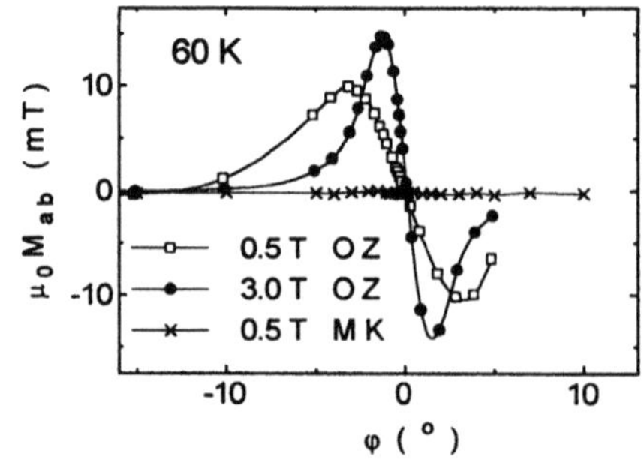

Fig. 4. The reversible magnetisation M_{ab} of the OZ crystal at 60 K when the magnetic field is rotated away from the *c*-axis (φ=0) perpendicular to the TB planes. For comparison, the behaviour of the det-winned MK crystal is shown also.

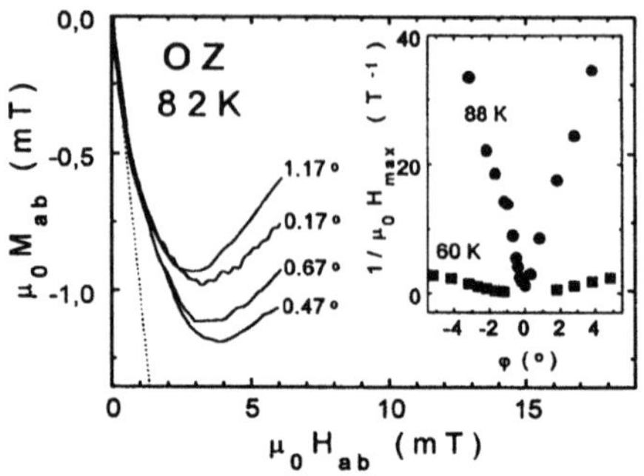

Fig. 5. Dependence of the reversible *ab*-plane magnetisation M_{ab} on the *ab*-plane magnetic field component H_{ab} at 82K; the dotted line has slope -1. Inset: angular dependence of the peak value H_{max}^{-1} (rotation plane perpendicular to the majority TB).

W Further on we consider the phenomena in more details on the basis of the vortex lock-in induced by TB. Because in this case the magnetic moment parallel to the *ab*-plane, m_{ab} , shows nearly reversible behaviour, hereafter we refer only to the reversible component of this moment. For a given field, both twinned crystals show maxima in m_{ab} as a function of φ (Fig.4). With increasing magnetic field this peak becomes sharper and shifts closer to the *c*-axis. At somewhat larger angles, m_{ab} becomes small, and thereafter hardly changes until the direction of $\boldsymbol{H}$ approaches the *ab*-plane. Both de-twinned crystals have vanishingly small m_{ab} (Fig.4), which supports the connection of this transverse magnetic moment with TB.

Theory [2] predicts that when φ is less than the critical locking angle φ_L, vortices are locked to the TB, so that as $\boldsymbol{H}$ is tilted out of the TB plane (rotating from the crystalline *c*-axis toward the *ab*-plane), $\boldsymbol{B}$ remains aligned in the TB plane. At larger angles $\varphi_L<\varphi<\varphi_T$ there is a kinked vortex structure. The tilted vortex segments make an angle φ_T with the TB, and link segments that remain locked to the TB plane. With increasing angle φ, the length of the locked segments decreases and vanishes at the trapping angle $\varphi=\varphi_T$. Finally, at large angles $\varphi>\varphi_T$, the TB cease to influence the vortex structure, so that M_{ab} falls to the (small) thermodynamic magnetisation. The behaviour of $M_{ab}(\varphi)$ shown in Fig.4 corresponds qualitatively to theory, which predicts increasing misalignement between $\boldsymbol{B}$ and $\boldsymbol{H}$ up to the locking angle, and appearance of the collinear phase at the trapping angle. Consequently the low-field and low-angle regime should correspond to the phase with vortices locked into the TB planes. For the locked state the transverse induction is zero. Therefore the transverse magnetisation should compensate the transverse magnetic moment. Fig. 5 shows a quantitative test of perfect shielding of the in-plane field component (H_{ab}) that is a signature of the locked vortex state. When plotted against H_{ab}, the initial slope of the in-plane magnetisation M_{ab} of crystal OZ is indeed close to -1 (note that demagnetising effects are insignificant for this direction). At higher fields, M_{ab} drops well below the full-shielding value. Even within a locked vortex state, a number of factors can suppress the shielding: (i) the presence of TB with another direction; (ii) penetration of vortices into the bulk away from the TB; (iii) ther-

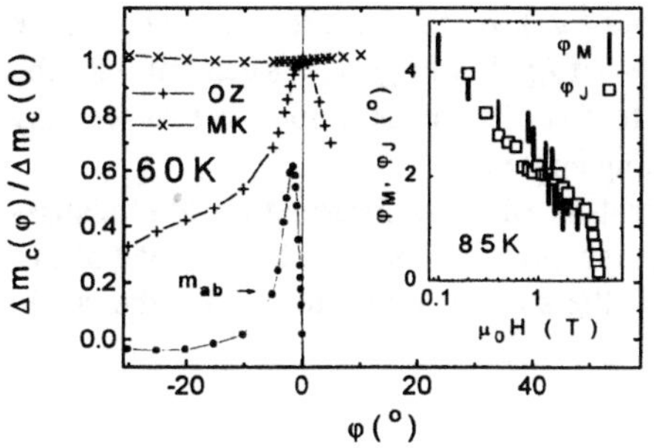

Fig. 6. The angular dependences of the irreversible magnetic moment Δm_c for the OZ (twinned) and MK (de-twinned) crystals at 60 K at constant $\mu_0 H_c$ ($=\mu_0 H \sin\varphi$) of 1.5 T. For comparison of the angular dependences, the reversible m_{ab} of crystal OZ is included too (arbitrary units). Inset: dependence of the characteristic angles φ_J and φ_M (see text) on field at 85 K, where the irreversibility field of ~4 T is within the experimental window.

mally-activated vortex tilt. Consequently, it is not surprising to observe a continuous and gradual reduction in the slope (dM_{ab}/dH_{ab}), rather than a discontinuity that would indicate a sharp transition from locked to kinked states.

Theory [1,2] predicts maximum misalignment between ***B*** and ***H*** at the transition from locked to kinked states, and that the position of this transition is determined by H_{ab} alone; consequently it is reasonable to identify the peak of m_{ab}(H) with this transition. For low fields (which correspond to larger angles), at a fixed temperature the maximum in m_{ab} indeed occurs at a value of the transverse field component H_{ab} that is to a first approximation independent of φ (to Fig.5). Because $H_{ab} \approx H\varphi$ this corresponds to a linear dependence of the inverse of the applied field H_{max} vs. φ (inset to Fig. 5).

The influence of the TB on the shielding currents j_s (measured by the *irreversible* part of the magnetic moment Δm_c) that circulate in the *ab*-plane is shown in Fig.6; to take account of the intrinsic angular dependence of the parameters, the data are plotted at fixed *c*-component of the field $H_c = H\sin\varphi$ [2,3]. The detwinned MK sample with dominant point-like disorder pinning has j_s independent of φ up to ~60°. In contrast, the OZ crystal exhibits a very sharp angular dependence, which flattens within the locking angle range (note that, for the same reasons as may account for the deviation from ideal shielding of H_{ab}, current is not angle-independent in that range). For $\varphi > \varphi_L$, there is a rapid drop in j_s.

As Fig. 6 shows, the angular width of the peak in the *irreversible* screening currents j_s associated with the vortex-TB interaction is close to the width of the peak in the transverse *reversible* moment m_{ab}. We therefore identify this angle as the boundary of the kinked state, which is the trapping angle φ_T; again, the transition is likely to be smeared, because of thermal and other factors.

In the kinked vortex state, a linear dependence of j_s on φ may be expected. The pinning force perpendicular to the TB is much larger than the longitudinal TB pinning force [2]. Therefore the irreversible magnetic moment is governed by the shielding currents flowing transverse to the TB. If bulk pinning can be neglected, [7] the pinning force is that acting on the locked segments of vortex line, so that $F = f_{TB}L(1-\tan\varphi_B/\tan\varphi_T)$, where f_{TB} is pinning force per unit length of locked segment, φ_B defines the average direction of ***B***, and L is the crystal thickness. On the other hand, for fixed j and $|\boldsymbol{B}|$, the Lorentz force component parallel to the *ab*-plane does not change with φ_B. Therefore the expected angular dependence is $j_c(\varphi) = j_c(0)(1-\tan\varphi_B/\tan\varphi_T) \approx j_c(0)(1-\varphi/\varphi_T)$ (we have used the

relation [2] $\varphi_B=\varphi_T(\varphi-\varphi_L)/(\varphi_T-\varphi_L)\approx\varphi$, valid for $\varphi_T>>\varphi_L$). As the inset to Fig. 6 shows, the value of φ_J determined from the maximum slope of $j_s(\varphi)$ is in good agreement with the angle φ_M, the angle at which m_{ab} has dropped to 5% of its peak value, so that we may regard them both as good measures of the trapping angle φ_T.
Behaviour qualitatively similar to shown in Figs.4-6, was also observed for the localisation induced by columns and *CuO*-planes. The new vortex phenomena described above can be useful for application in certain types of the suspension system when the stable direction of the levitating object is required.

ACKNOWLEDGEMENTS

This work was supported by the Russian Foundation for Basic Research Grant No 96-02-18376a, the joint RFBR (Grant 96-02-00235G) and DFG (Grant 436Rus-113/417/O(R)), NATO Linkage Grant No.HT931241, the UK Engineering and Physical Sciences Research Council, and INTAS Grant No. 94-3562.

REFERENCES

1. Feinberg, D. and Villard, C. (1990) *Phys.Rev.Lett.* **65** 919.
2. Blatter, G. et al. (1994) *Rev. Mod. Phys.* **66**, 1125.
3. Zhukov, A.A. et al. (1996) *J.Low Temp.Physics* **105**, 1105; (1997) *Phys.Rev. B***56** 3481.
4. Zhukov, A.A. et al. (1997) *Physica C***282-287** 2151.
5. Zhukov, A.A. et al. (1997) *Physica C***282-287** 2155.
6. Wolf, Th. et al. (1989) *J. Cryst. Growth* **96**, 1010.
7. Zhukov, A.A. et al. (1995) *Phys.Rev. B***52**, R9871.
8. Zhukov, A.A. et al. (1997) *Phys.Rev. B***56** 2809.

MAGNETO-OPTICAL CHARACTERISATION OF $Bi_2Sr_2CaCu_2O_x$ SINGLE CRYSTALS

M. ZAMBONI[♥], S. KOISHIKAWA and M. MURAKAMI
ISTEC-SRL, Div. VII
1-16-25 Shibaura, Minato-ku Tokyo 105 Japan
G.YANG and J.S. ABELL
University of Birmingham, Birmingham B15 2TT UK
[♥] Also at *University of Birmingham*

The flux density distribution in $Bi_2Sr_2CaCu_2O_x$ (Bi-2212) crystals with different annealing conditions has been investigated by magneto-optical (M-O) technique using an iron garnet film with in plane magnetisation. An unusual flux dynamics have been visualised. The vortices move across the sample preferentially along diagonal lines. The distance between the lines changes with the processing conditions. Although the sample's weak-links can be clearly visualised in the M-O images, these lines can not be associated directly to any visible defect. We found that these particular flux dynamics can be associated with changes in the anisotropy factor γ, as revealed by the magnetisation loops.

1. Introduction

The structure of the flux lattice and its evolution with the applied magnetic field in superconductor materials is of considerable importance both in terms of fundamental research and potential applications. In layered high T_c superconductors many types of structural defects can affect the flux behaviour since they may act as pinning centres or weak-links, depending also on the temperature and on the values of the external magnetic field. The M-O technique [1], based on the Faraday rotation in an indicator film, is a very powerful method to analyse the flux density distribution and its correlation with the sample microstructure since it can provide detailed information of local flux dynamics at any temperature of interest.
In this work we present the flux distributions obtained in Bi-2212 single crystals with different oxygen annealing using the M-O technique The results are correlated with those obtained by the analysis of magnetisation hysteresis measurements.

2. Experimental

The Bi-2212 samples were cut from a crystal grown by the large temperature gradient technique described in detail elsewhere [2]. Some crystals were investigated as-grown, others were annealed in air or in fluent oxygen. To avoid the size effect on the flux

G. Van Tendeloo et al. (eds.), High-Temperature Superconductors and Novel Inorganic Materials, 189–192.

dynamics [3], all the observed samples have comparable dimensions, approximately $1\times0.5\times0.05$ mm^3. The magnetisation curves were measured at different temperatures using a vibrating sample magnetometer VSM Oxford 3001.
The M-O flux observations were performed using the facility described in ref 4. The magnetic flux distribution in the crystals was visualised using a ferromagnetic garnet indicator (I-G) film with in-plane magnetisation, placed directly on top of the sample surface. In this kind of indicator film, the local magnetic field value and its changes are imaged as colour maps. In the black and white photographs presented in this paper that results in a grey scale image in which the brighter intensity corresponds to the higher value of the magnetic field. The external magnetic field was applied perpendicular to the a-b plane up to 3000 Oe at different temperatures. Firstly, after zero-field-cooling (ZFC) to the temperature of interest, the external field was gradually increased to visualise the flux penetration, then slowly it was decreased to observe the flux trapping mechanisms.

3. Results and discussion

Figures 1, 2 and 3 show the M-O images of the magnetic induction development in the a-b plane of three differently annealed Bi-2212 crystals. The external magnetic field H is applied perpendicular to this surface, that is parallel to the c-axis, after ZFC to 20 K.
It can be clearly observed that the flux moves across the samples mainly along diagonal lines. These lines can not be related to any visible defects in the crystals. It should be also noted that these samples are expected to be free of twins.

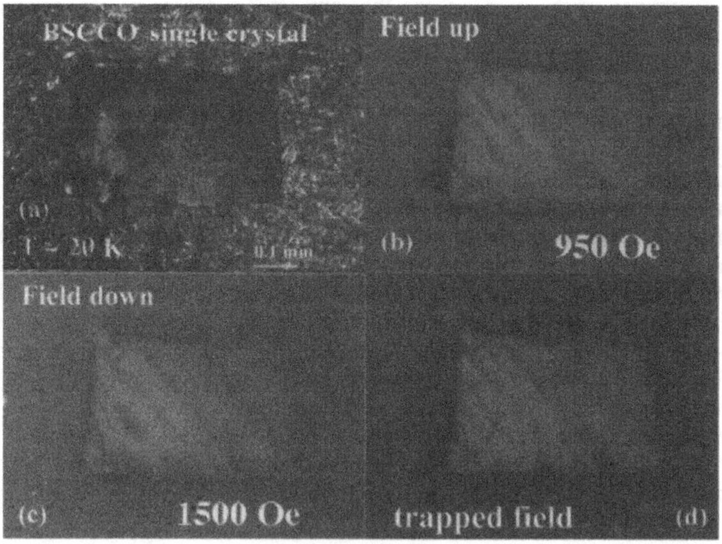

Figure 1. Magnetic induction development in the a-b plane of a Bi-2212 crystal annealed in oxygen. (a) sample surface without I-G film. (b) flux penetration increasing H. (c) flux trapping decreasing H and remnant state in (d).

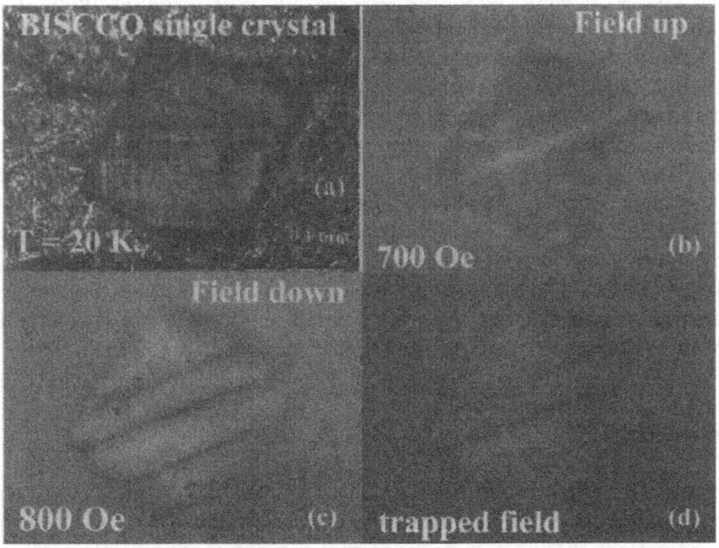

Figure 2. Magnetic induction development in the a-b plane of a Bi-2212 crystal annealed in air. (a) sample surface without I-G film. (b) flux penetration increasing H. (c) flux trapping decreasing H and remnant state in (d).

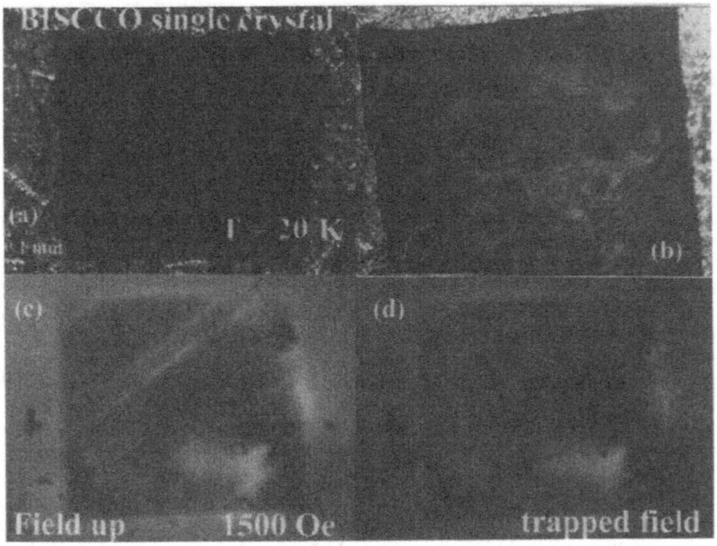

Figure 3. Magnetic induction development in the a-b plane of a Bi-2212 crystal as-grown. (a) and (b) optical images of the crystal surface. (c) flux penetration increasing H. (d) remnant state after turning off a field of 3000 Oe.

Since no visible defects can be ˙related to these penetration lines, one possible explanation [5] can be the presence of sub-grain boundaries, formed during the crystal growth. Nevertheless, according also with the VSM measurements [6], we believe that this particular flux dynamics is more likely to be connected to the anisotropy of the samples, rather than to some structural defects.
The variation of the anisotropy factor γ results in changes in the surface barrier values [7]. In our samples, the VSM curves show that γ can be affected by the annealing conditions, hence the surface barrier values will be different. In less anisotropic materials, such as YBCO or NdBCO [8, 9] this diagonal-type behaviour has not been osberved. On the contrary, it is common in our samples, but the distance between the lines changes in the different crystals. We have found, as will be discussed in detail in a separate publication, that the lines become closer in the crystals where, according to the VSM measurements, the anisotropy factor γ is expected to be smaller.

4. Conclusions

The magnetic flux dynamics in Bi-2212 crystals with different annealing conditions has been investigated both by integral magnetisation and by direct optical methods. The M-O investigations show a very peculiar flux behaviour. The vortices move across the sample along diagonal lines with no visible weak-links associated with this direction. A comparative analysis of the magnetisation curves and the M-O images suggests a relation between flux behaviour and processing conditions through the variation of the anisotropy parameter γ and the surface barrier.

ACKNOWLEDGEMENTS
This work was partially supported by NEDO for R&D of Industrial Science and Technology Frontier Programme.

5. References

1. Koblischka, M.R. and Wijngaarden, R.J. (1995) Magneo-optical investigations of superconductors. *Supercond. Sci. Technol.* **8**, 199-213
2. Yang,G. Sutton,S.D. Shang,P Jones,L.P. Gough,C.E. and Abell,J.S. (1993) Growth and microstructure in $Bi_2Sr_2CaCu_2O_y$ single crystals, *IEEE Trans. Appl. Superconduc.* **3**,1663-1666
3. Yang,G. Abell,J.S. and Gough,C.E. (1997) Effects of a geometrical length scale on the arrowhead magnetisation of $Bi_2Sr_2CaCu_2O_y$ single crystals, *Physica C* **282-287**, 2007-2008
4. Zamboni, M. Yoo, S.I. Higuchi, T. Waki, K. Koishikawa, S. and Murakami, M. (1997) Magneto-optical characterisations of flux dynamics in Nd-Ba-Cu-O bulk superconductors. *Physica C* **281**, 218-227
5. Indembom, M.V. van der Beek, C.J. Berseth, V. Wolf, T. Berger, H Li, T.W. and Benoit, W (1996) Morphology og growth of $Bi_2Sr_2CaCu_2O_8$ *Proc. of MOS'96*, Karlsruhe (Germany).
6. Yang,G. Abell,J.S. and Gough,C.E. (1994) Magnetic identification of 3D to 2D decomposition transition in BSCCO single crystals, *Proc. of 7th Int. Workshop on Jc*, Austria 264-267.
7. Chikumoto,N. Konczykowski,M. Motohira,N. and Malozemoff,A.P. (1992) Flux-creep cross over and relaxation over surface barrier in BSCCO crystals, *Phys. Rev. Lett.* **69**, 1260-1263
8. Dorosinskii,L.A. Indembom,M.V. Nikitenko,V.I. Polyanskii, A. and Prozorov,R.L. (1993) Magneto optic observation of Meissner effect in YBCO single crystals *Physica C* **206**, 360-366
9. Zamboni, M. Koishikawa, S. Mochida, T and Murakami, M. (1997) Magneto-optical characterisation of NdBCO single crystals. *Proc. of 3rd Int. Workshop on Superconductivity*, Hawaii (USA), 355-356

EFFECT OF Pb AND Sr SUBSTITUTIONS ON IRREVERSIBILITY AND FISHTAIL LINES OF Hg-1223 PHASE: POWDER, CERAMIC AND SINGLE CRYSTAL STUDIES

S.R.LEE, T.AKAO, H.SUEMATSU, H.YAMAUCHI
Materials and Structures Laboratory, Tokyo Institute of Technology, Yokohama 226, Japan
N.P.KIRYAKOV, M.S.KUZNETSOV, D.A.EMELYANOV, Yu.D.TRETYAKOV
Department of Chemistry, Moscow State University, Moscow 119899, Russia

1. Introduction

$HgBa_2Ca_2Cu_3O_y$ phase (Hg-1223) possess the highest values of critical temperature T_c = 134 K [1], but still does not considered as an attractive candidate for application in high magnetic fields. The main reason attributed to the relatively weak magnetic flux pinning characteristics of this phase. It was shown that irreversibility line (IL) of the Hg-1223, which set the upper limit for application of this compound in the H-T phase space, located between those of Y-123 and Bi,Tl-2212 phases [2], but still lower than that for Tl-1223. The irreversible properties of HTS can be essentially improved by chemical substitutions and Sr-containing Hg-1223 phases look very attractive for this purpose [3]. Obviously, the substitution of Ba by smaller Sr cation in $HgBa_2O_{2-\delta}$ «insulating» block should provide stronger coupling between $Ca_2Cu_3O_6$ «conducting» blocks and decrease anisotropy of Hg-1223 which is important for the increasing of irreversibility fields (H_{irr}) [4]. However the synthesis of Ba-free Hg-1223 was reported only under high pressure. Moreover, in order to stabilize such compounds, the partial substitution of Hg by Re, Mo or Pb is used [3,5].

In the present paper we report the synthesis of $(Hg,Pb)(Sr,Ba)_2Ca_2Cu_3O_z$ powder, ceramic and single crystal samples by an encapsulation and a high-pressure techniques using homogeneous precursors and influence of the substitutions on superconducting properties, irreversibility and fishtail lines.

2. Experimental

Precursor powders with a different ratio of Pb:Ba:Sr:Ca:Cu cations were synthesized by spray drying of the nitrates solutions and further decomposition of the obtained salt mixtures in a dynamic vacuum at 600-700 °C. After mixing with HgO the compacted pellets were placed into Al_2O_3 crucible and sealed in an evacuated quartz tube. Heat

G. Van Tendeloo et al. (eds.), High-Temperature Superconductors and Novel Inorganic Materials, 193–198.

treatment of the ampoule was performed at 850-870°C for 10-20 h. The details of the ampoule synthesis were described previously [6]. Single crystals with a size of 200x200x20 μm were isolated from the bulk samples obtained by intimate control of oxygen content in the precursors and green density of the pellets used for heat treatment in the sealed quartz ampoules. The samples obtained by the encapsulation technique were utilized as a precursors for high pressure (HP) experiments. Ceramic sample was grounded, pressed in a gold capsule and placed in an assembly in which NaCl used as the pressure media. Cubic-anvil-type apparatus was used for experiment at 5 GPa and 1000°C for 1 h. Finally the sample was quenched to room temperature. Post annealing was performed in a thermobalance in O_2, Ar or Ar/H_2 flow at 300-350°C for 10-20 hours. Magnetization measurements were carried out using 5 T SQUID magnetometer (MPMS-5, Quantum Design). Irreversibility lines were determined for the grain-aligned powdered samples in the H // c geometry and for ceramic samples with random orientation of crystallites. Fishtail effect was studied on the ceramic and single crystal samples. The single crystals were directly glued on the holder and magnetic measurements were performed in H//c and H//*ab* geometry.

3. Results and discussions

Recently we have reported the synthesis of $Hg_{1-x}Pb_xBa_{2-y}Sr_yCa_2Cu_3O_z$ compounds (x=0.0-0.5 and y=0.0-2.0) by encapsulation technique [6,7] with the T_c values of 110-130 K depended on the Pb and Sr content. Here we present the results of irreversibility lines (IL) study of these compounds obtained on the grain-aligned samples (fig.1).

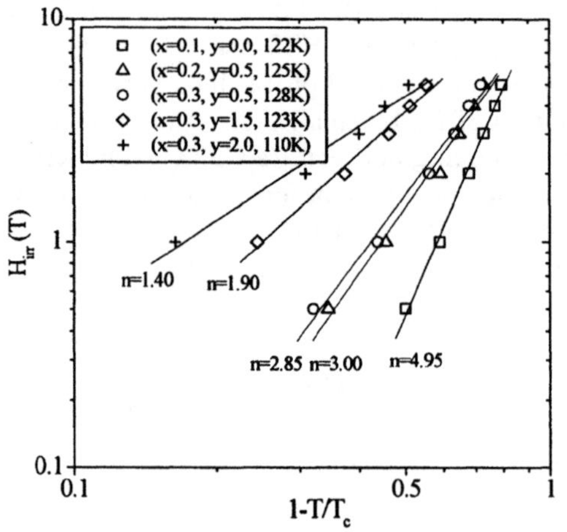

Figure 1. The irreversibility lines for grain-aligned $Hg_{1-x}Pb_xBa_{2-y}Sr_yCa_2Cu_3O_z$ powders (H II c).

It is clear that IL of the Hg-1223 samples shifted to a higher position with the substitutions of Hg by Pb and Ba by Sr. The dependence of H_{irr} was described as a power law $(1-T_{irr}/T_c)^n$ with the n varied from 1.4 to 4.95. For the Sr-free (HgPb)-1223 sample the n=4.95 was close to Bi(Tl)-2212 (n=5.5) and for Ba-free one the n=1.4 became similar to that reported for Y-123 (n=1.3-1.5) that can be explained by change of mechanism of flux motion from 2D independent pancake-like vortices to 3D vortex lines respectively. Such speculations are consistent with the results of lattice parameters study [6] which gradually decreased with the substitutions of Hg by Pb and Ba by Sr in the Hg-1223 phase that provide stronger coupling between CuO_2 planes and reduce anisotropy of substituted Hg-1223 crystals. It is important to notice that IL of the samples with the nominal compositions $Hg_{0.7}Pb_{0.3}Sr_{1.5}Ba_{0.5}Ca_2Cu_3O_z$and $Hg_{0.7}Pb_{0.3}Sr_2Ca_2Cu_3O_z$ located much higher than that reported for grain-aligned and single crystal samples of conventional HgBa-1223 phase [8] and that with the substitutions of Hg by Pb or Bi [9,10]. Such compounds could be the promising candidates for application in the high magnetic fields.

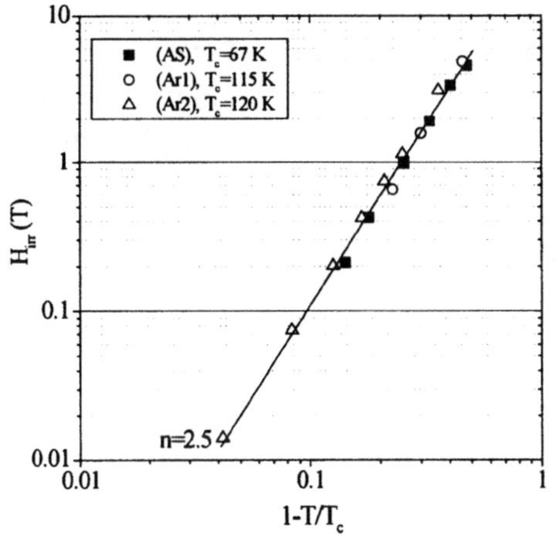

Figure 2. Irreversibility lines of $Hg_{0.7}Pb_{0.3}Sr_{1.5}Ba_{0.5}Ca_2Cu_3O_z$ ceramic sample prepared by high-pressure technique: (AS)-as-prepared, (Ar1)- after annealing in Ar gas flow at 300°C for 20 h, (Ar2)-after annealing in Ar at 350°C for 20 h.

The effect of doping, due to the different oxygen content, on the IL was studied on the ceramic samples with the nominal composition of $Hg_{0.7}Pb_{0.3}Sr_{1.5}Ba_{0.5}Ca_2Cu_3O_z$. Utilization of highly oxidized powder, prepared by ampoule method, as a precursor for HP experiment allowed us to obtain strongly overdoped samples with T_c of 67 K. After post-annealing in the Ar gas flow at 300 °C for 10 hours the sample with T_c=115 K was obtained. Annealing at 350°C for 10 hours resulted to the sample with T_c=120 K and further treatment in Ar for 20 hours does not changed the T_c value.
The IL were determined using criterion $j_c=30\Delta M/d=10^3 A/cm^2$ calculated by Bean's model for plate-like grains with d=10 μm (fig.2). The same piece of the ceramic sample was used for all the annealing experiments and magnetic measurements. It was found that location of the IL does not depends strongly on the post-annealing treatments. The different results were reported for over-, optimally- and underdoped samples of the

conventional Hg-1223 phase obtained by HP technique which showed a strong doping dependence of IL [11]. However almost similar to our data were reported for overdoped Tl(BaSr)-1223 crystals where position of IL does not depend much on the oxygen content controlled by annealing in reduced (Ar/H_2) atmosphere while for the Tl-2223 such variation in the location of IL is always exists [12].

The analysis of magnetization data in the high magnetic fields showed the significant variation in the shape of the M-H loops for samples prepared by different methods. We found the "fishtail"- or peak-effect (PE) for the ceramic samples prepared by encapsulation technique. It was reported that "fishtail" effect usually observed for the crystals, but no such feature has been observed for the ceramic samples, grain-aligned or random oriented powders [10]. However in our experiments the clear peak effect was observed in ceramic and single crystal samples as well. The typical M-H data for single crystal samples with nominal composition of $Hg_{0.7}Pb_{0.3}Sr_{1.5}Ba_{0.5}Ca_2Cu_3O_z$ at 60 K are shown in fig.3.

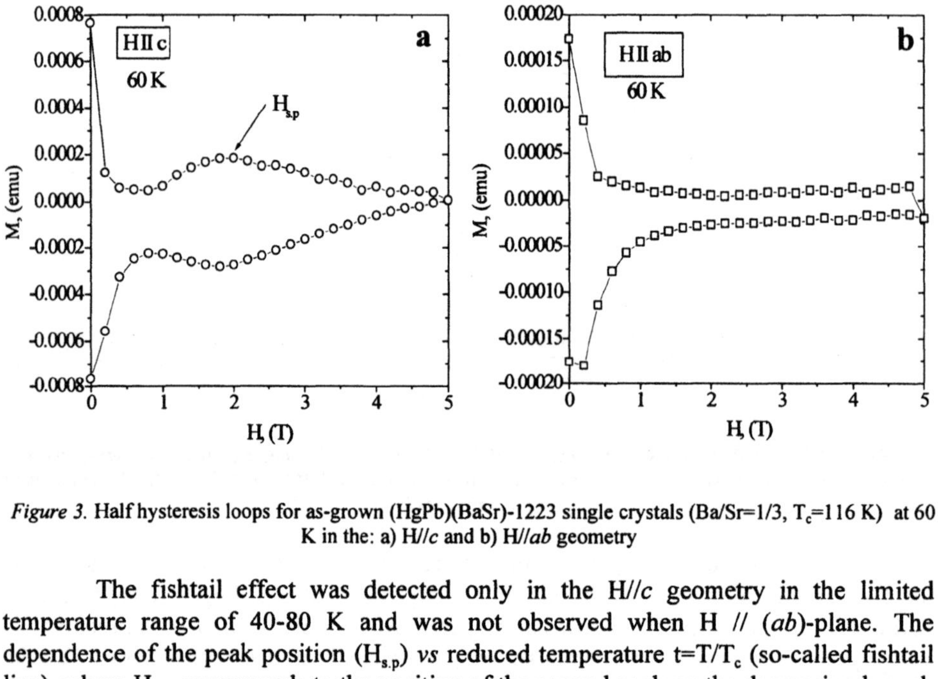

Figure 3. Half hysteresis loops for as-grown (HgPb)(BaSr)-1223 single crystals (Ba/Sr=1/3, T_c=116 K) at 60 K in the: a) H//*c* and b) H//*ab* geometry

The fishtail effect was detected only in the H//*c* geometry in the limited temperature range of 40-80 K and was not observed when H // (*ab*)-plane. The dependence of the peak position ($H_{s.p}$) *vs* reduced temperature $t=T/T_c$ (so-called fishtail line), where $H_{s.p.}$ corresponds to the position of the second peak on the decreasing branch of the hysteresis loop are shown in fig.4 for the ceramic samples with different Ba/Sr ratio. The position of fishtail line strongly depends on composition and correlated with the electronic anisotropy (γ) of studied compounds, obey the formula $H_{s.p.}=Af(t)\gamma^{-2}$ that was reported previously for Tl-based single crystals [13]. Therefore the higher location of the fishtail line corresponds to the Ba-free HgPb-1223 sample that has a lower γ value. In fig.5 our results of the fishtail line study for the single crystals presented in comparison with the published data for different Hg–1223 crystals. Obviously the $H_{s.p}(t)$

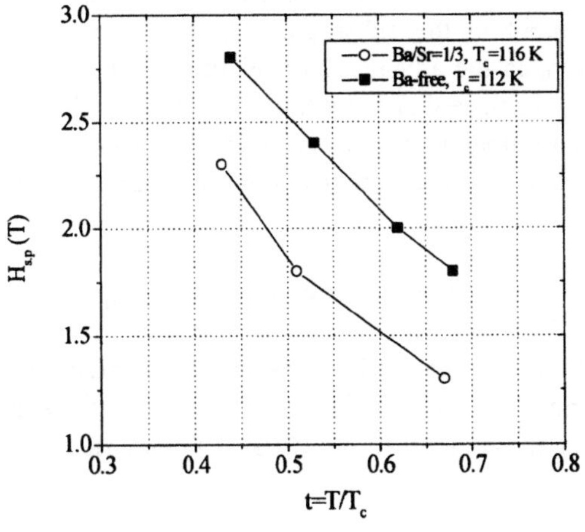

Figure 4. Fishtail lines for (HgPb) -1223 ceramic sample with Ba/Sr =1/3 and Ba-free one.

fishtail line position for as-prepared (HgPb)(BaSr)-1223 crystals stands higher then that reported for others Hg-1223 crystals [8,11]. Moreover the further comparison of the fishtail line with those reported for different Tl-based single crystals [13,14] confirmed the advantage of our samples. The annealing experiments with our crystals showed that they are overdoped in as-prepared state and T_c increased from 116 to 122 K after

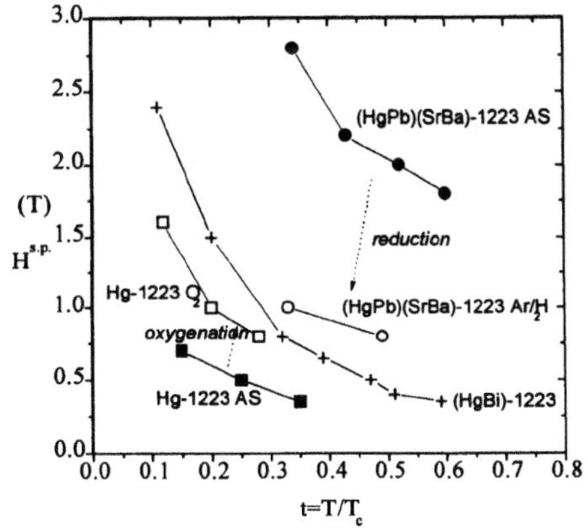

Figure 5. Fishtail lines of as-grown (AS, T_c=116 K) and reduced (Ar/H_2, T_c=122 K) (HgPb)(SrBa)-1223 single crystals [our results] compared with those of as-grown Hg-1223 (AS, T_c=105 K), Hg-1223 after oxygen annealing (O_2, T_c=133 K) [ref.8] and optimally doped (HgBi)-1223 (T_c=130 K) single crystals [ref.10].

annealing in Ar/H_2 gas mixture at 300°C for 20 hours. After such treatment the location of the fishtail line shifted to the low field region and became close to that for optimally doped (HgBi)-1223 crystals with T_c=130 K.

It was reported previously, for the conventional Hg-1223 single crystals that fishtail line for optimally doped sample with the T_c=133 K stands higher than that for as-grown, underdoped one with the Tc=105 K [8] Our results confirmed the strong influence of oxygen content in the substituted Hg-1223 crystals on the location of fishtail line and demonstrated the advantage of overdoped crystals. However the origin of the defects responsible for the enhanced pinning at a intermediate field is not clear yet for our samples. Local inhomogeneity in oxygen, Hg/Pb or Ba/Sr distribution, could be the possible reasons and more detailed studies of these samples are in progress.

Acknowledgments

S.Lee would like to thank the JSPS Postdoctoral Fellowship Program and Monbusho Grant in Aids for the financial support of his research in Japan. N.Kiryakov and D.Emelyanov are grateful to International Science Foundation for providing the scholarships in 1997 and 1998 years. The work was partially supported by Russian Scientific Council for Superconductivity and Russian Foundation for Basic Research.

4. References

1. Schilling, A., Cantoni, M.J., Guo, D., Ott, H.R. (1993) Superconductivity above 130 K in the Hg-Ba-Ca-Cu-O system, *Nature* **363**, 56-58.
2. Welp, U., Crabtree, G.W., Wagner, J.L., and Hinks, D.G. (1993) Flux pinning and the irreversibility lines in the $HgBa_2CuO_{4+\delta}$, $HgBa_2CaCu_2O_{6+\delta}$ and $HgBa_2Ca_2Cu_3O_{8+d}$ compounds, *Physica C* **218**, 373-378.
3. Chamaissem, O., Jorgensen, J.D., Yamaura, K., Hiroi, Z., Takano,M., Shimoyama, J., Kishio K. (1996) Crystal structures of Hg-Sr-Ca-Cu-O superconductors with enhanced flux pinning: $Hg_{1-x}Re_xSr_2Ca_{n-1}Cu_nO_{2n+2+\delta}$ (n = 2,3; x about 0.2-0.25), *Phys. Rev. B* **53**, 14647-14655.
4. Kim, D.H., Gray, K.E., Kampwirth, R.T., Smith, J.C., Richeson, D.S., Marks, T.J., Kang, J.H., Talvacchio, J., Eddy M. Effect of Cu-O layer spacing on the magnetic field induced resistive broadening of high-temperature superconductors, (1991) *Physica C* **177**, 431-437.
5. He, Z.H., Lin, Q.M., Gao, L., Sun, Y.Y., Xue, Y.Y., Chu, C.W. (1995) High-pressure synthesis of single-phase superconductors $Hg_zSr_xBa_{2-x}Ca_2Cu_3O_{8+\delta}$ and (Hg, M)$Sr_2Ca_2Cu_3O_y$ with M =Pb and Mo, *Physica C* **241**, 211-218.
6. Lee, S., Kuznetsov, M., Kiryakov, N., Emelyanov, D., Tretyakov, Yu. (1997) Synthesis and properties of $Hg_{0.7}Pb_{0.3}(BaSr)_2Ca_2Cu_3O_z$ superonductors, *Physica C* **290**, 275-280.
7. Lee, S.R., Petrykin, V.V., Kakihana, M., Tretyakov, Yu.D. (1997) Synthesis and raman characterization of $Hg_{1-x}Pb_xBa_{2-y}Sr_yCa_2Cu_3O_z$ superconductors, *Key Engineering Materials* **132-136**, 1257-1260.
8. Maignan, A., Putilin, S.N., Hargy, V., Simon, Ch., Raveau, B. (1996) Magnetic study of $HgBa_2Ca_2Cu_3O_{8+x}$ single crystals: effect of doping on irreversibility and fishtail lines,*Physica C* **266**, 173-177.
9. Isawa, K., Higuchi, T., Machi, T., Tokiwa-Yamamoto, A., Adachi, S., Murakami, M., Yamauchi, H. (1994) Irreversibility line for a Pb-doped Hg-Ba-Ca-Cu-O superconductor,*Appl. Phys. Lett.* **64**, 1301-1303
10. Pelloquin, D., Hardy, V., and Maignan, A. (1996) Synthesis and characterization of single crystals of the superconductors $Hg_{0.8}Bi_{0.2}Ba_2Ca_{n-1}Cu_nO_{2n+2-\delta}$, *Phys.Rev.B* **54**, 16246-16253.
11. Fujinami, K., Suematsu, H., Karppinen, M., and Yamauchi, H., (1997) Effect of overdoping on the irreversibility field and critical current density of the $HgBa_2Ca_2Cu_3O_{8+z}$ superconductor, to be published in *Phys.Rev.B.*
12. Wahl, A., Maignan, A., Hardy, V., Provost, J., Simon, Ch., Raveau B. (1995) Magnetic study of an overdoped "1223" system: properties of $TlBa_{0.8}Sr_{1.2}Ca_2Cu_3O_{9-\delta}$, *Physica C* **244**, 341-348.
13. Hardy, V., Wahl, A., Ruyter, A., Maignan, A., Martin, C., Coudrier, L., Povost, J., Simon, Ch. (1994) The fishtail effect in different Tl based single crystals. A possible interplay with the electronic anisotropy, *Physica C* **232**, 347-358.
14. Zeng, D.N., Johnson, J.D., Jones, A.R., Campbell, A.M., Liang, W.Y., Doi, T., Okada, M, Higashyama, K. (1995) Magnetic and transport measurements of Tl-1223 superconductors,*J.Appl.Phys.* **77**, 5287-5292.

CATION SUBSTITUTION, SECOND PHASE PRECIPITATION, SUPERCONDUCTIVITY, AND FLUX PINNING IN $LRE_{1+x}Ba_{2-x}Cu_3O7$

R. W. McCallum[*#], H. Wu[*#], K. W. Dennis[*], and M. J. Kramer[* #]
[*]Ames Laboratory and [#]Department of Materials Science and Engineering
Iowa State University, Ames, Iowa 50011, U.S.A.

1. Introduction

Unlike $YBa_2Cu_3O_{7+d}$ which forms only a stoichiometric compound, the light rare earth (LRE) elements form a solid solution of $LRE_{1+x}Ba_{2-x}Cu_3O_{7+d}$ (LRE123ss) with increasing substitution of the rare earth for Ba^{2+} as the ionic radii of the rare earth increases. The effects of oxygen partial pressure (Po_2) on the solubility limits of LRE123ss (LRE = Pr, Nd, Sm, Eu, and Gd) were studied by differential thermal analysis, X-ray diffraction and measurement of superconducting transition temperature (T_c). An understanding of the Po_2 and temperature dependence of the solubility limits will not only provide guidelines for obtaining high T_c in LRE123ss but also clarify the flux pinning mechanism in these systems.

The precipitation of second phases from a uniform solid solution is an effective method of introducing a uniform distribution of fine precipitates which may pin flux on a very fine scale. This process would appear to be particularly attractive in making bulk monolithic pieces for such applications as magnetic bearings since these are large domains of single crystal and grain boundaries do not play the determining role in the critical current density. Unfortunately in the $REBa_2Cu_3O_{7-d}$ (RE123) [*] superconductors (RE = rare earth elements), the most commonly investigated material Y123 is a line compound with respect to non-oxygen atoms. Except for Ce, Pr, Pm and Tb, the rare earth elements can be substituted for Y in $YBa_2Cu_3O_{7-d}$ (Y123) without destroying superconductivity ($T_c \approx 90K$). While 123 phase does not exist for Ce and Tb and has not been synthesized for radioactively unstable Pm, Pr123, although isostructural to Y123, is nonsuperconducting. For the light rare earth elements (LRE, including La, Pr, Nd, Sm, Eu, Gd), an extensive solid solution of $LRE_{1+x}Ba_{2-x}Cu_3O_{7-d}$ (LRE123ss) is formed with partial substitution of RE for Ba. This is attributed to the large ionic radii of these ions which approaches that of Ba. The upper solubility is small for Gd123ss but can be larger than x = 0.6 for larger rare earth elements (La, Pr, Nd and Sm). In air or 100% oxygen, the lower solubility limit for Sm, Eu and Gd is within experimental accuracy x=0, while there is a lower solubility limit of $x > 0$ for La, Pr and Nd.

---

[*] All compounds will be labeled by the first cation and the cation ratio in the order of rare earth, barium and copper, i.e., Gd211 for Gd_2BaCuO_5, 011 for $BaCuO_2$.

G. Van Tendeloo et al. (eds.), High-Temperature Superconductors and Novel Inorganic Materials, 199–204.

The substitution of a LRE^{3+} for Ba^{2+} must be compensated for by the addition of extra oxygen to maintain charge balance [1]. Since extra oxygen is required during the formation of the tetragonal material during high temperature reaction, the limits of solubility might be expected to be a function of Po_2. For the La123ss, there does not appear to be a Po_2 dependence of the solubility limit over a span of four orders of magnitude in Po_2. On the other hand, Pr123ss and Nd123ss show a strong influence of Po_2 on the lower solubility limit. For Pr123ss the lower solubility limit is x = 0.04 for 1 bar Po_2 but this decreases to x = 0.01 at 0.001 bar Po_2 [2]. Nd123ss also shows that the lower solubility limit is decreased in low Po_2 (x from 0.04 to 0) but the upper solubility limit was not changed (x = 0.6) over 3 orders of magnitude in Po_2. For Sm123ss, Eu123ss and Gd123ss, it is found that the upper solubility limit decreases with decreasing Po_2 [3].

For the superconducting compounds, the formation of solid solution causes the superconducting transition temperature (T_c) to depressed for x > 0.1. The simplest assumption for the location of the excess RE^{3+} is that it enters the Ba^{2+} sites in a random fashion. From neutron diffraction data, two facts are known. First the total O content increases to balance the charge of the excess Nd and second, the number of O atoms on the chain sites decrease by approximately 0.5x while the anti-chain O site occupancy increases approximately as x. This suggests that a RE on a Ba site always causes an O to occupy an adjacent O_5 site. In order to maintain charge balance, half of these O are additional atoms incorporated in the structure. The other half are relocated from O_1 sites. Due to the excess charge on the Nd, it is safe to assume that the O_1 sites vacated in this manner are not adjacent to the excess RE. For each occupied O_5 site created, two five-coordinated Cu atoms are produced. These five-fold-sites are similar to the five-fold-sites on the Cu-O plane. If the states arising from the hybridization of the O 2p and Cu states for these sites have slightly higher energy than the plane sites the excess electron from the O will occupy a lower lying state leaving a vacant state associated with the anti-chain oxygen site. As these sites are not periodically occupied, they are more localized than the plane sites and thus make a greatly reduced contribution to the conductivity and superconductivity of the materials as shown by XANES [4].

For the RE123ss which extend to large values of x, T_c may be reduced to zero. In addition, since there is very little driving force for the material to have a constant value of x within the solid solution region so that there may be significant variations in x through out a bulk sample. As T_c is dependent on x, broad superconducting transitions are observed for these materials when prepared using standard methods. For La123ss, a special treatment in oxygen and low Po_2 at temperatures somewhat below the decomposition temperatures results in the highest T_c for x = 0.18 [5]. Low Po_2 processing also gives rise to enhanced T_c in Nd123ss [6].

There has been increasing interest in the light rare earth LRE123 materials since high T_c and large J_c were reported for Nd123 melt-textured in low Po_2 by Yoo et al. [7]. An understanding of the Po_2 dependence of the solubility limits will not only provide guidelines for obtaining high T_c in LRE123ss but also clarify the flux pinning mechanism for large critical current density in the systems. In this paper we will review the solubility limits of Nd123ss, Sm123ss, Eu123ss, and Gd123ss. We will also discuss the ranges where the limits of solid solubility may be varied by changing process parameters.

2. Results

2.1. Nd123ss

The upper solubility limit of Nd123ss has been reported at x = 0.6 to 1.0 while the lower solubility limit has been reported at x = -0.1 to 0.05 [8-10]. Most determinations of the solubility limits are by the appearance of second phases in the XRD patterns. As depending on its composition, structure, size and distribution, a minor second phase may be difficult to detect in XRD, an over-estimate of the solid solution range may occur. On the other hand, if an equilibrium condition is not reached in the experiment, the solid solution range may be under-estimated. Differences in processing conditions (temperature and Po_2) also contributes to the discrepancies in the solubility limits reported in the literature. Currently the commonly accepted value for the upper solubility limit is x ~ 0.6. We find that this value is unchanged for all values of Po_2 for which Nd123 is stable. Since the upper solubility limit extends well into the non-superconducting range, the details of the dependence of the upper limit of the solid solubility as a function of Po_2 were not investigated. From the point of superconductor processing, it is more interesting to investigate whether the lower solubility limit will change with Po_2. Detailed work was conducted on the lower solubility limit of Nd123ss as a function of temperature in 1 bar, 0.2 bar, 0.01 bar and 0.001 bar Po_2 by a combination of DTA, XRD and lattice parameter measurements. The results for the lower solubility limits for Nd123ss for different Po_2 values are shown in Table 1.

Table 1 Lower Solubility limit for $Nd_{1+x}Ba_{2-x}Cu_3O_{7-d}$

Po_2, bar	Temperature, °C	Lower solubility limit x
	1070	0
1	1030	0
	900	0.05
	1050	0
0.2	998	0
	900	0.05
	1010	
0.01	937	0
	900	0
	990	0
0.001	915	0
	900	0

DTA data also provides information about the relative stability of different Nd123ss. The composition for the maximum melting temperature shifts from x = 0.1 in 1 bar Po_2 towards x = 0 with decreasing Po_2. It is also evident that in 1 bar Po_2 the stability changes slowly with x while the stability drops faster with decreasing Po_2. When the melt is cooled below the melting temperature, the composition of Nd123ss will follow the solidus line. Therefore, the composition obtained by liquid phase reactions would be around x = 0.1 in oxygen and the composition shifts to lower x value in a lower Po_2. Since there is a maximum in the stability, the direction of the composition shift with temperature depends on the overall composition.

From the DTA and the lower solubility limit data, a phase diagram of Nd-Ba-Cu-O in 1 bar Po_2 is proposed (Fig. 1). As Nd123 initially processed in low Po_2 will have a stochiometry close to x = 0, the narrowing of the solubility range in 1 bar Po_2 will result in the precipitation of second phases during the oxygen annealing required to convert a large sample from the non-superconducting tetragonal phase to the superconducting orthorhombic phase. The equilibrium second phase is $BaCuO_2$ for temperature above 608°C, and at lower temperatures the equilibrium second phases are $Ba_2CuO_{3.3}$ and $Ba_2Cu_3O_{5+y}$ [11]. Annealing at a low temperature can also produce a fine metastable transition phase. This can explain the enhanced flux pinning observed in Nd123 melt-textured in low Po_2, as discussed more thoroughly by Wu et al.

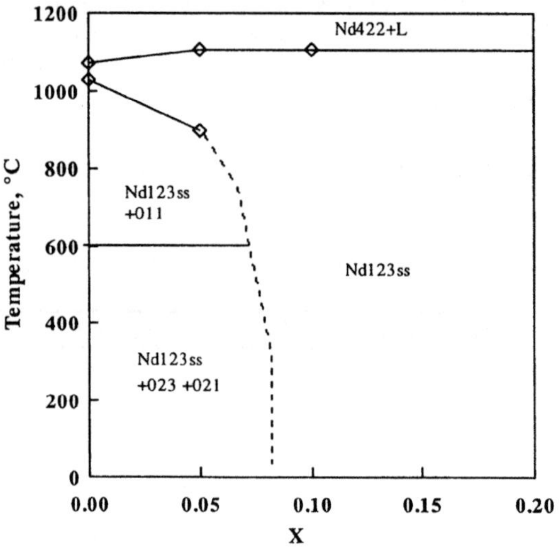

Fig. 1 Phase diagram of Nd-Ba-Cu-O in oxygen

2.2. Sm123ss

In O_2 Sm123ss forms of $x < 0.8$ where a minor peak of Sm201 appears in x-ray diffraction. For samples the upper solubility limit is x = 0.15. In low oxygen partial pressure, not only is the solubility limit decreased from x = 0.7 to x = 0.15, but also the equilibrium phase relations are changed. In oxygen, the phases in equilibrium with

Sm123ss are Sm201 and CuO. In 0.01 bar Po_2 at 950°C, Sm201 is replaced by Sm211 and CuO by Cu_2O. Investigation of the samples treated in low Po_2 in the multiphase region produced some interesting results. There are some nonsystematic changes in the lattice parameters and peak intensities with $x > 0.2$ for the samples treated in 0.01 bar Po_2 [3]. There are also some anomalies in the T_c curves (Fig. 9). For samples treated in oxygen, T_c decreases continuously with increasing x. For samples treated in 0.01 bar Po_2, T_c follows the same pattern up to x = 0.15. At x = 0.2, although the zero field cooled curve is similar to that of the sample with x = 0.15, the field cooled susceptibility becomes nearly zero, indicating possible strong flux pinning. As x increases beyond the solubility limit, we would expect that T_c should remain constant at the value determined by the solubility limit. Surprisingly, T_c increases and reaches an onset temperature of about 92 K. Initially it was suggested that this might be attributed to a new superconducting phase [3]. Later it was recognized that under these conditions the Sm123ss is part of a complex multiphase relationship so that the composition of the Sm123ss formed is not a simple function of the overall composition. As a consequence, neither the lattice parameters nor the T_c of the Sm123ss are well behaved when the overall composition is outside the region of solid solution [12].

2.3. Eu123ss

Eu123ss extends to about x=.6 in O_2. However, the upper solubility limit in 0.01 bar Po_2 is reduced to x = 0.05 as shown by DTA data. This is further verified by the T_c measurements of Eu123ss treated in 1 bar and 0.01 bar Po_2. In 1 bar Po_2, T_c continues to decrease with increasing x. All the samples treated in 0.01 bar Po_2 show the same T_c onset of about 94 K. From DC magnetization measurements it appears that the superconducting fraction decreases with increasing x in 1 bar Po_2 which suggests the presence of a second phase, however, we failed to detect any second phase or phase separation in the samples by XRD at room temperature and the unit cell volume seems to decrease smoothly with x as expected. The reason for the apparent decrease in superconducting volume fraction is not currently known. It may be due to a phase transition at a lower temperature or a charge separation. Normal state magnetization measurements of the samples are entirely consistent with Eu^{3+}. It should be noted that despite the J = 0 Hund's rule ground state of Eu^{3+}, the magnetic susceptibility is non-zero due to the presence of a low lying excited state.

2.4. Gd123ss

The solubility limit of Gd123ss is about x = 0.2 in oxygen. For the samples treated in 0.01 bar Po_2 at 950°C, XRD was used to determine the solubility limit by the disappearing phase method and DTA. The solubility limit was determined to be x = 0.04. T_c measurements provide further evidence of decomposition. For samples treated in 0.01 bar Po_2 at 900°C, T_c becomes independent of x when x > 0.04 since the composition of the superconducting phase has is at the solubility limit. The change of upper solubility limits provides another possible route to introduce flux pinning centers in LRE123ss [12].

3. Conclusions

Unlike Y123 which forms only a stoichiometric compound, the light rare earth elements form a solid solution of $LRE_{1+x}Ba_{2-x}Cu_3O_{7+d}$. For La, Pr and Nd, the lower limit of solid solubility is a function of both temperature and oxygen partial pressure. For Sm, Eu and Gd the upper limit of solid solubility is a function of oxygen partial pressure. The limits of solid solubility can be understood by considering both the energy associated with creating defects on respective sublattices within the RE123ss and the free energy of the phases in equilibrium with the solid solution phase. There are significant compositional regions for various RE123ss where, by changing temperature and/or Po_2 it is possible to move from a homogeneous solid solution to a multiphase mixture. By controlling the manner in which this transition occurs, it is possible to introduce defect which significantly enhance flux pinning and increase J_c of these high temperature superconductors.

4. Acknowledgments

This work was supported by United States Department of Energy under contract No. W-7405-ENG-82.

5. References

1. M.J. Kramer, S.I. Yoo, R.W. McCallum, W.B. Yelon, H. Xie, and P. Allenspach, Physica C, **219**, 145 (1994).
2. M. Park, M.J. Kramer, K.W. Dennis and R.W. McCallum, Physica C, **259**, 43 (1996).
3. M.J. Kramer, H. Wu, K. W. Dennis, B.I. Polzin, D.K. Falzgraf and R.W. McCallum, in Advance in Superconductivity VIII, H. Hayakawa and Y. Enomoto eds., p385, Springer, Tokyo (1996).
4 J.M. Chen, R.G. Liu, S.C. Chung, R.S. Liu, M.J. Kramer, K.W. Dennis, R.W. McCallum, Phys. Rev. B, 55, 3186 (1997).
5. T.B. Lindermer, B.B. Chakoumakos, E.D. Specht, R.K. Williams, and Y.J. Chen, Physica C **231**, 80 (1994).
6. R.W. McCallum, M.J. Kramer, K. W. Dennis, M. Park, H. Wu, and R. Hofer, J. Electronic Mater., **24**, 1931 (1995).
7. S.I. Yoo, N. Sakai, H. Takaichi, T. Higuchi, and M. Murakami, Appl. Phys. Lett. **65**, 633 (1994).
8. T.B. Lindemer, E.D. Specht, P.M. Martin, M.L. Flitcroft, Physica C **255**, 65 (1995).
9. E.A. Goodilin, N,N, Oleynikov, E.V. Antipov, R.V. Shpanchenko, G. Yu. Popov, V.G. Balakirev, Yu.D. Tretyakov, Physica C **272,** 65, (1996).
10. K. Osamura and W. Zhang, Z Metallkd., **84**, 523, (1993)
11. G.F. Voronin, S.A. Degterov, J. Solid State Chem. **110**, 50 (1994).
12. H. Wu, M.J. Kramer, K. W. Dennis, and R.W. McCallum, IEEE Trans. Appl. Superconductivity, **7**, 1731 (1997).

THALLIUM-BASED HTS THIN FILMS FOR MICROWAVE APPLICATIONS

DAVID DEW-HUGHES
Oxford University
Department of Engineering Science
Parks Road, Oxford OX1 3PJ, UK.

1. Introduction

Thin films of the thallium-based high temperature superconductors are particularly suitable for microwave applications. The Tl-2212, Tl-2223 and Tl-1223 phases have critical temperatures in the range 100-120K; thus at the proposed operating temperatures of ~80K, readily achievable with single-stage cryocoolers, their superconducting properties exhibit a slow variation with temperature, thus obviating the need for precise temperature control in maintaining device performance [1]. Device quality films of the above phases have been deposited on a variety of substrates, with and without buffer layers, by a number of groups and a range of practical microwave components have been successfully demonstrated [2]. Critical current densities at 77K and in zero field in excess of GAm^{-2}, and surface resistance values, normalised to 80K and 10GHz, ~0.5mΩ have been achieved in these films.

At Oxford University a collaborative programme between the Departments of Materials and Engineering Science has, for the past six years, investigated the deposition of thallium-based HTS thin films, studied the relations between deposition conditions, film structure, and superconducting properties, and has fabricated microwave devices from these films [3-5]. This paper describes the chemistry of the thallium compounds, the techniques used for thin films deposition and the relation between film structure and superconducting properties.

2. Chemistry of the Thallium-based HTS phases

There are three series of thallium-based high temperature superconducting mixed oxides. The double layer compounds, structurally analogous to the Bi-Sr-Ca-Cu mixed oxides, contain Tl, Ba, Ca and Cu, with the general formula $Tl_2Ba_2Ca_{n-1}Cu_nO2_{n+3}$. The phases of interest for practical applications in microwave devices are the Tl-2212 (n=2), with T_c up to 118K, and Tl-2223 (n=3), with T_c up to 125K, compounds. There is also a series of single layer phases containing the same cation constituents, with the general formula

G. Van Tendeloo et al. (eds.), High-Temperature Superconductors and Novel Inorganic Materials, 205–210.

$TlBa_2Ca_{n-1}Cu_nO_{2n+3}$. A third, also single layer, series is formed when Sr is substituted for Ba; this series has the general formula $TlSr_2Ca_{n-1}Cu_nO_{2n+3}$. This latter series is stabilised by the partial substitution of Tl by Pb or Bi. The single layer phase of practical interest is the Tl-1223 (n=3) compound with T_c up to 123K.

The chemistry and fabrication of the Tl-based mixed oxides is governed by the extreme volatility of Tl_2O. In other systems the superconducting compounds can be synthesised by mixing the appropriate oxides, sintering and then annealing in air or oxygen to establish the correct degree of oxygenation to ensure optimum superconducting properties. Thus the Tl-2212 phase should be formed by the reaction:

$$4BaCuO_2 + 2CaO + 2Tl_2O_3 \rightleftharpoons 2Tl_2Ba_2CaCu_2O_8 \quad \textbf{(1)}$$

However at elevated temperatures thallic oxide decomposes according to the equation:

$$Tl_2O_3(s) \rightleftharpoons Tl_2O(g) + O_2(g) \quad \textbf{(2)}$$

The equilibrium partial pressure of Tl_2O vapour, in 1 atmosphere of oxygen at 777°C is about 10^{-2} atmospheres. The mixed oxide similarly decomposes:

$$2Tl_2Ba_2CaCu_2O_8 \rightleftharpoons 4BaCuO_2 + 2CaO + 2Tl_2O(g) + 2O_2(g) \quad \textbf{(3)}$$

This reaction is exothermic, and the Tl-2212 phase is stabilised by high temperatures and high partial pressures of Tl_2O and O_2. The other Tl-containing mixed oxides decompose in a similar fashion. In the three Tl-based HTS compounds of practical interest, the proportion of Tl present as a fraction of the overall constituents decreases in the order 2212, 2223, 1223, and as the partial pressure of Tl_2O is lowered Tl-2212 will decay to Tl-2223, and this in turn will decay to Tl-1223, by the loss of thallium. The equilibrium between Tl-2212 and Tl-2223 as a function of oxygen partial pressure and temperature has been studied by Ahn, Lee and Beyers [6]; at 750°C the oxygen partial pressure must exceed 10^{-2} atmospheres, and at 900°C it must exceed 2 atmospheres, to favour the formation of the 2212 phase.

3. Thin Film Deposition Techniques.

The need to control the partial pressure of the Tl_2O vapour during film deposition and growth dictates the techniques employed in producing Tl-based thin films. There are two approaches to this problem. One is to carry out deposition in the presence of a separately controllable source of Tl_2O vapour; the *in-situ* process [7,8]. In this process material is deposited by off-axis laser ablation from a BaCaCuO target in the presence of Tl_2O vapour generated by heating a container of thallic oxide, all in an atmosphere of 3-5 mbar of oxygen. This results in the formation of a film of Tl-1223 by the reaction:

$$4BaCuO_2 + 2CuO + 4CaO + Tl_2O(g) + \frac{3}{2}O_2(g) \Rightarrow 2TlBa_2Ca_2Cu_3O_9 \qquad \textbf{(4)}$$

Films of up to 400nm thick were produced in this way on substrates of $LaAlO_3$, with T_c~99K, and of $SrTiO_3$, with T_c~109K [8]. The *in-situ* method seems capable of producing only the single layer thallium compounds.

In the *ex-situ* method, a precursor film, which may or may not contain thallium, is deposited by, *eg* thermal evaporation [9] or sputtering [10], and subsequently annealed in the presence of Tl_2O vapour. The reaction between the precursor film and Tl_2O generates the required superconducting phase. The requirements of the precursor film are that it should be of the correct stoichiometry and thickness; structural changes during the thalliation process, which may involve partial melting, render the structure of the precursor irrelevant to the quality of the final film. This latter is influenced by the substrate, the thalliation time and temperature, and the partial pressures of oxygen and thallous oxide. The source of the thallous oxide may be either Tl_2O_3 or one of the Tl-based mixed oxides.

There are two variants to thalliation stage of the *ex-situ* process. The precursor film and the thallium source may be sealed in the same crucible [11]. Both source and film will be maintained at the same temperature and in the same atmosphere. By having as the thallium source the same compound as that desired in the film, the correct vapour pressure should be ensured. Both single- and double-sided films of 2212, 2223 and 1223 have been grown by this method on a variety of substrates and substrate/buffer layer combinations [12, 13].

The alternative process is that using a two-zone furnace. The precursor film and thallium source are placed in separate, temperature controlled, zones of the furnace. In this way the temperature of the film, and the partial pressures of Tl_2O and O_2 can be independently varied, resulting in greater control of the thalliation process. Tl is transported to the film by molecular diffusion [14] or by a stream of flowing gas [15,16]. It is expected that the production, on a commercial scale, of large area double-sided device quality films with reproducible properties, will be carried out by this latter process.

4. Substrates and Buffer Layers

An HTS thin film suitable for microwave devices must have a low value of microwave surface resistance. This can only be achieved in films which are essentially single crystal, or quasi single crystal, with the c-axis normal to the plane of the film and the a,b axes lying in the plane of the film. Such films are grown epitaxially on the appropriate substrate, the prime duty of the substrate being to act as a crystallographic template for the growing film. There must therefore be a close crystallographic match between substrate and HTS compound. The substrate must also be inert with respect to the compound at the thalliation temperature. The substrate forms a part of the microwave circuit, and its dielectric properties must be compatible with the performance of the device. In particular the dielectric loss tangent must be small, $<\sim 10^{-5}$, and the relative dielectric constant should ideally be ~10, be isotropic, and invariant with respect to temperature, electric and magnetic fields, at the level of these quantities encountered under operating conditions.

The three Tl-based compounds of interest all have tetragonal structures, with a-axis parameters of 0.38550nm (2212), 0.38487nm (2223) and 0.38429nm (1223). Strontium titanate, $SrTiO_3$, with a lattice parameter of 0.3905nm, is the ideal substrate from the structural point of view. However it is a ferroelectric, with a relative dielectric constant >1000 at 80K, and is also very lossy. Thus it is totally unsuitable to form a significant part of the microwave circuit. In addition it is found that growth on $SrTiO_3$ favours the single-layer compound. This is believed to be due to the diffusion of Sr into the film, substituting for Ba and stabilising the Tl-1223 phase [13]. Lanthanum aluminate, $LaAlO_3$, is a pseudocubic material with a lattice parameter of 0.3792nm. It has a loss tangent of less than 10^{-5} at 77K, does not react with the HTS compounds, and the highest quality Tl-based films have been grown on substrates of this material. Its disadvantages are that it undergoes a structural transformation at about 500°C resulting in twinning on cooling and an anisotropy in its relative dielectric constant; this latter is, at 24, rather larger than desirable for many applications. Magnesium oxide, MgO, is cubic, has a relative dielectric constant of 10 and a loss tangent less than 10^{-5} at 77K, and from the dielectric point of view is an excellent substrate. Its lattice parameter, at 0.421nm, does not provide a good epitaxial match. Nevertheless, device quality films have been deposited on MgO substrates. The ideal dielectric substrate would be sapphire, αAl_2O_3. This has the lowest loss tangent of any material, 10^{-7} at 77K, and though having a rhombohedral structure, the R-plane cut has a pseudo-cubic parameter of 0.348nm which results in a good epitaxial match. The relative dielectric constant is slightly anisotropic, varying from 9.4-11.6. Unfortunately at the thalliation temperature Al diffuses into the film, degrading its superconductivity. The use of a ceria, CeO_2, buffer layer can block the diffusion of Al. Ceria is cubic, and the 45° lattice parameter is 0.3826, giving an excellent epitaxial match. However a new problem now manifests itself, in that during thalliation a layer of barium cerate forms, interfering with the epitaxial relation between the ceria layer and the HTS film [17]. This problem has been overcome by the development of a low temperature thalliation process; reducing the oxygen partial pressure lowers the melting point of the Tl-2212 phase, and thalliation can be carried out at ~730°C instead of the more usual ~850°C required at atmospheric pressure [18].

5. Processing, Properties and Microstructure.

The HTS thin film, substrate and buffer layer combinations studied at Oxford are listed in Table 1, together with the superconducting properties of the most favourable examples. Critical current density is that measured over the entire film, and is an order of magnitude less than measurements carried out on selected microbridges. There is no strong correlation between high J_c and low R_s. Both are measures of film quality, but R_s is dominated by surface condition, and J_c by the best current paths. Films on $LaAlO_3$ have both excellent c-axis texture normal to the plane of the film together with good in-plane alignment of the a,b-axes; those on MgO also have good c-axis texture but poor in-plane alignment and are

TABLE 1. Summary of best results for various HTS compound/substrate combinations

HTS phase	Substrate	Buffer Layer	Thalliation conditions	T_c, K	Large area J_c, GAm^2	R_s, mΩ at 10GHz & 80K
Tl-2212	$LaAlO_3$		855°C	105	3.0	<0.2
Tl-2212	$LaAlO_3$		730°C	98.5	3.4	0.4
Tl-2212	Sapphire	CeO_2	730°C	101	1.25	3.02
Tl-2212	MgO	CeO_2	740°C	96.5	1.94	2.98
Tl-2212 /2223	$SrTiO_3$	$LaAlO_3$	852°C	115	0.25	0.65
Tl-2223	$LaAlO_3$		855°C	112	0.50	0.20
Tl-2223	MgO		855°C	117	0.33	0.69
Tl-1223	$SrTiO_3$		868°C	103	3.27	2.0
Tl-1223	MgO	$SrTiO_3$	868°C	104	2.94	3.0 (<0.2 @55K)
Tl-1223	MgO	$SrTiO_3$/ $LaAlO_3$	854°C	100.5	0.83	39.2 (5.0 @ 30K

polycrystalline [13]. Thalliation at low temperatures produces slightly inferior films. The poorer performance of buffered films is due to the difficulty of depositing single crystal buffer layers; the generally high mosaic spread in the buffer layers cases some loss of epitaxy. The optimum film thickness for low R_s is 700nm; in thicker films the epitaxy is lost, for thinner films at microwave frequencies the field penetrates through the film into the substrate. The poor performance of the last two films in the table is due to their being of less than optimum thickness; the improvement in R_s at low temperatures results from the decrease in the penetration depth with temperature so that the field no longer samples the substrate.

The relation between film structure and microwave properties has been studied by SEM, plan view TEM, and cross-section TEM and HREM, and is found to be complex [19]. The necessity for c-axis texture and in-plane alignment has already been mentioned. Surprisingly a smooth surface does not seem to be of prime importance, and up to 10% of porosity and outgrowths of other superconducting phases does not degrade the surface resistance. Nor do defects such as dislocations have an effect on R_s. Internal planar defects, stacking faults, grain boundaries, a-axis textured material, and outgrowths of non-superconducting phases all result in considerable degradation of superconducting properties and microwave performance. A study of *in-situ* Tl-1223 demonstrates that continuity in the a,b planes is essential for good superconducting performance [20].

6. References

1. Dew-Hughes. D.(1997) Microwave properties and applications of high temperature superconductors, in H.Groll and I.Nedkov (eds.), *Microwave Physics and Techniques*, Kluwer Academic Publishers, Dordrecht, pp 83-114.
2. Hein, M.A.(1997) Progress, properties and prospects of passive high-temperature superconductive microwave devices in Europe, in H.Rogalla and D.H.A.Blank (eds), *Applied Superconductivity 1997*, IOP Publishing, Bristol, pp 261-266.
3. Kale, K.S. *et al.*(1996) in U.Balachandran, P.J.McGinn and J.S.Abell (eds), *High Temperature Superconductors: Synthesis, Processing and Large-Scale Applications*, The Minerals, Metals and Materials Society, Warrendale, pp 345-354.
4. Dew-Hughes, D. (1997) Application of high temperature superconductors in passive microwave devices, *Australian Journal of Physics* **50**, 363-379.
5. Jenkins, A.P. *et al.* (1998) Thin film Tl-based HTS microwave devices, *Journal of Superconductivity*, accepted for publication.
6. Ahn, B.T., Lee, W.Y. and Beyers, R (1992) Effect of oxygen pressure on $Tl_2Ca_2Ba_2Cu_3O_{10\pm\delta}$ formation, *Applied Physics Letters* **60**, 2150-2152.
7. Betz, J. *et al.* (1992) *In situ* preparation of high-T_c TlBaCaCuO thin films by a combination of laser ablation and thermal evaporation, *Journal of Applied Physics* **71**, 2478-2479.
8. Reschauer, N. *et al.* (1996) *In situ* preparation of epitaxial $TlBa_2Ca_2Cu_3O_9$ films by laser ablation in the presence of Tl_2O vapor, in U. Balachandran, P.J.McGinn and J.S.Abell (eds.), *High Temperature Superconductors: Synthesis, Processing and Large Scale Applications*, The Minerals, Metals and Materials Society, Warrendale, pp 339-344.
9. Lai, H.C. *et al.* (1991) Deposition of thallium-based thin films by a simple thermal evaporation method, *Superconductor Science and Technology* **4**, 306-311.
10. Morley, S.M. *et al.* (1993) Synthesis and microwave characterisation of DC sputtered Tl-Ba-Ca-Cu-O thin films, *Transactions on Applied Superconductivity* **3**, 1753-1756.
11. Ginley, D.S. *et al.* (1989) Morphology control and high critical currents in superconducting thin films in the Tl-Ca-Ba-Cu-O system, *Physica C* **160**, 42-48.
12. Bramley, A.P. *et al.* (1996) in R. Moerman (ed) *Proceedings of the 3rd HTS Workshop on Digital Applications, Josephson Junctions and 3-terminal Devices*, University of Twente, Enschede, pp 24-30.
13. Bramley, A.P. *et al.* (1997) The fabrication of Tl-Ba-Ca-CuO films on $SrTiO_3$ and $LaAlO_3$ buffered MgO substrates for microwave applications, in H. Rogalla and D.H.A.Blank (eds) *Applied Superconductivity 1997*, IOP Publishing, Bristol, pp 201-204.
14. DeLuca, J.A. *et al.* (1991)The synthesis of superconducting Tl-Ca-Ba-Cu-oxide films by the reaction of spray deposited Ca-Ba-Cu-oxide precursors with Tl_2O vapor in a two-zone reactor, *Journal of Materials Research* **6**, 1415-1424.
15. Kampwirth, R.T. *et al.* (1995) Properties of Tl-1223 superconducting thick films fabricated from precursor inks using a thallium vapor transport process, *Transactions on Applied Superconductivity* **5**, 1954-1957.
16. Pluym, T.C. *et al.* (1995) Superconducting $Tl_2Ba_2CaCu_2O_8$ thin films prepared by post-annealing in a flow-through multiple-zone furnace, *Transactions on Applied Superconductivity* **5**, 1339-1342.
17. Grovenor, C.R.M. *et al.* (1994) Deposition of TlBaCaCuO films on sapphire and CeO_2/sapphire substrates, *Physica C* **235-240**, 715-716.
18. O'Connor, J.D. *et al.* (1995) The processing at low temperature of Tl-2212 films in argon atmospheres for microwave devices, in D.Dew-Hughes (ed) *Applied Superconductivity 1995*, IOP Publishing, Bristol, pp 1055-1058.
19. O'Connor J.D. *et al.* (1998) Microstructure and electrical properties of $Tl_2Ba_2CaCu_2O_x$ thin films processed at low temperatures, *Superconductor Science and Technology* **11**, 207-216.
20. O'Connor J.D. *et al.* (1998) Transmission electron microscopy and electrical properties of epitaxial Tl-1222 and Bi-substituted Tl-1223 thin films grown by *in-situ* laser ablation, accepted by *Physica C*.

SUPERCONDUCTIVITY SUPPRESSION IN $(Nd_{1.05-x}Pr_x)Ba_{1.95}Cu_3O_7$ STUDIED BY X-RAY ABSORPTION SPECTROSCOPY

J. M. Chen
Synchrotron Radiation Research Center (SRRC), Hsinchu, Taiwan, R.O.C.
R. S. Liu
Department of Chemistry, National Taiwan University, Taipei, Taiwan, R.O.C.
M. J. Kramer, K. W. Dennis and R.W. McCallum
Ames Laboratory, Iowa State University, Ames, Iowa 50011, USA

O K-edge x-ray-absorption near-edge-structure (XANES) spectra for the series of $(Nd_{1.05-x}Pr_x)Ba_{1.95}Cu_3O_7$ compounds with x = 0.0 - 0.5 were measured to investigate how the variation of hole states related to the superconductivity. The prepeaks at ~ 527.6 and ~ 528.3 eV in O 1s x-ray absorption spectra are assigned to transitions into O 2p holes located in the CuO_3 ribbons and CuO_2 planes, respectively. As deduced from O 1s x-ray absorption spectra, both the O 2p hole concentration in the CuO_2 planes and CuO_3 ribbons decrease monotonically with increasing the Pr doping. The present XANES results clearly demonstrate that the suppression of superconductivity with Pr-doping results predominantly from the hole depletion effect.

1. Introduction

Chemical substitution of Pr for Y in the $(Y_{1-x}Pr_x)Ba_2Cu_3O_7$ system reduces the T_c value with superconductivity disappearing for x > 0.55, although $PrBa_2Cu_3O_7$ forms the same orthorhombic structure as $YBa_2Cu_3O_7$. This is in contrast to other earth-earth (R) substitutions (Nd, Sm, Eu, Gd, Dy, Ho, Er, Tm, Yb, and Lu) which remain superconducting with T_c near 90 K. In addition, it has been demonstrated that the depression of T_c by the Pr substitution in $(R_{1-x}Pr_x)Ba_2Cu_3O_7$ strongly depends on the rare-earth elements.[1] The suppression of superconductivity in $(R,Pr)Ba_2Cu_3O_7$ has been a remarkable puzzle in the cuprate superconductors.

An enormous amount of experimental investigations have been carried out on the $(Y_{1-x}Pr_x)Ba_2Cu_3O_7$ system in an attempt to understand the origin of the quenching of superconductivity in $(R,Pr)Ba_2Cu_3O_7$.[2] The models for explaining the depression of superconductivity in $(R,Pr)Ba_2Cu_3O_7$ include hole-filling, hole localization, percolation, magnetic pair-breaking, and hybridization. However, no model allows for a consistent interpretation of all the experimental data. Recently, it has been proposed by Fehrenbacher and Rice (FR) that the crystal structure of $PrBa_2Cu_3O_7$ favors a strong hybridization between Pr 4f states and conduction holes in the CuO_2 planes with a mixture of $4f^2L$ (Pr^{3+}) and $4f^1$ (Pr^{4+}) configuration, where L denotes a ligand hole in the

G. Van Tendeloo et al. (eds.), High-Temperature Superconductors and Novel Inorganic Materials, 211–216.

O 2p orbitals.[3] This hybridized state is competitive in energy to the hole states in the CuO_2 planes. The FR model assumed hole depletion in the CuO_2 planes, not because of higher Pr valence but because of transfer of the holes from the CuO_2 planes into the FR state which binds doped holes to the Pr sites. However, the FR model still can not explain the R-dependence on the suppression of T_c in $(R_{1-x}Pr_x)Ba_2Cu_3O_7$. Recently, Liechtenstein and Mazin (LM) calculated the electronic structure of $(R_{1-x}Pr_x)Ba_2Cu_3O_7$ using *ab initio* local density approximation plus Hubbard parameter including Coulomb correlation in the f shell.[4] They found that, in $PrBa_2Cu_3O_7$, there forms an additional hole-depleting band which crosses the Fermi level and consequently grabs holes from the CuO_2 band. On doping different rare-earth elements in $(R_{1-x}Pr_x)Ba_2Cu_3O_7$, the position of this hole-depleting band shifts with the atomic number of the rare earth, through the energy level of 4f orbitals, leading to the R-dependence of the suppression of T_c.

On the basis of band-structure calculations[5] and x-ray absorption measurements[6], it has been shown that the holes near the Fermi level in $YBa_2Cu_3O_7$ are distributed between the CuO_2 planes and CuO_3 ribbons. If the LM model is valid, it is expected that, in the $(Y_{1-x}Pr_x)Ba_2Cu_3O_7$ system, both the hole concentration originating from CuO_2 planes and CuO_3 ribbons should decrease with increasing the Pr doping. However, there is still a lack of direct experimental evidence in support of the LM model.

To improve the understanding on the mechanism of T_c-suppression, a detailed study on other $(R_{1-x}Pr_x)Ba_2Cu_3O_7$ systems would be helpful. We therefore measure systematically the variation of electronic structure near the Fermi level for a series of $(Nd_{1.05-x}Pr_x)Ba_{1.95}Cu_3O_7$ compounds. The chemical substitution of the Pr ion for the Nd ion in $(Nd_{1.05-x}Pr_x)Ba_{1.95}Cu_3O_7$ leads to a linearly decrease of the superconducting temperature from $T_c = \sim 92$ K for $x = 0$ to $T_c = 20$ K for $x = 0.3$. The compound becomes an semiconductor at $x \geq 0.4$.[7] X-ray absorption near edge structure is a direct probe of the character and local density of hole states responsible for high-T_c superconductivity. In this study, O K-edge x-ray absorption measurements in $(Nd_{1.05-x}Pr_x)Ba_{1.95}Cu_3O_7$ for $x = 0.0$ - 0.5 were performed in order to investigate how the variation of hole states near the Fermi level related to the superconductivity

2. Experimental

Details on the preparation of $(Nd_{1.05-x}Pr_x)Ba_{1.95}Cu_3O_7$ for $x = 0.0$ to 0.5 were reported elsewhere.[7] X-ray diffraction (XRD) showed that all samples were single phase with an orthorhombic structure. DC SQUID results showed that T_c decreases linearly with x, samples becoming normal for $x = 0.4$.

Using the 6-m high-energy spherical grating monochromator (HSGM) beamline, the x-ray absorption measurements were performed at the Synchrotron Radiation Research Center (SRRC) with an electron beam energy of 1.5 GeV. X-ray-fluorescence-yield spectra were recorded using a microchannel plate (MCP) detector.[8] The incident photon intensity (I_o) was measured simultaneously by a Ni mesh located after the exit slit of the monochromator. All the absorption spectra were normalized to I_o. The energy resolution of the monochromator was set to ~ 0.22 eV for the O K-edge x-ray absorption measurements.

3. Results and Discussion

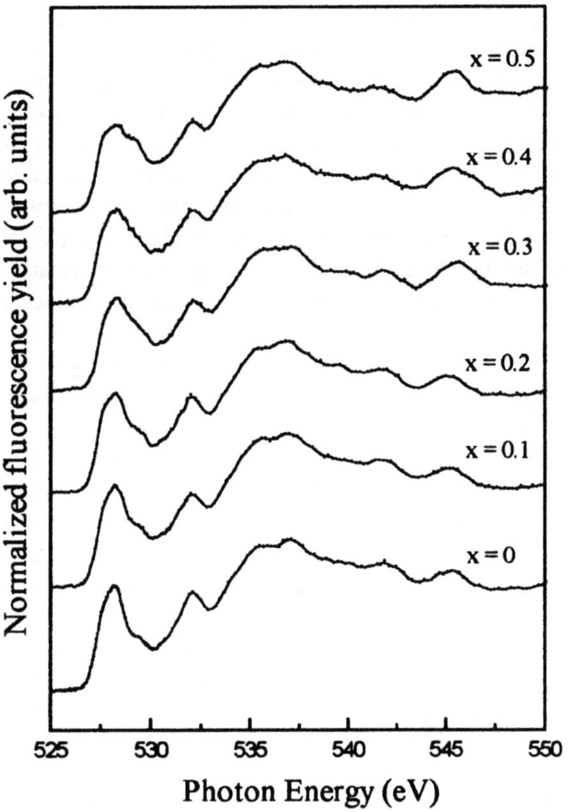

Figure 1 O K-edge x-ray absorption spectra for the series of $(Nd_{1.05-x}Pr_x)Ba_{1.95}Cu_3O_7$ samples with x = 0.0 - 0.5. The absorption spectra for various compounds with different x values were normalized to same height at the main peak of ~ 537 eV.

In Fig. 1, the O K-edge x-ray absorption spectra for a series of $(Nd_{1.05-x}Pr_x)Ba_{1.95}Cu_3O_7$ samples with x = 0.0 - 0.5 are shown in the energy range of 525 - 550 eV obtained using a bulk-sensitive total x-ray fluorescence yield method. The major features in the O 1s x-ray absorption spectrum for sample with x = 0.1 are two distinct prepeaks at ~ 528.3 eV and ~ 529.2 eV with a shoulder at ~ 527.6 eV, and a broad peak at ~ 537 eV. The low-energy prepeaks with energy below 532 eV are ascribed to photoexcitation of the O 1s core electrons to holes with predominantly 2p character on the oxygen sites. Unoccupied states related to the Ba 4d, Pr 5d or Pr 4f states hybridized with O 2p states could be the origin for the enhanced peaks above 532 eV in $Nd_{1.05-x}Pr_x)Ba_{1.95}Cu_3O_7$. The O K-edge x-ray absorption spectra for various compounds with different x values in Fig. 1 were normalized to the same height at the main peak of ~ 537 eV.

The crystal structure of $NdBa_2Cu_3O_7$ is composed of two Cu(2)O(2)O(3) layers separated by a Nd plane. The unit of CuO_2 and Nd planes are separated by a CuO_3 ribbon consisting of a BaO(4) plane, a Cu(1)O(1) chain along the b axis, and another BaO(4) plane. Therefore, there exists four nonequivalent oxygen sites in $NdBa_2Cu_3O_7$, O(2) and O(3) within the $Cu(2)O_2$ layers, O(4) in the BaO planes, and O(1) in the Cu(1)O chains. According to the LDA band-structure calculations in $YBa_2Cu_3O_7$ by Krakauer *et al.*,[9] the O(2,3) atoms in the CuO_2 planes and the O(1) atom in the CuO chain are predicted to have the largest and the lowest binding energy of the O 1s level, respectively. Based on the polarization-dependent x-ray absorption measurements on single-crystal $YBa_2Cu_3O_{7-\delta}$, it was found that O(2) 1s level is about 0.8 eV lower in energy than the O(4) 1s level.[6] The O(1) 1s binding energy is in between those of O(4) and O(2,3). These results are consistent with the theoretical predictions. In O K-edge x-ray absorption spectra of $YBa_2Cu_3O_{7-\delta}$, the prepeaks at ~ 527.8 eV are attributed to transitions into O 2p holes in the CuO_3 ribbons (apical oxygen sites and CuO chains). The high-energy prepeak at ~ 528.5 eV is ascribed to transitions into O 2p hole states within the CuO_2 planes.[6]

The orthorhombic $NdBa_2Cu_3O_{7-\delta}$ compound is isomorphic with $YBa_2Cu_3O_{7-\delta}$ (space group: Pmmm). When oxygen is removed from the $NdBa_2Cu_3O_{7-\delta}$ compounds, the system behaves analogously to $YBa_2Cu_3O_{7-\delta}$ with a reduction in total hole concentrations and T_c.[10] In addition, when oxygen is removed from $PrBa_2Cu_3O_{7-\delta}$, a strong reduction of the prepeak structure in O 1s x-ray absorption spectrum is observed,[11] as seen for the $YBa_2Cu_3O_{7-\delta}$ compounds. As shown in Fig. 1, the O 1s x-ray absorption spectrum of $(Nd_{1.05-x}Pr_x)Ba_{1.95}Cu_3O_7$ for x = 0 exhibits similar features as observed in $YBa_2Cu_3O_{7-\delta}$ with $\delta = 0$. We therefore adopt the same assignment scheme for the O 1s x-ray absorption spectra of $(Nd_{1.05-x}Pr_x)Ba_{1.95}Cu_3O_7$. The high-energy prepeak at ~ 528.3 eV is attributed to the excitation of O 1s electrons to O 2p holes in the CuO_2 planes. The low-energy prepeaks at ~ 527.6 eV in Fig. 1 are due to the superposition of O 2p hole states in the apical oxygen sites and the CuO chains.

Based on x-ray absorption studies in $La_{2-x}Sr_xCuO_4$, the absorption peak at ~ 529.2 eV in Fig. 1 can be ascribed to transitions between O 1s core states and the empty upper Hubbard Cu 3d conduction band.[12] Due to the strong on-site correlation effects on the copper sites in the cuprate superconductors, an upper Hubbard band (UHB) has always been assumed to exist. As noted from Fig. 1, the prepeak at ~ 528.3 eV originating from the O 2p hole states in the CuO_2 layers decreases in spectral weight with increasing the dopant concentration of Pr. This result clearly reveals that the chemical substitution of Pr for Nd in $(Nd_{1.05-x}Pr_x)Ba_{1.95}Cu_3O_7$ leads to a reduction in hole concentration in the CuO_2 planes.

In order to investigate the variation of the hole states among different oxygen sites as a function of the Pr doping, these pre-edge features shown in Fig. 1 were analyzed by fitting Gaussian functions to each spectrum. In Fig. 2 the integrated intensity of each prepeak, normalized against that of main peak at ~ 537 eV, is plotted as a function of Pr content x in $(Nd_{1.05-x}Pr_x)Ba_{1.95}Cu_3O_7$.

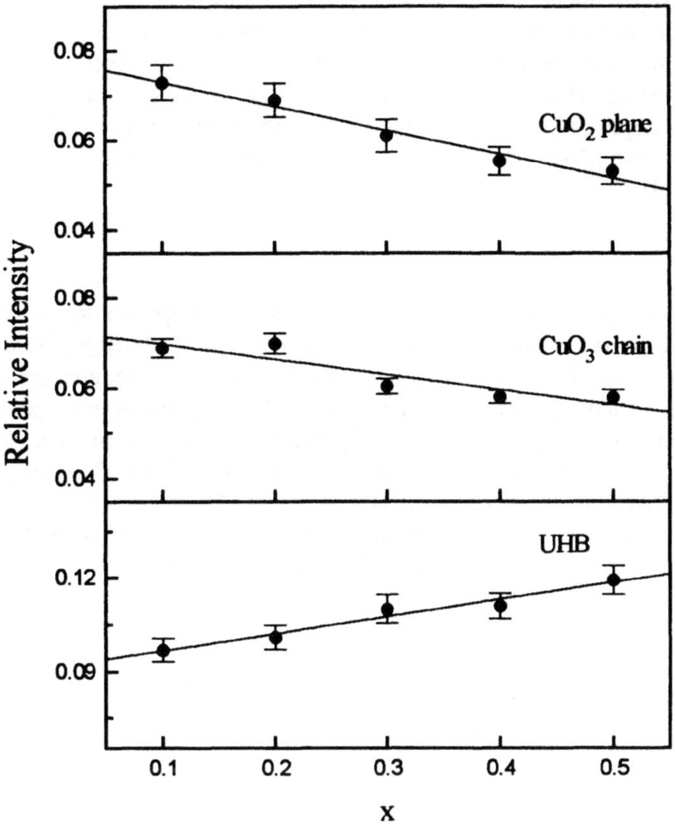

Figure 2. X-dependence of the relative intensity of hole states in $(Nd_{1.05-x}Pr_x)Ba_{1.95}Cu_3O_7$ on oxygen sites originating from the (a) CuO_2 planes, (b) CuO_3 ribbons, and (c) upper Hubbard band (UHB). The solid curves are drawn as a guide for the eyes.

As noted from Fig. 2a and 2b, the hole content from the CuO_2 planes decrease monotonically with increasing the Pr doping, while that from apical oxygen sites and CuO chains decreases slowly. According to the LM model, the hole reduction upon substituting R by Pr in $(R_{1-x}Pr_x)Ba_2Cu_3O_7$ originates from the hole transfer from the CuO_2 planes and CuO_3 ribbons into the LM band. As shown, the depletion in hole concentration observed in the $(Nd_{1.05-x}Pr_X)Ba_{1.95}Cu_3O_7$ system provides an evidence in support of LM model. It has been experimentally shown that the concentration of O 2p holes in the CuO_2 planes is strongly correlated with T_c in the p-type cuprtaes. Accordingly, the T_c value in the $(Nd_{1.05-x}Pr_X)Ba_{1.95}Cu_3O_7$ system should decrease with increasing the dopant concentration of Pr. Our experimental results clearly reveal that the quenching of superconductivity with Pr-doping in $(Nd_{1.05-x}Pr_x)Ba_{1.95}Cu_3O_7$ results predominantly from the hole depletion. In addition, as shown in Fig. 2c, the peak at ~ 529.2 eV originating from the UHB shows a monotonic increase in spectral weight as the Pr doping increases. This change is related to a spectral weight transfer of states from the UHB to doping-induced hole states near the Fermi level.[13] The similar behavior has also been observed in O 1s absorption spectra of other p-type cuprate superconductors.[14]

4. Conclusion

In this study, we reported the O K-edge x-ray absorption spectra for a series of $(Nd_{1.05-x}Pr_x)Ba_{1.95}Cu_3O_7$ compounds with x = 0.0 - 0.5. Near the O 1s edge, the multiple prepeaks is related to the different O 1s binding energies of the nonequivalent oxygen sites in the crystal structure. As deduced from the O 1s x-ray absorption spectra, the hole concentration originating from the CuO_2 planes and CuO_3 ribbons decrease monotonically in $(Nd_{1.05-x}Pr_x)Ba_{1.95}Cu_3O_7$ with the increase of the Pr doping. This result is consistent with the predictions by the LM model. The present XANES results clearly show that the superconductivity suppression with the Pr substitution in $(Nd_{1.05-x}Pr_x)Ba_{1.95}Cu_3O_7$ arises predominantly from the hole depletion effect.

Acknowledgment

This research was financially supported by SRRC and National Science Council of the Republic of China. Work at Ames Laboratory was supported by the Director of Energy Research, U.S. Department of Energy under contract No. W-7405-ENG-82.

References

1. Xu, Y. and Guan, W. (1991) *Appl. Phys. Lett.* **59**, 2183.
2. Radousky, H. B. (1992) *J. Mater. Res.* **7**, 1917, and reference therein.
3. Fehrenbacher, R. and Rice, T. M. (1993) *Phys. Rev. Lett.* **70**, 3471.
4. Liechtenstein, A. I. and Mazin, I. I. (1995) *Phys. Rev. Lett.* **74**, 1000.
5. Olés, A. M. and Grzelka, W. (1991) *Phys. Rev. B* **44**, 9531.
6. Nücker, N. Pellegrin, E. Schweiss, P. Fink, J. Molodtsov, S. L. Simmons, C. T. Kaindl, G. Frentrup, W. Erb, A. and Müller-Vogt (1995) *Phys. Rev. B* **51**, 8529.
7. Kramer, M. J. Dennis, K. W. Falzgraf, D. McCallum, R. W. Malik, S. K. and Yelon, W. B. (1997) *Phys. Rev. B.* **56**, 5512.
8. Chen, J. M. Chung, S. C. and Liu, R. S. (1996) *Solid State Commun.* **99**, 493.
9. Krakauer, H. Pickett, W. E. and Cohen, R. E. (1988) *J. Supercond.* **1**, 111.
10. Shanked, H. Veal, B. W. Faber, J. Hitterman, R. L. Balachandran, U. Tomlins, G. Shi, H. Morss, L. and Paulikas, A. P. (1988) *Phys. Rev. B* **37**, 5158.
11. Fink, J. Nücker, N. Romberg, H. Alexander, M. Maple, M. B. Neumeier, J. J. and Allen, J. W. (1990) *Phys. Rev. B* **42**, 4823.
12. Romberg, H. Alexander, M. Nücker, N. Adelmann, P. and Fink, J. (1990) *Phys. Rev.B* **42**, 8768.
13. Meinders, M. B. J. Eskes, H. and Sawatzky, G. A. (1993) *Phys. Rev. B* **48**, 3916.
14. Elp, J. van Eskes, H. Kuiper, P. and Sawatzky, G. A. (1992) *Phys. Rev. B* **45**, 1612.

INTERGRAIN PROPERTIES OF PRESSURE-OXYGENATED $REBa_2Cu_3O_{7-\delta}$

T.PLACKOWSKI, C.SUŁKOWSKI AND D.WŁOSEWICZ
Institute of Low Temperature and Structure Research
Polish Academy of Sciences
P.O.Box 937, 50-950 Wrocław 2, Poland

1. Introduction

The high-T_c superconductors belonging to the $REBa_2Cu_3O_{7-\delta}$ family may be synthesized basing on almost all rare earths (RE) and Y. The superconducting transition temperature T_c of these compounds shows only a weak increase with the size of the RE ion (from $T_c \approx 92.5$ K for smaller RE's up to $T_c \approx 98.5$ K for La-123 [1]). However, some clear dependences of different physical and chemical properties on RE ionic size have been revealed. The activation energy of oxygen within RE-123 group is higher for compounds with smaller RE's [2], whereas higher oxygen pressure is necessary to oxygenate compounds with larger RE's [3]. For the latter compounds a higher disorder was revealed within oxygen [4] and cation [5] sublattices.

There is also another factor influencing the properties of RE-123 with larger RE's. If the ionic radius of RE^{3+} ion is large enough to allow the partial substitution of Ba rare earth's ions (*i.e.* for $r \geq r_{Gd}$), then the 123 compounds reveal a tendency to create a solid solution of $RE_{1+x}Ba_{2-x}Cu_3O_y$ type, where Ba sites may be occupied by RE elements. In these solid solutions transport and superconducting properties are progressively degraded with increasing x [6, 7].

In case of some RE's the processing of the $REBa_2Cu_3O_{7-\delta}$ compounds under elevated oxygen pressure (hundreds of bars at temperatures up to 450 °C) may result in surprising changes of their properties while the oxygen content increases only slightly. Whereas several superconducting and normal state properties of Sm-123 (namely the specific heat jump at T_c, the thermoelectric power [8] or the thermal conductivity [9]) seem to be enhanced in the sense of additional charge carries doping, the electrical resistivity of this compound significantly increases and changes its temperature dependence into a semiconducting-like one.

The initial aim of this work was to find out the reason of the resistivity increase found previously for Sm-123 after elevated pressure oxygenation [8, 9]. We investigated the influence of normal and high pressure oxygenating on properties of $REBa_2Cu_3O_{7-\delta}$ compounds with rare earths of smaller (RE = Y, Dy, Ho) and larger (RE = Nd, Sm, Gd) ionic radii. The $Sm_{1+x}Ba_{2-x}Cu_3O_y$ solid solution with $x \leq 0.45$ was also oxygenated at elevated pressure for comparison. We present the results of electrical resistivity and thermopower measurements.

G. Van Tendeloo et al. (eds.), High-Temperature Superconductors and Novel Inorganic Materials, 217–222.

2. Experiment

Samples of $REBa_2Cu_3O_{7-\delta}$ were prepared by the conventional solid state reaction procedure from high purity RE_2O_3 (99.9 %), $BaCO_3$ (spectroscopically pure) and CuO (99.99 %) powders. The preparation of $Sm_{1+x}Ba_{2-x}Cu_3O_y$ solid solution was described elsewhere [7]. The obtained samples were heated at 400 °C in flowing oxygen for 20 h, and then cooled to room temperature. Such samples will be named LP-$REBa_2Cu_3O_{7-\delta}$. A few pellets of LP samples were then annealed in a closed reaction tube in pure oxygen under pressure of 250 bar (as measured at 450 °C) for 40 hours in Morris Research HPS-5015E furnace (these samples will be named HP-$REBa_2Cu_3O_{7-\delta}$). Powder X-ray diffraction analysis indicated that all samples were single-phase.

The thermopower (S) was measured at temperature 300 K using a differential technique with respect to pure copper. The resistivity (ρ) measurements were performed by standard four-probe method in the temperature range 4.2 - 300K.

3. Results and discussion

Ionic radius (r) of the RE^{3+} ions (for eight-fold coordination) changes from 0.977 Å (Lu) to 1.160 Å (La) [10]. In this work we present the results of investigation of $REBa_2Cu_3O_{7-\delta}$ compounds with rare earths elements of larger ionic radii (*i.e.* $r > 1.04$ Å): RE = Nd, Sm and Gd, and smaller ionic radii ($r < 1.04$ Å): RE = Dy, Ho and Y. Only the first group reveals a tendency to form a solid solution of

TABLE 1. Basic data for the investigated $REBa_2Cu_3O_{7-\delta}$ samples: r – RE^{+3} effective ionic radii for eight-fold coordination [10]; LP, HP – low (1 bar) or high (250 bar) pressure oxygenation; S_{300} – room temperature thermopower; oxygen uptake due to HP processing estimated basing on δ *vs.* S_{300} literature data (for Nd- and Sm-samples data for $SmBa_2Cu_3O_{7-\delta}$ were used [7, 8], for the rest we used data for $YBa_2Cu_3O_{7-\delta}$ [11]; T_c and T_c' (intergrain) were assessed from maxima in dρ/dT curves (in case of near lying peaks a numerical deconvolution was applied, dash denotes that the second peak can not be distinguished).

RE	R [Å]	type	S_{300} [μV/K]	oxygen uptake	T_c [K]	T_c' [K]	ρ_{300} [mΩcm]	ρ_{300}/ρ_{100}
Nd	1.109	LP	+11.5		89.9	87.3	1.7	1.74
		HP	+7.2	0.03	93.0	86.7	2.0	**0.72**
Sm	1.079	LP	+6.1		95.2	93.7	2.5	1.4
		HP	–1.5	0.05	95.3	~55	15.0	**0.3**
Gd	1.053	LP	+3.5		93.8	93.3	1.0	2.4
		HP	+1.2	0.01	93.6	~66	6.0	**0.75**
Dy	1.027	LP	+1.0		92.1	–	0.9	2.2
		HP	–0.4	0.03	92.3	91.6	1.3	1.4
Y	1.019	LP	+0.2		92.0	–	1.1	2.7
		HP	0.0	0.00	92.3	–	1.8	2.5
Ho	1.015	LP	+1.6		92.4	91.8	1.1	2.1
		HP	–1.4	0.07	92.9	91.7	1.7	1.6

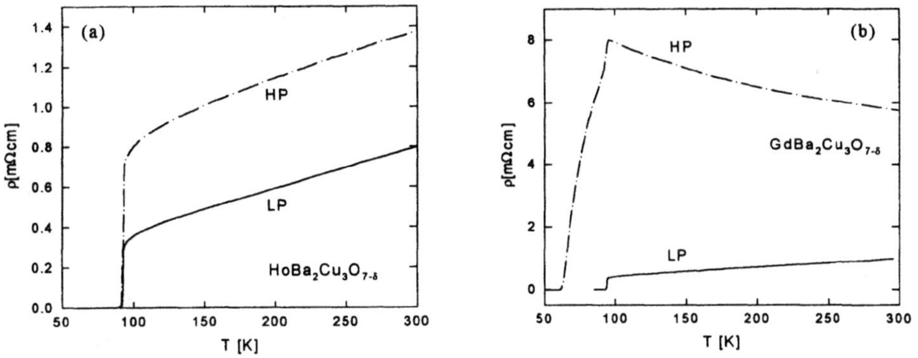

Fig.1 The electrical resistivity of the $HoBa_2Cu_3O_{7-\delta}$ (a) and $GdBa_2Cu_3O_{7-\delta}$ (b) before (LP, solid line) and after (HP, dash-dotted line) the oxygenation at 250 bar

$RE_{1+x}Ba_{2-x}Cu_3O_y$ type. Table 1 shows basic data for these samples processed at low (LP, 1 bar) and high (HP, 250 bar) oxygen pressure. The oxygen concentration was estimated basing on the S_{300} and T_c values and was found to be $(7–\delta) > 6.8$ for all LP samples. Since the slope of the (7-δ) vs. S_{300} dependence varies with RE, for Nd, Sm and Gd samples we use the literature data for Sm-123 [7, 8] and for Dy, Ho and Y samples we use data for Y-123 [11]. Although the above method of oxygen content determination has poor absolute accuracy, its high sensitivity allowed the detection of slight oxygen uptakes due to HP process. The uptakes did not exceed the value of 0.07 (Table 1).

The difference between the two groups of different RE's sizes is visible in the resistivity data after the HP treatment (Table 1). The results may be summarized as follows: (i) bulk (intragrain) T_c increased or remained constant; (ii) intergrain T_c' remained constant or decreased, the decrease is much more substantial for compounds with larger RE^{3+} ions; (iii) the value of resistivity (see ρ_{300}) increases, the increase is more evident for compounds with larger RE^{3+} ions; (iv) the character of the ρ(T) dependence remained metallic for compounds with smaller RE^{3+} ions, whereas for that with larger RE's the character became semiconductor-like (see the ρ_{300}/ρ_{100} values).

Fig. 1 shows the exemplary results of the resistivity for Ho-123 (smaller ionic RE^{3+} radius) and for Gd-123 (larger radius) before (LP) and after (HP) elevated pressure oxygenation. An increase of the resistivity was observed for both samples, but only for the Gd-sample the ρ(T) dependence is no longer metallic. Fig.2 presents the temperature dependences of the derivative dρ(T)/dT near T_c. Superconducting transition temperatures were determined from these derivatives as temperature, where the derivative has a maximum. Most dρ/dT curves show two maxima: the main maximum (T_c) is superconducting bulk transition (close to the T_c obtained from ac susceptibility), the second maximum (T_c') is the superconducting transition of intergrain material and it is equal or lower than bulk (intragrain) T_c (Table 1). Here the difference between two groups of samples is also evident; namely, the overall transition width is much wider for

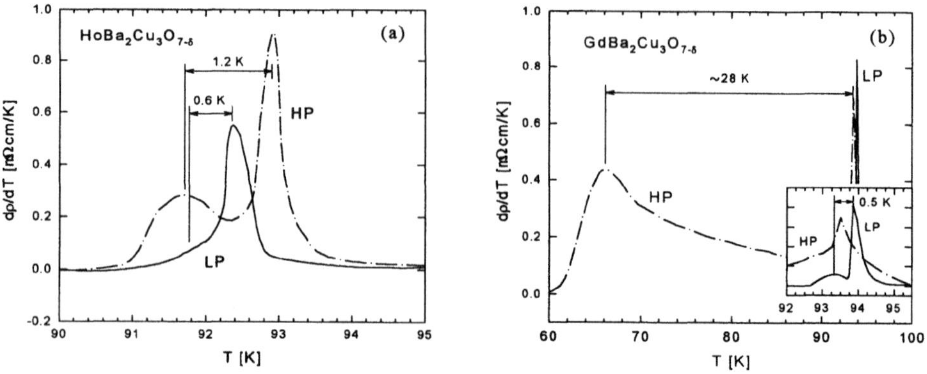

Fig.2 The derivative of the electrical resistivity of the $HoBa_2Cu_3O_{7-\delta}$ (a) and $GdBa_2Cu_3O_{7-\delta}$ (b) before (LP, solid line) and after (HP, dash-dotted line) the oxygenation at 250 bar. The bulk (intragrain), T_c, and intergrain, T_c^i, critical temperatures were assessed as the temperatures of the maxima in dρ/dT curves at higher and lower temperatures, respectively.

HP samples with larger RE ions. Moreover, the bulk (intragrain) transition temperature was found to be very sensitive to the value of the electrical current.

Fig. 3 presents the resistivity of $Sm_{1+x}Ba_{2-x}Cu_3O_y$ ($x \geq 0$) solid solution before and after HP treatment. The series of LP samples demonstrate a progressive destruction of the superconductivity with increasing additional Sm content x. It is visible that the increase of the resistivity due to HP processing, observed for the sample with $x = 0$ (Fig. 3a), is almost compensated by the effect of oxygen uptake for the composition $x = 0.15$ (Fig. 3b). For the samples with higher x values the effect of oxygen content increase dominates: for $x = 0.20$ ρ decreased and T_c increased (Fig. 3c), for $x = 0.45$ the HP treatment induced traces of the superconductivity in the nonsuperconducting LP sample (Fig. 3d).

Summarizing, the changes of the temperature characteristics induced by the HP treatment occurs only for 123 compounds with RE ions large enough to form solid solutions of $RE_{1+x}Ba_{2-x}Cu_3O_y$ type. However, this effect could not be observed in $RE_{1+x}Ba_{2-x}Cu_3O_y$ samples with $x > 0$, because it is overridden by the results of the additional oxygen uptake. The resistivity increase should be obviously attributed to the worsening of the grain boundaries, since the volume properties (like specific heat jump or the thermopower [8]) and even the thermal conductivity [9] have been improved due to the elevated pressure oxygenation.

The mechanism of the grain boundary deterioration seems to be connected with their microstructural properties and/or impurity phases formed at the boundary. The decomposition of the 123 phase hardly could be expected in the conditions of performed processings (T ≤ 450 °C, p ≤ 250 bar), but some impurities might appear during the high temperature synthesis, especially for 123 phases based on rare earths of high ionic radii. The possibility of creation of solid solution may impede the synthesis of pure RE-123 phases, resulting in the appearance of other impurity phases, as $BaCuO_{2+x}$:

$$(1+x)\ REBa_2Cu_3O_{7-\delta} \rightarrow RE_{1+x}Ba_{2-x}Cu_3O_y + 3x\ BaCuO_{2+x}$$

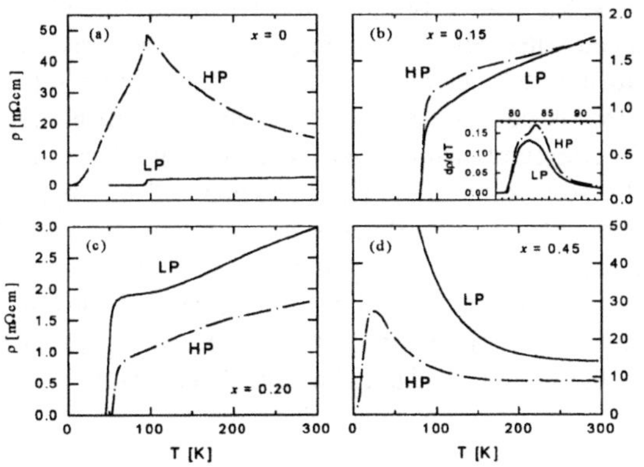

Fig.3 The electrical resistivity of the $Sm_{1+x}Ba_{2-x}Cu_3O_y$ solid solution with x = 0 (a), x = 0.15 (b), 0.20 (c) and 0.45 (d) before (LP, solid line) and after (HP, dash-dotted line) the oxygenation at 250 bar. The insert in Fig 3b. shows the derivative of the electrical resistivity for x = 0.15.

As the matter of fact, using the inclusions of the $BaCuO_{2+x}$ have been already observed in the $SmBa_2Cu_3O_{7-\delta}$ samples by EDX method [8]. We would like to suggest that the intensified presence of the impurity phases in 123 compounds with RE of higher radii might lead to the observed effect of the elevated pressure oxygenation. Namely, as it was shown in Ref. 12, the inclusions of the impurity phases lead to the formation of numerous structural defects during the normal pressure (1 bar) oxygenation, while the compound undergoes the tetragonal-orthorhombic phase transition. It seems that the mechanism of the discussed effect may be based on the kind of zone refinement, i.e. the migration of the defects towards grain boundaries that occurs during the elevated pressure oxygenation. After that treatment most defects are located near the boundaries, resulting in the worsening of their electrical properties. On the other hand, the refined grain interiors exhibit the enhancement of their superconducting properties, as revealed in Refs. 8 and 9.

4. Summary

The effect of RE^{+3} ionic radius on the resistivity of $REBa_2Cu_3O_{7-\delta}$ (RE = Ho, Y, Dy, Gd, Sm, Nd) after elevated pressure oxygenation (250 bar) has been investigated. For all compounds the following changes were observed in respect to the samples treated at 1 bar O_2: (i) the bulk (intragrain) T_c was slightly shifted towards higher temperatures or remained unmoved, whereas the intergrain $T_c^?$ remained unmoved or shifted toward lower temperatures; (ii) the normal state resistivity apparently increased. Both changes were significantly stronger for 123 compounds based on rare earths of high radii, *i.e.* large enough to partially substitute on Ba sites and form solid solutions of the $RE_{1+x}Ba_{2-x}Cu_3O_y$ type. In this case, despite the increased or constant bulk (intragrain) T_c, not only the raise of resistivity was observed, but also a turn of the ρ(T) curves into semiconductor-like dependences. The volume properties like the height of specific jump at T_c and the thermoelectric power [8] or even the thermal conductivity [9] of such ceramics were enhanced after HP processing. Hence, the observed changes in the resistivity should be attributed to the deterioration of the grain boundaries. The proposed mechanism of this effect is connected with the higher overall amount of the impurity phases present in 123 compounds with larger RE's (among other, due to the tendency to create $RE_{1+x}Ba_{2-x}Cu_3O_y$ solid solutions). The inclusions of the impurities also force the creation of large amounts of structural defects due to the tetragonal-orthorhombic phase transition, which occurs during the oxygenation [12]. We suggest that during the elevated pressure oxygenation these defects may be moved toward grain boundaries. Such zone refinement results in the enhancement of the intragrain properties, while the electrical properties of the grain boundaries became deteriorated.

5. Acknowledgments

The work was supported by the Polish State Committee for Scientific Research under contract No. 2PO3B 11613.

6. References

1. Lindemer, T.B., Chakoumakos, B.C., Specht, E.D., Williams, R.K., and Chen, Y.J. (1994) Effects of composition and processing on the superconductivity of $La_{1+z}Ba_{2-z}Cu_3O_y$, *Physica C* **231**, 80-90.
2. Buchgeister, M., Herzog, Hosseini, P.S.M., Kopitzki, K., and Wagener, D. (1991) Oxygen evolution from $ABa_2Cu_3O_{7-\delta}$ high-T_c superconductors with A=Yb, Er, Y, Eu, Sm, Nd and La, *Physica C* **178**, 105-109.
3. Tetenbaum, M., Tumidajski, P., Brown, D.L., and Blander, M. (1992) Some aspects of the thermodynamic behavior of high-T_c oxide systems via EMF and related measurements, *Physica C* **198**, 109-117.
4. Lutgemeier, H., Schmenn, S., Meuffels, P,. Storz, O., Schollhorn, R., Niedermayer, Ch., Heinmaa, I., and Baikov, Yu. (1996) A different type of oxygen order in $REBa_2Cu_3O_{6+x}$ HT_c superconductors with different RE ionic radii, *Physica C* **267**, 191-203.
5. MacManus-Driscoll, J.L., Alonso, J.A., Wang, P.C., Geballe, T.H., and Bravman, J.C. (1994) Studies of structural disorder in $ReBa_2Cu_3O_{7-x}$ thin fils (Re=rare earth) as a functin of rare-earth ionic radius and film deposition conditions, *Physica C* **232**, 288-308.
6. Singh, K.K., Morris, D.E., and Sinha, A.P.B., (1994) Stability of Ba, Sr and Ca cuprate phases as a function of oxygen pressure, *Physica C* **231**, 377-382.
7. Plackowski, T., Sułkowski, C., Bukowski, Z., Włosewicz, D., and Rogacki, K. (1995) Electronic structure parameters of $Sm_{1+x}Ba_{2-x}Cu_3O_y$ solid solution of orthorhombic and tetragonal structure, *Physica C* **254** 331-341.
8. Bukowski, Z., Plackowski, T., Sułkowski, C., Włosewicz, D., and Zaleski, A.J. (1996) $SmBa_2Cu_3O_{7-\delta}$ ceramics processed under oxygen pressure of 250 bar: Enhancement of intragrain superconducting properties, *Phys.Rev.B* **53,** 2781-2785.
9. Plackowski, T., Jeżowski, A., Bukowski, Z., Sułkowski, C., and Misiorek, H. (1997) Thermal conductivity of $SmBa_2Cu_3O_{7-\delta}$ oxidized at 250 bar: A comparison of the phonon and electron models, *Phys.Rev.B* **56**, 11267-11272.
10. Shannon, R.D. (1976) Revised effective ionic radii and systematic studies of interatomic distances in halides and chalcogenides, *Acta Cryst.* **A32**, 751-772.
11. Ouseph, P.J., and O'Brian, M.R. (1990) Thermoelectric power of $YBa_2Cu_3O_{7-y}$, *Phys.Rev.B* **41**, 4123-4125.
12. Kim, C.-J., Lee, Y.-S., Park, H.-S., Kuk, I.-H., Sung, T.-H., Kim, J.J., and Hong, G.-W. (1997) Oxidation induced formation of *a-b* planar defects in melt-textured $YBa_2Cu_3O_{7-y}$ containing Y_2BaCuO_5 inclusions, *Physica C* **276,** 101-108.

TRACER DIFFUSION OF OXYGEN IN 123YBCO FROM MONTE-CARLO SIMULATIONS OF THE ASYNNNI MODEL

A.S. ILCHEV, O.L. DE LANGE AND S. GUMEDE
Department of Physics, University of Natal, Private Bag X01, Scottsville, Pietermaritzburg 3209, South Africa

Abstract. The self-diffusion of oxygen in 123YBCO was studied by means of computer simulations of a two-dimensional lattice model. The tracer diffusion coefficients were calculated from samples of lattice configurations at equilibrium for given values of the temperature and the oxygen concentration. In obtaining these samples both random and ordered initial lattice configurations were used. Regardless of the fact that the simulations from ordered initial configurations do lead to different samples from those obtained for a random start (notably for lower temperatures), the diffusion coefficients do not change significantly. In particular, the strong dependence of the diffusion coefficients on the oxygen concentration, found by previous workers for more uniform configurations, persists even for our nonuniform configurations.

1. Introduction

Recent results obtained for oxygen diffusion in $YBa_2CU_3O_{6+x}$ from Monte Carlo simulations in the antisymmetric next-to-nearest-neighbour Ising (ASYNNNI) model show unsatisfactory features, namely the strong dependence of the tracer diffusion coefficient D^* on the oxygen concentration $c = x/2$ [1, 2, 3]. Because of this, the model has been extended from one based on jumps of oxygen atoms to nearest neighbour sites, to include as well jumps of next-nearest-neighbour sites [3]. However, a strong dependence of D^* on c still remains; for example, at a temperature 0.06eV the value of D^* decreases by about three orders of magnitude as c increases from 5% to 50% [3].

The motivation for the present work stems from the following. In the preceding paper in this book (hereafter referred to as I) we showed that the free energy of the ASYNNNI model possesses multi-minima and that,

G. Van Tendeloo et al. (eds.), High-Temperature Superconductors and Novel Inorganic Materials, 223–226.

as a consequence, nonuniform equilibrium configurations of the oxygen ions occur. We therefore consider the following question: how does this property of the ASYNNNI model affect the diffusion of oxygen and, in particular, is the strong dependence of D^* on the oxygen concentration modified for the nonuniform equilibrium configurations?

2. Calculations and Results

The tracer diffusion coefficients were evaluated from the mean-square displacement of tagged oxygen ions hopping to vacant nearest neighbours and next-to-nearest neighbours sites in a sample of lattices from the canonical ensemble at temperature T and concentration x. As in previous applications of the model, jumps over copper ions were disallowed. We have used the standard Metropolis algorithm both for preparing the sample of theramalized lattices (see I) and for moving the tracers. In particular, we have taken care to ensure that the destinations for the trial jumps are chosen uniformly from all available vacant sites in the neighbourhood of a given oxygen ion. This is essential for conformity of the algorithm with the detailed balance condition which, in turn, ensures the convergence of the corresponding Markovian process to a member of the canonical ensemble [4]. The results reported here correspond to $Q = 0$, Q being the energy barrier for a jump to a next-to-nearest neighbour vacant site.

The quality of the sample is of crucial importance for the reliability of the results. In particular, one has to ensure that the sample contains lattices which are at equilibrium for the chosen values of the temperature and concentration and that the correlations between the members of the sample are minimized. In the ideal case each member of the sample should be produced as the end-point of a separate Markovian process starting from significantly different initial configurations. Because this turns out to be prohibitively expensive in terms of computer time, one has to settle for a sample of only reasonable quality. Usually only one process is generated and after allowing it to develop for a certain number of Monte Carlo steps it is assumed that the process has converged to a configuration belonging to the ensemble. From this moment on one has two choices: either to count each consecutive lattice configuration as a member of the sample and to use it for evaluating the statistical averages of the observables of interest or one could decide to skip a number of Monte Carlo steps before including a configuration in the sample thus reducing the correlations between lattices. In our simulations we monitored the Metropolis acceptance rate and according to its current value have set the number of decorrelation steps (typically about a hundred MC steps).

The tracer diffusion coefficients are evaluated for the nonuniform equi-

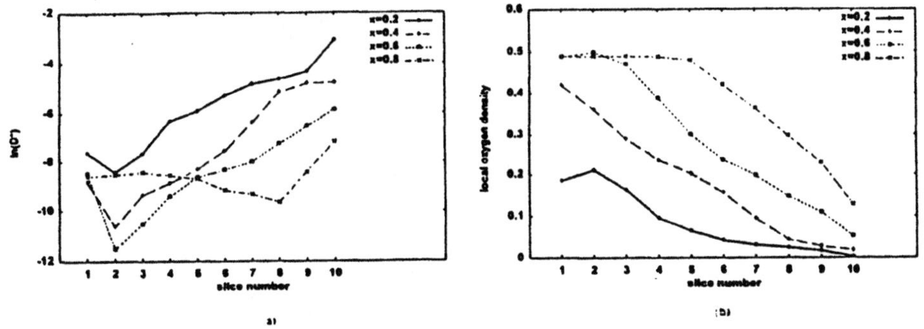

Figure 1. (a) Tracer diffusion coefficients in various lattice slices for a nonuniform equilibrium oxygen configuration, and various concentrations of oxygen. $T = 0.02\text{eV}$ and the lattice is 100×52. (b) Oxygen density profiles in the lattice slices after 200kMC steps.

librium configurations obtained from ordered starts (see I). For these nonuniform configurations the values of the tracer diffusion coefficients and other observables would exhibit a local dependence within the lattice - a feature which is absent from the simulations beginning with a random and uniform start. In order to detect any local dependence of the observables we divided the lattice into a number of slices and for each slice we measured the corresponding values of the tracer diffusion coefficient and the oxygen density. The local densities in the slices were used, after normalization, as weights for evaluating the global tracer diffusion coefficients from the corresponding local (slice) values. This procedure is trivial for homogeneous configurations but could, in principle, lead to a difference between the tracer diffusion coefficients obtained from disordered (random) start and from an ordered start. In Fig. 1 (a) we show the local values of the tracer diffusion coefficients corresponding to temperature $T = 0.02\text{eV}$ and different oxygen concentrations. The fluctuations are considerable due to the low number of oxygen atoms in the slices although lattices of size 100 by 52 were used. The corresponding local oxygen densities are given in Fig. 1 (b).

When the local (slice) values of the tracer diffusion coefficients are summed up with the corresponding statistical weights as obtained from the local density distribution, one gets values which are only slightly different from the corresponding values obtained by simulating the model from a completely random initial configuration. In fact, the difference is noticeable only for very low temperatures. In Fig. 2 the global values for the tracer diffusion coefficients obtained from an ordered (block) start at temperature $T = 0.02\text{eV}$ are shown together with the results published in Ref. [3].

As the temperature increases, even this slight difference between the tracer diffusion coefficients measured by Monte Carlo simulations from different starts disappears. Therefore the different phases observed in the

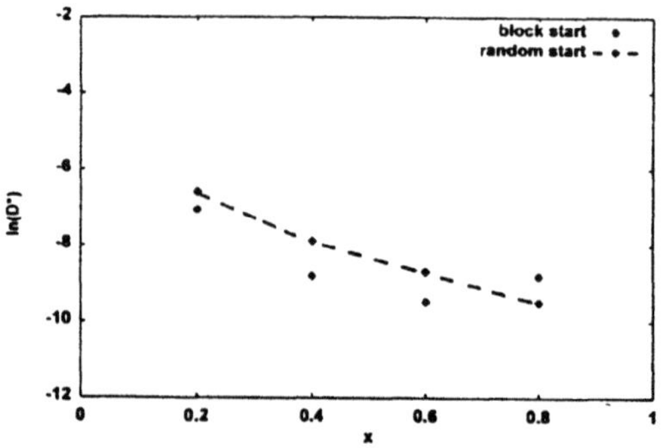

Figure 2. Tracer diffusion coefficients for a block start compared with those for a random start, calculated as described in the text.

ASYNNNI model (see I) are not characterized by different values of the tracer diffusion coefficients. A possible explanation of this fact could be that the difference between the configurations reached from random and ordered starts is in the long-range ordering with very little effect on the mobility of the oxygen ions.

A weak dependence of the diffusion coefficients on the oxygen concentration should follow from any realistic model of self-diffusion in YBCO. The negative result reported here, together with results reported before, casts doubt on the reliability of the ASYNNNI model. In a way this is surprising since once expects a two-dimensional Ising type system to be as adequate for this model as it is in many other similar situations. This makes it worthwhile to attempt other modifications of the characteristics of the ASYNNNI model and of the way it is exploited in an attempt to reconcile it with the experimental situation. The present work was an attempt in this direction. Another possibility is to depart from the simple application of the Metropolis algorithm and to, for instance, simulate the micro-canonical ensemble for the same system. An indication that such an approach could lead to different results for the tracer diffusion coefficients stems from the observation of extended structures in the various phases of this model (see I).

References

1. Salomons, E. and de Fontaine, D. (1990) Monte Carlo study of tracer and chemical diffusion of oxygen in $YBa_2Cu_3O_{6+2c}$, *Phys. Rev.* **B41**, 11159–11167.
2. Pekalski, A. and Ausloos, M. (1994) Monte Carlo simulation of oxygen diffusion in planar model of 123YBCO, *Physica* C**226**, 188–198.
3. Ausloos, M. and Pekalski, A. (1995) Model of oxygen diffusion in $YBa_2Cu_3O_{6+x}$ with next-nearest-neighbor hopping: A Monte Carlo study, *Phys. Rev.* **B52**, 4577–4582.
4. Binder, K. (editor) (1979) *Monte Carlo Methods in Statistical Physics*, Springer, New York.

ELECTRON DOPED MANGANITES

$Ca_{1-x}A_xMnO_3$ (A = Ln, Th).

B. RAVEAU, C. MARTIN, A. MAIGNAN and M. HERVIEU.

Laboratoire CRISMAT, CNRS/ISMRA, 6 boulevard du Maréchal Juin,14050 CAEN - FRANCE. <raveau@crismat.ismra.fr>

Abstract

The discovery of colossal magnetoresistance (CMR) in perovskite manganites has been the starting point of numerous investigations. Most of the compounds that have been studied to date are hole doped manganites, i.e., Mn(III) rich phases. In the present paper, we demonstrate that Mn(IV) rich manganites, $Ca_{1-x}Ln_xMnO_3$ (Ln = Ce, Pr, Nd, Sm, Eu, Gd, Ho) and $Ca_{1-x}Th_xMnO_3$ ($x \leq 0.20$) also exhibit CMR properties. These electron doped manganites exhibit ferromagnetism for low x values and a competition between ferromagnetism and antiferromagnetism as x increases, leading to the appearance of a peak on the M(T) curves. It is shown that the electron concentration is a crucial factor for the appearance and the amplitude of the CMR effect. In contrast to hole doped manganites, the CMR effect in electron doped manganites is observed for a small size of the interpolated cation (Ca, Ln). The size of the A site cations and their mismatch effect also influence the magnetotransport properties but at a lesser degree than the electron concentration.

1 - Introduction.

Numerous studies have been devoted to the perovskite manganites since the discovery of colossal magnetoresistance (CMR) in these oxides [1]. Most of the results have been obtained for Mn(III) rich manganites, i.e., hole doped compounds. The various investigations have shown that the CMR effect is strongly dependent on the hole concentration and on the average size of the interpolated cations. For instance, the phase magnetic diagrams established for the hole doped manganites $Ln_{0.7}A_{0.3}MnO_3$ [2] and $Ln_{0.5}A_{0.5}MnO_3$ [3] have shown that there exists a critical size of the interpolated cation below which CMR tends to disappear. The critical radius $<r_A>$ is of 1.18 Å for the $Ln_{0.7}A_{0.3}MnO_3$ series, corresponding to $Pr_{0.7}Ca_{0.3}MnO_3$, whereas it is of 1.19Å for the $Ln_{0.5}A_{0.5}MnO_3$ series.

G. Van Tendeloo et al. (eds.), High-Temperature Superconductors and Novel Inorganic Materials, 227–232.

In the present study, we explore the opposite side of the phase diagram, corresponding to the Mn (IV) - rich manganites, involving small A-site cations, i.e., $<r_A>$ smaller than 1.18 Å. This work is based on the fact that magnetoresistances with resistance ratio (RR) of about 3 at 50 K have been observed for the Mn(IV) rich manganites $Ca_{1-x}Bi_xMnO_3$ [4-5] and $Ca_{1-x}Eu_xMnO_3$ [6]. We demonstrate the existence of CMR properties for electron doped ceramics $Ca_{1-x}Ln_xMnO_3$ with Ln = Ce, Pr, Nd, Sm, Eu, Gd, Ho and $Ca_{1-x}Th_xMnO_3$ and we show that the electron concentration is a very crucial factor for the appearance of CMR in these materials.

2 - Anomalous magnetic behavior.

The systematic investigation of all this series of oxides $Ca_{1-x}Ln_xMnO_3$ (Ln = Pr to Ho) and $Ca_{1-x}Th_xMnO_3$ [7-8] allows the following features to be emphasized :

(i) For low x values, i.e. close to $CaMnO_3$, the doping with electrons induces a ferromagnetic (FM) state at low temperature. The shape of the M(T) curve is similar for all the lanthanides and for thorium, but it must be emphasized that the maximum magnetic moment at 5 K does not overpass 1.3 μ_B in a magnetic field of 1.45 T. But the most important point deals with the fact that the extension of the FM state depends on the valency of the doping element, corresponding to $0 < x < 0.14$ for trivalent elements (Ln = Pr, Nd, Sm, Eu, Gd) and to $0 < x < 0.07$ for tetravalent elements (Ce, Th). This behavior is illustrated by the M(T) curves of the oxides $Ca_{1-x}Sm_xMnO_3$ (Fig. 1a) and $Ca_{1-x}Th_xMnO_3$ (Fig. 1b).

(ii) For higher x values, a quite unusual behavior is observed on the M(T) curves (Fig. 1), characterized by a peak shape curve so that an antiferromagnetic (AFM) state is obtained at low temperature. Again, the composition range corresponding to this peak is closely related to the valency of the doping element, with $0.14 \leq x \leq 0.20$ for the trivalent elements (Ln = Pr, Nd, Sm, Eu, Gd), and $0.07 \leq x \leq 0.12$ for Ce and Th, as shown for Ln = Sm (Fig. 1a) and Th (Fig. 1b) respectively.

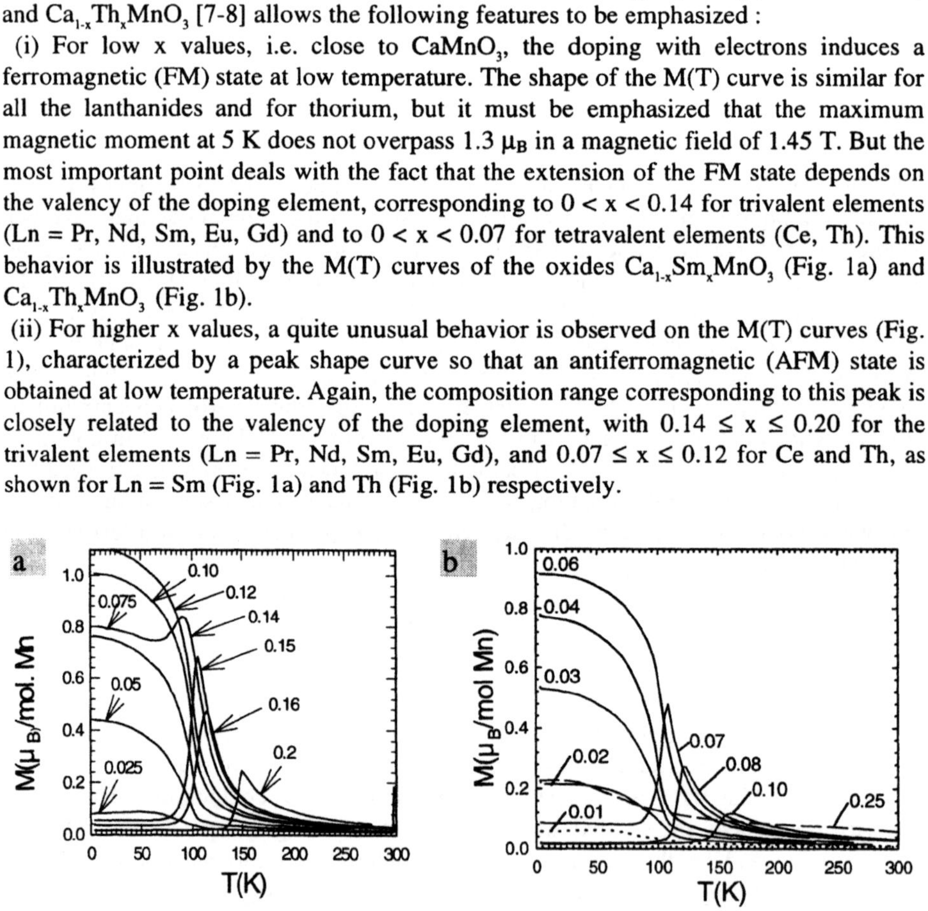

Figure 1 : Temperature dependence of the magnetisation M registered under 1.45 T for (a) $Ca_{1-x}Sm_xMnO_3$ and (b) $Ca_{1-x}Th_xMnO_3$ (x values are labelled on the graph).

3 - Transport properties.

The transport properties of the electron doped manganites are correlated to the magnetic behavior as shown from the ρ(T) curves of the Sm-phases (Fig. 2a) and Th-phases (Fig. 2b). For the low x values, the doping leads to a semi-metallic behavior which coincides with the ferromagnetic state as shown for Sm with x = 0.10 - 0.12 (Fig. 2a) and for Th with x = 0.04 (Fig. 2b). For higher x values a dramatic increase of the resistance is observed (Fig. 2). In fact a transition from the semi-metallic to the insulating state is observed as T decreases, around 100-150 K. Moreover, it is quite remarkable that the transition temperature T_R (see inflection point on the ρ(T) curves) coincides with the appearance of the peak on the M(T) curves (Fig. 1) so that $T_{peak} \approx T_R$.

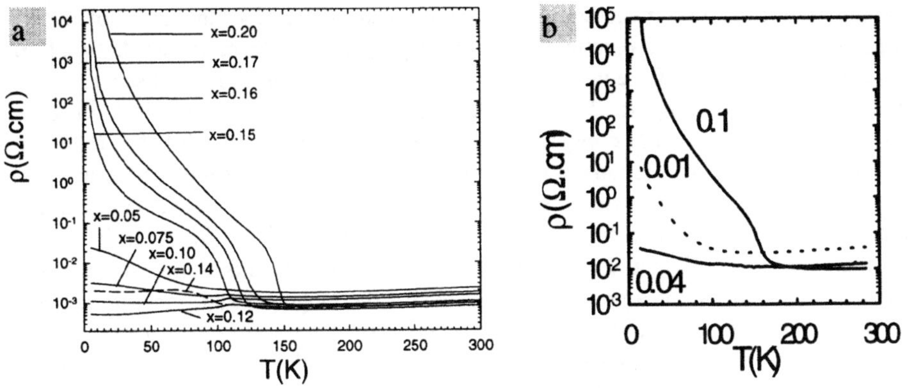

Figure 2 : Temperature dependence of the resistivity ρ for (a) $Ca_{1-x}Sm_xMnO_3$ and (b) $Ca_{1-x}Th_xMnO_3$ (x values are labelled on the graph).

4 - The crucial influence of electron concentration upon magnetic and transport properties.

The comparison of the magnetic and transport properties of the electron doped manganites can be explained on the basis of electronic localisation / delocalisation phenomena. For low x values (x < 0.14 for trivalent dopants and x < 0.07 for tetravalent dopants), introduction of electrons on the e_g level of manganese leads to an electronic delocalisation and consequently favours the ferromagnetic metallic (FMM) state. For higher x values (0.14 ≤ x < 0.20 for trivalent dopants and 0.07 ≤ x < 0.12 for tetravalent dopants) a localisation of electrons appears (charge ordering or local charge ordering in the form of clusterisation) , so that antiferromagnetism is favored at the expense of ferromagnetism. Consequently, ferromagnetism cannot fully develop and the saturation corresponding to a moment of 3.6 μ_B cannot be reached. Finally, the localisation predominates for higher x values, and antiferromagnetism expands rapidly as x increases, so that the intensity of the peak of the M(T) curves decreases abruptly as x increases (Fig. 1).

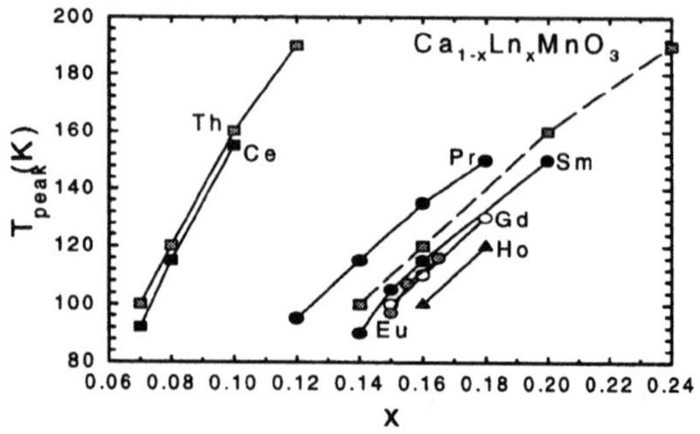

Figure 3 : T_{peak}, obtained from the M(T) curves, versus x for $Ca_{1-x}Ln_xMnO_3$ and $Ca_{1-x}Th_xMnO_3$.

At this point of the investigations it is easy to determine the role of electron concentration upon the magnetic and transport properties of these materials by plotting T peak vs x for the different series of oxides (Fig. 3). One indeed observes that the transition corresponding to T_{peak} (T_R) appears for much lower thorium or cerium contents (x ~ 0.07 - 0.12) compared to trivalent lanthanides (x ~ 0.135 - 0.20). This difference results from the tetravalent character of thorium or cerium, which introduces an electron concentration twice higher than that introduced by trivalent lanthanides. In fact, by plotting T_{peak} vs 2x for Th or Ce phases (dashed line Fig. 3), it can be seen that all these oxides exhibit a very similar behavior. These results demonstrate that the electron concentration is a predominant factor for the magnetic and transport properties of the electron doped manganites. This behavior is also confirmed from the graph of magnetization at 5 K versus x (Fig. 4) which shows that the M_{5K} (x) curves for Ln (III) doped manganites are shifted by a factor two along x with respect to the Ce(IV) or Th(IV) substituted phases.

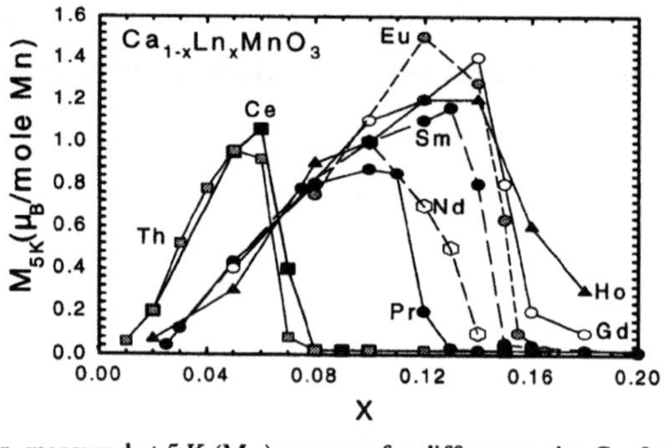

Figure 4 : Magnetization, measured at 5 K (M_{5K}) versus x for different series $Ca_{1-x}Ln_xMnO_3$ (Ln = Ce, Pr, Nd, Sm, Eu, Gd, Ho) and $Ca_{1-x}Th_xMnO_3$.

5 - CMR properties.

The ρ(T) curves registered in a magnetic field of 7 T show the existence of a negative magnetoresitance in the whole domain 0 < x < 0.20, whatever the valence of the dopant. Nevertheless, for the low x values, corresponding to the FMM region, the RR is low (< 3), comparable to that obtained for the bismuth manganites $Ca_{1-x}Bi_xMnO_3$ [4]. This can be explained by the fact that the resistance is rather low, so that the application of a magnetic field cannot modify dramatically the resistance. In fact, there exists an optimal x value for which the CMR effect is maximum. At the latter value, the competition between ferromagnetism and antiferromagnetism reaches it maximum and as a result the resistance ratio (RR) is high. Thus, the $x_{Opt.}$ value corresponds to the peak of highest intensity on the M(T) curve, i.e., just close to the ferromagnetic region and is closely related to the valency of the dopant. This is illustrated by the optimal CMR characteristics obtained for the Sm phase (Fig. 5a) and the Th phase (Fig. 5b) for x_{Opt} values of 0.15 and 0.08 respectively. For each series, the x_{Opt} values and the corresponding RR at three temperatures (T = 5 K, 50 K and 100 K) are listed in table 1.

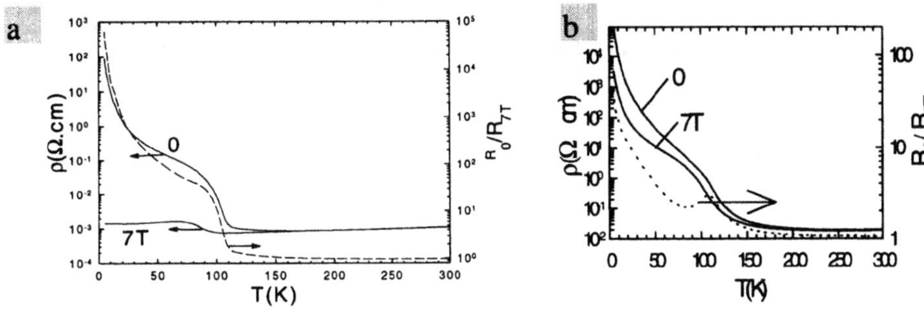

Figure 5 : Temperature dependence of ρ in 0 and 7T (left axis) and magnetoresistance ratio ρ_0/ρ_{7T} (right axis) for (a) $Ca_{0.85}Sm_{0.15}MnO_3$ and (b) $Ca_{0.92}Th_{0.08}MnO_3$.

It is worth pointing out that the trivalent lanthanides exhibit always a higher RR value than the tetravalent dopants (Ce, Th), the highest ratios being obtained for Sm and Eu : RR = 100-110 at 50 K to be compared to RR = 4-5 at 50 K for Th and Ce. But another interesting feature concerns the evolution of the properties of the Ln(III) doped manganites. One observes that x_{opt} increases slightly from 0.135 for Pr to 0.16 for Eu as the size of the lanthanide decreases, and then remains constant down to Ho. Correlatively, as the size of the interpolated cation decreases, the RR increases from 8 at 50 K for Pr to 110 for Eu, and then decreases down to 40 and 12 for Gd and Ho respectively. It is worth pointing out that the transition temperature T_{peak} is slightly affected by the size of the lanthanide, as shown from the diagram T_{peak} (x) for different Ln series (Fig. 3). It can be seen that for a fixed electron concentration x, the transition temperature T_{peak} decreases with the Ln size. In other words, as the mean ionic radius $<r_A>$ of the interpolated cation increases, a higher electron concentration is necessary to replace the FMM state at the benefit of the AFMI state. Thus, in contrast to the hole doped manganites, the stability of the FMM state increases as the size of the interpolated cation decreases. Note, however, that the size effect alone cannot explain the changes of properties that are observed, since the series $Ca_{1-x}Pr_xMnO_3$ - for which $<r_A>$ is kept constant - exhibit also the transition from a FMM to an AFMI state as x increases. It must also be emphasized that the mismatch due to the size difference between the A site cations may also influence these properties. The latter, characterized by the variance σ^2 defined as $\sigma^2 = \Sigma y_i r_i^2 - \langle r_A \rangle^2$ according to Attfield et al. [9], although it increases from Pr to Ho, remains rather small for x < 0.20, so that its influence cannot be evaluated easily (table 1).

TABLE 1. $\langle r_A \rangle$ (according to Shannon), σ^2 and resistance ratio ρ_0/ρ_{TT} for x_{opt} in each series $Ca_{1-x}Ln_xMnO_3$

				ρ_0/ρ_{TT}		
Ln	x_{opt}	$\langle r_A \rangle$ (Å)	σ^2 (Å^2)	at 5 K	50 K	100 K
Th	0.08	1.173	$5.96\ 10^{-4}$	50	4	3
Ce	0.08	1.167	$1.88\ 10^{-3}$	6000	5	4
Pr	0.135	1.18	$1.168\ 10^{-7}$	400	8	4.5
Nd	0.135	1.178	$3.37\ 10^{-5}$	1200	20	5
Sm	0.15	1.173	$6.7\ 10^{-3}$	1000	100	10
Eu	0.16	1.170	$4.84\ 10^{-4}$	3500	110	15
Gd	0.16	1.168	$7.16\ 10^{-4}$	500	40	10
Ho	0.16	1.163	$1.57\ 10^{-3}$	14	12	5

References.

1. Von Helmolt, R. Wecker, J. Holzapfel, B. Schultz, L. and Sanwer, K. (1993) *Phys. Rev. Lett.* **71**, 2331
2. Hwang, H.Y. Cheong, S.W. Radaelli, P.G. Marezio, M. and Batlogg, B. (1995) *Phys Rev. Lett.* **75**, 914.
3. Barnabé, A. Maignan, A. Hervieu, M. Damay, F. Martin, C. and Raveau, B. (1997) *Appl. Phys. Lett.*.**71**, 3907.
4. Chiba, H. Kikuchi, M. Kusaba, K. Muraoka, Y. and Syono, Y. (1996) *Solid State Comm.* **99**, 499.
5. Murakami, T. Shindo, D. Chiba, H. Kikuchi, M. and Syono, Y. (1997) *Phys. Rev. B*, **55**, 1.
6. Troyanchuk, I.O. Samsonenko, N.V. Szymczak, H. and Nabialek, A. (1997) *J. Solid State Chem.* **131**, 144.
7. Martin, C. Maignan, A. Damay, F. and Raveau, B. (1997) *J. Solid State Chem.* **134**, 198.
8. Maignan, A. Martin, C. Damay, F. and Raveau, B. *Chem. Mater.* accepted.
9. Rodriguez-Martinez, L.M. and Attfield, J.P. (1996) *Phys. Rev. B*, **54**, 15622.

$(La_{1-x}Pr_x)_{0.7}Ca_{0.3}MnO_3$ COLOSSAL MAGNETORESISTIVE MATERIALS: STRUCTURE AND ELECTRONIC PROPERTIES

O.YU.GORBENKO[1], A.R.KAUL[1], A.A.BOSAK[1], V.A. AMELICHEV[1], I.E.GRABOY[1], N.A.BABUSHKINA[2], L.M.BELOVA[2], H.W.ZANDBERGEN[3], B.GUETTLER[4], V.L.SVETCHNIKOV[5]
1) Chemistry Department, Moscow State University, 119899 Moscow, Russia;
2) RRC Kurchatov Institute, 123182 Moscow, Russia;
3) National Center for HREM, TU Delft, Rotterdamsweg 137, 2628AL Delft, The Netherlands;
4) PTB, Bundesallee 100, 38116 Braunschweig, Germany;
5) Institute for Metal Physics, Vernadskiy 36, 252680, Kiev, Ukraine

1. Introduction

$R_{1-y}M_yMnO_3$ perovskites, where R^{3+} is a rare earth cation, M is a doubly charged cation of large ionic radius, with both R and M occupying A sites of ABO_3 perovskite lattice, are a subject of renewed interest due to the recent discovery of colossal magnetoresistance (CMR) in such compounds [1]. The CMR is observed in the temperature range of the diversified phase transitions resulting in the formation of a half-metallic ferromagnetic phase stabilized by the mechanism of the double exchange. It was suggested that trapping of the itinerant electrons, which is assisted by phonons related to the dynamic Jahn-Teller distortion, occurs simultaneously at the ferromagnetic transition [2]. Forming three dimensional array in the structure, nearly invariant MnO_6 octahedra are tilted to reduce the space around the A cation if tolerance factor t (t=d(A-O)/ $\sqrt{2}$ d(Mn-O), where d denotes a interatomic distance) is less than 1. The smaller t, the higher the tilt of the octahedra. The buckling results in a decrease of Mn-O-Mn bond angle θ, which disrupts the overlapping of $2p$ orbitals of O and e_g orbitals of Mn ions. If the overlapping is insufficient then double exchange is no longer possible in Mn-O-Mn chains. The localized carriers tend to formation of charge ordered state (CO) [3].

In the current work we have studied the cumulative effect of chemical pressure and lattice strain in the sequence of $(La_{1-x}Pr_x)_{0.7}Ca_{0.3}MnO_3$ solid solution and demonstrated the importance of the lattice dynamics by colossal isotope effect for the critical value of t.

2. Experimental

Aerosol MOCVD used to prepare thin films includes deposition from precursors vapor produced by the evaporation of their organic solution nebula in the carrier gas flow [4]. Volatile precursors were as follows: $La(thd)_3$, $Pr(thd)_3$, $Sr(thd)_2$, $Ca(thd)_2$ and $Mn(thd)_3$, where thd is 2,2,6,6-tetramethylheptan-3,5-dionate. Single crystalline $LaAlO_3$

G. Van Tendeloo et al. (eds.), High-Temperature Superconductors and Novel Inorganic Materials, 233–238.

(pseudocubic cell) and (001) $SrTiO_3$ were used as substrates. Deposition runs were carried out at 750°C with total pressure of Ar - O_2 mixture 6 mbar (O_2 partial pressure 3 mbar). $(La_{1-x}Pr_x)_{0.7}Ca_{0.3}MnO_3$ films prepared were 60 and 300 nm thick. Directly after the deposition at 750°C the rector was filled with oxygen up to the pressure of 1 bar and films were annealed for 0.5 h at the deposition temperature. The ceramic samples were prepared as follows: ashfree paper filters were soaked with the water solution of the metal nitrates, then the ashes formed by burning of the dried paper were pressed in pellets and sintered in air at 1200°C for 12h.

SEM was accomplished by CAMSCAN equipped with EDAX system for quantitative chemical analysis. HREM study of the films was performed with a Philips CM30ST electron microscope operating at 300 kV. XRD with four circle diffractometer Siemens D5000 with secondary graphite monochromator (Cu K_α radiation) was used to determine phase composition, orientation and lattice parameters. θ-2θ (including measurements of the tilted samples over the reflection poles found by φ- and χ-scanning), φ scans and out-of-plane rocking curve measurements were used. Reflections of the substrates were used as inner standard. The study was repeated after the isotope exchange. For isotope exchange pairs of 1x8 mm^2 strips were cut from the samples and put into alumina boats which were placed in two quartz tubes mounted in the furnace. The quartz tubes formed parts of two identical closed loops where enforced circulation of gas was applied. The parallel samples were treated simultaneously: one part of the each sample was heated in $^{16}O_2$ atmosphere, the other part of the each sample was heated in $^{18}O_2$ (molar fraction of $^{18}O_2$ 85%). The diffusion annealing was carried out under the oxygen pressure 1 bar for 4h at the temperature 800°C (films) and for 48h at 950°C (ceramics). The completion of the exchange was controlled by gravimetry (ceramics) and by the measurement of the phonon bands shift in the Raman spectra (both ceramics and thin films).

3. Results and discussion

3.1 PECULIARITIES OF THE THIN FILM STRUCTURE

Perovskite substrates ($LaAlO_3$, $SrTiO_3$) permited deposition of layers showing the epitaxy of the 'cube-on-cube' type. HREM revealed clear and coherent film-substrate intefaces. The rocking curves for $(00l)_p$ reflections were rather narrow (FWHM 0.17-0.25°). Thin films $(La_{1-x}Pr_x)_{0.7}Ca_{0.3}MnO_3$ have a different structure as compared to that of the ceramics. The buckling is expected to increase gradually with the increase of Pr content. The buckling is the main cause of the variation of lattice parameters and especially of the molar volume both for the ceramics and films. On the other hand, in the whole series we have found a discrepancy between d-values of pseudocubic $(00l)_p$ reflections of the films on $LaAlO_3$ and $SrTiO_3$. The latter demonstrated lower d-values than the former with difference increasing with x (Fig.1). The reflections were narrow unsplit peaks with the profile fitted well by the Voigt function for $\alpha_1+\alpha_2$ spectral doublet. By measurements of θ-2θ scans under the tilting we have found that the unit cell is close to tetragonal.

The tetragonal distortion has opposite direction for the films on $LaAlO_3$ and $SrTiO_3$ in agreement with the mismatch between lattice constants of the substrates and pseudocubic lattice constant for the films a_p. The lattice constant of $SrTiO_3$ is larger, consequently the tensile stress demands in-plane expansion in the film. The deformation of the lattice normal to the substrate (ε_3) is proportional to the in-plane strain components (σ_1,σ_2): $\varepsilon_3= -(\sigma_1+\sigma_2)\eta/E$, where η - Poisson coefficient, E - Young modulus. The films on $SrTiO_3$ are contracted along the normal. The opposite is valid for the films on $LaAlO_3$. Values of a_p are close for the films on the both substrates and ceramics of the same composition suggesting that $\eta \sim 0.5$.

The tetragonal distortion depends on the film thickness (Table 1). The gradual character of the transition implies that the local lattice symmetry can deviate significantly from the mean tetragonal metrics found by XRD. On the other hand, no significant strain gradient is present in the thicker films since no line broadening is observed in the XRD patterns. Thus, we can assume that the relaxation is related to the appearance and gradual increase of the density of the misfit dislocation at the film-substrate interface but not in the film depth. HREM study of the mechanism now is in progress.

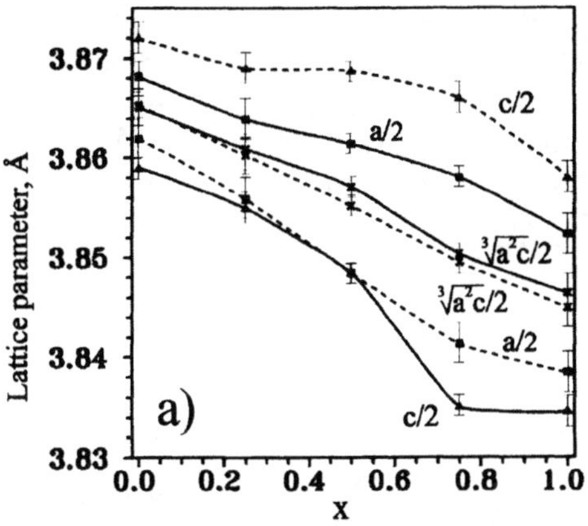

Figure 1. Tetragonal lattice parameters of $(La_{1-x}Pr_x)_{0.7}Ca_{0.3}MnO_3$ films 300 nm thick: on $LaAlO_3$ (dashed lines) and on $SrTiO_3$ (solid lines). The tetragonal distortion of the opposite direction is observed but the mean volume of the perovskite cube ($a_p^3 = a^2c$) remains nearly constant.

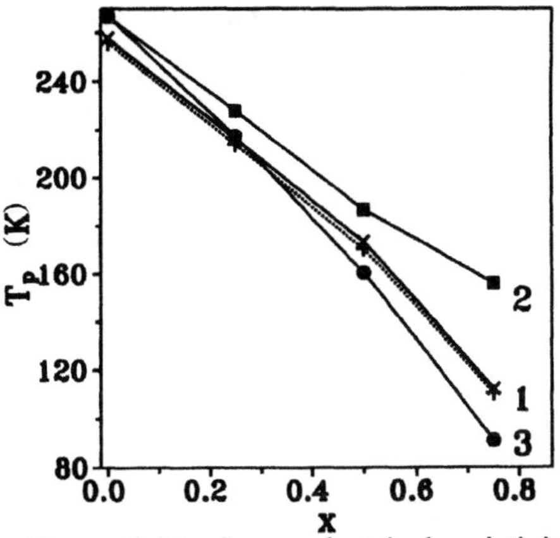

Figure 2. Maximum electrical resistivity temperature T_P (x) for $(La_{1-x}Pr_x)_{0.7}Ca_{0.3}MnO_3$ ceramics (1) and thin films grown on $SrTiO_3$ (2) and $LaAlO_3$ (3). The dotted line shows the variation of the Curie temperature T_c (x) for ceramic samples.

3.2 ELECTRICAL PROPERTIES

In the series of $(La_{1-x}Pr_x)_{0.7}Ca_{0.3}MnO_3$ solid solutions tolerance factor t passes through its critical value 0.91 for $x \sim 0.75$ [5]. Starting with ~260 K for $La_{0.7}Ca_{0.3}MnO_3$, the Curie temperature (T_c) and nearly coincident T_p drop rapidly with increase of Pr content for ceramics. All forms of the material, both with and without epitaxy-related stress, show the similar dependence on x (Fig.2). The variation of the sample form (film or ceramics) and substrate material manifest itself in the change of the buckling anisotropy in the MnO_6

octahedra array. The metal-insulator phase transition is very sensitive to a change of the Mn-O-Mn bond angle. Hence, we could expect a peak of sensitivity near $t \sim 0.91$.

In fact, T_p values become divergent approaching the critical t value. In doing so T_p varies from the higher value for the film on $SrTiO_3$ to the lower value for the film on $LaAlO_3$. In the film on $SrTiO_3$ the bending of the Mn-O-Mn chains is decreased in the plane of the film. On the contrary, in the plane of the film on $LaAlO_3$ the bending of the Mn-O-Mn chains is increased. The ceramic samples occupy an intermediate position - the buckling is more uniform in their structure. On the other hand, we have found no drastic effect of the magnitude of the lattice strain on T_p for $La_{0.35}Pr_{0.35}Ca_{0.3}MnO_3$ (Fig.3): the thinner films with higher lattice strain reveal only few K shift of T_p as compared to the thicker films with lower lattice strain. More extended experimental data are necessary for a detailed understanding of such small shifts.

TABLE 1. Structural characteristics of the $La_{0.35}Pr_{0.35}Ca_{0.3}MnO_3$ films on the perovskite substrates

Substrate	Film thickness, nm	Parameters*		Tetragonal
		$a/2, b/2$	$c/2$	Distortion, %
$LaAlO_3$	60	3.846±0.003	3.872±0.002	-0.7
	300	3.848±0.003	3.869±0.002	-0.5
$SrTiO_3$	60	3.879±0.003	3.823±0.002	1.4
	300	3.862±0.003	3.848±0.002	0.4

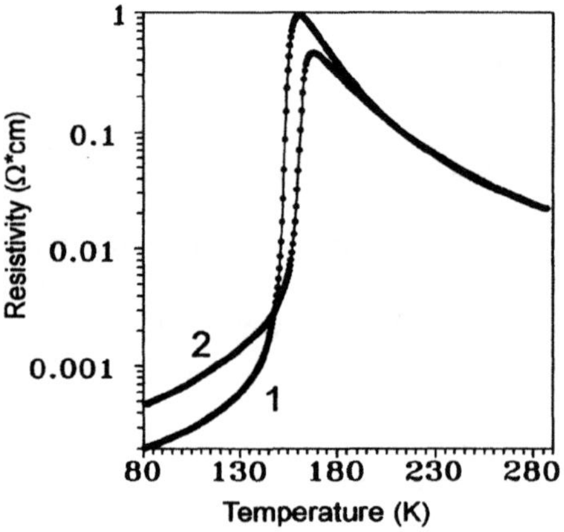

Figure 3. Effect of the film thickness on the temperature dependence of the electical resistivity of $La_{0.35}Pr_{0.35}Ca_{0.3}MnO_3$ films on $LaAlO_3$: (1) - 300 nm, (2) – 60 nm. Analogous results were obtained for the films on $SrTiO_3$.

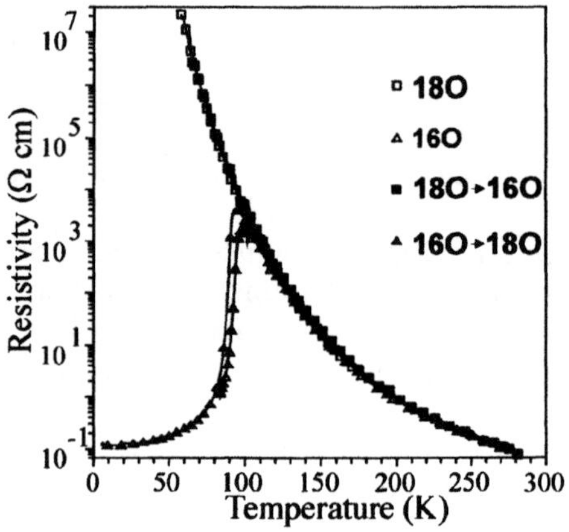

Figure 4. The temperature dependence of resistivity for $La_{0.175}Pr_{0.525}Ca_{0.3}MnO_3$ ceramic samples after the first isotope exchange (^{18}O and ^{16}O) and after the isotope back-exchange ($^{18}O \rightarrow ^{16}O$ and $^{16}O \rightarrow ^{18}O$, an arrow means 'instead of').

3.3 ANNEALING AND OXYGEN ISOTOPE EFFECT

The isotope effect observed by us in the ceramics of the critical composition $La_{0.175}Pr_{0.525}Ca_{0.3}MnO_3$ has been far in the excess of our expectations, as not merely a shift of T_p but a metal-insulator transition due to the isotope exchange was established (Fig.4). The ^{16}O sample showed a maximum of resistivity at 95 K typical for a ferromagnetic phase transition in $R_{1-x}M_xMnO_3$, the ^{18}O sample was insulating down to 4.2 K. After the isotope back-exchange, the sample, which had been firstly saturated by ^{18}O, became metallic below 95 K after subsequent annealing in $^{16}O_2$. Respectively, the sample, which had been firstly treated in $^{16}O_2$, became insulator after the treatment in $^{18}O_2$.

The application of magnetic field is known to provide melting of CO state in $(La_{1-x}Pr_x)_{0.7}Ca_{0.3}MnO_3$ resulting in a drastic decrease of the electric resistivity. We have found that the magnetic field of 1T was not enough to suppress the CO state in $La_{0.175}Pr_{0.525}Ca_{0.3}Mn(^{18}O_{0.85}{}^{16}O_{0.15})_3$ in spite of the more than 30 K upward shift of T_p in the case of $La_{0.175}Pr_{0.525}Ca_{0.3}Mn(^{16}O)_3$. Nevertheless in the field of 2T the melting of charge ordered state was clearly detected (Fig.5). The isotope shift of T_p comprised 63 K in the field of 2T and 54 K in the field of 3T, which is considerably more than isotope effects whenever reported for $R_{1-x}M_xMnO_3$. The resistivity curves at the temperature higher than T_p are practically independent of the isotope exchange.

Also, the long-term annealing in $^{16}O_2$ atmosphere results in a shift of the maximum resistivity temperature T_p (Table 2). In contrast to the small shift of T_p upwards after annealing of the $La_{0.35}Pr_{0.35}Ca_{0.3}MnO_3$ films in $^{16}O_2$ a drastic decrease of T_p was observed after the parallel annealing in $^{18}O_2$. The effect was higher for the film on $LaAlO_3$: no transition to the metallic state was detected down to 4.2 K for ^{18}O sample (Fig.6). We would remind that due to the strain the film on $LaAlO_3$ is actually closer to the critical state for localization than the film on $SrTiO_3$. One can learn it from the variation of T_p for $(La_{1-x}Pr_x)_{0.7}Ca_{0.3}MnO_3$ films as a function of x: T_p decreases with the growth of x more rapidly for the films on $LaAlO_3$.

TABLE 2 The comparison of the maximum resistivity temperatures T_p (K) for the 60 nm thick $La_{0.35}Pr_{0.35}Ca_{0.3}MnO_3$ films

Substrate	T_p (K)		
	As-grown film	After 4h annealing in $^{16}O_2$	After 4h annealing in $^{18}O_2$
$LaAlO_3$	168.0	182.3	no T_p down to 4.2 K
$SrTiO_3$	176.6	190.1	125.2

The high temperature annealing could cause the relaxation of the film structure. The XRD study after the isotope exchange showed the relatively small relaxation for the films 300 nm thick but no relaxation for the films 60 nm thick. On the other hand, the annealing effect is observed in the strain-free ceramics too. Thus, the shift of T_p is due to other mechanism not related to the variation of the lattice parameters (in particular, no oxygen nonstoichiometry or Mn^{3+}/Mn^{4+} variation is involved in the process). Such a mechanism could be a clustering [6]. The enthalpy gain originates from the fact that Mn^{4+} containing MnO_6 octahedra tend to be surrounded by A^{2+} cations providing the more effective screening of the Coulomb potential.

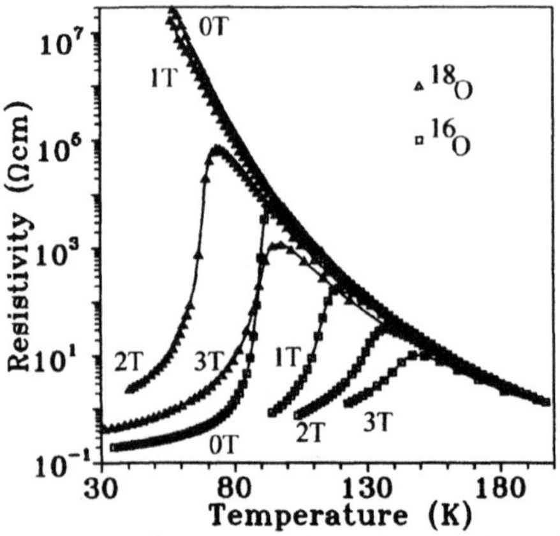

Figure 5. The temperature dependence of resistivity for $La_{0.175}Pr_{0.525}Ca_{0.3}MnO_3$ under various magnetic fields.

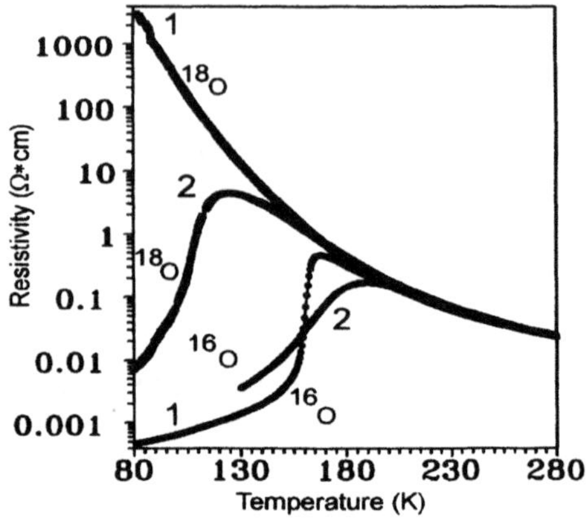

Figure 6. The oxygen isotope effect in the 60 nm thick films of $La_{0.35}Pr_{0.35}Ca_{0.3}MnO_3$ on $LaAlO_3$ (1) and $SrTiO_3$ (2).

4. Conclusions

To sum up, we have found a colossal oxygen isotope shift of the maximum resistivity temperature for $(La_{1-x}Pr_x)_{0.7}Ca_{0.3}MnO_3$ near critical tolerance factor value 0.91. The magnitude of the effect is a vivid demonstration of the importance of the electron-phonon coupling in $R_{1-y}M_yMnO_3$ waiting for a theoretical comprehension. Also, the lattice strain in thin films becomes important for electrical properties when t is close to 0.91. On the other hand, the shift of T_p during high temperature annealing is not due to the strain relaxation because it happens even without such relaxation. A plausible reason for the shift of T_p is a clustering influenced by the annealing.

5. Acknowledgements

This study was partly supported by VW Foundation (grant No. I/73628) and RFBR (grant No. 97-03-32979a).

6. References

1. Rao, C.N.R., Cheetham, A.K. and Mahesh, R. (1996) Giant magnetoresistance and related properties of rare earth manganites and other oxide systems, *Chem. Mater.* **8**, 2421-2432.
2. Millis, A.J., Mueller, R., and Shraiman, B.I. (1996) Fermi-liquid-to-polaron crossover, *Phys. Rev. B* **54**, 5389-5417.
3. Yoshizawa, H., Kawano, H., Tomioka, Y. and Tokura, Y. (1996) Magnetic field-induced metal-insulator transition in $Pr_{0.7}Ca_{0.3}MnO_3$, *J. Phys. Soc. Jpn.* **65**, 1043-1052.
4. Gorbenko, O.Yu., Fuflyigin, V.N., Erokhin, Y.Y., Graboy, I.E., Kaul, A.R., Tretyakov, Y.D., Klippe, L. and Wahl, G. (1994) YBCO and BSCCO thin films prepared by wet-CVD, *J. Mater. Chem.* **4**, 1585-1589.
5. Hwang, H.Y., Cheong, S.W., Radaelli, P.G., Marezio, M. and Batlogg, B. (1995) Lattice effects on the magnetoresistance in doped $LaMnO_3$, *Phys. Rev. Lett.* **75**, 914-917.
6. Hervieu, M., Van Tendeloo, G., Caignaert, V., Maignan, A. and Raveau, B. (1996) Monoclinic microdomains and clustering in the colossal magnetoresistive manganites $Pr_{0.7}Ca_{0.25}Sr_{0.05}MnO_3$ and $Pr_{0.75}Sr_{0.25}MnO_3$, *Phys. Rev. B* **53**, 14274-14284.

HREM STUDY OF DOMAINS IN THIN FILMS OF $La_{0.73}Ca_{0.27}MnO_3$ AND $(La_{1-x}Pr_x)_{0.7}Ca_{0.3}MnO_3$

H.W.ZANDBERGEN[1], J. AARTS[2], S. FREISEM[2], I.E.GRABOY[2], O.YU.GORBENKO[2] AND A.R.KAUL[2]

2) Kamerlingh Onnes Laboratory, Leiden University, P.O. Box 9506, 2300 RA Leiden, The Netherlands.

1) National Centre for HREM, TU Delft, Rotterdamsweg 137, 2628AL Delft, The Netherlands;

3) Chemistry Department, Moscow State University, 119899 Moscow, Russia;

1. Introduction

Ever since the discovery of oxide high T_c superconductors (HTS), oxide materials are playing a prominent role in both fundamental and applied materials research. Lately, Colossal Magnetoresistance (CMR) materials [1], in which the resistance is changed by orders of magnitude upon applying a magnetic field, have received widespread attention. The superconductivity is related to the details of charge doping of the CuO_2 planes on an atomic scale, in which the microstructure does not play a major role. In contrast, the CMR effect depends on the mobility of electrons, combined with their spin memory, and is extremely sensitive to the microstructure [2]. Even in films of the same compound, different deposition methods, annealing treatment and substrate material lead to serious differences in the magnitude and temperature dependence of the magnetoresistance (MR) effect, while it is usually also different from the MR in the equivalent bulk material. The latter fact is partly fundamental in nature. Both the magnetic ordering temperature and the MR effect are intrinsically sensitive to strain and pressure, as for instance imposed by a substrate. This also means that the fundamental properties of films are often masked by inhomogeneities due to (partial) relaxation effects, which may even be enhanced by post-anneal treatments. The question of fundamental film properties constitutes one reason to investigate very thin films, where relaxation has not yet taken place. Another is the possibility of combining e.g. CMR and HTS materials, with the purpose of using the magnetism or the MR effect to change (or switch) the properties of the superconductor. In that case, high quality hetero-epitaxy requires typical film thicknesses of 10 nm rather than of 100 nm.

The growth of the thin films is performed in a temperature range where the structure is not yet orthorhombic. This implies that upon cooling down several variants can be formed. At the boundaries of the domains of different variants a structure will be present which differs from the bulk structure. Since the properties are very dependent on the Mn-O-Mn bond angle, such domain boundaries can have a strong effect on the overall properties. For this reason we have started an electron microscopy study on the

G. Van Tendeloo et al. (eds.), High-Temperature Superconductors and Novel Inorganic Materials, 239–243.

domains occurring in thin films and in particular on the domain boundaries. The first results are reported in this paper.

2. Experimental

The $La_{1-x}Ca_xMnO_3$ films were sputter deposited [3] from ceramic targets $La_{1-x}Ca_xMnO_3$ with a nominal composition of x=0.33, on [001] oriented substrates of $SrTiO_3$ with a miscut of less than 0.1°. Deposition was performed in a pure oxygen atmosphere of 300 Pa, with substrate and source in line and about 2 cm apart. The high pressure causes a very low growth rate of the order of 0.9 nm/min. The growth temperature was chosen at 840° C, in order to be able to grow multilayer films with $YBa_2Cu_3O_7$ under the same conditions. After deposition, the film was cooled to room temperature under the same oxygen pressure, without further annealing. The chemical composition of thick films was determined by microprobe analysis, which showed a Ca content of x=0.27.

$(La_{1-x}Pr_x)_{0.7}Ca_{0.3}MnO_3$ films were prepared by aerosol MOCVD deposition from precursors vapour produced by the evaporation of their organic solution nebula in the carrier gas flow [4]. Further details are given in ref. [5].

Electron transparent areas in cross-section were obtained by ion milling with the film facing away from the ion guns. High resolution electron microscopy (HREM) and electron diffraction (ED) were performed with a Philips CM30UT electron microscope with a field emission gun operated at 300 kV.

Figure 1. HREM image taken along the [100] direction of $SrTiO_3$. The 30 nm film of $La_{0.73}Ca_{0.27}MnO_3$ shows three orientations: i) c axis in vertical direction, ii) c axis in horizontal direction and iii) c axis coming out of the figure.

Figure 2. HREM image of the $SrTiO_3/(La_{1-x}Pr_x)_{0.7}Ca_{0.3}MnO_3$ interface taken along the [110] direction of $SrTiO_3$. The $(La_{1-x}Pr_x)_{0.7}Ca_{0.3}MnO_3$ film is about 200 nm thick. One can observe two domains corresponding to the [100] and [001] directions of LCMO. The boundary between the two domains is not sharp indicating that the boundary is inclined to the viewing direction.

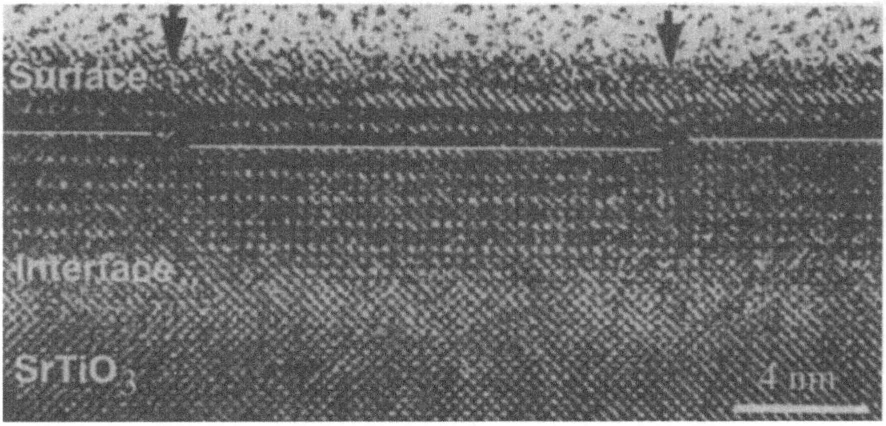

Figure 3. HREM image of a 5 nm $La_{0.73}Ca_{0.27}MnO_3$ film viewed along [101] of LCMO. One can observe (101) mirror twin boundaries (TBs), which are indicated by black arrows and by the white horizontal lines.

3. Results and discussion

An important feature of the structure of bulk $(La,Pr)_{0.7}Ca_{0.3}MnO_3$, $La_{1-x}Ca_xMnO_3$, and many other CMR oxides is the tilting of the MnO_6 octahedra which lead for $La_{1-x}Ca_xMnO_3$ to an $a_p\sqrt{2},2a_p,a_p\sqrt{2}$ superstructure [6,7] where a_p is the unit cell axis of the simple perovskite structure. The reason for this tilt is the too small size of the La and Ca atoms for the 12 coordination of oxygen atoms which is present in the simple perovskite structure; the tilts of the MnO_6 octahedra allow a reduction of the volume of the unit cell and thereby the space for the (La,Ca) atoms. These tilts do hardly change the octahedral coordination of the Mn atoms and the MnO_6 units can - in first approximation - be considered as rigid bodies. In this case the tilt of one MnO_6 octahedron determines the tilts of all other MnO_6 octahedra, since all octahedra are coupled by corner sharing. The size of the unit cell depends the La/Ca ratio, the Mn^{3+}/Mn^{4+} ratio which is obviously linked to the La/Ca ratio and the tilt of the MnO_6 octahedra, which again is linked to the La/Ca ratio in the case of bulk materials, but which can be quite different for a thin epitaxial film. In the case of $La_{1-x}Ca_xMnO_3$ Radaelli et al [6] report for bulk $La_{0.75}Ca_{0.25}MnO_3$ that a=0.5481 nm, b=0.7751 nm, c=0.5499 nm whereas Dai et al [7] report a=0.5454 nm, b=0.7704 nm, c=0.5468 nm for bulk $La_{0.65}Ca_{0.35}MnO_3$. In the case of very thin films of $La_{0.7}Ca_{0.3}MnO_3$ on $SrTiO_3$, the unit cell parameters depend on the orientation of the unit cell to the [001] oriented substrate. If the b axis is along the interface normal, the unit cell parameters are a=0.5516 nm, b= 0.764 nm, c=0.5516 nm, whereby the a and c axes adopt the lattice parameter of the substrate resulting in an enlargement and the b is shortened [8] to compensate for the otherwise large increase in the unit cell volume. For other epitaxial orientations a similar substrate-imposed increase and decrease of cell parameters occur. For thicker films, dislocations do occur at the interface, resulting in change of the cell parameters towards the bulk values.

A change in the tilts of the octahedra leads to a change in the Mn-O-Mn bond angle. Since such a change results also directly in a change in the overlap of $2p$ orbitals of O and e_g orbitals of Mn ions, the double exchange which governs the magnetic interaction between the Mn ions will change, resulting in a different magnetic behaviour and the CMR. In addition, the MnO_6 octahedra can be distorted due to the Jahn-Teller (elongation along one of the axes) effect as is reported for $LaMnO_3$ [9] and $La_{0.5}Ca_{0.5}MnO_3$ at low temperatures [10].

As is obvious from the HREM images, domains occur which are due to the ordering of the MnO_6 tilt. The domains are the result of the six possible variants of the MnO_6 tilts. Relating the tilts to the $SrTiO_3$ lattice, which adopts the basic perovskite structure, and taking the c axis of $SrTiO_3$ as the normal of the interface the domain orientations are:

i) c(LCMO) // c($SrTiO_3$) and a (LCMO) // [110]($SrTiO_3$)
ii) c(LCMO) // c($SrTiO_3$) and a (LCMO) // [1-10]($SrTiO_3$)
iii) c(LCMO) // a($SrTiO_3$) and a (LCMO) // [011]($SrTiO_3$)
iv) c(LCMO) // a($SrTiO_3$) and a (LCMO) // [1-10]($SrTiO_3$)
v) c(LCMO) // b($SrTiO_3$) and a (LCMO) // [110]($SrTiO_3$)
vi) c(LCMO) // b($SrTiO_3$) and a (LCMO) // [1-10]($SrTiO_3$)

During the growth the CMR material is still cubic or rhombohedral, the phase transformation upon cooling can lead to each of the six variants. The fraction of each of the domains occurring in the thin film will depend on the strain. For thin films the highest fraction is expected for that orientation which has the lowest misfit energy with the substrate. We observe that the two domains with the b axis along the interface normal are the major fraction.

If one images the thin film along the [100] direction of $SrTiO_3$ only three variants (the interchange between the a and the c axes of LCMO cannot be observed) can be distinguished as is clear from Figure 1. The three are:

i) c(LCMO) // c($SrTiO_3$)
ii) c(LCMO) // b($SrTiO_3$)
iii) c(LCMO) // a($SrTiO_3$)

If one images the thin film along the [110] direction of $SrTiO_3$ one can clearly distinguish the interchange between the a axis and the c-axis of LCMO, but one can only see the b axis if it is parallel to the interface normal, as can be seen from Figure 2.

No sharp domain boundaries across which the a and the c axes interchange can mostly be observed in images taken along the [110] direction of $SrTiO_3$. On the other hand, HREM images taken along the [100] direction of $SrTiO_3$ show sharp boundaries as can be seen from Figure 3. This implies that the domain boundary is normally parallel to a {110} plane of LCMO as is schematically represented in Figure 4.

The orientation of the LCMO and LPCMO films can be determined easiest from the d(010) spacing, which can also be observed in Figure. However, this spacing cannot be observed if the crystal in oriented exactly along the [110] because this reflection is forbidden in this space group Pnma. To image this fringe one has to break the symmetry either at the crystal (by a misorientation) or in the imaging by allowing the beam not to be on the optic axis of the microscope. The symmetry can also be broken by lowering the symmetry of the space group of the LCMO, as Hervieu [11] has suggested (space group $P2_1/n$). It is still an open question whether a lowering of the symmetry occurs locally. However, since no indication has been obtained from X-ray diffraction for such a lowering, it is not likely that this lowering occurs at a significant scale.

4. Conclusions

Six variants of the orientation of LCMO occur in the thin films. The dominant orientations are those with the b axis along the interface normal. Deviations from this orientation occur predominantly in the thicker films.

The boundary at which the a and c axes are interchanged runs along the {101} planes of LCMO, which is logical because this type of boundary has a low energy (other boundaries would lead to a frustration of the tilts of the octahedra near the boundary.

5. Acknowledgements

This study was partly supported by INTAS (grant 93-936-ext) and by the 'Stichting voor Fundamenteel Onderzoek der Materie (FOM), which is part of NWO.

EFFECT OF TEMPERATURE ON $Ca_2Fe_2O_5$ AND CRYSTAL STRUCTURES IN THE $(Ca,Sr)Fe_{1-y}Ti_yO_{3-\delta}$ SYSTEM STUDIED BY POWDER NEUTRON DIFFRACTION

P. BERASTEGUI[1], S.-G. ERIKSSON[2] and S. HULL[1]
[1]*ISIS Facility, Rutherford Appleton Laboratory, Chilton, Didcot, Oxfordshire OX11 0QX, UK*
[2]*Dept. of Inorganic Chemistry, Chalmers University of Technology and University of Göteborg, S-412 96 Göteborg, Sweden*

1. Abstract

A study of the phases present in the $(Ca,Sr)Fe_{1-y}Ti_yO_{3-\delta}$ system has been carried out by powder neutron diffraction. Data were collected for a series of samples with composition $CaFe_{1-y}Ti_yO_{3-\delta}$ ($0\leq y\leq 0.83$) and the structures of the single phase materials with y= 0, 0.33 and 0.80 were refined by Rietveld analysis. The temperature dependence of the brownmillerite phase $Ca_2Fe_2O_5$ was investigated and a structural phase transition from *Pcmn* to *Icmm* was observed at about 700°C. Samples with $SrFe_{1-y}Ti_yO_{3-\delta}$ composition annealed in different atmospheres were also studied. An orthorhombic symmetry was confirmed for the superstructure with $SrFeO_{2.86}$ stoichiometry and a distortion of the perovskite structure was observed for samples with low Ti contents annealed in air and nitrogen.

2. Introduction

The perovskite structure can accomodate a large range of metal ions with mixed valences within its basic framework of a mixed metal/oxygen close packing. Ordered or random anion vacancies can be stabilised in the structure and may lead to the formation of superstructures which are intergrowths of perovskite and oxygen deficient perovskite layers. A sequence of idealised structures has been proposed by Grenier *et al.* [1] and can be viewed as composed of $A_nB_nO_{3n-1}$ phases where (n-1) layers of BO_6 octahedra alternate with one layer of BO_4 tetrahedra.
Electron microscopy studies have revealed a number of ordered and disordered phases in the $CaFe_{1-y}Ti_yO_{3-\delta}$ system [2]. At high Ti contents, the symmetry of the $CaTiO_3$ phase is retained and the vacancies are disordered but at y= 0.5, an order-disorder transition was observed as a discontinuous variation in the unit cell volume. The ordered phases below y= 0.5 were idientified as consisting of ordered sequences of tetrahedra and octahedra, with the number of octahedra decreasing until the brownmillerite structure is obtained. It was also proposed that for low vacancy ratios, the oxygen vacancies are ordered along rows of different lengths, that become progressively more ordered until at y=0.5, long range order can be observed.

G. Van Tendeloo et al. (eds.), High-Temperature Superconductors and Novel Inorganic Materials, 245–250.

The known ordered phases in the $Ca_n(Fe,Ti)_nO_{3n-1}$ model have thus been observed to take values of n= 2, 2.5, 3, 4 and ∞. On the other hand, $SrFe_{1-y}Ti_yO_{3-\delta}$ forms a solid solution, and the synthesis procedure, i.e. atmosphere, heat treatment and rate of cooling, has been shown to have a large effect on the degree of oxidation of iron, high oxygen pressures being necessary in order to crystallise samples with the perovskite structure and containing only Fe^{4+}. Studies carried out on samples in this series have reported a simple perovskite cell with the cell volume increasing as the Ti content is increased. They have also shown a discontinuity in magnetic and electrical behaviour at a Ti content of y= 0.8, $SrTiO_3$ being paramagnetic and an insulator [3]. Furthermore, oxygen vacancies are created in the case of samples containing Fe^{3+}, but no long range ordering was found in samples with e.g. Sr_2FeTiO_{6-x} composition [4]. However, two superstructures have been found in the $SrFeO_{3-\delta}$ system with $0.0 \leq \delta \leq 0.5$, which has been reported to consist of an ideal series of compositions with the nominal stoichiometry $SrFeO_{3-1/n}$ where n= 2, 4, 8 and ∞. The oxygen content can be varied by annealing at different oxygen pressures and temperatures to form $SrFeO_3$, $SrFeO_{2.86}$, $SrFeO_{2.73}$ or $SrFeO_{2.50}$ phases, thus changing the Fe^{3+}/Fe^{4+} ratio [5].

3. Experimental

The samples under study include $CaFe_{1-y}Ti_yO_{3-\delta}$ with y= 0-0.83 and $SrFe_{1-y}Ti_yO_{3-\delta}$ with y= 0-0.75. Polycrystalline samples were prepared by mixing stoichiometric amounts of CaO, $SrCO_3$, Fe_2O_3 and TiO_2 and grinding in ethanol. The mixtures were then sintered at 1000°C for 24 h, pressed into pellets and sintered at 1100-1300°C for 5-8 days with several intermediate grindings. The reactions were carried out in air, oxygen or nitrogen atmosphere.

Time-of-flight powder neutron diffraction data were collected using the backscattering banks of the POLARIS powder diffractometer at the ISIS spallation source, Rutherford Appleton Laboratory. Rietveld analysis was carried out using the program GSAS.

4. Results and Discussion

4.1. $Ca_2Fe_2O_5$

The brownmillerite structure has a unit cell related to the basic perovskite cell by $a \approx 2\sqrt{a_p}$, $b \approx 4a_p$ and $c \approx 2\sqrt{a_p}$. Depending on metal composition, brownmillerite compounds are known to crystallise in space groups *Ibm2* (no. 46), *Pcmn* (no. 62) or *Icmm* (no. 74). The alternating layers of octahedra and tetrahedra are elongated along the *b*-axis, with O(2) as a link between the two layers. Two short Fe(1)-O(1) and one long Fe(1)-O(2) bond form the octahedra, while two long Fe(2)-O(3) and one short Fe(2)-O(2) bond form the highly distorted tetrahedra. The tetrahedra have different orientations along the *b*-axis and build a stacking sequence according to -OTOT'-. In this work, the crystal structure of $Ca_2Fe_2O_5$ has been studied as a function of temperature up to 1000°C, see Fig. 1. The magnetic structure was refined in space group *Pcm'n'* (T_N =725K). A structural phase transition from *Pcmn* to *Icmm* takes place at about 700°C, above which temperature reflections with $h+k+l \neq 2n$ are no longer observed, while the cell parameters vary linearly with temperature across the transition, see Fig. 2.

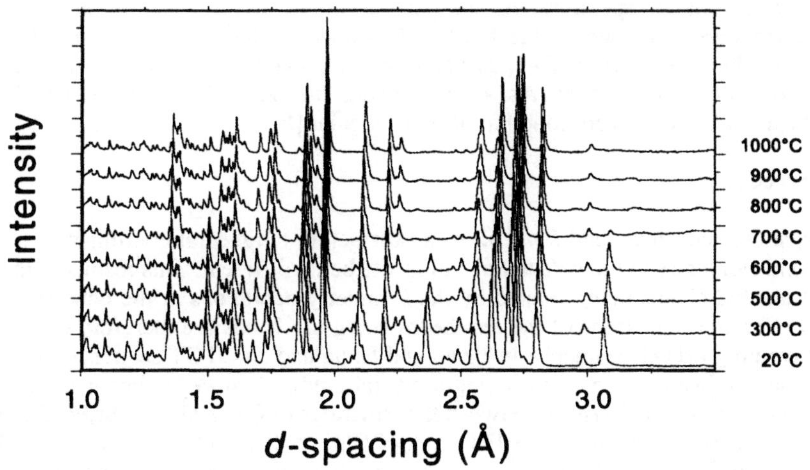

Figure 1. Neutron powder diffractograms of $Ca_2Fe_2O_5$ from room temperature up to 1000°C.

The polyhedra around the metal ions are highly distorted by an elongation along the *b*-axis and the transition induces a change in the tilting of the octahedra and displacements of the oxygen ions in the tetrahedra. In the high temperature structure, increasing disorder along the chains result in highly irregular tetrahedra and an ordering of the octahedra and coordination polyhedron around Ca are observed.

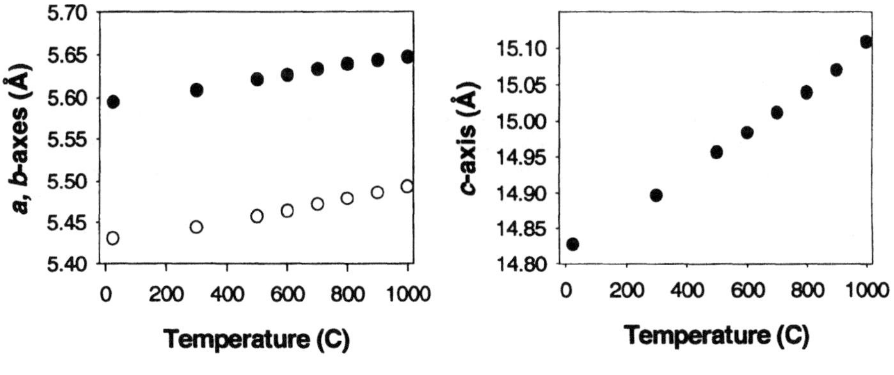

Figure 2. Change in unit cell parameters with temperature in $Ca_2Fe_2O_5$ ($a < b$).

The elongation of the octahedra along the *b*-axis increases continuously with temperature and across the structural transition. The mean Fe(1)-O(1) distance increases from 1.965 to 1.982 Å at 1000°C, while the Fe(1)-O(2) bond distance increases from 2.131 to 2.180 Å. The mean octahedral O-O bond distances expand linearly with temperature, while the tetrahedral distances are the most affected at the transition. The tetrahedra are also highly elongated along the *b*-axis, as can be seen from the O(2)-O(2) distance which is 3.237 Å compared to a mean distance of 3.021 Å for the two O(2)-O(3) bonds. At 1000°C, the O(2)-O(2) distance is 3.441 Å compared with 3.581 Å in

$Sr_2Fe_2O_5$ at room temperature. As the position of the O(3) ion shifts on the *ac*-plane above the transition, one of the Fe(2)-O(3) bonds in the tetrahedra along the *c*-axis shortens while the other Fe(2)-O(3) bond becomes longer, resulting in highly irregular tetrahedra along the chains. As a comparison, the bond distances in the chains of tetrahedra in $Sr_2Fe_2O_5$ were found to be more regular [6].

4.2. $SrFeO_{3-\delta}$

In $Sr_2Fe_2O_5$, both *Pcmn* and *Ibm2* allow a satisfactory tetrahedral coordination with only the point symmetry at the octahedral cation site being different. This results in the layers of tetrahedral chains being related by the symmetry operations at the octahedral cation site. Different orientations of the tetrahedra along the *b*-axis are therefore observed, i.e. an alternating OTOT' sequence in *Pcmn* and *Icmm* and an OTOT sequence in *Ibm2*, the latter having been observed in a single crystal study of $Sr_2Fe_2O_5$ For a more detailed discussion on the relationships between the structures of $Ca_2Fe_2O_5$ and $Sr_2Fe_2O_5$, see [6]. In the mixed-valence phases $SrFeO_{2.75}$ ($a=2\sqrt{2}a_c$, $b=2a_c$ and $c=\sqrt{2}a_c$), and $SrFeO_{2.875}$ ($a=2\sqrt{2}a_c$, $b=2a_c$ and $c=2\sqrt{2}a_c$), the ordering has been described as similar to the one in brownmillerite. In the first case, only half of the vacancies are present so as to generate equal numbers of five and six-coordinated sites in every $(001)_c$ plane with Fe^{4+} and Fe^{3+} in equal proportions, the two types of polyhedra alternating along $[001]_c$. In the second case, the vacancies are further separated from each other and the symmetry is orthorhombic. A small orthorhombic distortion was observed in the powder pattern of $SrFeO_{2.875}$ and reflections could be indexed in space group *Immm* with cell parameters a= 10.9186(1), b= 10.9041(1) and c= 7.6852(1) Å. However, higher resolution data is necessary to resolve the overlapping peaks and in order to successfully refine the atomic positions.

4.3. $CaFe_{1-y}Ti_yO_{3-\delta}$

For compositions with $0.16< y\leq 0.33$, $Ca_3Fe_{3-3y}Ti_{3y}O_8$ (n= 3) was observed to have only a small solid solution. Refinements were performed in space group $Pcm2_1$ starting from a model from a previous X-ray study [7]. Calculated cell parameters were a= 5.5266(1), b= 11.2123(2), c= 5.4367(1) and accurate positions for the oxygen atoms were obtained. Samples with the same composition annealed in nitrogen or oxygen consisted of a mixture of $Ca_3Fe_2TiO_8$ and a $CaFe_{1-y}Ti_yO_{3-\delta}$ phase with higher Ti content. Results from these refinements will be presented in a forthcoming paper.

TABLE 1. Cell parameters and Fe-O bond distances for $CaFe_{1-y}Ti_yO_{3-\delta}$ samples.

y	a	b	c	3-δ	$Fe-O_{ac}$	$Fe-O_b$
0.85	5.3951(3)	7.6548(5)	5.4308(3)	2.7	1.951(1)	1.963(2)
0.89	5.3944(3)	7.6529(5)	5.4321(3)	2.8	1.952(1)	1.962(2)
0.94	5.3893(1)	7.6482(2)	5.4368(1)	3.0	1.952(2)	1.960(1)

In the range $0.75< y\leq 0.83$, the distorted perovskite phase of $CaTiO_3$ (*Pcmn*) was obtained. With increasing Ti content, the unit cell volume was observed to decrease while the octahedra become more regular and the amount of vacancies decreases as shown in Table 1.

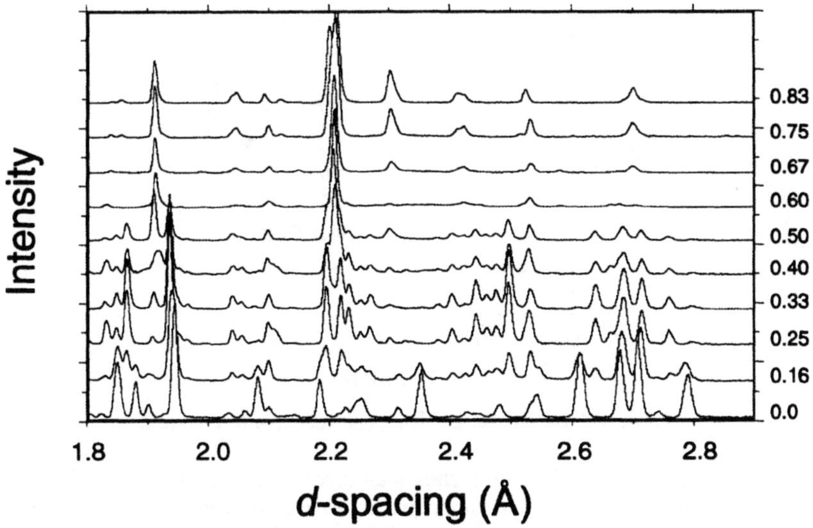

Figure 3. Neutron powder diffractograms of $Ca_{1-x}Fe_{1-y}Ti_yO_{3-\delta}$ samples at room temperature.

The structure of the n=4 member in the $Ca_n(Fe,Ti)_nO_{3n-1}$ series, $Ca_4Fe_2Ti_2O_{11}$, has also been reported [8], but in order to obtain a single phase sample, longer annealing times than the ones used in the preparation of our samples are necessary.

4.4. $SrFe_{1-y}Ti_yO_{3-\delta}$

Our studies on $SrFe_{1-y}Ti_yO_{3-\delta}$ samples annealed in air and nitrogen have revealed distortions of the simple perovskite unit cell with $Pm\bar{3}m$ symmetry. For samples sintered in air, a rhombohedral symmetry with space group *R3m* was found for compositions $0.25 \leq y < 0.66$ while for samples annealed in nitrogen, a doubling of the unit cell axis to give $Fm\bar{3}m$ symmetry was observed, both symmetries being subgroups of $Pm\bar{3}m$ and involving no octahedral tilts ($a^0a^0a^0$ according to Glazer's notation). Calculated cell parameters for each series are shown in Table 2.

TABLE 2. Cell parameters for $SrFe_{1-y}Ti_yO_{3-\delta}$ samples annealed in air and nitrogen.

y		0.25	0.33	0.50	0.66	0.75
air	*a*	5.479(1)	5.490(1)	5.494(1)	3.9006(8)	3.9032(6)
	c	6.757(1)	6.759(5)	6.760(3)		
N_2	*a*	7.8337(7)	7.8302(6)	7.8207(6)	3.9082(2)	3.9087(3)

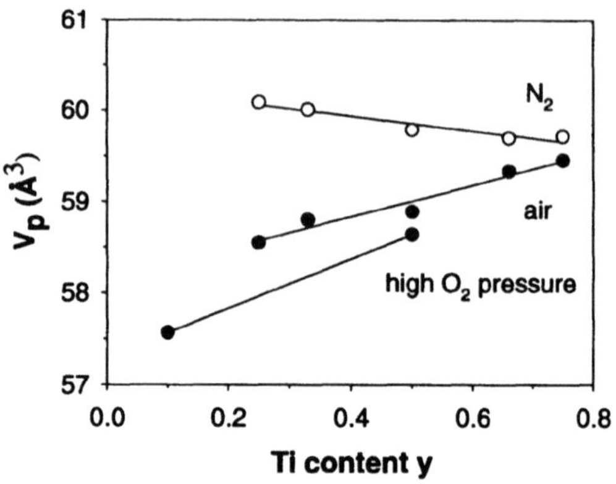

Figure 4. Change in the $SrMeO_3$ unit volume with Ti content.

A plot of the reduced unit cell volume shows an increase for samples annealed in air and a decrease for samples annealed in nitrogen, as expected from the difference in ionic radii of the substituted iron ions (Fe^{3+} hs r= 0.64 Å; Fe^{4+} r= 0.58 Å; Ti^{4+} r= 0.60Å). The two series converge at a composition close to y=0.85, see Fig. 4 where data for two samples annealed at high oxygen pressure have been included for comparison.

5. Acknowledgements

One of the authors (P.B.) wishes to thank the Swedish Foundation for International Cooperation in Research and Higher Education for financial support.

6. References

1. Grenier, J.-C., Pouchard, M. and Hagenmuller, P. (1981) Vacancy ordering in oxygen-deficient perovskite-related ferrites, *Structure and Bonding* **47**, 1-25.
2. Grenier, J.-C., Schiffmacher, G., Caro, P., Pouchard, M. and Hagenmuller, P. (1977) Etude par diffraction et microscopie electronique du systeme $CaTiO_3$-$Ca_2Fe_2O_5$, *J. Solid State Chem.* **20**, 365-379.
3. Clevenger, T.R. (1963) Effect of Fe^{4+} in the system $SrFeO_3$-$SrTiO_3$, *J. Am Ceram. Soc.* **46**, 207-210.
4. Greaves, C. and Buker, R.A. (1986) The defect structure of Sr_2FeTiO_{6-x}, *Mat. Res. Bull.* **21**, 823-833.
5. Takeda, Y., Kanno, K., Takada, T., Yamamoto, O., Takano, M., Nakayama, N. and Bando, Y. (1986) Phase relation in the oxygen nonstoichiometric system, $SrFeO_x$ ($2.5 \leq x \leq 3.0$), *J. Solid State Chem.* **63**, 237-249.
6. Berastegui, P., Eriksson, S.G. and Hull, S. (1997) A neutron diffraction study of the temperature dependence of $Ca_2Fe_2O_5$,submitted to *Mat. Res. Bull.*
7. Rodriguez-Carvajal, J., Vallet-Regi, M. and Gonzalez Calbet, J.M. (1989) Perovskite threefold superlattices: A structure determination of the $A_3M_3O_8$ phase, *Mat. Res. Bull.* **24**, 423-430.
8. Gonzalez Calbet, J.M. and Vallet-Regi, M. (1987) A new perovskite-type compound: $Ca_4Fe_2Ti_2O_{11}$, *J. Solid State Chem.* **68**, 266-272.

STRUCTURE AND PROPERTIES OF SOME THIOSPINELS

H.M. PALMER AND C. GREAVES
University of Birmingham
School of Chemistry, University of Birmingham, Edgbaston, Birmingham, B15 2TT, UK

Abstract. Colossal magnetoresistance has recently been observed in the chromium chalcogenide spinels, $FeCr_2S_4$ and $Fe_{0.5}Cu_{0.5}Cr_2S_4$ and the effect in a 6 Tesla field was found to be comparable to that shown by the perovskite material, $La_{0.67}Sr_{0.33}MnO_3$. A structural study of $Fe_{0.5}Cu_{0.5}Cr_2S_4$ has been performed and Rietveld refinement based on ambient temperature powder neutron diffraction data has revealed that $Fe_{0.5}Cu_{0.5}Cr_2S_4$ unusually displays both magnetic and cationic ordering.

1. Introduction

Colossal magnetoresistance is the dramatic change in the resistance of some materials caused by large magnetic fields. It may be defined at fixed temperature and applied field, H, as the percentage difference between resistance at zero field and resistance at applied field. This effect is technologically important for devices such as sensors, actuators and in magnetic recording (*i.e.* high capacity hard-disks for computers). To date the only structural families known to show CMR (colossal magnetoresistance) behaviour are the manganese oxide perovskite and Ruddlesden-Popper phases (*e.g.* $Ln_{1-x}A_xMnO_3$ (Ln = rare earth, A = alkaline earth)), the pyrochlore, $Tl_2Mn_2O_7$ (and some doped variants) and two chromium thiospinels [1]. The recent report of CMR behaviour in the chromium thiospinels was the first time CMR behaviour had been observed in either this structure type or in a non manganese oxide phase. It was reported that $FeCr_2S_4$ and $Fe_{0.5}Cu_{0.5}Cr_2S_4$ show a CMR effect comparable to that observed in the perovskite material, $La_{0.67}Sr_{0.33}MnO_3$ [2]. These phases display colossal magnetoresistance at 170K and 340K, respectively in the presence of a 6 Tesla field.

This study focuses on the structure of latter phase, $Fe_{0.5}Cu_{0.5}Cr_2S_4$. The normal spinel structure consists of a cubic close packed array of anions (X^{2-}) with an eighth of the tetrahedral holes filled by a divalent cation (A^{2+}) and half of the octahedral holes filled by a trivalent cation (B^{3+}) giving the stoichiometry AB_2X_4. A wide variety of spinels can be prepared due to the relative ease of substitution on both the cation sites. As a consequence of this, thiospinels show great diversity in their magnetic properties, *e.g.* $Li_{0.5}Ga_{0.5}Cr_2S_4$ is antiferromagnetic (T_N = 14K) [3], $HgCr_2S_4$ is ferromagnetic (T_c = 36K) [4], Fe_3S_4 is ferrimagnetic (T_c = 580K) [5] and $In_{0.5}Ag_{0.5}Cr_2S_4$ is

G. Van Tendeloo et al. (eds.), High-Temperature Superconductors and Novel Inorganic Materials, 251–256.

helimagnetic [6]. Although $Fe_{0.5}Cu_{0.5}Cr_2S_4$ itself is known to be a ferrimagnetic semiconductor [1] there appear to be no reports of a magnetic structure refinement in the literature. The magnetic structure of this phase is now of considerable interest due to the reported CMR behaviour.

A possible variation on the spinel structure is the substitution of a mixture of trivalent and monovalent cations for the divalent tetrahedral cations. Cationic order is known to occur in spinel phases of this type when there is an equal quantity of monovalent and trivalent cations. $Fe_{0.5}Cu_{0.5}Cr_2S_4$ is such a phase and indeed it has been suggested by Lotgering *et al.* [7] that this phase shows cationic order. This ordering takes the form of each monovalent cation being surrounded by a tetrahedral arrangement of trivalent cations (and vice versa). This is shown in Figure 1 (the octahedral cations (Cr^{3+}) and anions have not been drawn for clarity). However, a structural refinement has not been performed to confirm the order.

Here a Rietveld refinement of both the nuclear and magnetic structure of this phase is described.

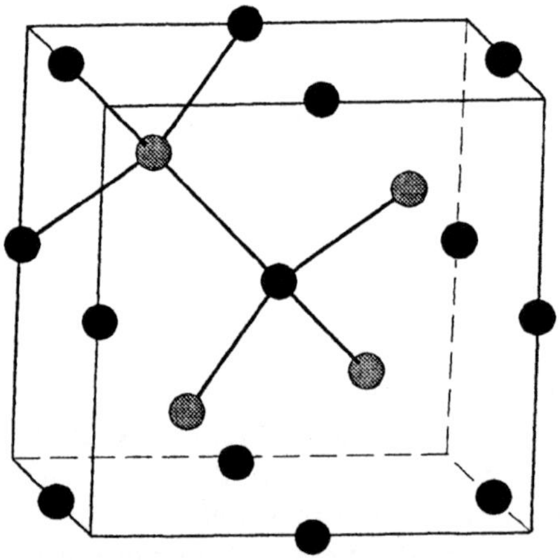

Figure 1. Cation Order in the Tetrahedral Sublattice of $Fe_{0.5}Cu_{0.5}Cr_2S_4$
(Fe^{3+} are dark spheres, Cu^{+} are the light grey spheres)

2. Experimental

Standard solid state techniques were used to prepare $Fe_{0.5}Cu_{0.5}Cr_2S_4$. The appropriate high purity metal powders (Fe(99.99%), Cu(99.999%) and Cr(99.5%)) were intimately ground together with S(99.998%) powder. The mixture formed was pressed into a 7mm pellet and fired in a sealed evacuated (~1×10^{-4} atm) quartz tube at 750°C for 12 hours. The sample was then reground, repressed and resealed in an evacuated quartz tube and fired for a further 12 hours at 750°C to ensure phase purity.

Phase purity of the sample was checked by X-ray powder diffraction (Siemens D5000, $CuK_{\alpha 1}$). Ambient temperature powder neutron diffraction data were collected on the POLARIS diffractometer at the Rutherford Appleton Laboratory. The Rietveld refinement was carried out using the GSAS suite of programs [8]. Room temperature magnetisation versus magnetic field data were collected using an Oxford Instrument 12 Tesla VSM (Vibrating Sample Magnetometer) while magnetic susceptibility versus temperature data were collected using a pendulum magnetometer (built in house [9]).

3. Results and Discussion

The magnetic and nuclear structure were considered to be two phases for the purposes of the refinement. Both the low angle (A) and backscattering (C) bank data were refined simultaneously. The cationic order has the effect of lowering the structure symmetry from the normal spinel symmetry, Fd3m (227) to F-43m (216). Therefore reflections are now seen in the diffraction patterns that were previously symmetry forbidden *(i.e.* (200) and (420)). An example of this is the peak at approximately 5Å (Fig. 2) which is the (200) reflection. The refinement was therefore carried out in F-43m and the results are shown in Figures 2&3 and Table 1.

It can be seen that the lower symmetry leads to several structural effects. The tetrahedral site is split into two distinct positions each of which is fully occupied by either Fe or Cu (this is a single site in Fd3m which is shared by Fe and Cu). The sulphur site has also been split into two positions leading to each type of cation being bonded to one "type" of sulphur ion. The Cr cations have moved from their special position. The bond distances and angles are listed in Table 1.

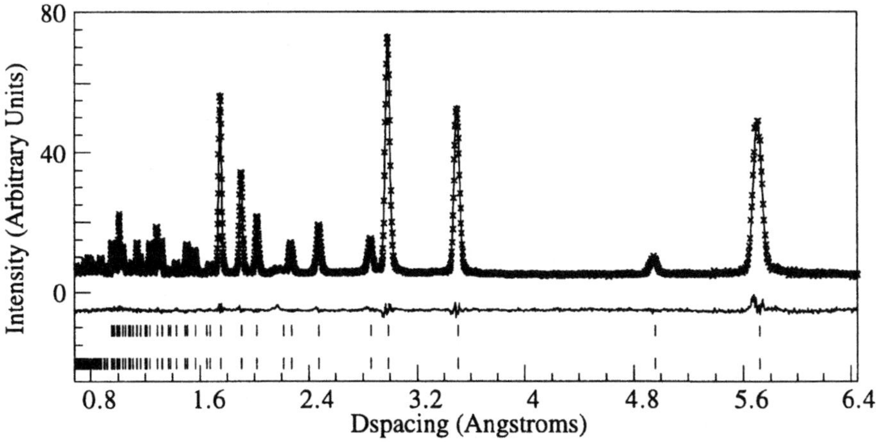

Figure 2. Low Angle (A) Bank Powder Neutron Diffraction Data, $Fe_{0.5}Cu_{0.5}Cr_2S_4$
(Reflections due to the magnetic structure are indicated by the upper tick marks)

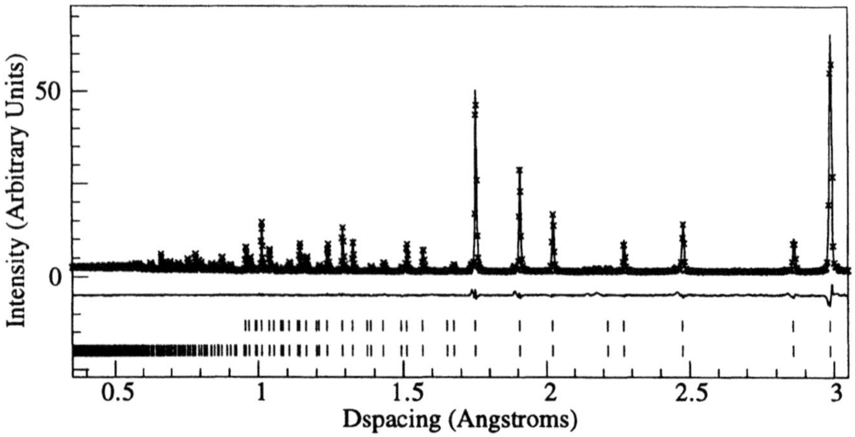

Figure 3. Backscattering (C) Bank Powder Neutron Diffraction Data, $Fe_{0.5}Cu_{0.5}Cr_2S_4$
(Reflections due to the magnetic structure are indicated by the upper tick marks)

TABLE 1. Refined structural data for $Fe_{0.5}Cu_{0.5}Cr_2S_4$

Atom	Position	x	y	z	100*U (Å^2)	μ (BM)
Fe	4a	0	0	0	0.67(2)	-1.97(5)
Cu	4d	0.75	0.75	0.75	1.11(3)	
Cr	16e	0.3749(3)	0.3749(3)	0.3749(3)	0.65(1)	1.49(3)
S(1)	16e	0.1317(1)	0.1317(1)	0.1317(1)	0.70(4)	
S(2)	16e	0.6148(1)	0.6148(1)	0.6148(1)	0.61(5)	

Unit cell dimensions: a = 9.90705(5) Å: space group F-43m (216):
R factors low angle bank; R_{wp} = 3.54%, R_{exp} = 1.68%. R factors high angle bank; R_{wp} = 2.63%, R_{exp} = 1.25%. Bond distances(Å); Fe-S(1) 2.261(3) [x4], Cu-S(2) 2.320(3) [x4], Cr-S(1) 2.411(3) [x3], Cr-S(2) 2.382(3) [x3]. Bond angles(°); Cr-S(1)-Cr 93.1(2) [x3], Cr-S(2)-Cr 94.8(1) [x3].

The Fe-S(1) bond distance is less than the Cu-S(2) distance as expected from the Fe^{3+} and Cu^{+} ionic radii (0.74Å and 0.63Å, respectively [10]). The bond angles also show that there is no significant distortion from the ideal cubic structure. The occupancies of both Fe and Cu were allowed to vary, but as they tended towards their ideal values they were constrained to be 4.0.

The magnetic structure has the effect of increasing the relative intensity of some of the reflection peaks. This is most obvious in the low angle bank pattern (Figure 2). In particular the massively increased intensity of the lowest angle peak at approximately 5.7Å is unambiguous evidence of the magnetic order. The refined magnetic moments are shown in Table 1. The antiferromagnetic coupling of the Fe^{3+} and Cr^{3+} to give a ferrimagnetic material is confirmed. The overall magnetic moment is calculated to be 1.99(8) BM. This compares favourably with the magnetic moment determined from ambient temperature VSM data (Figure 4). The VSM data gives a moment of 1.73 BM, therefore the two values are within experimental error of each other.

As previously mentioned $Fe_{0.5}Cu_{0.5}Cr_2S_4$ has a transition temperature of approximately 340K [1]. The transition temperature of our sample was determined using a pendulum magnetometer under a field of 0.2 Tesla. This gives a transition temperature of 345(5)K (Figure 5) which again agrees well with the previously published values (340K [1] and 349(5)K [11]).

The R_{exp} and R_{wp} values for both the low angle and backscattering banks are shown in the table. It can be seen that a very good fit to the data was achieved. Thus the Rietveld structure refinement has unequivocally shown that $Fe_{0.5}Cu_{0.5}Cr_2S_4$ displays both cationic and magnetic order.

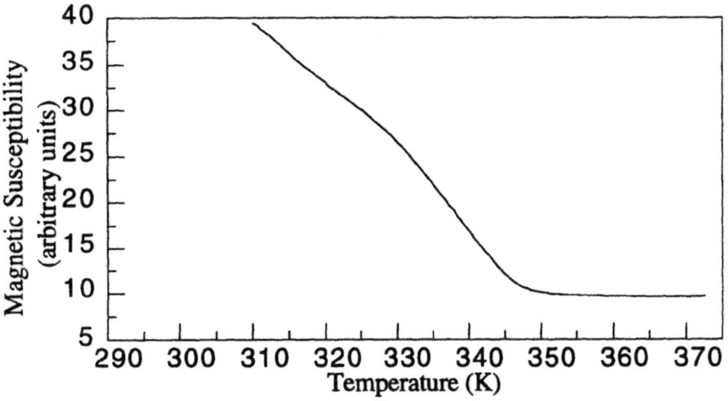

Figure 4. Magnetic Susceptibility against Temperature Data, $Fe_{0.5}Cu_{0.5}Cr_2S_4$

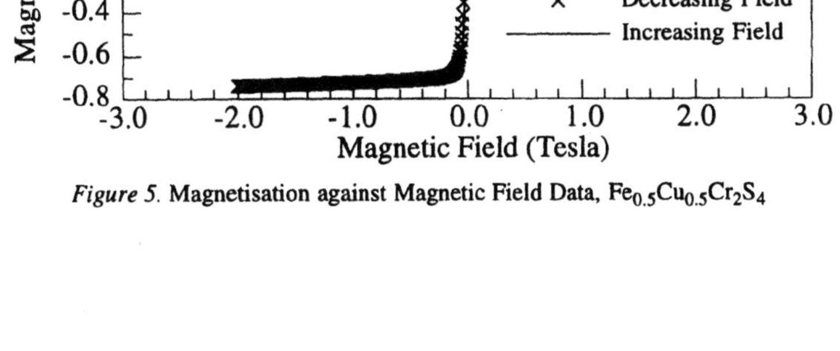

Figure 5. Magnetisation against Magnetic Field Data, $Fe_{0.5}Cu_{0.5}Cr_2S_4$

4. Conclusions

A structural study of the thiospinel, $Fe_{0.5}Cu_{0.5}Cr_2S_4$ which shows colossal magnetoresistance has been reported. A Rietveld refinement based on ambient temperature neutron diffraction data has confirmed the magnetic structure of this ferrimagnetic material. For the first time the presence of cationic order on the tetrahedral sites has been shown. The importance of these structural features for the observed CMR behaviour and indeed for other properties of the material is not clear at present. Further study on these systems is required if the role of the ordering is to be understood.

5. Acknowledgements

We would like to thank Dr M Slaski at the School of Physics and Astronomy, University of Birmingham for carrying out the VSM measurements and Mr R Mottram at the School of Materials and Metallurgy for his help in carrying out the measurements on the pendulum balance. Finally we would like to thank EPSRC for funding.

6. References

1. Ramirez, A.P., Cava, R.J. & Krajewski, J. (1997) Colossal magnetoresistance in Cr-based chalcogenide spinels, *Nature* **386**, 156-159
2. Hwang, H.Y., Cheong, S.W., Ong, N.P. & Batlogg, B. (1996) Spin-polarised intergrain tunneling in $La_{2/3}Sr_{1/3}MnO_3$, *Phys. Rev. Lett.* **77**, 2041-2044
3. Pinch, H.L., Woods, M.J. & Lopatin, E. (1970) Some New Mixed A-site Chromium Chalcogenide Spinels, *Mat. Res. Bull.* **5**, 425-430
4. Baltzer, P.K., Woijtowicz, P.J., Robbins, M. & Lopatin, E. (1966) Exchange Interactions in Ferromagnetic Chromium Chalcogenide Spinels, *Phys. Rev.* **151**, 367-377
5. Robbins, M., Wolff, R., Kutzig, A.J., Sherwood, R.C. & Mikovsky, M.A. (1970) Ferrimagnetic Compositions in the System $Fe_{1+x}Cr_{2-x}S_4$, *J. Appl. Phys.* **41**, 1086-1087
6. Plumier, R. & Sougi, M. (1971) Mis en Evidence Par Diffraction Des Neutrons D'un Terme D'Echange Biquadratique Dans L'Helimagnetique $Ag_{1/2}In_{1/2}Cr_2S_4$, *Solid State Comm.* **9**, 413-416
7. Lotgering, F.K., Van Stapele, R.P., Van Der Steen, G.H.A.M. & Van Wieringen, J.S. (1969) Magnetic Properties, Conductivity and Ionic ordering in $Fe_{1-x}Cu_xCr_2S_4$, *J. Phys. Chem. Solids* **30**, 799-804
8. Larson, A.C. & Von Dreele, R.B. (1985-1990) General Structure Analysis System, University of California
9. Burkhardt, C., Mottram, R., Dimc, F., Kobe, S. & Harris, I.R. (1998) Design and Application of a Pendulum Magnetometer, *J. Appl. Phys.*, submitted
10. Shannon, R.D. (1976) Revised Effective Ionic Radii in Halides and Chalcogenides, *Acta Crystallogr.* A **32**, 751-767
11. Ok, H.N., Baek, K.S., Lee, H.S. & Kim, C.S. (1990) Mössbauer Study of $Cu_{0.5}Fe_{0.5}Cr_2S_4$, *Phys. Rev. B* **41**, 62-64

PHASE TRANSITIONS IN TERNARY OXIDES

H. EHRENBERG[a,b], H. FUESS[a], H. WEITZEL[a]

[a] *Darmstadt University of Technology, Materials Science, Petersenstr. 23, D-64287 Darmstadt, Germany*
[b] *Interdisciplinary Research Centre in Superconductivity, Madingley Road, Cambridge CB3 0HE, UK*

The interest in physical properties of oxide materials has experienced a renascence with the discovery of high temperature superconductivity and giant magnetoresistance in such compounds. From a crystallographic point of view the recent investigations are strongly concentrated on compounds with crystal structures derived either from the perovskite-, spinel- or garnet-type structure. In this contribution we want to point out some less intensively studied ternary oxides of general composition ABO_4 or AB_2O_6, where A is a 3d-transitional metal from Mn to Cu and B is a 4d- or 5d-transitional metal Nb, Mo or Ta, W. All compounds under consideration crystallize at least for one modification in the very similar wolframite- (ABO_4) or columbite-type structure (AB_2O_6). This presentation will report on phase transitions induced by the application of pressure or magnetic fields, summarized in p-T and H-T phase diagrams.

The wolframite and columbite structure are based on a hexagonal-close package of oxygens, where half the octahedral interstices are filled by cations with a characteristic distribution of the cations resulting in zig-zag chains of filled octahedra, and these chains are arranged in layers of either A-type or B-type ions only. The difference between both structure types is the sequence of such layers: in the wolframite-type structure A- and B-layers are alternating, while in the columbite one the A-layers are separated by two B-layers. The compounds α-$AMoO_4$ crystallize at ambient pressure in a structure with no close packing of oxygens, but if sufficently high pressure is applied, the wolframite structure-type is formed instead and remains (meta-)stable at ambient conditions. These high-pressure modifications are labelled $AMoO_4$-II for A = Mn to Ni and $AMoO_4$-III for A = Cu. The magnetic properties of both the α- and the high-pressure modifications will be summarized and compared to the tungstates AWO_4, which also crystallize in the wolframite structure-type: Although

G. Van Tendeloo et al. (eds.), High-Temperature Superconductors and Novel Inorganic Materials, 257–262.

the compounds α-$AMoO_4$ are isotypic for A = Fe, Co and Ni, essentially different antiferromagnetic structures are observed at low temperature as indicated by the propagation vectors $\vec{k} = (0, \frac{1}{2}, \frac{1}{2})$ for Fe, $\vec{k} = (\frac{1}{2}, 0, \frac{1}{2})$ for Co and $\vec{k} = (0, 0, \frac{1}{2})$ for Ni. While nearly collinear structures are observed for the Co- and Ni-compounds, two perpendicular sublattices are formed in α-$FeMoO_4$ according to the two different crystallographic sites for the A-ions. The magnetic phase diagram of α-$NiMoO_4$ is characterized by a spin-flop phase for applied fields nearly parallel to the orientation of the magnetic moments and a direct transition from the antiferromagnetic ground state into the forced ferromagnetic state for an applied field perpendicular to the easy direction [1]. The magnetic ordering in α-$CoMoO_4$ is exceptional as two magnetic phase transitions occur even in the absence of an external field. The following figure shows the temperature dependence of specific heat and magnetization of a powder sample:

Figure 1: Specific heat and magnetization versus temperature for α-$CoMoO_4$.

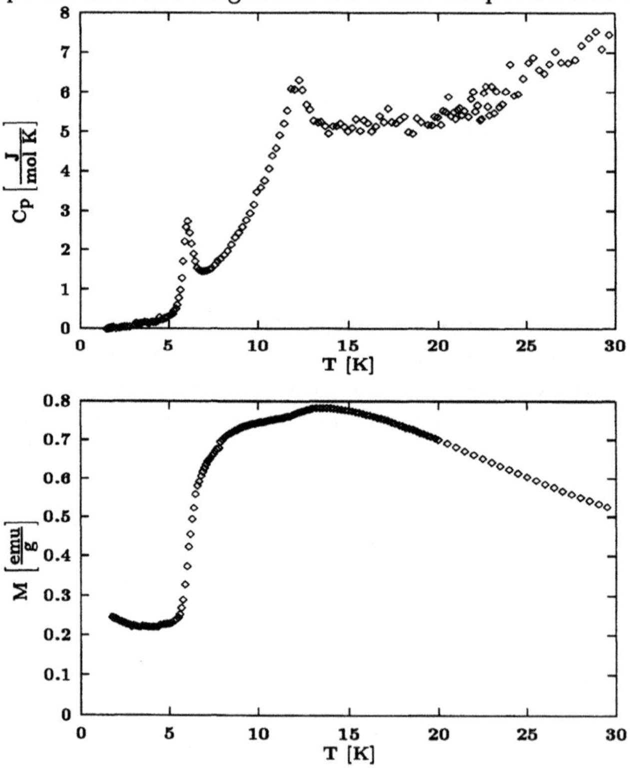

The intermediate phase is incommensurate with propagation vectors $\vec{k} = (k_x, 0, k_z)$, and both k_x and k_z varying with temperature. Although the magnetic reflections are significantly shifted with respect to those of the commensurate ground state, there is a one-to-one correspondence with no magnetic reflection splitted. The detailed magnetic structure of the incommensurate phase is not solved yet.

The antiferromagnetic structures of AWO_4, A = Fe, Co and Ni, are characterized by ferromagnetic zig-zag chains in agreement with the Kanamori rules for the case of 90° superexchange. As respresentative, $CoMoO_4$-II was studied by neutron powder diffraction, and the same antiferromagnetic structure as in $CoWO_4$ was confirmed [2]. A totally different situation is realized in the case of A = Mn and Cu, where the sign of the 90° superexchange coupling cannot be deduced from symmetry considerations, and the magnetic structures depend severely on supersuperexchange couplings via two oxygens, which are more sensitive to details of the underlying crystal structure. If only supersuperexchange couplings via two oxygens, which are forming one edge of a coordination octahedron, are taken into account, there are still 8 such couplings in addition to the one superexchange within the zig-zag chains, resulting in a pronounced competition of interactions. As a consequence, the very complicated phase diagrams of $MnWO_4$ result, with three different phases even in the absence of an external field [3]. Depending on the direction of an applied field one further high-field phase is induced, the determination of its magnetic structure is in progress to complete the description of the topology of the H-T phase diagrams. The analysis of experimental magnon dispersion curves, based on a bilinear spin Hamiltonian and treated by Holstein-Primakoff and Bogoliubov transformations, confirm the relative weakness of the superexchange coupling [4]. The dominant antiferromagnetic supersuperexchange coupling is the one corresponding to the largest Mn-Mn distance of all pairs included, reflecting the general tendency of increasing antiferromagnetic coupling strength as the bridging Mn-O-O angles approach the 180° limit. From magnetic excitation studies on single-crystals of $CuWO_4$ two models for relevant couplings have been proposed [5]. From a comparison with the isostructural compound $CuMoO_4$-III one of these models can be rejected, while the other one is strongly supported [2]. Although of very similar crystal structures, the magnetic unit cells of $CuWO_4$ and $CuMoO_4$-III are doubled with respect to different axes, *a* for W and *c* for Mo. However, the relative orientation in spin pairs coupled by the dominant interactions should be the same for both compounds. An analysis of the geometry of all possible super- and supersuperexchange paths confirms this assumption for one of the two proposed models, and the following criteria can be applied to predict dominant supersuperexchange couplings in networks of Jahn-Teller distorted CuO_6 octahedra: i) Both Cu-O distances in the Cu-O-O-Cu path must be short bonds, i.e. not belong to an elongated O-Cu-O axis due to the Jahn-Teller effect. ii) The antiferromagnetic coupling strength increases with both bonding angles Cu-O-O and O-O-Cu. The relative strong antiferromagnetic supersuperexchange couplings result in identical spin arrangements for sublattices in $CuWO_4$ and $CuMoO_4$-III, but these sublattices are orientated in different ways with respect to each other, determined by any combination of weaker couplings. By cooling α-$CuMoO_4$ a first-order phase transition takes place at about 190 K into γ-$CuMoO_4$ with a crystal structure similar to that of $AlNbO_4$, but a more complicated cation distribution. The same phase transition can also be induced at room temperature by the application of 0.2 GPa pressure, and the p-T phase diagram has been determined from in-situ synchrotron diffraction experiments [6]. The phase transition from α- to γ-$CuMoO_4$ is called *pseudoreconstructive*, because on one hand bonds are broken and the topology is changed, on the other hand no pronounced diffusion of atoms takes place and a one-to-one correspondence between atomic sites in both structures exist. Both high-pressure modifications γ-$CuMoO_4$ and $CuMoO_4$-III can also be synthesized at ambient pressure, if some of the Mo-ions are replaced by W, e. g. the γ-phase is stabilized by chemical pressure for 15% of W and -III by 75%. It is an interesting task to study how the magnetic structure changes from $CuMoO_4$-III to $CuWO_4$ with an increasing amount of Mo in $CuMoO_4$-III being replaced by W. Down to 2 K no magnetic ordering takes place in γ-$CuMoO_4$ as only slight deviations from a Curie-Weiss behaviour are observed in magnetization studies and no extra peaks occur in neutron powder diffraction.

Further information about the competing interactions in wolframites can be deduced from a comparison with the magnetic properties of compounds which crystallize in the related columbite structure. In this case the couplings between magnetic bc-layers are considerably weaker due to a second diamagnetic layer between them. These additional diamagnetic layers do not only result in lower Néel temperatures, but also emphasize the role of the ferromagnetic chains parallel b in ANb_2O_6 for A = Fe, Co and Ni. As in α-$AMoO_4$ the Co-compound is exceptional again as magnetic ordering takes place via an intermediate incommensurate phase, too. The magnetic phase diagram has been determined by neutron diffraction on a single crystal in applied fields and is shown for an external field parallel to the c-axis, see figure 2. The magnetic structure of the incommensurate phase is sinusoidally modulated with a temperature- and field-dependent propagation vector $\vec{k} = (0, k_y, 0)$, and k_y becomes rational for the magnetic ground state AF ($\frac{1}{2}$) and for the spin-flip phases SF1 ($\frac{1}{3}$) and SF2 ($\frac{1}{2}$). For other field directions in the ac-plane a further spin-flip phase SF3 with $k_y = \frac{1}{4}$ occurs, e. g. for angles of 45° to the a- and c-axes [7]. The spin-flip transitions are due to the switching of whole chains, and therefore plateaus of characteristic fractions of the saturation value are observed in the field dependence of magnetization. This effect is shown in figure 3 for the compound $NiNb_2O_6$ and an applied field in the ac-plane, forming an angle of 15° with the c-axis.

Figure 2: Magnetic phase diagram of $CoNb_2O_6$ for an applied field parallel to the c-axis.

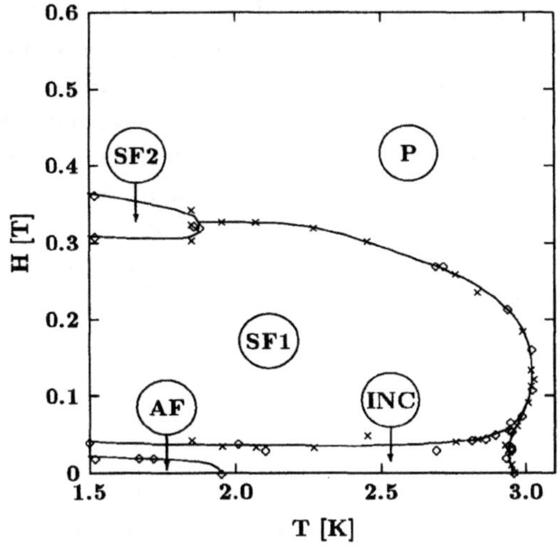

Figure 3: Field induced phase transitions in $NiNb_2O_6$ for an applied field direction forming an angle of 15° with the c- and 75° with the a-axis.

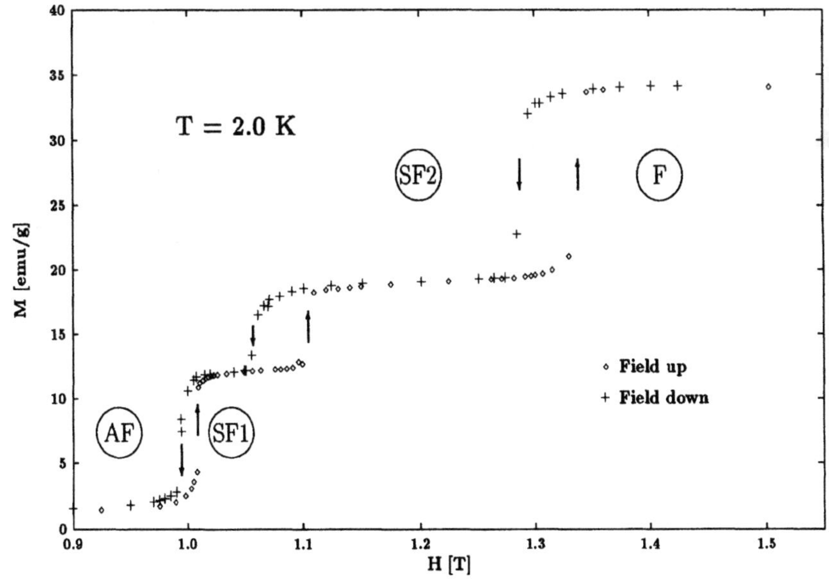

The plateaus appear at $\frac{1}{3}$ (SF1) and $\frac{1}{2}$ (SF2) of the saturation magnetization (F), and pronounced hystereses effects can be seen. The stability of intermediate spin-flip phases is very sensitive to the specific field direction. If the field is applied parallel to c, SF2 does not appear, and SF1 vanishes for an increased angle of 25° to the c-axis in the ac-plane. There is also in important effect of temperature: At 3.5 K the three phase transitions, shown in figure 3 for the 15° case, are thermally washed out, and even two transitions can hardly be resolved. With respect to the Néel temperature of 5.7 K speculations about the existence of further plateaus at lower temperature with very small field ranges of stability are supported. The compounds ANb_2O_6 are therefore promising canidates to exhibit a pronounced stair-case behaviour as predicted by theoretical considerations, e. g. the ANNNI model.

References:

[1] Ehrenberg, H., Wiesmann, M., Paulus, H., and Weitzel, H. (1998) Magnetic phase diagrams of α-$NiMoO_4$, *J. Magn. Magn. Mater.* , in print.

[2] Ehrenberg, H., Wiesmann, M., Garcia-Jaca, J., Weitzel, H., and Fuess, H. (1998) Magnetic Structures of the High-Pressure Modifications of $CoMoO_4$ and $CuMoO_4$, *J. Magn. Magn. Mater.* **182**, 152-160.

[3] Ehrenberg, H., Weitzel, H., Heid, C., Fuess, H., Wltschek, G., Kroener, T., van Tol, J., Bonnet, M. (1997) Magnetic phase diagrams of $MnWO_4$, *J. Phys.: Condens. Matter* **9**, 3189-3203.

[4] Ehrenberg, H., Weitzel, H., Fuess, H. (1997) Magnon dispersion and magnetic phase diagram of $MnWO_4$, *Physica B* **235-237**, 560-563.

[5] Lake, B., Tennant, D.A., Cowley, R. A., Axe, J. D., and Chen, C. K. (1996) Magnetic excitations in the ordered phase of the antiferromagnetic alternating chain compound $CuWO_4$, *J. Phys.: Condens. Matter* **8** 8613-8634.

[6] Wiesmann, M., Ehrenberg, H., Miehe, G., Peun, T., Weitzel, H., and Fuess, H. (1997) p-T phase diagram of $CuMoO_4$, *J. Sol. State Chem.* **132**, 88-97.

[7] Heid, C., Weitzel, H., Burlet, P., Winkelmann, M., Ehrenberg, H., and Fuess, H. (1997) Magnetic phase diagrams of $CoNb_2O_6$, *Physica B* **234-236**, 574-575.

MESOSCOPIC SCALE STRUCTURES OF A FERROELASTIC DOMAIN WALL

JURICA NOVAK AND EKHARD K.H. SALJE
Department of Earth Sciences, University of Cambridge, Downing Street, Cambridge, CB2 3EQ, United Kingdom

Abstract. We present a precise description of a twin domain wall and its surface mesoscopic structures. The bulk structure of the wall is described numerically, and analytically using the Landau-Ginzburg model extended do the fourth order spatial derivative of the order parameter. The surface topography of the domain wall is seen as a rounded ridge. In contrast, the elastic response of the surface is found to have an exotic form, exhibiting ridge-trough-ridge structure which should be visible in AFMTM images. The characteristic width of the domain wall is found to be the same at the surface and in the bulk.

1. Introduction

Almost all ferroelastic or co-elastic materials exhibit domain patterns when they go through a structural phase transition, and these domain patterns influence their physical properties[1]. The thermodynamic description of these processes is well understood[2, 3, 4, 5, 6]. This description, though, does not provide a precise determination of the structure on a mesoscopic scale. Although the twin domain walls, which are the building blocks of the mesoscopic structures, appear as distinct features in images obtained with TEM techniques[7], the preparation of the samples as thin slabs or wedges for use with TEM significantly alters any fine structure present. Recently, microstructure has been studied by high-resolution X-ray diffraction which directly observes the diffuse scattering around Bragg reflections[8]. Although the sensitivity of this technique is relatively good (identification of wall patterns with one wall in 10000 Å)[9, 10], the more complex details of the mesoscopic structures, such as the surface structures and needle tips are still not accessible, due to the fact that the X-ray diffraction studies give

G. Van Tendeloo et al. (eds.), High-Temperature Superconductors and Novel Inorganic Materials, 263–270.

bulk properties. Another possible mode of experimental investigation is by atomic force microscopy (AFM) or related techniques which are sensitive to the spatial variation of the surface topography or physical properties of material[11, 12, 13, 14]. Besides the fact that these techniques can concentrate on the surface properties exclusively, thus excluding the determination of sub-surface or even bulk properties, they have not been successful in determining even that in sufficient detail.

2. Simulation method

The model we use to simulate twin domains and mesoscopic structures of interest is a simple two-dimensional lattice that represents a hypothetical plane perpendicular to the twin walls that form the mesoscopic structures in the ferroelastic phase[15]. In order to generate surface relaxations, which were necessary to describe the surface behavior of mesoscopic structures, it is essential that interatomic interactions extend to, at least, two interatomic layers perpendicular to the surface[16]. The ground state of the lattice is characterized by the shear angle, which was chosen to be 3° in agreement with some representative real materials[17].

Domain structures and surface relaxation were then calculated by numerical minimization of the total energy functional:

$$\delta \int dr^3 \left\{ \sum_{i,j} \mathcal{U}(i,j) \right\} = 0, \tag{1}$$

where $\mathcal{U}_p(i,j)$ is the potential energy of a point and the sum extends over all lattice points. Free surface boundary conditions were used to facilitate surface relaxation where appropriate. We have not included any thermodynamic fluctuations, and have effectively simulated a system far away from the ferroelastic-paraelastic transition temperature T_c using a standard MD algorithm. 80000 points were used to simulate desired mesoscopic effects. The production runs of the simulation were carried out on the Hitachi S-3600/180 vector supercomputer, a part of the Cambridge High Performance Computing Facility.

3. Twin domain wall in the bulk

Analytically, the ferroelastic materials have been well described using Landau-Ginzburg phenomenological model, which expresses the Gibbs free energy as a polynomial in the order parameter and its spatial derivative[1], with a chosen strain component being the order parameter Q of the ferroelastic phase transition. A domain wall is a low energy excitation of the system, and the total energy of the domain wall is minimized, with the conditions that, in general, no secondary strain fields are excited, and that no lattice imperfections occur. In our model, neither of these can happen, so the minimization takes the following form:

$$\delta \int dr\, G(r) = 0, \tag{2}$$

where $G(r)$ is the energy functional of the Gibbs free energy, and r is the coordinate perpendicular to the wall. The boundary condition for minimization is the relation of the order parameters in the adjacent parts of the lattice, i.e. equal magnitude, opposite sign. The two adjacent domains are twin related, in our model, and in the most ferroelastics. The energy minimum can be obtained by solving Eq. (2) numerically, which we did in our simulation, or, analytically by solving the corresponding Euler-Lagrange differential equation:

$$g'\frac{d^4Q}{dr^4} - g\frac{d^2Q}{dr^2} + A\theta_s\left(\coth\frac{\theta_s}{T} - \coth\frac{\theta_s}{T_c}\right)Q + BQ^3 = 0, \tag{3}$$

. The solution to Eq. (3) for $g' = 0$ is well known:

$$Q(r) = Q_0 \tanh\left[\frac{r}{\mathcal{W}}\right], \tag{4}$$

with $\mathcal{W}$ being the characteristic thickness of the domain wall. The solution for $g' \neq 0$ does not have an analytical form, and can be approximately derived by perturbation methods. The coefficients can be renormalized[1, 6], leading to the dimensionless Euler-Lagrange equation:

$$-Q + Q^3 = \frac{d^2Q}{dr^2} - \gamma\frac{d^4Q}{dr^4}, \tag{5}$$

with Q redefined in units of Q_0, and $\gamma = g'|a|/g^2$. The solution to this equation, for $\gamma = 0$, is Eq. (4), with $Q_0 = 1$ and $\mathcal{W}^2 = 2$. As a first approximation for $\gamma \neq 0$, we can use the same functional form of Q and treat $\mathcal{W}$ as a variational parameter. In the limit of small r, Eq. (5) then yields:

$$\mathcal{W}^2 = 1 + \sqrt{1 + 16\gamma} \approx 2 + 8\gamma,$$

which points in the direction of an effective increase in the wall thickness compared to the minimum value of $\mathcal{W}^2 = 2$.

To get a better estimate for the effect of the fourth order gradient term, the trial function can be expanded for $r \ll \mathcal{W}$:

$$Q = \tanh\left[\frac{r}{\mathcal{W}}\right] + \delta(r), \tag{6}$$

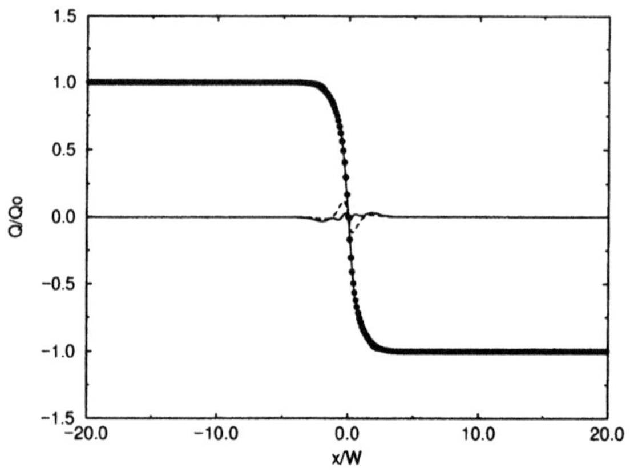

Figure 1. The numerically obtained data for the twin domain wall profile in the bulk is represented with dots. The best fit using the fourth order Landau-Ginzburg model Eq. (6) is represented by a dotted line. Solid line is the difference curve for the fourth order model, and the dashed line is the difference curve for the standard, second order model Eq. (4).

where $\delta(r)$ is a correction function close to the center of the domain wall. We still treat the wall width $\mathcal{W}$ as a variational parameter that is expected to increase with γ. The contribution of $\delta(r)$ in terms of the Eq. (5) is:

$$\delta - 3\delta \tanh^2 \frac{r}{\mathcal{W}} + \frac{d^2\delta}{dr^2} - \gamma \frac{d^4\delta}{dr^4} =$$

$$-\left(1 - \frac{2}{\mathcal{W}^2} - \frac{16\gamma}{\mathcal{W}^4}\right) \tanh \frac{r}{\mathcal{W}} - \left(1 - \frac{2}{\mathcal{W}^2} - \frac{40\gamma}{\mathcal{W}^4}\right) \tanh^3 \frac{r}{\mathcal{W}} - \frac{24\gamma}{\mathcal{W}^4} \tanh^5 \frac{r}{\mathcal{W}}. \tag{7}$$

The functional form of $\delta(r)$ is clear from the above expression if we write:

$$\delta(r) = (Ar^3 + Br)f(r), \tag{8}$$

with $f(r)$ being a concave function, flat for $r \ll \mathcal{W}$ and approaching 0 rapidly for $r \gg \mathcal{W}$. By Taylor expansion of the hyperbolic tangent, we can calculate $A(\gamma)$ and $B(\gamma)$ for $r \ll \mathcal{W}$. We have fitted the numerical data using both models, described by Eq. (4) and Eq. (6), with the $\delta(r)$ taking the form of Eq. (8), by looking at the value of the strain in the middle of the simulated sample, in order to avoid any surface relaxation effects.

Using the Eq. (4) we found the best fit for the wall width to be $\mathcal{W} = 10$, measured in unit cells. The fit is apparently very good, but the difference curve shows quite a noticeable discrepancy between the data and the model in the region close to the center of the domain wall [Fig. 1]. This discrepancy is a consequence of the fact that the fourth order spatial derivative in Eq.

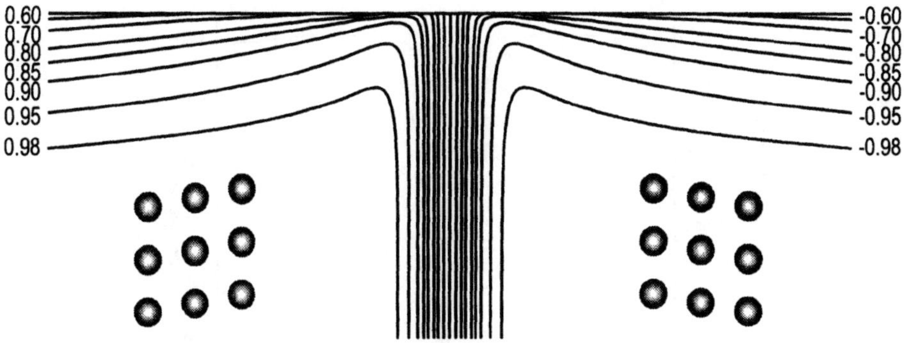

Figure 2. Distribution of the order parameter $\mathcal{Q}$ at the surface of the lattice (first 50 layers). Lines represent constant $\mathcal{Q}/\mathcal{Q}_0$. There are three lines in the middle of the twin domain wall that are not labeled, they represent the $\mathcal{Q}/\mathcal{Q}_0$ values of 0.40, 0.00, and -0.40 respectively. Notice the steepness of the gradient of $\mathcal{Q}/\mathcal{Q}_0$ through the twin domain wall. The two structures represent sheared twin atomic configurations in the bulk (far from the twin domain wall and surfaces).

(3) was neglected ($g' = 0$) in the derivation of the wall profile (Eq. (4)), as discussed above.

When the numerical data was fitted with Eq. (6) we got much better results, as can be seen from the shape of the difference curve. In this case, we found the wall width to be $\mathcal{W} = 7$, setting $\gamma = 0.3$ and using a Gaussian of appropriate width $\sigma = 6$ as $f(r)$. Hence, the wall profile was found to be:

$$\mathcal{Q} = \tanh\left[\frac{r}{\mathcal{W}}\right] + (Ar^3 + Br)\frac{\mathcal{H}}{\sqrt{2\pi}\sigma}e^{-\frac{1}{2}\left(\frac{r}{\sigma}\right)^2},$$

with all the symbols as defined above, and $\mathcal{H}$ the scaling factor for the height of the Gaussian.

The difference in the wall widths for the two different models are expected. The model that includes the fourth order derivative Eq. (6), yields a lower value for the width of the domain wall, and the effects of the additional factors from Eq. (8) account for the broadening evident from the fit of the simple model Eq. (4).

To conclude, the wall widths we obtained from the simulation are in general agreement with the values for real materials[17, 18, 19, 20].

4. Single domain wall and its surface features

The above presented analytical method based on the Landau-Ginzburg phenomenological model has not been employed to deal with the mesoscopic structure of the twin domain wall at the surface. The mesoscopic features at

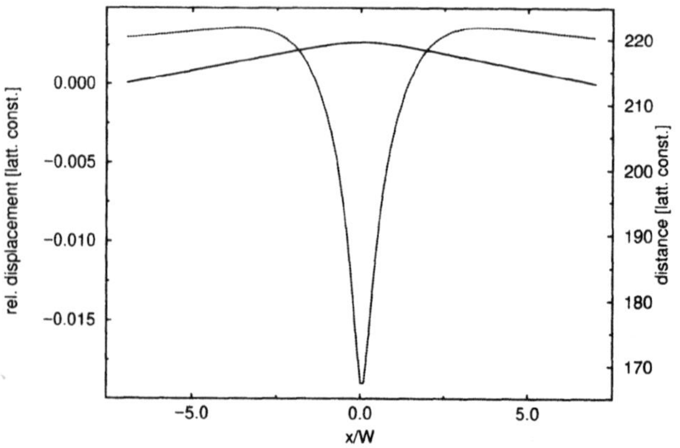

Figure 3. Surface topography as a solid line with the scale on the right and the groove due to the wall relaxation as a dashed line with the scale on the left. The order of magnitude of surface deformation is three times larger compared to that of the deformation due solely to the domain wall relaxation.

the surface arise from a solution to a double relaxational problem, consisting of both the relaxation originating in the effects of the free surface and those due to the effects of a domain wall. The two relaxational problems have not been solved concurrently within the Landau-Ginzburg framework, their complexity of the concurrent treatment arising from the inherent multi-dimensionality of the problem that includes both surface and domain wall relaxations[21].

We found that close to the surface the wall apparently widens in a characteristic trumpet-like shape, a consequence of both surface and domain wall origins of the relaxation effects [Fig. 2]. The calculated domain wall width at the surface, using the following expression:

$$\mathcal{Y}_s(x) = \mathcal{Y}_0 + \int_0^x \mathcal{Q}_s(t)dt,$$

($\mathcal{Y}_s$ is the real space position of the surface particles, $\mathcal{Y}_0$ is the position of the particle at the center of the domain wall, $\mathcal{Q}_s$ is the order parameter distribution at the surface) has indicated that there is no appreciable difference between $\mathcal{W}$ and $\mathcal{W}_s$.

The surface relaxation depth λ was obtained from the standard expression arising from the thermodynamic Landau-Ginzburg description[22]. Close to the twin domain wall interface with the surface, this description does not hold due to the influence of the twin domain wall. The magnitude of λ was found to be at a minimum close to the twin domain wall and increasing with distance away from it, as seen on the [Fig. 2], where the measure of the surface relaxation is any contour of constant $\mathcal{Q}$. At an

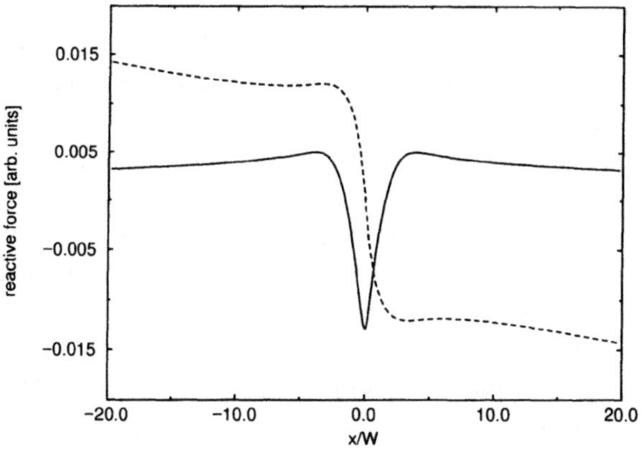

Figure 4. Distribution of the normal (solid line) and lateral (dashed line) response to normal displacement of the surface particles. Lateral response can be useful for observing the domain structure, while normal response can be useful for the observation of twin domain walls.

infinite distance λ would reach its maximum value $\lambda_{max} = \mathcal{W} = 7$, calculated from the simulation results, measured in lattice constants, which is the surface relaxation depth of the lattice if no twin domain walls are present. In materials with microstructure formed by an array of periodic twin domain walls[23], the depth of surface relaxation λ_{array} is suppressed proportionally to the domain wall density.

The real space topography of the surface is determined by both sources of relaxation - twin domain wall and the surface. These are distinct and when considered separately, we found the effect due to the wall relaxation larger by about three orders of magnitude then that due to the surface relaxation [Fig. 3]. In fact, the wall relaxation completely dominates the topography of the twin domain wall surface structure, entirely masking the groove-like effects originating in the surface relaxation, thus creating the surface rounding centered at the twin domain wall [Fig. 3]. A recent AFM-tapping mode study[11] has identified this feature. It reports a "nose" effect on the surface, that we have not been able to verify.

In the attempt to predict the possible experimental results of AFM investigations of the surface structure of the twin domain wall, we emulated the effect that the tip at the end of an AFM cantilever has on the surface of the material. This we did by displacing each particle in the surface layer by $10^{-8}a$ in the $-\hat{y}$ direction. We then calculated the lateral and normal components of the reactive force. The lateral force distribution shows a dependence similar to that of the order parameter $\mathcal{Q}_s$ [Fig. 4]. The normal force distribution has a profile similar to that of the square of the order parameter, $\mathcal{Q}_s^2$, with ridges on both sides of a groove [Fig. 4].

The change in the sign of the order parameter at the surface has been observed (for ferroelectrics) by using a mode of imaging developed for the detection of static surface charge[13]. For ferroelastics discussed here, this corresponds to the profile of the lateral reactive force. The SFM non-contact dynamic mode images[14] would correspond to the distribution of the normal reactive force. The divergence of the lateral force distribution away from the center of the wall can be attributed to the simulated infinite extension of the lattice. In the simulated array, the lateral component of the force reached a finite value between two adjacent domain walls.

These results can be used as a guidance for the future experimental work. In order to determine the twin domain wall width $\mathcal{W}$ in the bulk, one only needs to determine the characteristic width $\mathcal{W}_s$ of the surface structure of the domain wall. Previously, these features of the twinning materials were investigated using mainly X-ray techniques. In fact, the only necessary information for the determination of the twin domain wall width $\mathcal{W}$ are the real space positions of the particles in the surface layer.

References

1. E.K.H. Salje, Phase transitions in ferroelastic and co-elastic crystals, Cambridge University Press (1993)
2. S. Marais *et.al.*, Phys. Rev. Lett. **66**, 2480 (1991)
3. K. Parlinski *et.al*, J. Phys.: Condens. Matter **5** 497 (1993)
4. E.K.H. Salje *et.al*, J. Phys.: Condens. Matter **5** 4775 (1993)
5. A.M. Bratkovsky *et.al.*, J. Phys.: Condens. Matter **6**, 3679 (1994)
6. I. Tsatskis *et.al.*, J. Phys.: Condens. Matter **6**, 11027 (1994)
7. S. Amelinckx, D. van Dyck, J. van Landuyt, G. van Tendeloo, Handbook of Microscopy, VCH Verlagsgesellschaft mbH (1997)
8. E.K.H. Salje, Phase Trans **55**, 37 (1995)
9. J. Chrosch and E.K.H. Salje, Physica C **225**, 111 (1994)
10. D. Viehland *et.al.*, Phil. Mag. **71**, 205 (1995)
11. D. Bosbach *et.al.*, J. Phys.: Condens. Matter **9**, 8397 (1997)
12. C. Schönenberger, Physics World **10**, No. 9, 25 (1997)
13. F. Saurenbach and B.D. Terris, Appl. Phys. Lett. **56**, 1703 (1990)
14. R. Lüthi *et. al.*, J. Appl. Phys. **74**, 7461 (1993)
15. J.Novak and E.K.H. Salje, J.Phys.: Condens. Matter **10**, L359 (1998)
16. B. Houchmanzadeh *et.al*, J. Phys.: Condens. Matter **4**, 9779 (1992)
17. K.R. Locherer *et.al.* Phil. Trans. R. Soc. Lond. A **354**, 2815 (1996)
18. B. Wruck *et.al.*, Phase Transitions **48**, 135 (1994)
19. E.K.H. Salje, Phase Transitions **55**, 37 (1995)
20. S.A. Hayward *et.al.*, Eur. J. Mineral. **8**, (1996)
21. I. Rychetský, J. Phys.: Condens. Matter **9**, 4583 (1997)
22. B. Houchmanzadeh *et.al*, Phase Transitions **38**, 77 (1992)
23. E.K.H. Salje and K. Parlinski, Supercond. Sci. Technol. **4**, 93 (1991)

STRUCTURAL EFFECTS OF THE DIFFUSION ENHANCEMENT IN GROWING OXIDE FILMS

A.R. KAUL,[1] A.A. MOLODYK[1], S.V. SAMOYLENKOV[1], N.V. MARKOV[1], O.Yu. GORBENKO[1], I.E. GRABOY[1], I.E. KORSAKOV[1], G. WAHL[2], O. STADEL[2]

1) Department of Chemistry, Moscow State University, 119899 Moscow, Russia

2) IOPW, TU Braunschweig, Bienroder Weg 53, 38108 Braunschweig, Germany

1. Introduction

The growth of thin films from gas phase by various deposition techniques may result in different film morphologies, textures and crystallinity. Crystal growth theories regard these properties to be dependent on the interplay of condensation rate and surface diffusion mobility in the growing film. The homologous temperature, which is the ratio of deposition temperature T_d to melting temperature T_m, can be used as a measure of diffusivity in growing film. In general, high growth rate at low homologous temperature (low diffusivity) leads to amorphous deposits, while epitaxial films can be obtained at low deposition rates and high homologous temperatures if epitaxial relations between the substrate and the film are possible. Between these two opposite situations the growth of polycrystalline films consisting of chaotic or oriented grains occurs. When the diffusivity in the growing film is not high enough to accommodate the flow of the substance coming to the substrate from the gas phase, an admixture of the grains with kinetically preferred orientation appears in the equilibrium (thermodynamically stable) epitaxial matrix. In this work, the illustrations of these principles are given and some approaches to overcome the kinetical hindrance during the growth of thin epitaxial films of oxide materials are developed.

2. Competition of a- and c- growth in $RBa_2Cu_3O_{7-y}$ thin films.

Normally thin films of $RBa_2Cu_3O_{7-y}$ superconductors demonstrate the ability to grow in two orientations, (100) and (001), with the axes a - and c- perpendicular to the substrate respectively. Epitaxial films of c- orientation usually possess the best superconducting properties, while the presence of a-oriented grains always decreases

G. Van Tendeloo et al. (eds.), High-Temperature Superconductors and Novel Inorganic Materials, 271–278.

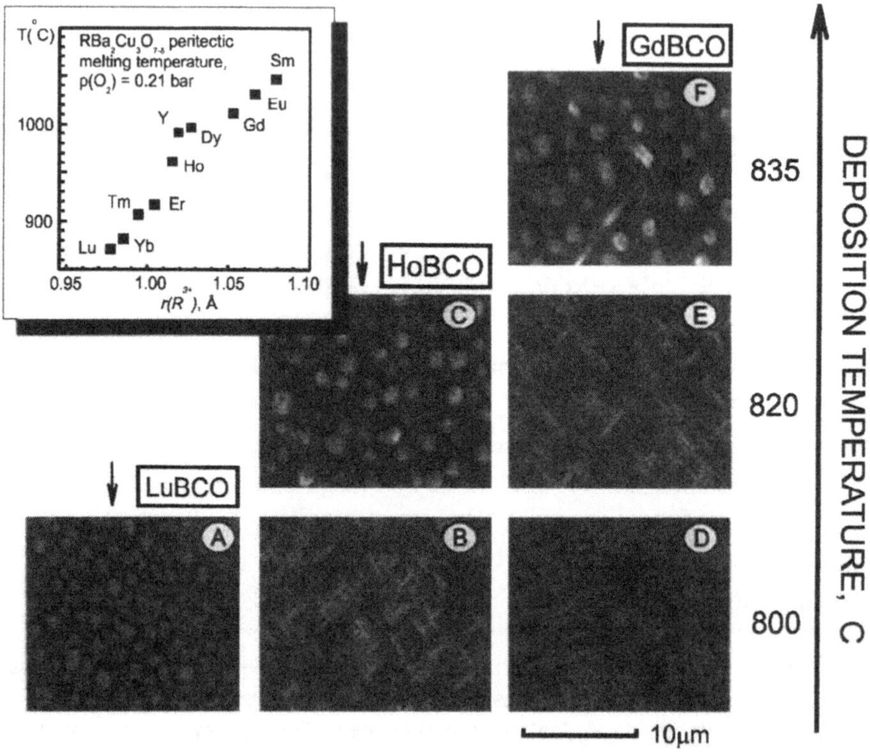

Figure 1. SEM images of the surface of $RBa_2Cu_3O_{7-y}$ films deposited under various conditions. Inset: the peritectic melting temperature of $RBa_2Cu_3O_{7-y}$ vs. R^{3+} ionic radius, according to *Janovsky et al., S.Ph.Ch.T., v2, 1989, p. 30 (in Russian).*

T_c and transport characteristics. The kinetics and thermodynamics of a- and c- growth of YBCO and BSCCO were discussed in details in [1] and the conclusion was made that c-orientation is thermodynamically preferable due to lower surface energy while a-orientation dominates at the kinetic hindrance of crystallisation. In this work, the ability to a- and c- growth is compared for various rare earths in $RBa_2Cu_3O_{7-y}$ series. For this purpose, nearly stoichiometric thin films of GdBCO, HoBCO and LuBCO were prepared using flash evaporation single source MOCVD technique, the experimental is described in details elsewhere [2]. Changing the oxygen partial pressure at the chosen temperature we have optimized (pO_2 - T) conditions of thin film deposition. The absence of a-oriented inclusions (determined by XRD and SEM) and the value of critical current density (determined from magnetic susceptibility), j_c, were used as optimization criteria. We found for all RBCO studied the optimal conditions to be along the same "epitaxy line" known for YBCO and coinciding with the equilibrium line CuO/Cu_2O [3].

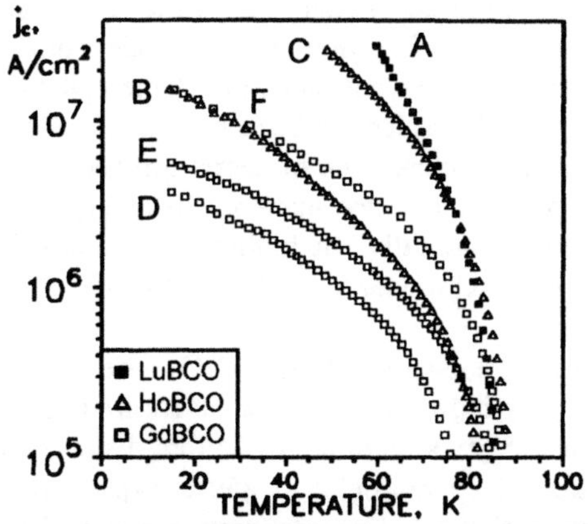

Figure 2. The temperature dependencies of the critical current density of $RBa_2Cu_3O_{7-y}/LaAlO_3$ films, showing optimization of the film quality and the influence of the rare earth cation on the slope $j_c(T)$ dependence. The indexes are the same as in Fig.1.

However, optimized conditions systematically vary from one RBCO phase to another: the higher deposition temperature is needed for larger R. In other words, the tendency to a-axis oriented growth of $RBa_2Cu_3O_{7-y}$ on the same substrate material, at the same deposition temperature, gradually decreases with the decrease of the size of R^{3+} cation. This is illustrated by Fig.1, where the morphology of the films deposited on (001) $LaAlO_3$ is compared. It can be seen that a-oriented grains, which look like light bars (B, D, and E in Fig.1), disappear in LuBCO, HoBCO and GdBCO films grown at 800, 820 and 835°C respectively. These results are naturally explained as manifestation of diffusivity decrease when going from LuBCO to GdBCO that can be deduced from systematic change of RBCO peritectic melting temperature (Fig. 1). Also the improved crystallization at the above mentioned temperatures results in the pronounced hillocks (Fig.1) which evidently grow through step-flow mechanism along the screw dislocations. Critical current densities of the films grown in identical conditions (optimal for LuBCO) correlate with the fraction of a-grains (see curves A, B, D in Fig.2). Nevertheless, the difference of j_c for various RBCO remains even for pure c-oriented films deposited under the optimized conditions (see curves A, C, F in Fig. 2). The difference is related to fundamental peculiarities of RBCO structures resulting in remarkable change of pinning potential, but not to variation of film morphology. The study of this phenomenon is in progress.

3. Effect of heterovalent doping on orientation and morphology of thin films.

Thin films of CeO_2 are used as buffer layers permitting the deposition of RBCO superconductors on $(1\bar{1}02)$ plane of Al_2O_3 (R-sapphire). Structural and morphological defects of buffer layer strongly influence the superconducting characteristics. For high frequency application, it is very desirable to obtain very smooth (001) oriented epitaxial buffer layers at high deposition rate. In the case of slow surface diffusion the growth of epitaxial CeO_2 films is complicated by appearance of (111) oriented grains embedded in (001) oriented thermodynamically stable epitaxial matrix. The growth in [111] direction is kinetically preferable because it is a direction of closest package of atomic planes in fluorite structure. The hindered diffusion results also in formation of coarse surface that is partly due to domination of island growth and partly due to cone-like form of (111) grains. Naturally, the higher is the deposition rate and/or the lower is the deposition temperature, the larger is the fraction of (111) oriented grains and the rougher is the film surface.

In our experiments on MOCVD of CeO_2 (the pulse single source version of MOCVD was used), very high quality buffer layers free of (111) oriented grains and characterised by R_t factor (maximum top-to-bottom distance) of 3.4 nm were obtained at the deposition rate ≅ 0.15 nm/pulse at 850°C [4]. It was demonstrated also that negative results of deposition rate increase (appearance of (111) grains and coarsening of the surface) could not be compensated by increase of the deposition temperature due to limited thermal stability of cerium precursor $Ce(thd)_4$. At higher deposition temperature gas phase nucleation with formation of fine oxide particles occurs in the reactor zone prior to the substrate resulting again in coarsening of the film surface.

We proposed an approach to get high quality films at high deposition rates consisting in heterovalent doping of the film that led to increase of the surface diffusion. The following quasichemical reaction explains the increase of oxygen vacancy concentration in CeO_2 structure when it is doped with oxide of a metal having the valence less than 4; Gd is taken here as an example:

$$Gd_2O_3\ (\rightarrow CeO_2) = 2Gd'_{Ce} + 3O^x_O + V^{\bullet\bullet}_O$$

In the first approximation, oxygen vacancy concentration, $[V^{\bullet\bullet}_O]$, is proportional to the dopant concentration and provides high ionic mobility to the substance. Doped ceria is known to be a solid electrolyte with high oxygen ionic conductivity. The correlation of ionic conductivity and diffusivity expressed by Nernst – Einstein equation is useful to predict a kinetical consequence of doping on the film deposition process. Taking into account that maximum ionic conductivity in $Ce_{1-x}R_xO_{2-x/2}$ corresponds to $x \approx 0.2$-0.3, we deposited thin films, with $x = 0.2$ for R = Y, Nd, Lu and $x = 0.1, 0.2, 0.3$ for R = Gd, at the deposition rates varied in the interval 0.15 – 2.5 nm/pulse. As it can be seen from Fig. 3, which presents the results of XRD and AFM characterisation of the films obtained, the orientation and surface roughness of doped films are nearly independent on the deposition rate in contrast to those of undoped ones. It is worthy to note that doping is very effective regardless to the type and the concentration of dopants. The reason is that even in low doped samples $[V^{\bullet\bullet}_O]$ value increases by a factor $10^4 - 10^5$ in comparison with equilibrium vacancy concentration in undoped state, while the variation of $[V^{\bullet\bullet}_O]$ value in the studied interval of dopant concentration should be well within one order of magnitude. According to literature data on conductivity for $Ce_{1-x}R_xO_{2-x/2}$, the variations of ionic mobility in the systems with various rare earth dopants R^{3+} should not exceed one order of magnitude too. Thus, heterovalent doping creating ionic vacancies is an effective approach giving the

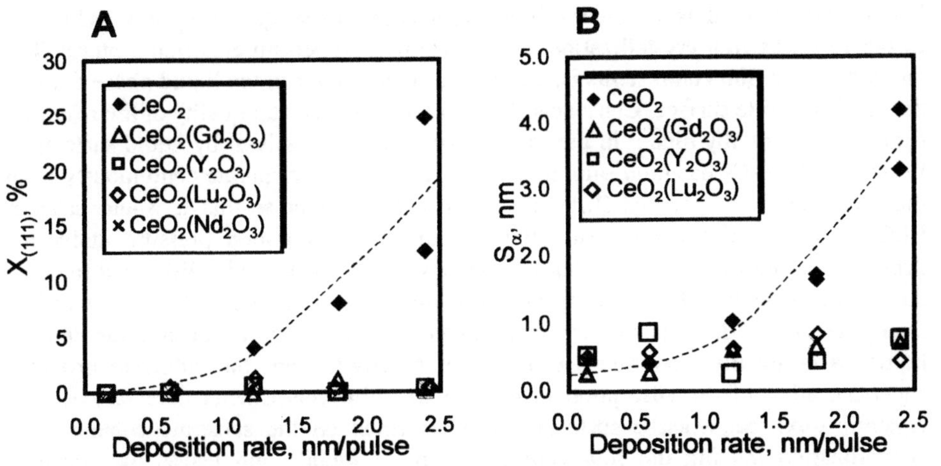

Figure 3. Dependences of (111) oriented grain fraction (**A**) and root mean square roughness calculated from AFM data (**B**) for $Ce_{0.8}R_{0.2}O_{1.9}$ films vs. deposition rate. $X_{(111)}$ was calculated as $0.3S_{111}/(0.3S_{111}+S_{002})$, where S is θ-2θ reflection integral intensity.

opportunity to attain higher deposition rates keeping high crystalline quality of epitaxial oxide films.

4. Volatile Surfactant Assisted MOCVD

Another promising buffer layer for HTSC's with even better match than CeO_2 would be lanthanum aluminate. It can be used also as a dielectric barrier in HTSC heterostructures. However, there have been few reports on successful deposition of $LaAlO_3$ films of high crystalline quality. High deposition temperatures (1000°C) are required to provide the epitaxy of $LaAlO_3$ by thermal MOCVD. Such difficulties with $LaAlO_3$ deposition are related to low diffusion mobility of the phase at the temperatures used for the deposition that correlates with very high melting point of lanthanum aluminate (T_m = 2100°C). The same problem is valid for MOCVD of $BaTiO_3$ (T_m = 1620°C) ferroelectric films. Therefore, to obtain the films of high crystalline quality, diffusion mobility in the growing films must be enhanced.

We tried to accelerate the surface diffusion by changing the film growth mechanism to one close to vapor-liquid-solid. It was accomplished by adding bismuth-containing precursor triphenylbismuth ($BiPh_3$) vapor into the reactor during the deposition. We supposed that bismuth oxide produced on the substrate would act as a surfactant facilitating the crystallization process. Thus, the following mechanism of Volatile Surfactant Assisted MOCVD (VSA MOCVD) was proposed. Precursor molecules (including $BiPh_3$) adsorbed on the substrate are oxidized in surface reaction producing corresponding oxides. Bismuth oxide possesses low melting point 820°C; furthermore, it forms even lower melting eutectics with oxides to be obtained. Liquid

Bi_2O_3 is widely used as a solvent of many oxides for crystal growth by flux technique. In our case, the film crystallization takes place from supersaturated liquid phase. The mobility of the film components is, obviously, much higher in the liquid phase than on the solid substrate surface. Consequently, we anticipated crystal quality of the resulting film to be essentially higher in the case of crystallization assisted by liquid surfactant. Besides that, Bi_2O_3 has high saturation vapor pressure to evaporate continuously from the substrate. We organized $BiPh_3$ transport to the reactor so that it would produce Bi_2O_3 with partial vapor pressure lower than Bi_2O_3 saturation pressure under the deposition conditions. It means that no condensation of bulk liquid Bi_2O_3 had to occur during the deposition. Hence, we believe it is more correct to speak about more or less thick quasiliquid Bi_2O_3 adsorption layer on the substrate than about bulk amounts of liquid. Additionally, continuous evaporation of Bi_2O_3 from the substrate provided supersaturation of the adsorption layer with the film components necessary for the crystallization. So, volatile surfactant Bi_2O_3 formed on the substrate enhances the crystallization making the film components more mobile and leaves the substrate completely by the end of the process.

Our experiments on VSA MOCVD were performed with single source aerosol MOCVD setup described in detail elsewhere [1]. First, it should be noted that there was no Bi detected by EDX analysis in $BaTiO_3$ and $LaAlO_3$ films deposited in the presence of Bi_2O_3 that correlated with the prediction of thermodynamics. The next important result we got was significant increase of the film deposition rate in the presence of Bi_2O_3 comparing to that for the films deposited without Bi_2O_3 with the same supersaturation in the reactor. For $LaAlO_3$ films the deposition rate increase of up to 5 times was observed. Higher sticking coefficient for precursor adsorption on liquid surface is the most possible explanation of this fact as it is the case for vapor-liquid-solid technique.

The most important, crystalline quality and surface morphology were found to improve remarkably for $BaTiO_3$ and $LaAlO_3$ films deposited in the presence of Bi_2O_3. Thus, $LaAlO_3$ films obtained at relatively low temperatures (800 – 900°C) without Bi_2O_3 were amorphous while by deposition in the presence of Bi_2O_3 crystalline films were obtained. According to XRD θ-2θ- and ϕ- scans, epitaxial $BaTiO_3$ (at 800°C) and $LaAlO_3$ (at 950 – 1000°C) films were obtained both with and without addition of $BiPh_3$ to the starting solution. Addition of $BiPh_3$ in the aerosol led to pronounced increase of the epitaxial quality of the films as indicated by FWHM values of XRD rocking curves, HREM and AFM data. The rocking curve widths were essentially smaller for the films grown with addition of $BiPh_3$ (Fig. 4). Higher mosaisity and less perfect interface structure were found in $LaAlO_3/NdGaO_3$ films grown without Bi_2O_3 by HREM investigation. AFM study revealed also a remarkable difference in the surface morphology pointing out to higher surface diffusion of the film components during the growth in the presence of Bi-containing vapor. It is worthy to note again that all these results were obtained for the films deposited in the presence of Bi_2O_3 at the growth rate 5 times higher than that for those deposited without Bi_2O_3. This fact emphasizes how high is the degree at which the surface diffusion is accelerated by Bi_2O_3 surfactant.

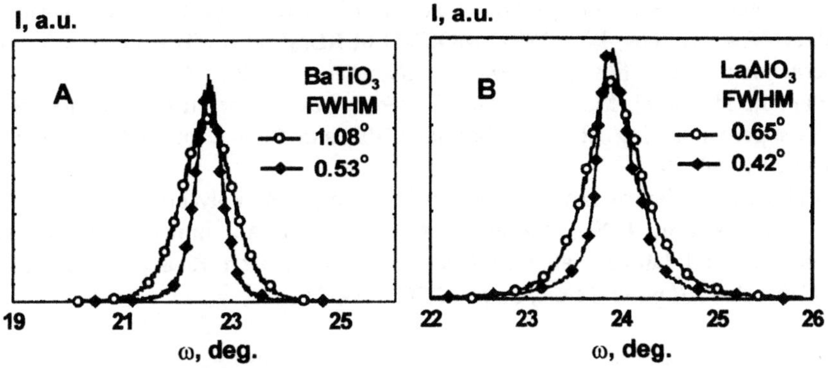

Figure 4. XRD rocking curves for (002) reflexes of epitaxial films (A) $BaTiO_3$/MgO grown at 780°C, (B) $LaAlO_3$/$NdGaO_3$ grown at 950°C. Depositions without $BiPh_3$ —o— , with $BiPh_3$ —◆—

5. Conclusions

Experimental results on MOCVD of various oxides are interpreted in regard to surface diffusion mobility during film growth.

Deposition conditions of RBCO films by MOCVD are optimized (for R = Lu, Ho and Gd), being correlated with systematic shifts in RBCO melting temperatures that provides differences in diffusion mobility in the growing films. Heterovalent doping is proposed as a method to enhance the diffusion mobility that leads to improved crystalline quality and surface morphology of CeO_2 films. A new approach to MOCVD based on use of volatile low melting oxide as a surfactant promoting the crystallization is proposed and examined in growth of $BaTiO_3$ and $LaAlO_3$ epitaxial films.

The approaches developed facilitate MOCVD of oxides to meet industrial requirements.

6. Acknowledgements

This work was partly supported by projects RFBR N 96-03-33027 and BMBF 13 N6947/5. A.A.M. is grateful to Robert Havemann Stiftung for a fellowship.

7. References

1. Gorbenko, O.Yu., Fuflyigin, V.N., Erokhin, Yu.Yu., Graboy, I.E., Kaul, A.R., Tretyakov, Yu.D., Klippe, L. and Wahl, G. (1994) YBCO and BSCCO thin films prepared by wet-CVD, *J. Mater. Chem.* **4**, 1585-1589.

2. Samoylenkov, S.V., Gorbenko, O.Yu., Kaul, A.R., Papucha, S.V., Mirin, N.A., Maleev, V.V., Graboy, I.E. (1997) MOCVD of $RBa_2Cu_3O_{7-y}$ (R = Lu, Ho, Y and Gd) superconducting thin films, *Electrochem. Soc. Proc.*, **97-25**, 990-997.
3. Yamane H., Hasei, M., Kurosawa H. (1991) Low temperature formation of Y-Ba-Cu-O superconducting films by thermal CVD and their j_c in high magnetic field, *Physica C*, **190**, 79.
4. Graboy, I.E., Markov,N.V., Maleev, V.V., Kaul, A.R., Polyakov, S.N., Svetchnikov, V.L., Zandbergen, H.W., Dahmen, K.-H. (1997) An improvement of surface smoothness and lattice match of CeO_2 buffer layers on R-sapphire processed by MOCVD, *J.of Alloys and Compounds*, **251**, 318-321.

PREPARATION OF FERROELECTRIC OXIDE FILMS BY SOL-GEL METHOD

VLADIMIR N.FUFLYIGIN, KEWIN.K.LI, FEILING WANG, HUA JIANG, SHIWEN LIU, JING ZHAO, PETER NORRIS,
NZ Applied Technologies
Woburn, MA 01801; USA
PEARL YIP
U.S. Air Force Rome Laboratory
Hanscom Field, MA 01731, USA

1. Introduction

Among other solid state media, materials in PLZT and $BaTiO_3$ families possess the strongest electrorefractive effect, i.e. field-induced change of the refractive index. It is well known that some relaxor compositions in the PLZT system possess a quadratic electro-optic coefficient as high as 38×10^{-16} $(m/V)^2$[1]. Oxide materials have some very desirable properties, such as fast electro-optic response speed, for use as light modulating media [2]. To implement light modulating devices using hot-pressed bulk PLZT materials, substantial efforts have been made to integrate the electro-optic materials with semiconductor materials, particularly silicon, to form smart pixels [3]. In these devices the light passes through a PLZT wafer of sufficient thickness in order to collect enough optical phase shift. High operating voltages are usually associated with devices of this kind. Their integration with semiconducting materials in high-density format has remained a substantial challenge [4]. In this regard, ferroelectric films represent a viable alternative to the bulk material.

Many fabrication methods such as sputtering, chemical vapor deposition (CVD) and a sol-gel process have been used to prepare ferroelectric thin films [4,5]. However, all of them failed to fabricate thick (> 1 μm) films with high optical quality. We have recently developed a processing technique which enabled us to deposit PLZT and $BaTiO_3$ thin and thick films on a variety of substrates, including silicon and highly reflective dielectric mirror [6].

2.Experimental

The deposition of the oxide films was performed using the sol-gel technique [7]. Carboxilates of barium, lead, lanthanum, and zirconium and titanium diisopropyl

G. Van Tendeloo et al. (eds.), High-Temperature Superconductors and Novel Inorganic Materials, 279–284.

bisacetylacetonate ($Ti(O\text{-}iPr)_2(acac)_2$) were used as precursors. The precursors were dissolved in a water-ethanol mixture. The solution thus prepared was used for the PLZT and $BaTiO_3$ film deposition. Composition of PLZT films was 5-10/65/35 (La/Zr/Ti). Barium titanate based solid solutions with composition of $Ba_{1-x}Pb_xTiO_3$ (x=0-0.2) were also prepared. Excess of lead was used to promote the phase formation. Precursor films were deposited by dip-coating and subjected to a thermal treatment. PLZT films were annealed at 650-800°C. Annealing temperature for barium titanate was in the range of 750-800°C.

R-sapphire, (100) and (110) oriented $SrTiO_3$ were used as substrate materials. Thickness measured by profilometer "Dektak" was in the range 0.5-20 μm, and the thickness of a single layer varied from 0.02 to 0.3 μm.

As deposited films of PLZT and $BaTiO_3$ were transparent, crack free and adhered well to the substrate.

A transmission differential ellipsometer was employed to measure the first order electro-optic properties of the oxide thin film materials. Transmission differential ellipsometry is a phase-sensitive detection method, especially designed for electro-optic measurements of thin film materials [8]. Coplanar interdigitated electrodes were fabricated on film samples. Light beam was perpendicular to the films surface, and electrical field was applied perpendicular to the light beam. Phase shift was measured at 633 nm.

Refractive index and optical loss in the oxide films were measured by a Metricon prism-film waveguide coupler.

3.Results and Discussion

There is a number of difficulties in fabrication of PLZT and $BaTiO_3$ films related to the ferroelectric phases as such. The main difficulty is the slow kinetics of phase formation. This results in a low growth rate and requires a growth temperature higher than 800°C. It is well known that solid state reactions proceed faster in the presence of additives having a low melting point or forming eutectics with film's components. Lead oxide has relatively low melting point (886°C) and might form eutectics with other oxide components of the film promoting a diffusion and thereby enhancing the phase formation process. $BaTiO_3$ and $PbTiO_3$ form a continuous solid solution, hence phase segregation can be potentially avoided. As a result of more intense phase formation and sintering processes, $BaTiO_3$ films with better crystallinity (and transparency) were expected. An excess of lead in PLZT composition was used to suppress a formation of zirconium rich non-ferroelectric pyrochlor phases [9]. This phases can build up at relatively low temperature (500-600°C) and transforms extremely slowly onto PLZT phase at a higher temperature. It is to be noted that it was impossible to prepare single phase PLZT films at reasonable annealing time unless excess of lead was introduced. The excess of lead was introduced into oxide system also to compensate the loss of lead during the thermal processing due to a volatility of lead oxide at elevated temperatures.

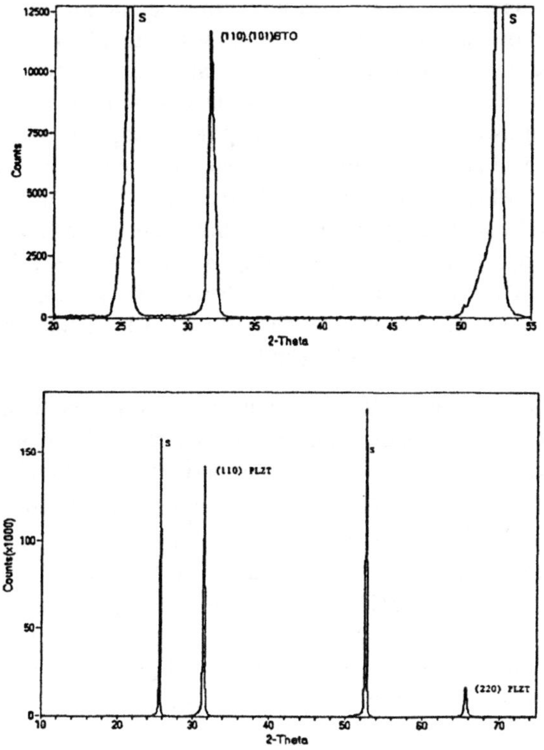

Figure 1. XRD patterns of the $Ba_{1-x}Pb_xTiO_3$ (a) and PLZT (b) films grown on Al_2O_3. Film thickness is about 1 μm.

We did not succeed in preparing films of barium titanate thicker than 1.5 μm on R-sapphire. Thicker films have cracked and delaminated. However, it was possible to grow thicker films on strontium titanate. PLZT films up to 25 μm thick were successfully deposited onto R-sapphire wafers.

Fig.1 shows θ-2θ X-ray diffraction patterns of $BaTiO_3$ films grown on R-Al_2O_3. The X-ray pattern of $Ba_{1-x}Pb_xTiO_3$ films on R-sapphire demonstrates the predominant (110) orientation of the films. It is to be noted that (101) and (110) peaks are not resolved. That allows us to suggest the structure of $Ba_{1-x}Pb_xTiO_3$ films to be pseudocubic. The lattice constant changes linearly from 3.999(±0.003)Å to 4.092 (±0.004)Å when lead content changes from 0 to 0.2. Films grown on (100) and (110)$SrTiO_3$ had orientation of (100) and (110) type respectively.

PLZT films deposited on R-plane sapphire were single phase and exhibited predominant (110) orientation. The contribution of other orientations was affected by the growth rate and the annealing temperature. Pseudocubic structure was suggested for PLZT films as well.

Optical transmission was measured in the range of 500-3000 nm. Transmission spectrum had a cut off at 250 nm, meanwhile, both PLZT and barium titanate films were transparent at 250-3000 nm. This suggests their possible applications in IR range. To accurately measure the optical loss in ferroelectric films and PLZT films grown on R-sapphire, a prism coupling experiments were performed. The optical loss coefficient of 1 μm barium titanate film measured at 633 nm was 3-5 dB/cm which is significantly lower than that measured for barium titanate films grown by metalorganic chemical vapor deposition (20-30 dB/cm for 1 μm thick film). The optical loss in PLZT films (3-5 μm thick) was 5-7 dB/cm.

Both materials have exhibited the quadratic electro-optic behavior (Fig.2). The value of second-order electro-optic coefficient (R) was calculated as follows,

$$R=\lambda\Delta\Gamma/(\pi n^3E^2L) \qquad (1)$$

where $\Delta\Gamma$ is the relative phase shift in radians, λ is a wavelength in nm, n is the refractive index of the film, E is the density of the electric field in V/m and L is the film thickness in nm. The R values measured for barium titanate and PLZT films are shown in the Table 1. A maximum quadratic EO coefficient $R=0.65\cdot10^{-16}$ $(m/V)^2$ was achieved in transmission mode for $Ba_{0.9}Pb_{0.1}TiO_3$ film. The higher lead content leads to the decrease of the relative phase shift. The R value of $0.45\cdot10^{-16}$ $(m/V)^2$ measured for 10 μm thick 7.5/65/35 PLZT film is the highest ever reported in the literature for PLZT films. This value is still lower than that of the bulk PLZT ceramic ($R=7\times10^{-16}$ $(m/V)^2$). However, a number of light controlling devices have been demonstrated using PLZT films thus prepared [6].

TABLE 1. Second-order electro-optic coefficients measured for $Ba_{0.9}Pb_{0.1}TiO_3$ and PLZT films.

Material	Orientation	R, $\times10^{-16}$ $(m/V)^2$
$Ba_{0.9}Pb_{0.1}TiO_3$*	(100) on (100)$SrTiO_3$,	0.35
	(110) on (110)$SrTiO_3$	0.65
	(110) on (110)$SrTiO_3$	0.63
	(111) on (111)$SrTiO_3$	0.09**
PLZT (7/65/35)***	$I_{(110)}/\Sigma_I=99\%$	0.42
	$I_{(110)}/\Sigma_I=65\%$	0.24
	$I_{(110)}/\Sigma_I=50\%$	0.16

* 1-1.5 μm thick films, ** MOCVD film [5], *** 6-10 μm thick films

The "structure-properties" relationships have been extensively studied in ferroelectric ceramics. The analogous study on thin and thick films is still in its infancy. This is particularly true for the electro-optic properties of the these materials.

Compositional effect was studied on PLZT films. Fig.3 shows the dependence of the second-order electro-optic coefficient on the lanthanum content.

This dependence goes through the maximum at app. 7.5 at.% of La. This composition lies near the boundary between the "rhombohedral" and "tetragonal" ferroelectric phases. This morphotropic phase boundary is especially important for the PLZT ceramics since many properties of interest are maximized along it [1]. The behavior of the PLZT films is, therefore, very similar to that of the bulk PLZT ceramics in this respect.

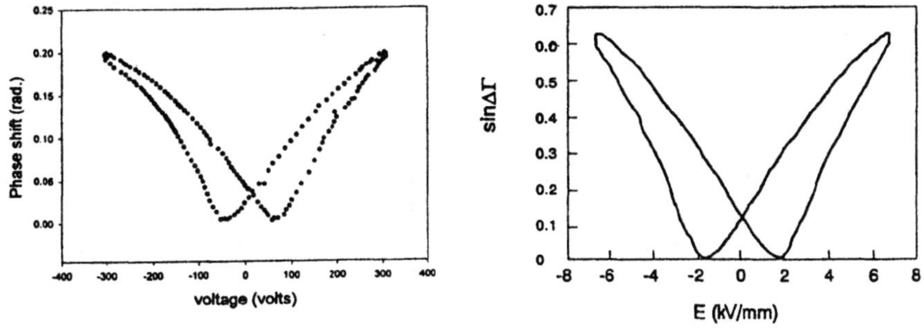

Figure 2. Dependence of the relative phase shift on applied voltage for the films of $Ba_{0.9}Pb_{0.1}TiO_3$/R-Al_2O_3 (a), and PLZT/R-Al_2O_3 (b).

Orientation of the ferroelectric films appears to be one of the most important factor affecting electro-optical properties of the PLZT and $BaTiO_3$ films. It was found out that $BaTiO_3$ films with orientation of (110) type demonstrated a larger EO effect compared to those having (100) and (111) orientation. The same phenomena was observed in PLZT films on R-sapphire: the higher the contribution of (110) orientation, the larger quadratic electro-optic coefficient (Table 1).

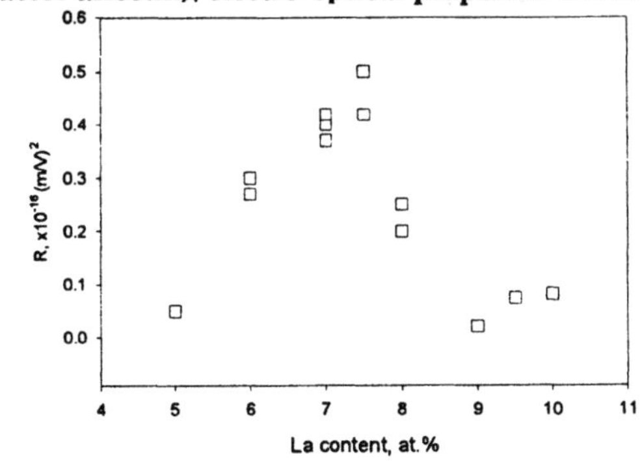

Figure 3. Dependence of quadratic EO effect on the La content in PLZT films. Film thickness is 3 μm.

This dependence for PLZT film is in a good agreement with a fact that polar axis in PLZT with composition 8/65/35 ("rhombohedral" phase) is parallel to [111] direction. This direction lies in the substrate plane and is, therefore, perpendicular to the incident light beam. In this arrangement, a maximum phase retardation between ordinary and extraordinary components of the light beam can be reached. However, the observation for barium titanate films can not be interpreted as straightforward as it has been done for PLZT films. The ferroelectric $BaTiO_3$ phase, which exists at 0-120°C has a tetragonal symmetry (*4mm*) and the spontaneous polarization parallel to [001] direction. The rhombohedral (*3m*) phase exists only at the temperatures lower than -90°C [1]. It should be also pointed out that if $SrTiO_3$ is used as substrate material, a contribution of the birefringence from the substrate might be noticeable.

A remarkable feature of the PLZT films is that their grains have a submicron size (Fig.4). This is in contrast to the sintered PLZT ceramic consisting of the grains of few microns in size. This might explain a lower EO coefficient measured for PLZT film compare to the bulk ceramic. The smaller birefringence observed in PLZT films can be attributed to the grain size effects since smaller grains are more subject to grain boundary pinning not allowing for flexible domain movement.

4.Conclusions

We have successfully demonstrated the growth of thick films of barium titanate and PLZT by sol-gel process. This method enabled the growth of up to 25 μm thick films which retained a high optical quality. The materials showed quadratic electro-optic behavior with large electro-optic coefficients. The relationship between the structure of PLZT and barium titanate films was studied as well. As a result of the optimization of the films optical and electro-optical properties the films of the device quality were obtained.

5. Acknowledgements

This work was supported by USAF Rome Laboratory (Contract F19628-93-C-0078).

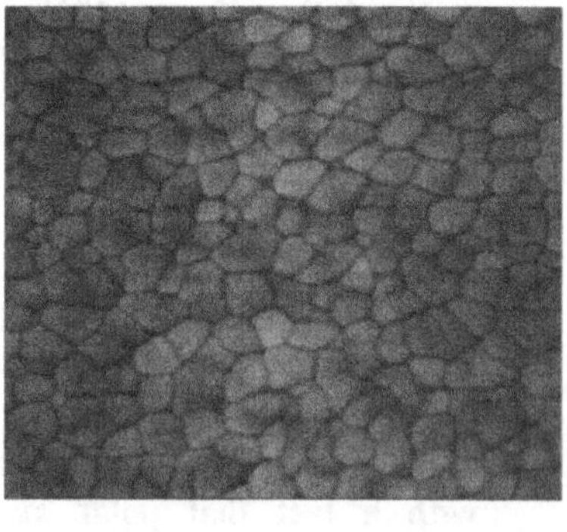

Figure 4. AFM image of the PLZT film surface.

6. References

1. Xu, Yuhuan. (1991) *Ferroelectric Materials and Their Applications*, Elsevier Science Publishers B.V., Amsterdam.

2.Wang F., Li., K.K., and Haertling, G.H. (1992) Transverse electro-optic effect of antiferrolelectric lead zirconate films, *Optics Lett.*, **17**, 1122-1124.

3.Jin, M.S., Wang, V., Ozguz, V., and Lee, S.H. (1994) Monolithic integration of a silicon driver circuit onto a lead lanthanum zirconate titanate substrate for smart spatial light modulator fabrication, *Appl.Optics* **14**, 2842-2848.

4.Lin, T.-H., Ersen, A., Wang, J.H., Dasgupta, S., Esener, S., and Lee, S.H. (1990) Two dimensional spatial light modulators fabricated in Si/PLZT, *Appl.Optics* **11**, 1595-1603.

5. Zhao, J., Fuflyigin, V., Wang, F., Norris, P., Bouthilette, L., and Woods, C.(1997) Epitaxial electro-optical $BaTiO_3$ films by single source metal organic chemical vapour deposition, *J.Mater.Chem* **7**, 933-936.

6. Wang F., Li, K.K., Fuflyigin, V., Jiang, H., Zhao, J., Norris, P., and Goldstein D. (1998) Thin ferroelectric interferometer for spatial light modulations, Appl. Optics, Paper No. AO 0150160.

7. Preston, K.D., Haertling, G.H., (1992) Comparison of electro-optic lead lanthanum zirconate titanate films on crystalline and glass substartes, *Appl. Phys. Lett.* **60**, 2831-2833.

8. Wang, F., Furman, E., and Haertling, G.H. (1995) Electro-optic measurements of thin films materials by means of reflection differential ellipsometry, *J.Appl.Phys.* **78**, 9-15.

STUDY OF THE REACTIVITY OF NANOCRYSTALLITES OF SnO_2 WITH H_2S BY COUPLED ELECTRICAL AND RAMAN MEASUREMENTS

M. BOULOVA[2] T. PAGNIER[1],M. RUMYANTSEVA[2], A. GASKOV[2], G. LUCAZEAU[1]
1 *Lab. d'Electrochimie et de Physico chimie des Matériaux et des Interfaces, UMR 5631, ENSEEG, BP 75 38402 Saint Martin d'Hères France*
2 *Laboratory of Diagnostic of Inorganic Materials, Department of Materials, Moscow State University, Leninskie Gory 119899 Moscow Russia*

1. Introduction

The present contribution consists in the characterization of the reactivity of nanoceramics of Cu doped tin oxide towards H_2S by Raman spectroscopy. It will be shown in the followings that the specific Raman signature due to the nanometer size of crystallites and the in-situ Raman spectroscopy are particularly useful to treat such a system:
The 2D dimensionnality achieved with thin epitaxial films leads to red or blueshifts of Raman bands depending the film is in tensile or compressive stress state. However no important band broadening seems to be necessarily associated with the reduction of the dimensionnality (1, 2) On the contrary when the three dimensions are reduced as i.e. for microcrystals smaller than 10nm one observes a broadening and a downshift of Raman bands (3). The spectra of SnO_2 nanocrystallites have been interpreted on the basis of bands slightly shifted and broadened representative of volume modes and of broad features representative of surface modes; the relative intensity of which varying from 1 to zero when the crystallite size varies from 4nm to 2nm (4). The reactivity of pellets of nanocrystallites (4 to 7 nm) of pure SnO_2 towards H_2S has been monitored by coupled electrical and Raman measurements (5). The following conclusions were obtained: insulating samples in their initial state react with H_2S and lead to the formation of surface "SO_4" groups and to SnS_x tin sulfide, while conducting samples lead to tin sulfide formation only. The corresponding thermodynamic stability diagram vs $P(O_2)$, $P(S_2)$ has been calculated and allows to check that the reversible formation of sulfide is normally forbidden for the partial pressures (10^{-7} and 10^{-3} bars respectively) used in the above experiments. The formation of SnS_x (metal) was strongly correlated with the decrease of electrical resistance.
The electrical response of Cu doped samples is much more sensitive towards H_2S than undoped samples. Actually it is considered that CuO-SnO_2 contacts are p-n junctions which make the tin oxide insulating under air and then increases the R variation when the material is in contact with H_2S. The aim of this Raman study is to indentify i):the

G. Van Tendeloo et al. (eds.), High-Temperature Superconductors and Novel Inorganic Materials, 285–290.

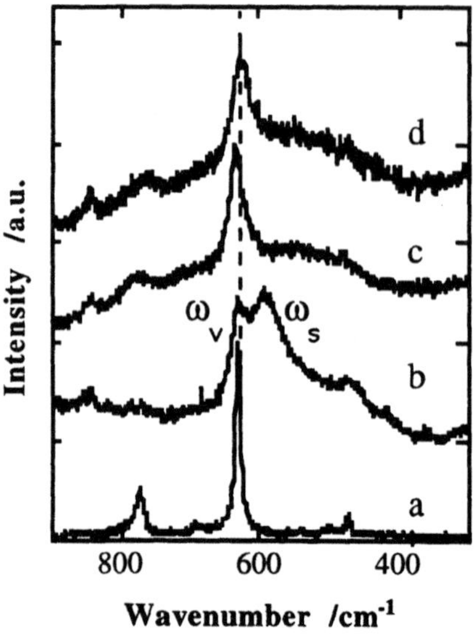

Figure 1. Effect of Cu doping (% at) on spectra of nanocrystallites of SK+x%Cu. b: x=0, c: x=5, d: x=10. For comparison, the spectrum of a pure microcrystalline SnO_2 (a) is given. Note the huge broadening of the s band when Cu is introduced.

nature of the chemical species that Cu can form under air and under H_2S, ii) their location (surface or bulk of SnO_2 crystallites or as independent particles.) iii) to correlate the different response of the samples with their structure.

2 Experimental

Powders were prepared following different methods: I and II routes are described in ref (4). the route I consists in the precipitation of stannic acid and in the treatment of sodium stannate in H^+ resin, the obtained colloidal solution is then freeze-dried in a sublimator and calcined at 600°C. The route II starts with the classical hydrolysis of $SnCl_4$ to obtain the precipitation of stannic acid folllowed by washing, drying at 100°C and calcination. The route III involves the hydrolysis of tin-isopropylate while route IV is identical to route II followed by freeze drying in sublimator. Copper is introduced in a second stage in the pure SnO_2 powder as copper ethylacetate, then the solution is evaporated and dried. Powders were then pressed at 700MPa and the obtained pellets were partially covered with silver electrodes. The crystallite sizes have been estimated from the intensity ratio of the 630 cm^{-1} line (volume mode) to the 550 cm^{-1} line (surface mode) see ref (4) In addition elemental analysis, XRD and SEM controls were performed.
Coupled electrical and Raman measurements were made in a heating (25-900°C) cell allowing atmosphere control. Flowing gas was either dry air (20% O_2) or 300ppm H_2S in nitrogen.
MicroRaman spectra were obtained on a Dilor XY spectrometer with 10mW of the 514.5nm exciting line of an argon laser.

TABLE 1. Main properties of SnO_2 nanoceramics presently studied

Sample	Synthesis route (1)	Crystallite size /nm	H_2S reactivity			Resistance /Ω 100°C		Conductor in initial state
			SO_4	SnS_2	Cu_2S	air	H_2S	
SV_6	II	4	+	+				
SK	II	6	+	+		$>10^8$	5.10^5	-
SX	I	7	-	+		4.10^2	2.10^2	+
SKCu5	II	10	+	+	+	$>10^8$	4.10^5	-
SXCu5	I	5	-	+	-	$2\ 10^5$	2.10^3	+
SACu5	III	2	+	-	+	$>10^8$		-
SSCu5	IV	4	-	+	-			+

(1): see text

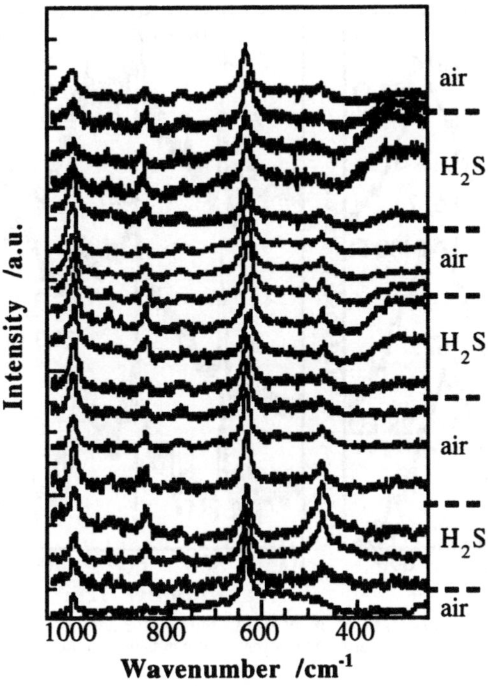

Figure 2. Raman spectra of SKCu5 at 100°C during atmosphere alternances

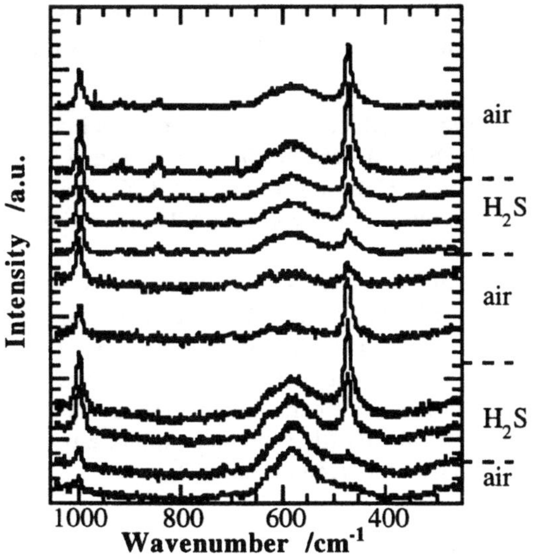

Figure 3. Raman spectra of SACu5 at 100°C during atmosphere alternances.

3. Results and Discussion

Table 1 reports the main properties measured in the different samples studied presently .
Fig 1 represents the spectral changes due to the micro-nanoscale size effect (a-b spectra) and to copper doping (c-d spectra). Note the appearance of a broad component noted ω_S and attributed to surface modes (4).while the narrow component due to volume modes is observable in all spectra. The intensity ratio of these bands is related to the size of the crystallites (4). Note the broadening of the surface band when the samples are Cu doped. This broadening can be attributed to the disorder induced by copper on the surface of the crystallites.

Fig 2 represents the evolution of the spectra of SKCu5 at 100°C versus atmosphere alternances
The band at $990cm^{-1}$ is attributed to sulfate species noted $SnO_2(SO_2)_{ad}$ to indicate that these groups are bridged on the surface and that it can explain the intensity increase when the sample is exposed to air. The growth of the Cu_2S band at $480cm^{-1}$ is clearly observed together with a decrease of the surface band of the tin oxide particle. This shows that Cu_2S is formed on the surface. The broad band at about $300cm^{-1}$ is likely due to SnS_x species which form slowly and at the expense of copper sulfide. At the difference of sulfate, the formation of both

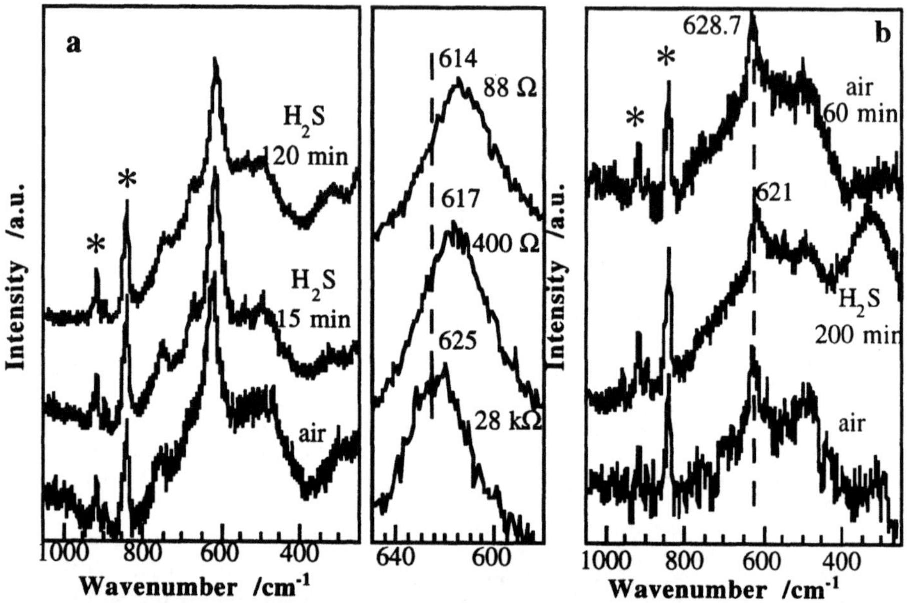

Figure 4.Effect of atmosphere on the spectrum of SXCu5. a: :at 100°C, the volume band is magnified in the right part. b: at 150°C. the asterisk indicates fluorescence bands due to the microscope..

sulfides is reversible. Fig 3 represents the spectra of SACu5 at 100°C. Note that the volume mode is absent and that this indicates that the size of the crystallite is close to 2nm. However the surface band is somewhat narrower than in SKCu5. The sulfate and Cu_2S bands are clearly observable, that of SnS_x doesn't show up. The initial resistance of SK and SA samples is very high and this can explain the particular reactivity of these compounds (5).

The two other samples SXCu5 and SSCu5 are conductors in their initial state and do not form neither sulfate nor Cu_2S with H_2S as it can be checked in Fig 4 and 5 respectively. On the contrary SnS_x forms under H_2S, and faster at 150°C. This slow kinetics is likely the indication that tin sulfide forms in the volume while the sulfate and copper sulfide species form at the surface

The main volume Raman peak at about 620cm^{-1} downshifts upon exposition to H_2S as well at 100 as 150°C. In the same time SnS_x forms and the electrical resistance decreases. The sulfides which form in the crystallite volume could decrease the density of the crystallites and induce tensile stresses. On the other hand this band upshifts when temperature is raised (compare fig 4a and 4b). Such a behaviour is unusual for bulk solids. Aggregation of crystallites could reduce the tensile state and explain a positive shift, however this shift being reversible such a phenomenon is unlikely.

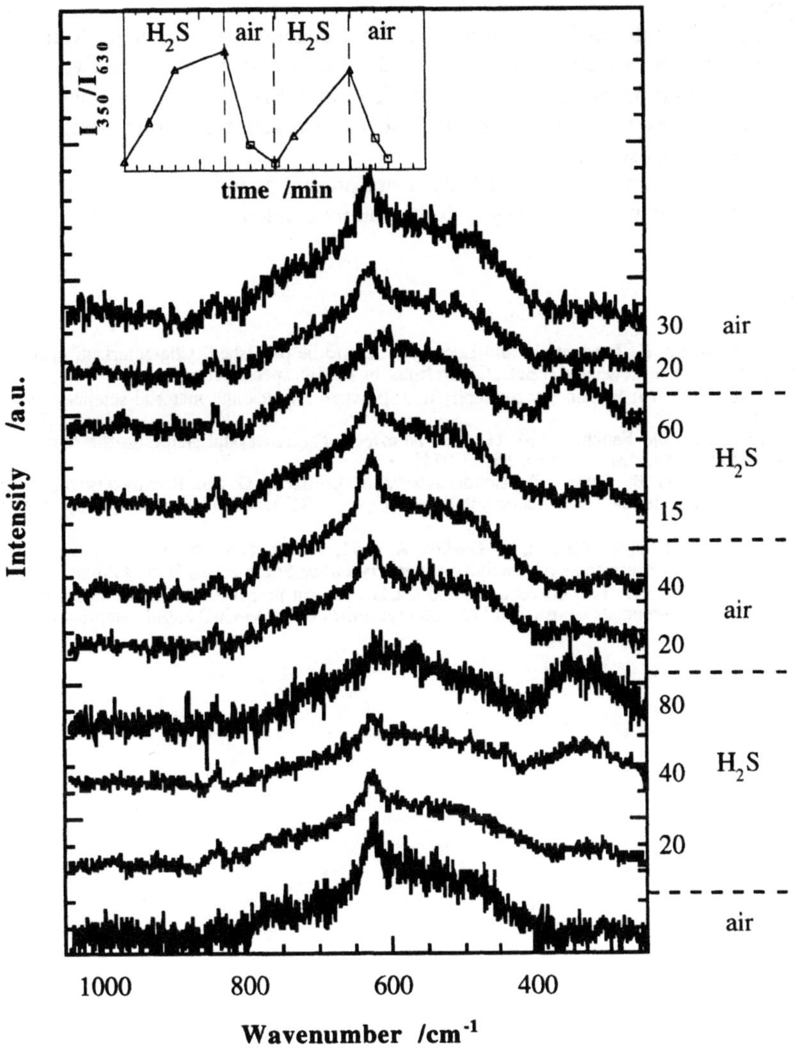

Figure 5. Raman spectra of SSCu5 at 100°C during atmosphere alternances. The variations of the relative intensity of the broad band at about 300cm^{-1} and attributed to SnS_x is plotted in the upper part.

Conclusions

-SnS_x, and Cu_2S are obtained under H_2S although their formation is not allowed from thermodynamic considerations.The formation of sulfides could be assisetd by the fact that nanocrystallites have a small enthalpy of formation compared to normal tin oxide

or copper oxide.
- In agreement with stability diagram, sulfate species are obtained. Actually these species are mainly located at the surface and are noted $SnO_2(SO_2)$ad. They are formed only with insulating pure and Cu dopedSK and SA samples. A mechanism for explaining the present reaction and the absence of reaction for conducting SS and SX compounds is proposed in (5).
-It seems that at least a fraction of CuO is not present as independent crystallites but is bonded to tin oxide particles and should be mainly located on the surface of crystallites.

References

1. Abello, L.,Weiss, F., Lindner, J., and Lucazeau , G. (to be published) Characterization of stresses in epitaxallialy grown CeO_2 and YBaCuO thin films by Raman spectroscopy
2. Lucazeau, G. (1995) Raman Spectroscopy in Solid state Physics and material science. *Analusis* , **23**, 301-311
3. Campbell, I.H. and Fauchet, P.M. (1986) The effect of microcrystal shape on the raman spectra of semiconductors *Solid State Comm.* **58**, 739-741
4. Abello, L., Bochu, B., Gaskov, A., Loudryavtseva, S., Lucazeau, G. and. Roumyantseva, M. (in press) Structural Characterization of nanocrystalline SnO_2 by DRX and Raman spectroscopy, **J. Solid State Chem**
5. Pagnier, T., Boulova, M., Galerie, A., Gaskov, A. and Lucazeau, G. (submitted) In situ coupled Raman and impedance measurements of reactivity of nanocrystalline SnO_2 versus H_2S. **J .Solid State Chem**
6. Boulova, M., Pagnier, T., Lucazeau, G. and Gaskov, A. (in preparation) In situ coupled Raman and impedance measurements of reactivity of nanocrystallites of SnO_2-CuO based ceramics in H_2S/N_2 and N_2/O_2

A DISCUSSION OF DIFFUSION AND SOLID STATE REACTIONS IN ALLOYS OF II-VI, III-VI AND IV-VI COMPOUNDS ON THE BASIS OF PHASE DIAGRAMS

V. LEUTE
Institut für Physikalische Chemie, Universität Münster
Schlossplatz 4, D-48149 Münster

1. Introduction

Interdiffusion processes in solid solutions of quasibinary and quasiternary systems are strongly correlated with the thermodynamic properties of these systems. The basic function describing the thermodynamic properties is the mean molar Gibbs energy. If this function is known in dependence on composition and temperature the phase diagram of the system can be calculated. On the other side, if essential parts of a phase diagram are known from experiment, the thermodynamic interaction parameters can be determined and therefrom the Gibbs energy can be calculated. In this paper we will show at certain examples of quasiternary systems how the interdiffusion processes and the paths of solid state reactions are influenced by the mean molar Gibbs energy of quasiternary systems.

The composition of a quasiternary system of type $(M_kN_{1-k})(X_lY_{1-l})$ can be described by two mole fractions, namely k for the metal sublattice and l for the chalcogen sublattice. The mean molar Gibbs energy for such a system consists of a standard term, an ideal mixing term and an excess term g^E that reflects the chemical properties of the quasibinary edge systems $M(X_lY_{1-l})$, $N(X_lY_{1-l})$, $(M_kN_{1-k})X$ and $(M_kN_{1-k})Y$.

$$g^E = k(1-k)(\alpha_1 + k\beta_1 + l\gamma_1 + kl\delta_1) + l(1-l)(\alpha_2 + l\beta_2 + k\gamma_2 + kl\delta_2) + k(1-l)\Delta_R G^0$$

The interaction parameters α_i, β_i, γ_i and δ_i can be calculated from the interaction parameters of the four quasibinary edge systems [1]. The only ternary term that occurs in g^E is the standard Gibbs energy, $\Delta_R G^0$, of the characteristic solid state reaction MX + NY $\rightarrow$ NX + MY.

2. Gibbs Energy and Thermodynamic Factor

The phase diagrams for the quasiternary system $(Sn_kPb_{1-k})(S_lTe_{1-l})$ and for its quasibinary edge systems were determined by X-ray phase analysis [2]. Therefrom, by adjusting calculated miscibility gaps to the experimental data, both the interaction parameters and $\Delta_R G^0$ were calculated.

G. Van Tendeloo et al. (eds.), High-Temperature Superconductors and Novel Inorganic Materials, 291–296.

On quasibinary sections the interdiffusion coefficient $\tilde{D}$ is correlated with thermodynamics according to the Darken equation [3]:

$$\tilde{D} = (x_1 D_2 + x_2 D_1)F$$

The thermodynamic factor F introduced by this equation depends, via the activity coefficient, on the second derivative of the excess Gibbs energy:

$$F = 1 + \frac{x_2(1-x_2)}{RT}\left(\frac{\partial^2 g^{\mathrm{E}}}{\partial x_2^2}\right)$$

The Sn/Pb-interdiffusion coefficient on sulfur rich quasibinary sections was calculated from Sn and Pb mole fraction profiles measured by an electron microprobe (Fig.1). Fig.2 shows that the thermodynamic factor F is nearly independent on the mole fraction l, but depends on the mole fraction k. The interdiffusion coefficient is reduced as long as $k < 0.78$ and is enhanced for $k > 0.80$. If $\tilde{D}$ is divided by F, one obtains the ideal interdiffusion coefficient reflecting only the mobility part, contributed by the tracer diffusion coefficients (Fig.1).

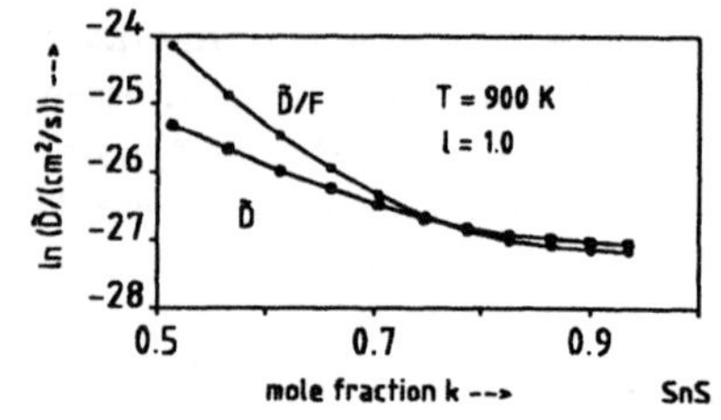

Figure 1: Interdiffusion coeff.

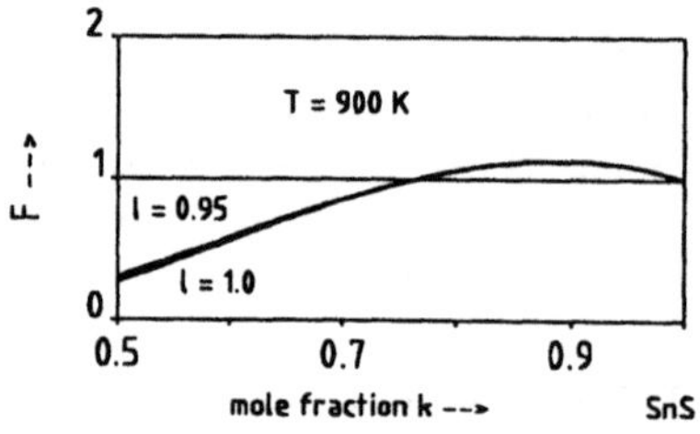

Figure 2: Thermodynamic factor

3. Reaction Path and Phase Diagram

The result of solid state reactions in quasiternary systems after a given annealing time can be described by a three-dimensional curve $G(k,l,x)$. The variables of this implicit function G are the mole fractions k and l and the space coordinate x describing the extension of the reaction zone. The mole fraction profiles $k(x)$ and $l(x)$, usually measured by an electron microprobe, are the projections of these curve onto the planes k-x and l-x. The so-called reaction path $\Gamma(k,l)$, calculated from these profiles by elimination of the space coordinate x, corresponds to the projection of G onto the l-k plane.

Fig.3 shows the phase diagram of the quasiternary system $(Cd_kPb_{1-k})(S_lTe_{1-l})$ [4] together with a reaction path, calculated from the Pb and Te profiles for a reaction couple consisting of PbS and CdTe crystals with the composition of minimal vapour pressure. This plot shows that the equilibrium tie lines, determine the possible jumps across the miscibility gaps. The course of the reaction path within the homogeneous regions of the phase diagram, however, is influenced by the diffusivities. If the components of the reaction couple were metal saturated, the reaction path would be different, because intrinsic doping changes the diffusivities. Nevertheless, the possible jumps across the miscibility gaps will always run along the equilibrium tie lines.

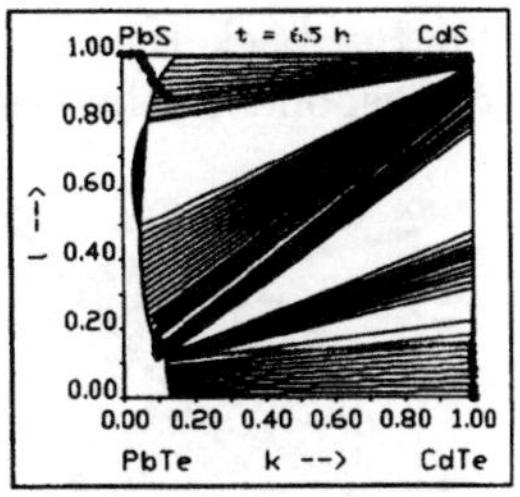

Figure 3: Reaction path for samples with minimal vapour pressure

4. Calculation of Reaction Paths

Obviously, a reaction path is determined by transport properties as well as by thermodynamic properties. Within the scope of linear irreversible thermodynamics the diffusion fluxes are linear functions of the gradients of chemical potentials of independent components, $J_i = \sum_j L_{ij}(\partial\mu_j/\partial x)$. These chemical potentials can be expressed as the first derivatives of the mean molar Gibbs energy with respect to the molar fractions k and l. If the flux equations are to be described in dependence on composition gradients, the matrix **D** of the partial diffusion coefficients has to be introduced. This diffusion matrix is proportional to the product of the Onsager Matrix **L** containing the mobilities with the Hessian matrix **H** consisting of the second derivatives of the mean molar Gibbs energy, $\mathbf{D} \sim \mathbf{L} \cdot \mathbf{H}$ [5].

The coefficients of the Onsager matrix are functions of the tracer diffusion coefficients. The cross coefficients $L_{12} = L_{21}$ are here taken as zero.

The Hessian matrix consists of an ideal mixing part:

$$M_{11}^{id} = \frac{1}{k(1-k)}, \quad M_{22}^{id} = \frac{1}{l(1-l)}, \quad M_{12}^{id} = M_{21}^{id} = 0,$$

and an excess part:

$$\begin{aligned} & M_{11}^{ex} = (2-6k)(\beta_1 + l\delta_1) - 2(\alpha_1 + l\gamma_1) \\ M_{12}^{ex} \quad = \quad & M_{21}^{ex} = (1-2k)\gamma_1 + (2k-3k^2)\delta_1 + (1-2l)\gamma_2 + (2l-3l^2)\delta_2 - \Delta_R G^0 \\ & M_{22}^{ex} = (2-6l)(\beta_2 + k\delta_2) - 2(\alpha_2 + k\gamma_2) \end{aligned}$$

The thermodynamic parameters $(\alpha_i, \beta_i, \gamma_i, \delta_i)$ can be derived from the interaction parameters of the subregularly treated quasibinary edge systems. The cross coefficients $M_{12}^{ex} = M_{21}^{ex}$ are determined predominantly by the standard Gibbs energy $\Delta_R G^0$ for the reaction between MX and NY.

From the diffusion matrix, the mole fraction profiles $k(x)$ and $l(x)$ are calculated by solving the diffusion equations using the method of finite differences:

$$\partial k/\partial t = \nabla \cdot (D_{11}\nabla k + D_{12}\nabla l), \qquad \partial l/\partial t = \nabla \cdot (D_{21}\nabla k + D_{22}\nabla l).$$

As the Onsager cross coefficients were taken as zero, any ternary effect, like uphill diffusion, is due to the influence of the Hessian matrix, i.e. due to thermodynamics.

Figs.4 demonstrates the effect of $\Delta_R G^0$ on the path of the reaction MX + NY $\rightleftharpoons$ MY + NX. For $\Delta_R G^0 > 0$ the path for the reaction between MX and NY runs nearly along a straight line, whereas one obtains an Z-shaped reaction path for the back reaction, for which $\Delta_R G^0 < 0$. The situations shown in Figs.4-6 correspond to the system $(Hg_k Cd_{1-k})(Te_l Se_{1-l})$ Fig.5 shows that the course of

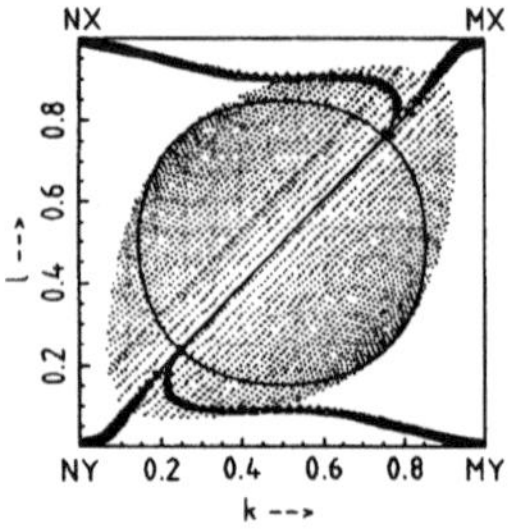

Fig.4: Reaction paths for $T = 723$ K

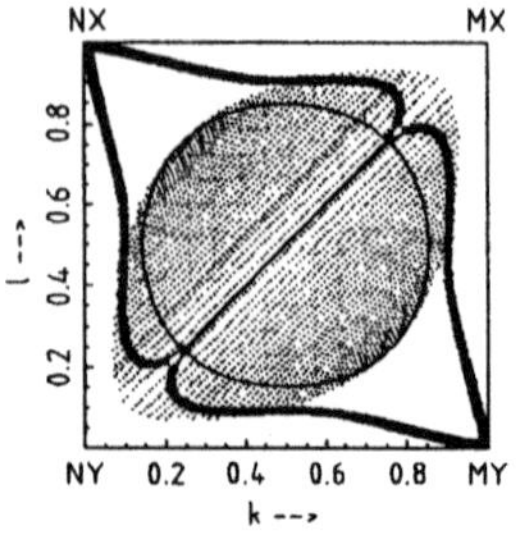

Fig.5: Influence of diffusion coefficients

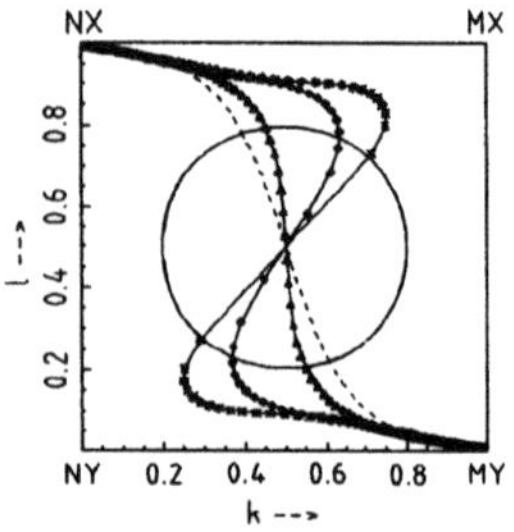

Fig.6: Temp. dependence of reaction paths

the reaction path within the homogeneous regions of the phase diagram is governed by the diffusivities. If the diffusion coefficients for the constituents of the chalcogen sublattice are higher then those for the constituents of the metal sublattice, then the reaction path will run predominantly along the chalcogen edges N(X,Y) and M(X,Y), in the opposite case along the metal edges (M,N)X and (M,N)Y.

The effect of thermodynamics becomes most evident in the temperature dependence of the reaction path (cf.Fig.6). In the quasiternary system $(Hg_k Cd_{1-k})$-$(Te_l Se_{1-l})$ there is a spinodal miscibility gap with a critical point at $T_{crit} = 984$ K. Thus, at 800 K ($*$) the reaction path crosses the miscibility gap, whereas at 1100 K ($\diamond$), above the critical temperature, the reaction path runs through regions of complete solid solubility. Nevertheless there is a distinct region of uphill diffusion, that can be caused only by thermodynamic influences, because the cross coefficients of the Onsager matrix have been taken as zero. Even at 2400 K (Δ), at a temperature where the system were already molten, still a remarkable deviation from ideal behaviour (- - -) would occur.

5. Systems with Order-Disorder Transitions

Another problem occurs, when systems are to be considered, that show regions with ordering tendencies. Special systems of this type are alloys of II-VI (CdTe or HgTe) with III-VI compounds ($Ga_2\underline{V}Te_3$ or $In_2\underline{V}Te_3$), both crystallizing in the zincblende lattice. In the III-VI compounds however, because of their stoichiometry, only 2/3 of the sites within the cation sublattice are occupied. The unoccupied sites $\underline{V}$ are called structural vacancies. In randomly mixed alloys these vacancies act as diffusion supporting defects, whereas in well ordered phases the structural vacancies behave as interstitial sites of the superstructure. Thus, one should expect that the ideal interdiffusion coefficient $\tilde{D}/F$ will be distinctly higher in disordered than in ordered phases with equal composition.

But real interdiffusion coefficients calculated from electron microprobe measurements will differ from ideal coefficients by the thermodynamic factor. To calculate the thermodynamic factor, we need an expression for an excess Gibbs energy including the contributions from ordering. To obtain this function we add a special ordering term g^s to the normal excess term $g^{E,n}$ [6]. The ordering term is formulated as a product of the Gibbs energy for the order-disorder transition $\Delta_{Tr}G^0$ with a distribution function of the Gauss type describing the extension σ_k of the ordering tendency on both sides of the ideal stoichiometric composition k^* of the superstructure phase.

$$g^E = g^{E,n} + g^s = g^{E,n} + \Delta_{Tr}G^0 \cdot \exp\left[-\frac{(k-k^*)^2}{2\sigma_k^2}\right]$$

Fig.7 shows $g^{E,n}$ and the complete g^E function including three ordering terms. The system $(Hg_{3-3k}In_{2k}\underline{V}_k)Te_3$ yields superstructures at the mole fractions k = 0.375, 0.75 and 1.0. Because of the regions with negative curvature in this Gibbs energy function one obtains regions that are unstable with respect to diffusion. In phase diagrams these regions appear as spinodal miscibility gaps.

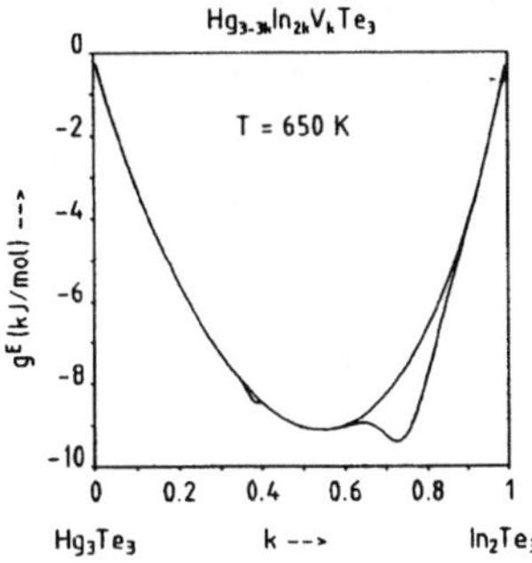

Figure 7: Gibbs energy with ordering tendencies

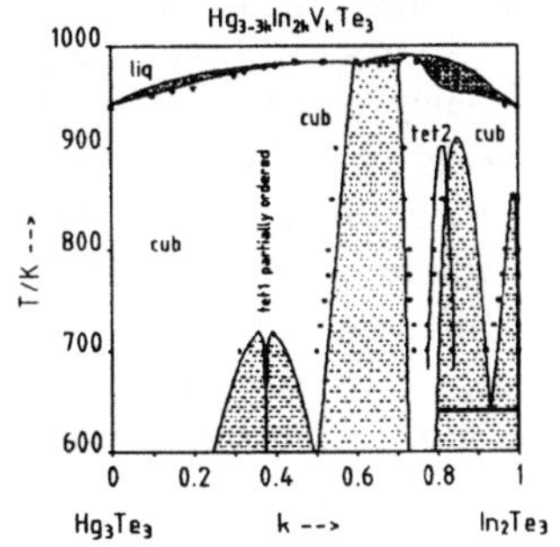

Figure 8: Calculated phase diagram

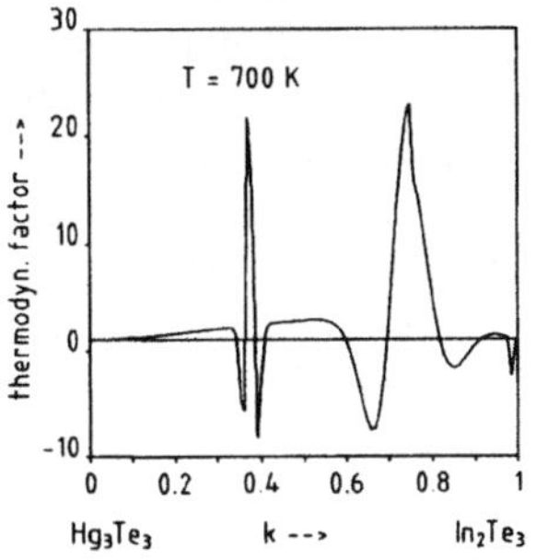

Figure 9: Thermodynamic factor

The phase diagram for this system, calculated from the complete function for the excess Gibbs energy, reflects exactly the topology of the experimentally determined phase diagram (cf.Fig.8). Generally, the quantitative agreement between the results from calculation and from measurement is also quite satisfying. Only the miscibility gap on the In_2Te_3-rich side of the tetragonally ordered chalcopyrite phase (k = 0.75) is calculated somewhat too broad and is shifted a little to higher k-values compared with the experimental data. But nevertheless this calculated miscibility gap reflects the characteristic feature of the experimental data in so far as its critical point lies within the sub-solidus region. This is the reason why there is a continuous transition from the tetragonal superstructure to the randomly distributed cubic alloy. This experimentally observed behaviour could not be explained by the usual Gibbs energy functions. From a single continuous Gibbs energy function one can also calculate a continuous function for the thermodynamic factor. In Fig.9 the thermodynamic factor is plotted as a function of composition. According to this figure the thermodynamic factor is drastically enhanced in the well ordered regions near the mole fractions k

= 0.375 and 0.75. The regions with negative thermodynamic factor, however, correspond to those regions of miscibility gaps that are unstable with respect to diffusion. If $F \geq 1$ holds generally for all ordered regions, then one will expect that on quasibinary sections the interdiffusion is favoured thermodynamically in all ordered regions compared to disordered regions. Fig.10 shows the phase diagram for the quasiternary system $(Cd_{1-y}Hg_y)_{3(1-k)}In_{2k}\underline{V}_kTe_3$. The section with a constant $In_2\underline{V}\ Te_3$ mole fraction of $k = 0.375$ shows a tetragonally ordered region for Hg/Cd-ratios $y \geq 0.5$. If on this section the Hg/Cd-interdiffusion were governed alone by thermodynamics, then the interdiffusion coefficient within the ordered region (tet1) should be increased compared to the disordered cubic region. But the experiment yields the opposite effect. Fig.11 shows the Hg/Cd-interdiffusion coefficient for $k = 0.375$ as a function of y at different temperatures. In the phase diagram for $T \geq 750$ K no ordered region occurs and consequently the interdiffusion coefficient is nearly constant. It increases only slightly with increasing lattice constant. At lower temperatures the interdiffusion coefficient decreases in those regions where ordering occurs. Therefore, in this case interdiffusion can not be determined by thermodynamics, i.e by the thermodynamic factor F, but must be dominated by the ideal interdiffusion coefficient. The observed behaviour of the interdiffusion coefficient can be explained by the properties of the structural vacancies. On the chosen section, at $k = 0.375$, 1/8 of the sites in the cation sublattice are vacant and these vacancies act as defects as long as the cations are randomly distributed. If however the vacancies occupy definite sites of a superstructure they will behave as interstitial sites of this structure and will contribute much less to the mobility of the cations. This is the reason why the Hg/Cd-interdiffusion is distinctly lower in the ordered region than in the disordered one.

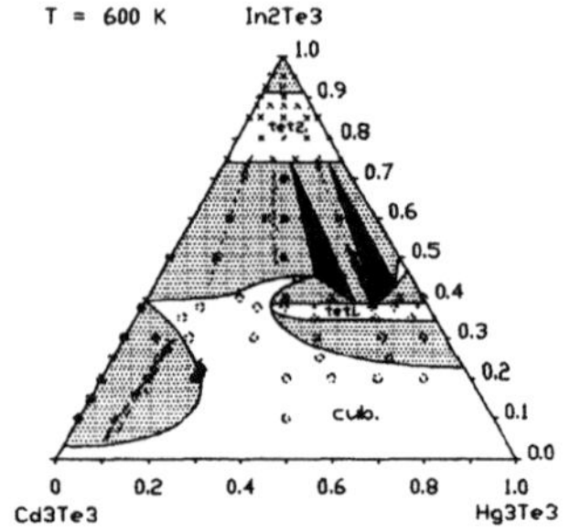

Fig.10: Phase diagram

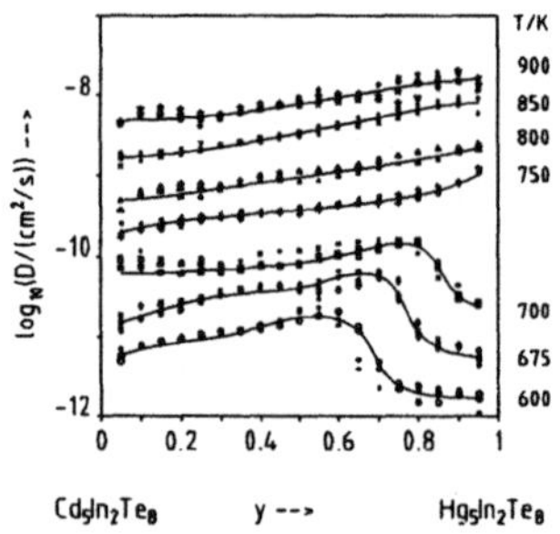

Fig.11: Interdiffusion coeff.

Acknowledgement: Financial support of this work by the Deutsche Forschungsgemeinschaft and by the Fonds der Chem. Industrie is acknowledged.

References

[1] Leute, V. (1989) *Ber.Bunsenges.Phys.Chem.* **93**, 7-13

[2] Leute,V. Brinkmann,S., Linnenbrink,J., and H.M.Schmidtke (1995) *Z.Naturforsch.* **50a**, 459-467

[3] Darken,L.S. (1948) *Trans. A.I.M.E.* **175** 184

[4] Leute,V., Haarmann,N.J., and Schmidtke,H.M. (1995) *Z.Phys.Chem.* **190**, 253-266

[5] Kokkonis,P.A. and Leute,V. (1997) *Defect and Diffusion Forum* **143-147**, 1159-1164

[6] Leute,V. (1996) *Calphad* **20**, 407-418

[7] Wensierski,v.H., Weitze,D., and Leute,V. (1997) *Solid State Ionics* **101-103**, 479-487

OPTICAL PROPERTIES OF GALLIUM-DOPED PbTe

N. ROMČEVIĆ[1], M. ROMČEVIĆ[2], D.R. KHOKHLOV[3],
I.I. IVANCHIK[3]

[1]Institute of Physics, 11001 Belgrade, P.O.Box 57, Yugoslavia
[2]Institute of Nuclear Sciences Vinča, 11001 Belgrade, P.O. Box 522, Yugoslavia
[3]Low-Temperature Physics Department, Moscow State University, 117234 Moscow, Russia

Abstract: We present Raman and far-infrared reflection spectra of PbTe single crystal doped with 0.5 at.% Ga at temperatures between 10 and 300K. In the Raman spectra, well-resolved peaks at about 104cm^{-1} (impurity induced PbTe LO mode), 166cm^{-1} and 187cm^{-1} were observed at all temperatures. One additional mode appears at about 117cm^{-1} at temperatures bellow 200K. These modes are local modes of Ga. The analysis of the far-infrared reflection spectra was made by a fitting procedure based on the model of coupled oscillators. At temperatures bellow T=200K, we also found an local mode of gallium at about 170cm^{-1}.

1. Introduction

Doping of the lead telluride-based alloys with the group III impurities such as In, Ga, Tl modifies substantially the electronic spectrum of the semiconductor. This modification implies formation of a deep (quasi)local level exhibiting quite unusual properties, such as Fermi level pinning [1], persistent photoconductivity or superconductivity [2].

When lead telluride is doped with gallium, the Fermi level becomes pinned only in some Ga content region. The pining position is inside the gap resulting in the semiinsulating state of a semiconductor [3]. Further doping leads to the increment of free electron concentration. In this situation the donor action of Ga becomes unstable with respect to the external factors, such as hydrostatic pressure or temperature [4]. In PbTe(Ga) the decrease of free electron concentration under pressure was detected [4, 5]. Lowering of temperature has the same effect as an application of pressure [5], so the model has been proposed, where the ability of Ga to act as a donor was assumed to be strongly coupled to the lattice [6]. Theoretical estimates based on the evaluation of ionic radii of different impurity atoms in PbTe lattice have shown that the conclusion concerning the donor action of Ga atoms depends critically on the lattice parameter, so

G. Van Tendeloo et al. (eds.), High-Temperature Superconductors and Novel Inorganic Materials, 297–302.

that even minor changes of this parameter may lead to transition to electrically neutral state of Ga [6].

In this paper we present Raman and far-infrared reflection spectra of 0.5 at.% Ga doped PbTe single crystal from 20 to 250cm^{-1} at temperatures of 10 to 300K. These far-infrared spectra analyzed using a fitting procedure based on the modified plasmon-phonon interaction model (but incorporating an additional oscillator corresponds the local mode of gallium). Together with the strong plasmon - phonon coupling we obtain the three local modes of gallium. We believe that observed local modes of Ga have the F_{1u} symmetry which energies depend on the impurity center charge.

2. Experimental

The sample under investigation was PbTe(Ga) single crystal with some excess concentration, so that the regime of unstable donor action of Ga was realized. That single crystal grown by the modified Bridgeman technique. The gallium impurity was introduced into the liquid zone.

The Raman spectra were excited by the 488nm and 514.5nm lines of an Argon laser (the average power was about 100mW). The geometry was of a bac-scatering type. For measurement of the Raman spectra we used the Jobin Yvon model U-1000 monohromator with conventional photocounting system. The samples were held in a closed-cycle cryostat, equipped with a low-temperature controller and evacuated by a turbo pump.

A Bruker 113v spectrometer with an Oxford cryostat was used to measure the far-infrared reflection spectra at low temperature.

3. Results and discussion

The nonpolarized Raman spectra of PbTe(Ga) in the spectral range from 80-210cm^{-1} at temperatures between 30 and 300K, are given in Fig.1.a.. At room temperatures the modes at 104, 166 and 187cm^{-1} are clearly observed. One additional mode appears at about 117cm^{-1} at temperatures bellow 200K.

For the analysis of Raman spectra a deconvolution technique was used, where it was accepted that all lines are of Lorentzian type. A smooth, solid line in Fig.1., was obtained by adding four different profiles:

$$I(\omega) = I_0 + \frac{2A}{\pi} \cdot \frac{W}{4(\omega - \omega_c)^2 + W^2} \qquad (1)$$

Here I - is a line intensity, I_o - the hight of the baseline, ω_c - position of the maximum, W - a half-width of the peak and, A - parameter that depends on W. A typical line shape obtained by deconvolution is presented in Fig.1.b. for T=150K. The results of the analysis were given in Fig. 2. Fig.2.a. shows the temperature changes of line half-widths (W) obtained by deconvolution technique. Two distinct types are clearly seen:

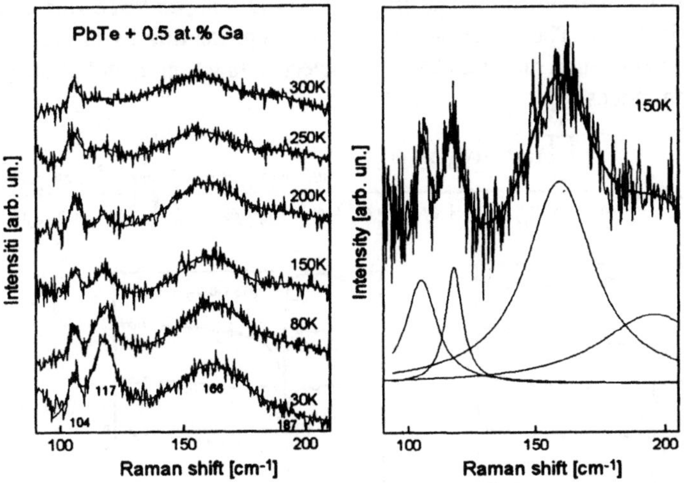

Figure 1. a. Nonpolarized Raman spectra of 0.5 at.% Ga-doped PbTe at different temperature.
b. Result of the application of the deconvolution on the spectrum at T=150K.

a) the first one comprises the lines at 104 and 117cm^{-1}. These are narrow phonon lines whose halfwidths practically do not change with a temperature decrease;
b) the second type is made of two broad lines at 166 and 187cm^{-1}. Here the line half-width increases for the 166 cm^{-1} line and decreases for the 187 cm^{-1} one with temperature lowering.

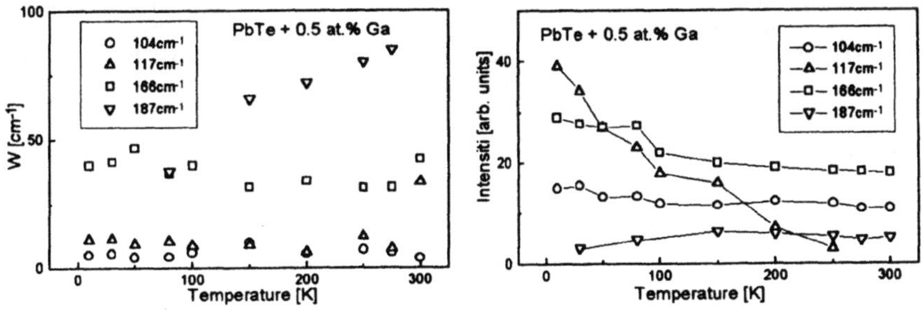

Figure 2. a. Temperature dependence of half-width. b. Intensity temperature dependence.

The temperature changes of line intensities $I(\omega_c)$ are shown in Fig.2.b. The mode at 187cm^{-1} has very low intensity which is almost independent of temperature. Similar situation was found for the clear mode at 104cm^{-1}. Under a temperature decrease the intensity of the mode at 117 cm^{-1} increases rapidly whereas the increase of the mode at 166 cm^{-1} is less pronounced. As will be seen latter the mode at 104cm^{-1} is the PbTe LO mode while the other modes represent the galium impurity local modes.

The far-infrared reflection spectra of the gallium doped PbTe single crystal sample is shown in Fig.3. The experimental data are representing by circles. The solid line were obtained using a modified factored dielectric function model of coupled plasmon-LO phonon modes [7]:

$$\varepsilon_f(\omega)=\varepsilon_\infty\frac{\prod_{j=1}^{2}(\omega^2+i\gamma_{jl}\omega-\omega_{jl}^2)}{\omega(\omega+i\gamma_p)(\omega^2+i\gamma_t\omega-\omega_t^2)}\,\frac{\omega^2+i\gamma_{Lo}\omega-\omega_L^2}{\omega^2+i\gamma_{to}\omega-\omega_0^2}\prod_{k=1}^{2}\frac{\omega^2+i\gamma_{kLO}\omega-\omega_{kLO}^2}{\omega^2+i\gamma_{kTO}\omega-\omega_{kTO}^2} \qquad (2)$$

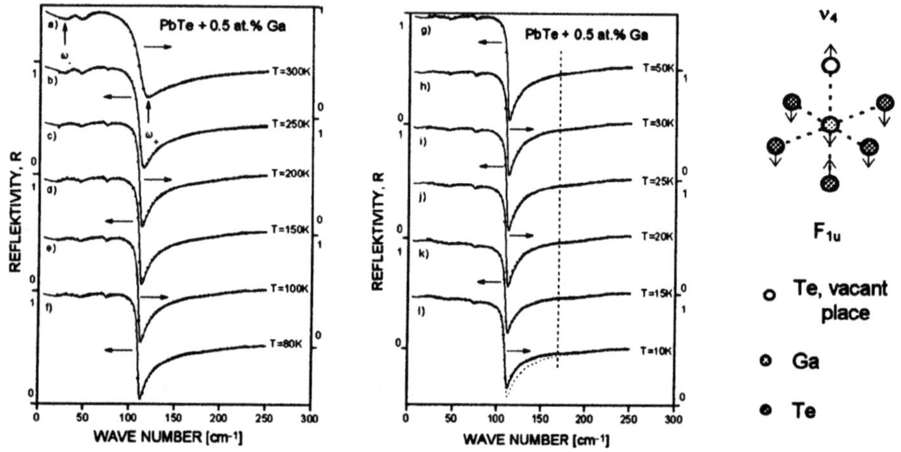

Figure 3. Far-infrared reflectivity spectra of PbTe(Ga) single crystal measured at different temperatures. The experimental spectra are represented by circles. The solid lines are calculated spectra obtained by using fitting procedure based on the model given by (2). All the parameters value have been given in Table 1. For T=10K, the calculated spectra are given with (solid line) and without (dashed line) additional oscillator at 170cm^{-1}. *Inset:* ν_4 local mode of XY_6 molecule.

The ω_{jl} and γ_{jl} parameters of the first numerator should be understood as the eigen-frequencies and damping coefficients of the longitudinal plasmons-phonon (LP+LO) waves, which arise as a result of the initial modes. The parameters of the first denominator correspond to similar characteristics of the transverse (TO) vibrations; the γ_P values describe the LP-mode damping coefficient in the limit of zero frequencies and ε_∞ constant denotes the contribution of the excitations at high relative to the spectral interval of interest. The second term in Eq.2. represent an Ga-impurity local mode, to be discussed latter. ω_{kLO} and ω_{kTO} are the longitudinal and transverse frequencies, and γ_{kLO} and γ_{kTO} stand for the damping of uncoupled modes of the host crystal. The best fit parameters have been listed in Table 1.

The oscillators denoted by j=1,2 in Eq.2. are the dominant structures in the far-infrared reflection spectra and represent the position of the coupled plasmon-LO phonon modes. The frequencies of these modes ($\omega_\pm$) are marked by arrows in Fig.3. The temperature dependence of ω_{TO}, γ_t, and ε_∞, given in Table 1, are characteristic for this type of materials in the case when there is no a phase transition [8] and the effect of persistent photoconductivity [9]. Oscillators at about 45cm^{-1} and 70cm^{-1}, are modes from the edge of the Brillouin zone, because the phonon density of PbTe [10] has a maximum at these frequency.

TABLE 1. Optical parameters (in cm^{-1}) obtained by an oscillator fitting of the PbTe(Ga) reflection spectra.

T	300K	200K	100K	80K	50K	30K	25K	20K	15K	10K
ε_∞	26	44	46	46	47	48	48	48	49	50
γ_p	21	15	14	10	8	7	6	6	6	5.5
ω_t	34	29	26	26	22	18	15	14	14	13
γ_t	9	12	12	12.5	13	13	13	13	13	15
ω_{1l}	34	28	26	26	22	18	18	18	14	8
ω_{2l}	115.5	110.7	110.5	110.4	110.4	110.3	110	110	110	110
γ_{1l}	22	22	18	18	18	18	18	18	18	18
γ_{2l}	16	3.2	2.5	2.5	2.2	2.3	2.3	2.6	4.5	4.5
ω_0	-	164	166	169	170	173	173	174	176	176.5
γ_{lo}	-	56	56	56	56	56	56	56	58	70
ω_L	-	246	266	338	340	519	519	522	1232	1240
γ_{lo}	-	15.6	16.3	18.4	18.4	22.7	22.9	22.8	35	35.2
ω_{1LO}	43	44	47	46	47	47	45	45	42	40
ω_{2LO}	73	73	72.8	72.8	72.9	72.9	73	73	73	73
γ_{1LO}	20	20	28	29	29	29	30	30	30	30
γ_{2LO}	8.5	8.5	8	7.8	7.8	7.8	7.8	7.8	7.8	7.8
ω_{1TO}	50.8	48.8	51	51	49.5	49	49	49	46	45
ω_{2TO}	73	73.5	73.3	73.5	73.1	73.5	73.7	73.7	73.5	73.5
γ_{1TO}	17	19	24	27	27	27	27	27	33	35
γ_{2TO}	8.5	9	8.7	8.7	8.7	8.7	8.7	8.7	8.7	8.7

A weak structure is noticeable at T=50K at about ω_o~170cm^{-1} (denoted with the vertical dashed line) is clearly observed at Fig.3. In order to obtain agreement between the measured and the calculated spectra, an oscillator of characteristic frequency ω_o, was added to Eq.2. G is the damping and ω_L depends of "strength" of this oscillator.

In order to demonstrate the existence of an additional mode at about 170cm^{-1}, the calculated spectra with (solid line) and without (dashed) this oscillator are shown in Fig.3.i. and l. A significant disagreement between the calculated and the experimental spectra data on this picture, led us to the conclusion that the mode at about 170cm^{-1} really exists. The temperature change of the position and intensity of the mode at around 170cm^{-1} is very similar to corresponding changes of the mode at 166cm^{-1} obtained from Raman measurements. The other modes in Fig.1. were not observed in the far-infrared reflection spectra, probably due to very low intensity. The mode at around 170cm^{-1} interacts with a plasmon. This is the reason for the changes of the hight of a global minimum in the far-infrared reflection spectra (see Fig3.j. k. and l). The plazmon- phonon interaction is also responsible for the line broadening of the lines at 166 i 187cm^{-1} (see Fig.1.a.).

The mode at 104cm^{-1} observed both in the Raman and the far-infrared reflection spectra represents the LO mode of PbTe due to impurities. We believe the mode at 117, 166 (170 IR) and 187cm^{-1} are the local mode of gallium for the following reasons: the first-order Raman-modes are not active in NaCl type of crystal structure (O_h space group symmetry), bat lead telluride grows with rather high concentration of internal defects (vacancies on Te places) which makes selection rules symmetry braching. Besides, gallium as substitution impurity ion has a consequence the fact that every ion in PbTe is no longer at the center of inversion symmetry and PbTe vibration modes could be Raman active (the impurity ion is still at the center of inversion, but its six nearest neighbors are not).

We can consider the local mode of gallium as a vibrational mode of the molecule XY_6 (X is gallium atom and Y is Te or vacant place) as it was done in [11] (see inset on Fig.3). The mode ν_4, F_{1u} symmetry, at around 187cm^{-1} was observed in the transmission measurements [11]. We think that the modes at 117 and 166cm^{-1} are also local modes of the gallium F_{1u} symmetry. Namely, the neutral state of gallium in the lattice PbTe (Ga^{++}) is unstable [6] and according to equation:

$$2Ga^{2+}=Ga^{+}+Ga^{3+} \tag{3}$$

transverse to the two-electron stable state (Ga^{+}), which is energy-wise more favorable than the metastable single electron (Ga^{++}) and state of empty center (Ga^{+++}). It is expected that the force constant of the different charged ion will be different and the impurity ion will displace from the regular action site (O_h symmetry) with local distortions due to its small ionic radius [6, 12]. In that way we observe three local modes of the F_{1u} symmetry, where the mode at 117cm^{-1} corresponds to a metastable one-electron state on an impurity center of gallium, the mode at 187cm^{-1} corresponds to a stable two-electron state, and the mode at 166cm^{-1} to the state Ga^{+++}.

Conclusion

In this paper we have shown non-polarized Raman and far-infrared reflection spectra of gallium doped PbTe single crystal sample at temperatures between 10 and 300K. In addition to the strong plasma - phonon interaction, we registered three impurity local modes of gallium. The position of these modes depends of impurity center charge.

References

1. Kaidanov, V.I., Mel'nik, R.B., and Chernik, I.A. (1973), *Fiz. Tekh. Poluprovodn.* **7**, 759
2. Kaidanov, V.I., and Ravich, Yu.I. (1985) *Usp. Fiz. Nauk.* **145**, 51
3. Bushmarina, G.S., Gruzinov, B.F., Drabkin, I.A., Lev, E.Ya., and Nelson, N.V. (1977), *Fiz. Tekh. Poluprovodn.* **11**, 1874
4. Akimov, B.A., Brandt, N.B., Gas'kov, A.M., Zlomanov, V.P., Ryabova, L.I., and Khokhlov, D.R. (1983), *Fiz. Tekh. Poluprovodn.* **17**, 87
5. Akimov, B.A., Brandt, N.B., Ryabova, L.I., Khokhlov, D.R., Chudinov, S.M., and Yatsenko, O.B. (1980), *Pis'ma v Zh. Eksp. Teor. Fiz.* **31**, 304
6. Weiser, K. (1981), *Phys. Rew. B* **23**, 2741
7. Kukharskii, A.A. (1973), *Solid State Commun.* **13**, 1761
8. Miljkovic, J.M., Romcevic, N., Popovic, Z.V., König, W., and Nikiforov, V.N. (1996), *Phys. Stat. Sol. (b)* **193**, 43
9. Romcevic, N., Popovic, Z.V., Khokhlov, D.R., and König, W. (1997), *Infrared Phys.& Tekh.* **38**, 117
10. Cochran, W., Cowley, R.A., Dolling, G., and Elcombe, M.M. (1966), *Proc. Roy. Soc. A* **293**, 433
11. Takaoka, S., and Murase, K. (1983) *J. Phys. Soc. Jpn.* **52**, 25
12. Katayama, S., and Murase, K. (1980) *Solid State Commun.* **36**, 707

PRECURSORS FOR BARIUM METAL-ORGANIC CHEMICAL VAPOR DEPOSITION. ADVANTAGES AND DISADVANTIGES IN IMPROVEMENT OF BARIUM HEXAFLUOROACETYLACETONATE

N. P. KUZMINA, S. I. TROYANOV, M.V.RYAZANOV
Moscow State University, Department of Chemistry
Moscow, 119899, Russia

1. Introduction

The intensive development of Metal-organic Chemical Vapor Deposition (MOCVD) technique has opened up opportunities for producing a wide variety of materials. The research on MOCVD is far from complete and still offers many challenges to a chemist. Some of them concern volatile starting compounds (precursors). The problem is to search and synthesize metal coordination compounds answering the criteria of metal-organic precursors [1]. Volatility, thermal and storage stability are the most important for MOCVD precursors. One of the most interesting and challenging tasks for a coordination chemist is synthesis of volatile precursors of alkaline earth elements (AEE). These elements form coordination compounds with predominantly electrostatic bond nature. It results in some special features of the complex formation with organic ligands:
(i) the AEE ions show high and varying values of coordination numbers that leads to formation of different mixed ligand complexes or complexes with oligomeric structure;
(ii) predominantly electrostatic metal-ligand bonds results in flexibility of complex geometry, that influences strongly on packing density in crystal structures and on volatility of complexes.
The formation of intermolecular bonds leads to polymeric or oligomeric structure and decreases volatility, especially in case of barium compounds. In spite of this, some β-diketonates of AEE are volatile as the introduction of bulky or fluorinated alkyl groups into ligands moderates intermolecular bonds, increases the volatility of metal β-diketonates and offers the possibility to use these compounds as a precursors [2]. Barium complexes with 2,2,6,6,-tetramethylheptane-3,5-dionato (thd^-) and 1,1,1,5,5,5-hexafluoropentane-2,4-dionato (hfa^-) ligands are the most known and widely used barium containing precursors. However, the practical use of these precursors is complicated by their low volatility and stability, arising from their polymeric or oligomeric structures [3]. Obviously the best precursors could be mononuclear complexes, coordinatively saturated. In order to achieve these desired properties two ways can be realized: – the design of new ligands, containing bulky groups and additional donor atoms, and the obtaining of new complexes with these ligands or the improvement of known complexes due to the formation of mixed ligand complexes

G. Van Tendeloo et al. (eds.), High-Temperature Superconductors and Novel Inorganic Materials, 303–307.

(adducts) with additional neutral donor ligands. Either of the two ways has its advantages and disadvantages, many famous chemists are working in these ways, however the best solution of barium precursors problem is not found yet. In this paper the structure and properties of some new and known adducts of barium hexafluoroacetylacetonate are discussed.

2. Adducts of barium hexafluoroacetylacetonate with neutral organic ligands

Adducts of $Ba(hfa)_2$ with neutral donor ligands would be appropriate for demonstration of mutual ligand influence on thermal stability and volatility of barium β-diketonates. As compared to non-fluorine containing complexes, like the well-known $Ba(thd)_2$, the Lewis acid character of $Ba(hfa)_2$ and hence its tendency to adduct formation is much higher. It is well-known that $[Ba(hfa)_2]_n$ has polymeric structure [3] and is not volatile below 205° C at 10^{-5} Torr [5], but coordination of the long polyether $CH_3O(CH_2CH_2O)_5CH_3$ (tetraglyme) dramatically enhances the volatility of this barium complex. So, $Ba(hfa)_2$(tetraglime) adduct sublimes readily at 120° c at 10^{-2} Torr [4]. Vapor pressure measurements show that it was well behaved and was considerably more volatile than the parent $[Ba(hfa)_2]_n$ [6]. The similar results are obtained in case of $Ba(hfa)_2$ adducts with crown ether ligands, containing oxygen or nitrogen donor atoms [2]. All these adducts sublime without decomposition into the neutral ligand and parent $Ba(hfa)_2$ complex. The significantly short intermolecular contacts are not found in the crystal structure of these adducts. However, it has not been clear whether or not intermolecular interaction influenced on thermal stability and volatility of mononuclear adducts of barium hexafluoroacetylacetonate.

The adduct of $Ba(hfa)_2$ with o-phenantroline (Phen) was chosen to study the influence of intermolecular interactions. Earlier it was shown [7], that in crystal structure of non-fluorinated adducts $M(thd)_2(Phen)_2$ (M = Ba, Sr, Ca) stacking interaction between aromatic fragments of Phen ligands played very important role and influenced on their volatility. The new $Ba(hfa)_2(Phen)_2$ adduct was synthesized and characterized [8]. The crystal structure of $Ba(hfa)_2(Phen)_2$ consists of monomeric molecules with CN of central barium atom 8. All donor atoms of ligands are coordinated: four oxygen atoms of hfa ligands and four nitrogen atoms of Phen ligands. This adduct is volatile at low pressure (0.01 Torr), but it sublimes into two stages:

$$Ba(hfa)_2(Phen)_2\,(s) \xrightarrow[180\,C]{} Ba(hfa)_2(Phen)\,(s) + Phen(g) \xrightarrow[220\,C]{} Ba(hfa)_2(Phen)\,(g),$$

that is incongruently in contrary to other known adducts of barium hexafluoroacetylacetonate with polyethers and crown ethers [4,6]. The difference in thermal behaviour of these adducts is attributable to stacking interaction in crystal structure of $Ba(hfa)_2(Phen)_2$. The packing mode of molecules in the crystal structure of $Ba(hfa)_2(Phen)_2$ can be described as zigzag chains due to π–π-stacking interaction between Phen ligands of the neighboring molecules, the Phen-Phen interplanar separation being about 3.5 A (Fig. 1).

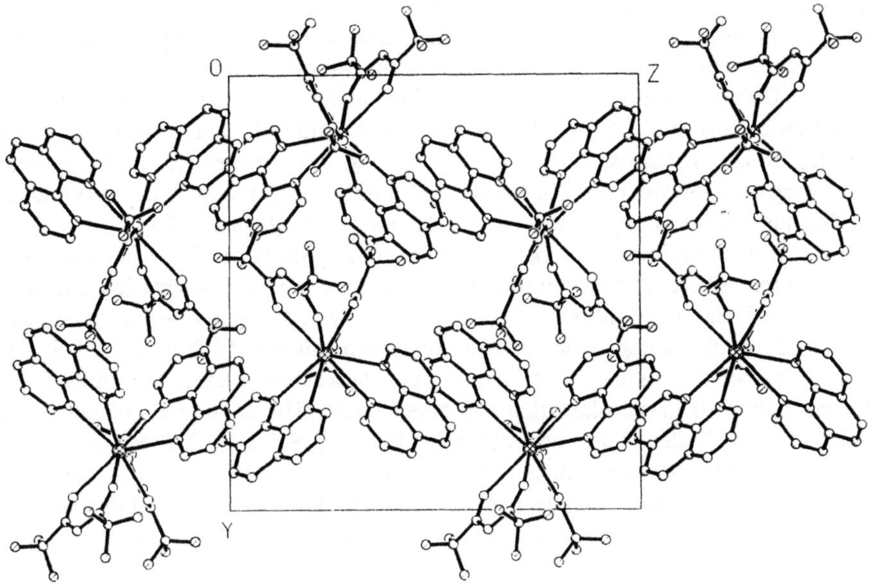

Figure 1. The arrangements of molecules in the zigzag chains in the crystal structure of $Ba(hfa)_2(Phen)_2$ [8].

The analysis of packing way of $Ba(hfa)_2(Phen)_2$ shows another interesting feature – CF_3-groups of hfa-ligands, which belong to molecules from neighboring chains, are situated in dihedral angles between two Phen ligands. Such packing way of chains corresponds to the short contacts between fluorine atoms and periphery part of Phen molecules, the distances between fluorine atoms and some atoms of Phen ligands being in the range of 3.07-3.38 A. It leads to increasing of the packing density and the decreasing of volatility in contrary to adduct with tetraglime [4]. Complex $Ba(hfa)_2(Phen)_2$ is the first example of monomeric barium hexafluoroacetylacetonate adducts with strong Lewis base ligand which sublime at rather high temperature with decomposition.

3. Adduct of barium hexafluoroacetylacetonate with metal complexes as ligands

The idea of using of metal complexes as ligands to prepare heterobimetallic complexes has long been developed for the purpose of magnetic studies. Recently it has been adapted to prepare volatile heterobimetallic compounds [9]. Schiff-base complexes of *d*-elements are used as terminal ligands to complete the coordination geometry about an *f*-element. The *d*-element is surrounded by a multidentate ligand with peripheral coordination sites that chelate the *f*-element. Two series of such complexes were prepared and investigated: Cu(saloph)Ln$(hfa)_3$ and Ni(salen)Ln$(hfa)_3$ (H_2saloph and

H_2salen – Schiff bases from salycilic aldehyde and o-phenylenediamine or ethylenediamine) [9]. It has been shown that some of them have the molecular structure and sublime without decomposition. To test the validity of this idea for s-d elements, complex $Ba(hfa)_2(Phen)_2$ was obtained by interaction between $[Ba(hfa)_2]_n$ and NiSalen in organic solvents. X-ray structure analysis showed that crystal structure of $Ba(hfa)_2(Nisalen)_2$ consists of discrete molecules, two Nisalen molecules acting as a neutral ligands toward Ba(II). Central barium atom coordinates four oxygen atoms of hfa-ligands and four oxygen atoms of Nisalen molecules (ligands) (Fig. 2).
There is a stacking interaction between aromatic fragments of Nisalen molecules in the coordination sphere of the central barium atom. Possibly, this interaction inluences the volatility of $Ba(hfa)_2(Nisalen)_2$. According to the thermogravimetric and elemental analysis data , this complex sublimes at high temperature (240-260° C, 0.01 Torr) with partial dissociation into parent barium and nickel complexes and futher thermal decomposition of barium hexafluoroacetylacetonate. The sublimate obtained becomes enriched with Nisalen. This data shows that $Ba(hfa)_2(Nisalen)_2$ does not satisfy general criteria of selection of precursors [1]. However, it should be noted that there is a rich variety of design of new complexes with metal complexes as additional ligand and some of these compounds can be a prospective precursors for MOCVD. So, the preliminary results show that complex $[Mg(hfa)_2(Nisalen)]$ sublimes congruently.

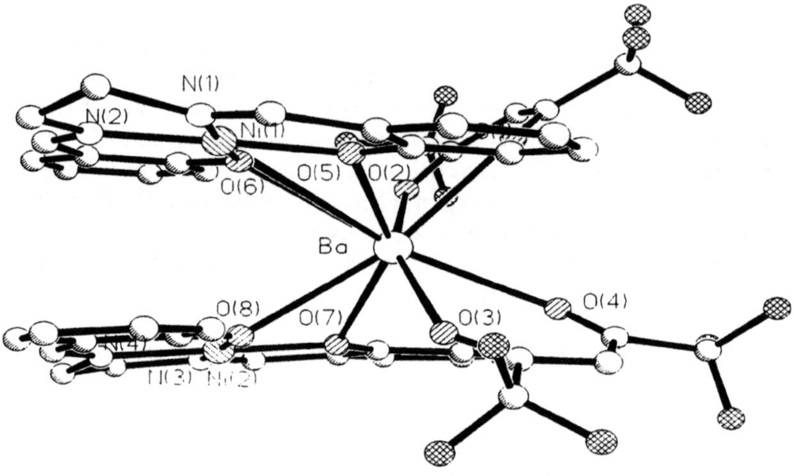

Fig. 2. Molecular structure of $Ba(hfa)_2(Nisalen)_2$.

4. Conclusions

The results presented in this paper were focused on some adducts of barium hexafluoroacetylacetonate. An attractive strategy for the improvement of volatility of this precursor is to minimize the intermolecular interaction in its crystal structure by the formation of adducts with neutral donor ligands. The transformation of polymeric $[Ba(hfa)_2]_n$ into monomeric complexes $[Ba(hfa)_2Q_m]$ (Q - neutral donor ligands) is a necessary condition for the improvement of volatility and stability of this complex, but it is inadequate condition for the desighn of excellent precursor.The properties of $[Ba(hfa)_2Q_m]$ can enhance if there are no short intermolecular contacts between periphery parts of ligands in the crystal structure of these adducts. It means that the use of ligands, containing π-aromatic fragments, can not be effective for enhancing of volatility. Examples of $Ba(hfa)_2$(tetraglime) and $Ba(hfa)_2(Phen)_2$ illustrate this statement. It should be taken into account when designing of new adducts and especialy searching of heterobimetallic volatile compounds with metal complexes as neutral ligand.

5. Acknowledgements

This work was partly supported by INTAS-RFBR (grant No. 95-0118) and RFBR (grant No. 96-03-32937a).

6. References

1. Maury, F. (1995) Recent trends in selection of metal-organic precursors for MOCVD process, *J. De Physique IV C5, suppl. J. De Physique II*, **5**, 449-463
2. Tiita, M. and Niinisto, L. (1997) Volatile metal β-diketonates: ALE and CVD precursors for electroluminescent device thin films, *Chemical Vapor Deposition*, **3**, 167-182
3. Drozdov, A. and Troyanov, S. (1996) The structural chemistry of IIA group metal diketonates, *Main Group Metal Chemistry*, **XIX**, 547-569
4. Gardiner, R., Brown, D.W. and Kirlin P.S. (1991) Volatile barium β-diketonates polyether adducts. Synthesis, characterization, and metalloorganic chemical vapor deposition, *Chem. Mater.*, **3**, 1053-1059
5. Purdly, A. P., Berry, A. D., Holm, R. T., Fatemi, M. and Gaskill, K. K. (1989) Chemical vapor deposition experiments using new fluorinated acetylacetonates of calcium, strontium and barium, *Inorg. Chem.*, **28**, 2799-2805
6. Timmer, K., Spee, K. I. M. A, Mackor, A., Meinema, H. A., Spek, A. L. and van der Sluis, P. (1991) Synthesis and characterization of volatile, thermally stable bis(1,1,1,5,5,5-hexafluoropentane-2,4-dionato)barium-, -strontium- and calcium polyether complexes, suitable as precursors in chemical vapor deposition, *Inorg. Chim. Acta*, **190**, 109-117
7. Troyanov, S. I., Kuzmina, N. P., Soboleva I. E. and Chugarov, N. V. (1997) Packing in crystal structure of metal β-diketonates used in MOCVD and its correlation to volatility, *Electrochem. Soc. Proceed. 14 th Intern. Conf. On CVD and EUROCVD11*, **97-25**, 852-856
8. Kuzmina, N., Troyanov, S. and Ryazanov, M. (in press) Synthesis, ctystal structure and thermal behaviour of $Ba(hfa)_2(Phen)_2$, *Inorg. Chim. Acta*
9. Gleizes, A., Julve, M., Kuzmina, N., Alikhanyan, A., Lloret, F., Malkerova, I., Sanz, J. L. and Senocq, F. (in press) Heterobimetallic d-f metal complexes as potential single-source precursors for MOCVD: structure and thermodynamic study of the sublimation of $[Ni(salen)Ln(hfa)_3]$, Ln = Y, Gd, *Eur. J. Inorg. Chem.*

GPSR Compliance

The European Union's (EU) General Product Safety Regulation (GPSR) is a set of rules that requires consumer products to be safe and our obligations to ensure this.

If you have any concerns about our products, you can contact us on ProductSafety@springernature.com

In case Publisher is established outside the EU, the EU authorized representative is:

Springer Nature Customer Service Center GmbH
Europaplatz 3
69115 Heidelberg, Germany

Batch number: 10755110

Printed by Printforce, the Netherlands

Religion and Law in South Africa

Religion and Law in South Africa

Charles Maimela

This book was originally published as a monograph in the International Encyclopaedia of Laws/Religion.

Founding Editor: Roger Blanpain
General Editor: Frank Hendrickx
Volume Editor: Rik Torfs

Wolters Kluwer

Published by:
Kluwer Law International B.V.
PO Box 316
2400 AH Alphen aan den Rijn
The Netherlands
E-mail: international-sales@wolterskluwer.com
Website: lrus.wolterskluwer.com

Sold and distributed in North, Central and South America by:
Wolters Kluwer Legal & Regulatory U.S.
7201 McKinney Circle
Frederick, MD 21704
United States of America
Email: customer.service@wolterskluwer.com

Sold and distributed in all other countries by:
Air Business Subscriptions
Rockwood House
Haywards Heath
West Sussex
RH16 3DH
United Kingdom
Email: international-customerservice@wolterskluwer.com

Printed on acid-free paper

ISBN 978-94-035-1151-1

e-Book: ISBN 978-94-035-1200-6
web-PDF: ISBN 978-94-035-1201-3

Printed in the United Kingdom.

The Author

Charles Maimela is a Senior Lecturer in the Department of Private Law at the University of Pretoria, South Africa. Dr Maimela teaches Legal Pluralism, Law of Damages and Law of Delicts to law students. He holds the following qualifications: LLB (Unisa), LLM (Unisa) and LLD (Unisa).

The Author

Table of Contents

List of Abbreviations

ACHPR	African Court on Human and Peoples' Rights
AIDS	Acquired Immune Deficiency Syndrome
ANC	African National Congress
APRM	African Peer Review Mechanism
AQIM	Al Qaeda of the Islamic Maghreb
AU	African Union
BCEA	Basic Conditions of Employment Act
CCMA	Commission for Conciliation, Mediation & Arbitration
COIDA	Compensation for Occupational Injuries and Diseases Act
CRL	Cultural, Religious and Linguistic Communities
DA	Democratic Alliance
EU	European Union
EEA	Employment Equity Act
HIV	Human Immunodeficiency Virus
HPCSA	Health Professions Council of South Africa
ICCPR	International Covenant on Civil and Political Rights
ICC	International Criminal Court
ICESCR	International Covenant on Economic, Social and Cultural Rights
ICJ	International Court of Justice
ICTY	International Criminal Tribunal for the former Yugoslavia
IMF	International Monetary Fund
JSC	Judicial Services Commission

List of Abbreviations

LAC	Labour Appeals Court
L&D	Laws and Disciplines (of the Methodist Church of Southern Africa)
LRA	Labour Relations Act
NPA	National Prosecuting Authority
NGOs	Non-governmental Organisations
NEPAD	New Partnership for Africa's Development
OHSA	Occupational Health and Safety Act
OAU	Organisation for African Unity
PAC	Pan African Movement
PCIJ	Permanent Court of International Justice
SADC	Southern African Development Community
SDA	Skills Development Act
TAC	Treatment Action Campaign
TRC	Truth and Reconciliation Commission
UDHR	Universal Declaration of Human Rights
UNESCO	United Nations Educational, Scientific and Cultural Organisation
UIA	Unemployment Insurance Act
UN	United Nations

General Introduction

§1. General Background of the Country

I. Geographical Background

1. South Africa is located in the southern part of the African continent. It comprises of nine provinces: Gauteng, Limpopo, Northern Cape, Western Cape, North West Free State; KwaZulu-Natal; Mpumalanga and Eastern Cape. The geographical position of South Africa is considered to be perfect because it is the producer of important minerals which include gold and diamonds among other natural resources.[1] This has resulted in making South Africa one of the leading economically developed states in the African continent.[2] Each province among the nine provinces of South Africa is unique and contributes to the overall vision and mission of making South Africa a dynamic and sound country in the African continent. For example, Gauteng province is the smallest province in South Africa and overly populated. This is attributed to the fact that most mining industries and economic activities of South Africa are vested in this province.[3] Based on this economic position of Gauteng it attracts people from all walks of life, particularly from different African countries and locals from other provinces in pursuit of wealth and better opportunities. This means that the people who reside in this province are diverse and share one common goal of making a better living.[4]

2. The province of Gauteng despite its mining and industrial activity is also known for big cities like Pretoria, which is the seat of government and the capital city of South Africa, and Johannesburg, which is the city that is regarded as the economic hub of South Africa because it is an industrial city and has the richest square mill which is Sandton city among other things.[5] The Limpopo province is one of the rural provinces of South Africa and contributes with regard to farming activities for the country and most of the people from this province move to Gauteng in search

1. B. Buscher, The *Political Economy of Africa's Natural Resources and the Great Financial Crisis*, 26 TESG 137 (2011).
2. *Ibid.*, at 138.
3. C. Fourie, *Why Is Johannesburg the Economic Hub of South Africa: A City of Opportunities for Entrepreneurs*, 3 Randburg Sun 4 (2017).
4. *Ibid.*
5. *Ibid.*

of better opportunities.[6] Western Cape is also one of the developed and economically sound provinces in South Africa. This is due to the fact that Table Mountain and Ocean in the city of Cape Town serve as a major boost to the tourism of the country. Cape Town is a well-known and regarded city across the world due to its beauty and Mediterranean climate. The Western Cape province experiences Mediterranean climate, which means that it experiences rain in winter season; however this is not the case for the other eight provinces of South Africa.[7] Furthermore, the Western Cape also plays a major role when it comes to food production and security through grape and apple farms as well as the production of tea such as Rooibos. In turn, this results in the production of wines and juices, which are well known in the world, and are considered for both imports and exports purposes in the country and beyond.[8]The Northern Cape is the largest province in South Africa and boasts of agricultural land and also various solar systems that are well known, for example, in connection with detecting weather conditions for the rest of the country.[9]

3. North West province is regarded as the platinum province because it is the world producer of platinum. It is also known for its rich soil for farming, and it plays a role in food production in the country, in particular vegetables and other staple foods.[10] The province of Mpumalanga, which means where the sun rises, is one of the provinces that is leading in coal mining and production. Tourism is also one of the strong factors which is dominating in this province. Mpumalanga prides itself as being home to the Kruger National Park, which is the biggest game reserve in the African continent, and consists of big fives (the Lion, Leopard, Buffalo, Elephant and Rhino) as the official residents of the park.[11]

4. KwaZulu-Natal is the warmest and friendly province in South Africa, and it is put on the map by the city of Durban for its beautiful and clean beaches as well as hotels and resorts, which are favourites among both locals and international visitors. However, tourism is not the only factor or blueprint that determines this province; this is because this province is the largest sugar producer due to the sugar plantations that rest on the belly of this province. This means that this province contributes for the economy of South Africa through sugar production and tourism.[12]

---

6. *The Great Limpopo Province, History of the Province*, http://www.dolimpopo.com/limpopo-province (accessed 6 Jul. 2018).
7. P. De Lille, *The Western Cape Government: History and Regional Development Profile for the City of Cape Town* 3 WCG 14 (2018).
8. Cape Town is also a city which hosts parliament where law is made and where the executive, which is the government, is called to account from the various political parties that are represented in parliament. *See* De Lille, *supra* n. 3 at 15.
9. J. Morris, *Northern Province and the Karoo Rock in South Africa*, 55 Journal of Human Evolution 11 (2016).
10. J. Nkosi *Investment Potential of the Platinum Province: Areas of Investment That Are Expected to Grow Fast Include Fuel Cell Technology, Machinery for Mining, Energy Manufacturing and Component Production*, 4 Global Africa 12 (2017).
11. *Mpumalanga and the Big Five Animals*, https://www.mpumalanga-info.co.za/provinces/article/378/big-five-animals-in-south-africa (accessed 17 Aug. 2018).
12. *A Warm Welcome to Kwazulu Natal in South Africa*, http://www.zulu.org.za/about/key-facts/welcome (accessed 16 Aug. 2018).

Free State province is also an active province in farming activities due to the vast land that is found in the province and is considered to be a beautiful province due to the city of Bloemfontein which is regarded as the city of roses because of the beautiful flowers which are found only in that city. This results in an increase in tourism in order to explore the beauty of this province, and there are other tourist attractions to see and enjoy for both locals and visitors.[13] Finally, the province of Eastern Cape is also one of the largest provinces second to the Northern Cape and has a rich history when it comes to the liberation of the South African community. This province is known as the home of legends, because struggle stalwarts in South Africa come from this province and this includes the well-known and regarded Nelson Mandela among others.[14] Furthermore, this province has rich agricultural land, which plays a major role when it comes to farming activities.[15]

5. Based on the geographical position or location of South Africa, it is very important to note that it is a tourist's hub and also a socio-economical country that is in a position to develop and prosper on the basis of its position. The South African economy is driven by mining and agriculture because South Africa is one of the leading exporters of raw materials like cold, coal, gold and platinum to name but a few.[16] South Africa enjoys these benefits because of its geographical location or position, which in a nutshell, it positions the country far better than other African countries, and as it is often argued or believed that indeed South Africa has been blessed and has seen the grace of God. This is due to the prosperity it enjoys and prides itself with, due to its geographical position or location. It is important for the reader to further bear in mind that each province in South Africa is unique and has a role to play with the aim of making sure that a country like South Africa becomes a world leader in terms of its economic position. One can sum up the role and position of each province in South Africa as unique with the aim of putting South Africa and the continent of Africa on the map, by ensuring that they complement and work together effectively in building the rainbow nation of South Africa.

II. Population

6. South Africa is still a developing country with a population size of about 57 million people, and this figure excludes people who have entered the country illegally or legally for different reasons as part of migration and bearing in mind the geographical location of South Africa as already discussed above.[17] The population of South Africa can best be described as pluralistic in nature, and this means that the population of South Africa is diverse. Different people from different cultural

---

13. F. Richardson, *The Orange Free State and Inhabitants* 5 (Cambridge University Press 11th ed. 1999).
14. *South African History Online*, www.sahistory.org.za (accessed on 5 Jun. 2018).
15. *South African History Online Place of Birth of Nelson Mandela*, www.sahistory.org.za (accessed on 5 Jun. 2018).
16. Nkosi, *supra* n. 10, at 12.
17. *Statistics of the Population of South Africa*, http://www.statssa.gov.za/publications/P0302/P030220 18.pdf (accessed 11 Jul. 2018).

and religious backgrounds make up the South African population. This is the reason why South Africa is regarded as the rainbow nation.[18] South Africa has eleven official languages: English, Afrikaans, IsiZulu, IsiXhosa, South Sotho, North Sotho, Setswana, Ndebele, Xitsonga, Tshivenda and Sepedi.[19] Out of the eleven official languages as outlined, both English and Afrikaans owe their historical roots or connection to the colonisers of South Africa. This is because South Africa was colonised by both the English and Dutch at some point and one cannot discuss the history and legal system of South Africa without referring to both the English and Dutch legal systems. The reason why both systems of English and Dutch are significant when one is discussing the position of South Africa is because the legal system of South Africa is built from both the English and Dutch legal systems, which makeup what is known as common law in South Africa.[20] This means that the systems of both English and Dutch did not only play a role in relation to languages but it played an important role when it comes to the development and shaping of the legal system of South Africa as it is known today. The other nine official languages form a part of the indigenous communities of South Africa, and the people who speak these languages are spread out across the nine provinces of South Africa and majority are black Africans.[21] However, it is important to bear in mind that because the South African community or population is pluralistic it does not mean that only black Africans are allowed to speak these indigenous languages. People from other ethnic groups like white Africans, for example, are not barred from speaking them, but they are encouraged to do so as part of nation building.[22] This is because freedom of association and expression is protected in the Constitution in terms of section 16 of the Constitution of the Republic of South Africa, 1996 (herewith referred to as the Constitution).[23]

7. The debate currently in South Africa is based on the fact that despite the recognition of various languages, still a lot of work needs to be done to add other languages which have not been recognised thus far, even after twenty-four years of democracy. This new language is sign language, which will accommodate people living with disability, and it is argued that the incorporation of this language into the recognised languages in the Republic will be one of the progressive and socially relevant moves with the aim of ensuring that people living with disabilities are

---

18. J. Bekker & C. Rautenbach, *Introduction to Legal Pluralism* 12 (LexisNexis, South Africa 4th ed. 2015).
19. *Ibid.*, at 13.
20. *Ibid.*, at 14.
21. Section 6(2) of the *Constitution of the Republic of South Africa*, 1996 (herewith referred as the Constitution) provides that in recognising the historically diminished use and status of indigenous languages of our people, the state must take practical and positive measures to elevate the status and advance the use of these languages.
22. Section 6(2) of the Constitution.
23. Section 16(1) of the Constitution states that everyone has the right to freedom of expression, which includes: freedom of the press and media; freedom to receive or impart information or ideas; freedom of artistic creativity; and lastly academic freedom and freedom of scientific research.

accommodated in the South African society.[24] Without a doubt to include sign language as one of the official languages in South Africa will surely put South Africa in the world map and a leader in the African continent to such an extent that it will be seen as a country that embraces diversity even though it is diverse by its nature. Furthermore, it will be regarded as a state that cares and recognises people who suffer from disabilities and in this case particularly people who are deaf and dumb.[25]

8. The population of South Africa, which is 57 million as outlined, comprises of people from different backgrounds based on their culture and colour, but it is important to bear in mind that they all make up the South African community or society irrespective of their diversity. The population comprises of black Africans who are the original inhabitants of South Africa and are the majority in the country.[26] They are considered the custodians of the nine indigenous languages as outlined above and indigenous culture and practices.[27] White Africans owe their historical roots to the colonial past of South Africa, and they are a minority in South Africa. This ethnic group speaks both English and Afrikaans.[28] Furthermore, the South African community is made up of Indians who owe their ancestral links to Gandhi as they were brought in by the colonisers to South Africa as slaves in the sugar plantations of KwaZulu-Natal.[29] South Africa also comprises of people from Asian background who form a part of the population of South Africa and are a minority in the country.[30]

9. Based on the above assessment it is very clear that the South African community or society is diverse and pluralistic in nature. To best sum up the population of South Africa, one can say that it is multicultural in nature due to the various cultural and legal systems, which will be explored later on in this research text.

III. Government Administration

10. When it comes to the government administration in South Africa, it is a democratically elected government and the country goes to elections every five

24. T. Reagan *South African Sign Language and Language in Education Policy in South Africa*, 38 Linguistics 166 (2008).
25. People who suffer from different physical and mental deformities are often neglected in both developed and developing countries, and the incorporation of sign language can be onc of the solutions to resolve and eradicate this problem encountered by the developing world. *See* Reagan, *supra* n. 24, at 167.
26. Bekker & Rautenbach, *supra* n. 18 at 20.
27. *Ibid.*
28. *Ibid.*, at 21.
29. K. Ma, *Control and Conscience: Positivist Approaches to Religion in India and South Africa*, 260 SOAS LJ 261 (2015).
30. Ma, *supra* n. 24, at 262.

years for the will of the people through free and fair elections.[31] This was unfortunately not the case in the past because during the colonial and apartheid era only white people were considered competent and eligible to vote and they enjoyed certain privileges. All of this happened at the exclusion of black people who were considered as sub-humans and were not eligible to vote, but were subject to white rule and oppression.[32] Black Africans did not have a voice in government formation and how they would want to be governed, even though they were the majority in the country, and they were the original inhabitants of South Africa. The oppression of black Africans ended in 1994 when the first election was held and the first president was democratically elected, which was the late President Nelson Mandela.[33] The reader will be taken through the process and history as to how apartheid and colonial rule were destroyed in South Africa, with religion in particular, in chapters that will follow, when discussing the role of religion in the liberation of South Africa. However, for now it is important to note that all the people of South Africa are eligible to vote and have a voice on who would govern and how they would like to be governed by their chosen government.[34] The period of 1994 is one of the foundational periods in the history of South Africa because it gave rise to the Constitution of the Republic of South Africa, which consists of a Bill of Rights that protects the fundamental rights of all people in South Africa.[35] Furthermore, multi-party democracy was and is still encouraged, and it is considered the order of the day in South Africa and in the area of politics as part of growing the young democracy of South Africa.[36]

11. From 1994 until 2018, the government of South Africa was run by the African National Congress (ANC), which was elected to power through the democratic processes of elections every five years.[37] The government of South Africa is divided into the following three tiers, which make up the government of the Republic of South Africa.[38] The national government is the first tier or branch, which comprises of the president and his or her cabinet. The second tier is the provincial government, which is run by the various premiers in the nine provinces with their respective cabinets and lastly the local government tier, which comprises of different

31. Section 1 of the Constitution states that the Republic of South Africa is one sovereign, democratic state founded on the following values: human dignity, the achievement of equality and the advancement of human rights and freedoms; non-racialism and non-sexism; Supremacy of the Constitution and the rule of law; universal adult suffrage, a national voters' roll, regular elections and a multi-party system of democratic government, to ensure accountability, responsiveness and openness.
32. Bekker & Rautenbach, *supra* n. 18, at 22.
33. J. Marrison, *Nelson Mandela, South Africa's First Black President, Dies at the Age of 95*, 5 The Washington Times 14 (2015).
34. Section 1 of the Constitution.
35. Chapter 2 of the Constitution is the Bill of Rights and it sets out the fundamental rights of all South Africans, which include the rights to human dignity and equality among other things. The Bill of Rights also make provision as to when rights can be limited.
36. Section 1 of the Constitution.
37. The ANC was formed in 1912 and it is the oldest liberation movement not only in South Africa but also in the entire continent of Africa. *See* s 1 of the Constitution.
38. Section 8 of the Constitution.

municipalities spread out across the nine provinces.[39] The local government ensures that essential services reach the people. These include water and electricity, and both the national and provincial governments respectively oversee with the aim of ensuring that people on the ground are provided with all the essential services.

12. All the three branches of government are essentially one, and this structure of division in government is aimed at ensuring that the government is close to the people and the needs of society are provided through the process of service delivery, with the much-needed urgency.

IV. Judicial Administration

13. South Africa is a constitutional democracy, section 2 of the Constitution states that the Constitution is the supreme law of the land, and any law found to be inconsistent with the Constitution must be declared null and void and the obligations imposed by it must be fulfilled.[40] This means that the rule of law is supreme in the Republic, and the government is duty bound to respect and uphold the Constitution in its function of providing services to the public, and the same can be said for the inhabitants of the Republic.

14. The Constitution as the supreme law of the land has set down mechanisms in place, which are aimed at ensuring that democracy in South Africa is protected and also maintained through the process of outlining how the state needs to function. Provision is made in the Constitution that the state for its functionality is divided into three spheres: the executive which is the government as outlined above; the legislature, which is parliament, that is responsible for the creation and formulation of laws. Furthermore, parliament is also a place where the executive is called to account in cases where they fail to execute their role towards the people of South Africa. In parliament multi-party democracy and representation is encouraged due to the diverse nature of the South African population. The judiciary, which is the last sphere of government, acts as a watchdog and ensures that the rule of law is upheld at all times. Independence of the judiciary is also protected by the Constitution to such an extent that courts are able to give decisions that do not favour the government with the aim of developing and protecting the democracy in South Africa.[41]

15. The judiciary plays a significant role in the South African democratic dispensation and justice must be accessible to all the people of South Africa irrespective of their socio-economic position, background, colour, religion or cultural beliefs. This is the reason why the structure of the judiciary through the various

39. *Ibid.*
40. Section 2 of the Constitution.
41. Section 165(2) of the Constitution states that courts are independent and subject only to the Constitution and the law, which they must apply impartially and without fear, favour or prejudice.

courts is spread out in the length and breadth of South Africa in terms of section 166 of the Constitution.[42] Section 166 of the Constitution provides for the following court structures.

A. *The Constitutional Court*

16. This is the highest court in the land and it deals with only constitutional matters. It is the highest court in handling constitutional matters and the highest Court of Appeal when it comes to constitutional matters. The Constitutional Court is not a Court of First Instance, meaning one cannot approach this court directly, except if one has compelling and substantive reasons to this effect.[43] This court consists of eleven justices who serve for a non-renewal term of twelve years each, and the head of this court is the chief justice who is assisted by the deputy chief justice.[44] This court currently sits in the city of Johannesburg, built in the area or vicinity of the former female prison where women lost their lives during the apartheid era. The Constitutional Court is considered the guardian of the South African democracy, and it plays an important part in the administration of justice and preservation of the hard-earned democracy of South Africa through the application and enforcement of the Constitution.

B. *The Supreme Court of Appeal*

17. The Supreme Court of Appeal is the highest Court of Appeal in terms of non-constitutional matters.[45] This court only listens to appeals from the various divisions of the high courts in South Africa, and the decisions it makes in such matters are considered final and binding. The head of this court is the judge president and deputy judge president and they work with other judges who are appointed by the court to ensure that the administration of justice is provided to all the people in South Africa. This court sits in the city of Bloemfontein, which is affectionately known as the city of roses, in the province of Free State.

C. *The High Courts*

18. High courts are spread throughout the country in each province for the administration of justice. High courts each have a judge president and deputy judge president who are responsible for the courts' administration and day-to-day operation of the courts and are assisted by the various judges who are appointed by that

42. Section 166 of the Constitution.
43. Section 167(3) of the Constitution.
44. Section 167(1) of the Constitution.
45. Section 168(3) of the Constitution.

particular division or court.[46] High courts listen to different matters, including criminal and civil matters; they also hear constitutional matters, but the final arbiter in such matters will remain the Constitutional Court as already outlined above.[47] However, it is important to note that the high court has inherent jurisdiction and is the upper guardian of all minors, and this means that this is the type of court that is vested with the best interests of minors as provided for in section 28 of the Constitution.[48]

D. Magistrate Courts

19. Magistrate courts are divided into two categories: regional courts and district courts. Regional courts have a higher jurisdiction than the district courts, and they can try a person who has been convicted with murder. Regional courts are stationed in the various major cities of South Africa, and they aim to alleviate the burden from the high courts.[49] The presiding officer in the regional courts is the chief magistrate, who will be assisted by other magistrates. In contrast to the regional courts, district courts are spread out in the townships and rural areas of South Africa.[50] The jurisdiction of the district courts is limited and the presiding officers are the chief magistrate and other magistrates who are stationed in that particular court.

E. Small Claims Court

20. Small claims court has a very limited jurisdiction, and the amount of dispute between the parties must not exceed ZAR 5,000.[51] Legal representation is not permitted in this court and only the parties of the dispute will appear before the presiding officer who is a commissioner from the ranks of advocate, attorney or legal academic.[52] The commissioners in the small claims courts are not remunerated, and they do the work as a community engagement or service with the aim of harnessing their skills as well as ensuring that the administration of justice is achieved through resolving the dispute of the parties without any delay.

F. Specialised Courts

21. In a country like South Africa with a developing democracy, litigation is increasing and the court rolls are high. The reader must bear in mind that as per the

---

46. Section 169 of the Constitution.
47. Section 167(3) of the Constitution.
48. Section 28 of the Constitution affirms the upper guardianship of the high court when it comes to determining the best interests of minor children.
49. Section 170 of the Constitution.
50. *Ibid.*
51. *Ibid.*
52. *Ibid.*

Constitution each and every person in South Africa has a right to fair and speedy trial, and in order to realise this right the government has set mechanisms in place to ensure that this materialises through the establishment of specialised courts.[53] Specialised courts include the Labour Court and Labour Appeals Court (LAC), which strictly listen to labour disputes between employer and employee or between a trade union and the employer.[54] Specialised commercial courts deal with offences like fraud, corruption and other white-collar crimes.[55] The minister of justice and constitutional development has powers that are vested in him by the Constitution to establish circuit courts in the various provinces to deal with, for example, divorce or any matter that the minister may deem fit with the aim of ensuring that the administration of justice is realised for all the people in South Africa as guaranteed by the Constitution.[56]

V. The Practice of Law in South Africa

22. The court structures as outlined above display how the judicial system of South Africa is structured and how it functions. However, it will be an incomplete picture if the reader is not given a brief background in relation to the role players or practitioners that appear before these various courts, as agents.

23. In South Africa for a person to be eligible to practice law and to be considered a fit and proper person, he or she must first be in possession of an LLB degree.[57] In case a person wants to be an admitted attorney, he or she must further train as a candidate attorney for two years and then write board exams after the training.[58] It is important to note that an attorney can appear in all the courts as outlined above, but must apply to be admitted to appear in the high court because attorneys are not litigation specialists.[59] In contrast to being an attorney, one can elect to be an advocate and such a person must do one year of in-house training at the bar council, which is pupillage, and write bar exams. Advocates work on briefs from attorneys and they are litigation specialists and have automatic entrance to appear in all the courts as discussed.[60] Furthermore, a candidate with an LLB degree can elect to be a prosecutor; he or she must go through the in-house training at the National Prosecuting Authority for a year, and will mostly work to prosecute criminal matters in both the high courts and magistrates courts.[61]

24. However, the practice of law in South Africa is not confined to the three career options that are outlined above; it goes further than that because a person

53. Section 35 of the Constitution.
54. Section 170 of the Constitution.
55. *Ibid.*
56. *Ibid.*
57. Section 3 of the Attorney's Act of 1979.
58. Section 4 of the Attorney's Act of 1979.
59. Section 13 of the Attorney's Act of 1979 and s. 3 of Right of Appearance in Court Act 62 of 1995.
60. Section 3 of the Advocate's Act 74 of 1964.
61. *Ibid.*

with an LLB can also choose to be an academic and embark on postgraduate studies.[62] Furthermore, one can also choose to be an adviser or researcher in the various state institutions, and this shows that the practice of law is not only confined to the courts but it is beyond that aspect. From the various courts as discussed above, it must be noted that the presiding officers in the courts are recruited from the ranks of advocates, attorneys and academics who are experienced and have been in the fraternity for a long time. Once a person is appointed as a judicial officer, they must be impartial and objective all the time with the aim of ensuring that justice is served. The judicial officers of South Africa owe their allegiance to the Constitution and the same applies to the legal practitioners. In case where a judicial officer is found to have acted inappropriately then the Judicial Service Commission (JSC), which is a body that deals with the appointment of judicial officers, will embark on an investigation and disciplinary processes on the responsible judicial official.[63] The same is the case with a legal practitioner who has abused his or her position of trust with clients. That practitioner will be disciplined by the law society in his or her area of jurisdiction or the relevant bar council.[64] These statutory bodies aim to ensure that the integrity of the legal profession is protected at all times, while ensuring that justice is administered to the people of South Africa without fear or favour.

VI. Conclusion

25. South Africa is a developing democracy with a very vibrant Constitution, which champions the fundamental rights of all people in the Republic, and this is evident from the various mechanisms it has set in place with the aim of protecting and preserving democracy. A lot still needs to be done in ensuring that the rule of law is preserved and championed in this young constitutional democracy. The next chapter will focus on the history and social aspect of South Africa with particular reference to religion, which is the main focus of this research endeavour.

§2. Social and Historical Facts About Religion in South Africa

I. Introduction

26. South Africa like all the countries in Africa did not escape the colonial rule and apartheid system. The apartheid era in South Africa can best be described as the dark period for the country due to the atrocities that were committed by the apartheid government on black people.[65] Millions of lives were lost and the degrading as well as painful oppression of the majority people by minority white people form part of this rich history of South Africa. The question the reader is probably

62. *Ibid.* and s 3 of the Attorney's Act of 1979.
63. Section 177 of the Constitution sets out the grounds on which the JSC, after it has done an investigation, can remove a judicial officer from office on the basis of gross misconduct and incapacity.
64. Section 44 of the Attorney's Act of 1979 and s. 36 of the Advocate's Act 74 of 1964.
65. Bekker & Rautenbach, *supra* n. 18 at 13.

asking is how South Africa got out from the painful past of oppression and discrimination for it to be the country that it is today, a leader in the continent of Africa. Without a doubt religion was the solution apart from other political and economic pressures that were put on the then government of South Africa. Religion kept the torch burning for the liberation of South Africa.

27. Religion was one of the instruments that were used to bring social cohesion, and it still is one of the factors that keep a pluralistic community like South Africa together to this very day. This is attributed to the fact that religion brings peace and stability without taking into account the background or colour of an individual. The diverse nature of the South African society means that different religious and cultural practices are recognised in South Africa. These include the Christian religion, Muslim religion, Hindu religion, African customary culture and practices, and all of them are recognised and are all equal as provided by the Constitution.[66]

II. The Socio-Economic Aspect of Religion in South Africa

A. During Apartheid and Colonial Period

28. Religion was one of the driving forces and instruments that have played a major role in the liberation of South Africa, with the aim of fighting the unjust practices of apartheid at the time. When looking in particular at the Christian religion there is a debate that this religion was imposed by colonisers on the South African communities and this was done through the process or work of missionaries from European countries.[67] These missionaries established independent schools in South Africa and educated the South African community in particular black children the Christian way of life. In addition, churches that preached the Christian Gospel were established in various parts of the country. Churches like the Methodist Church, Roman Catholic Church, Anglican Church, Dutch Reformed Church and Lutheran Church among others were formed due to the work of missionaries, and it is important to affirm that these churches are still functioning to this very day in South Africa.[68]

29. During this period, the South African community was educated and the doctrine of Christ formed part of their lives, particularly black people, as they were the recipients of missionary education. Despite the Christian indoctrination being introduced during the missionaries around the 1600s, still there was a great deal of oppression from the colonisers that was taking place against black people.[69] The

66. Section 15 of the Constitution states that everyone has a right to freedom of religion, conscience, thought, belief and opinion.
67. P. Coertzen, *Constitution, Charter and Religions in South Africa*, 14 AHRLJ 127 (2014).
68. *Ibid.*, at 128.
69. W.R. Kauuova, *The Religions of Southern Africa: A Synopsis of Their Basic Beliefs*, 325 Institute for Reformed Studies 3 (1997).

indigenous people of South Africa continued to suffer and receive inferior education and in most instances were denied the right to a fair and quality education. This was attributed to the fact that Afrikaans was imposed as a medium of instruction to all public schools and this was to the detriment of black people who did not understand the language.[70] This move by the apartheid government of excluding black people from mainstream educational opportunities through the use of language as a barrier and also to deny black people better opportunities of employment was unjust. This resulted in black people, particularly children, resorting to missionary schools as refuge and the redeeming feature to access better education, whose foundation, was based on the Christian way of life.[71] However, it is important to note it was not all black people who were accessing missionary education. The pupils who received missionary education were also members of that particular church that offered missionary education. This is because the foundation was largely based on the Christian methodology of that particular church formation or group.[72] The work of missionary's schools and the different church formations, which were established, played a significant role when it comes to the liberation of South Africa. This is because the liberation movement like the ANC, which is currently the governing party in South Africa, and its members were the custodians or recipients of missionary education.[73] For example, the struggle stalwart Nelson Mandela received missionary education and was a member of the Methodist Church, including other struggle stalwarts.[74]

30. These struggle stalwarts due to their Christian background and education used the church to mobilise and garner the necessary support to fight the oppressive government. This included even the various ministers of the above-mentioned church formations assisting liberation movements like the ANC and the Pan African Movement to gather at the church premises and hold secret meetings and to even provide financial means to these organisations in fighting the government.[75] The aim was to expose the ill effects of apartheid and to garner all the support from both the African continent and the globe. Church ministers in addition to the support they offered to the struggle stalwarts and organisations in South Africa, they also provided various other support such as spiritually, emotionally and financially and in some instances the church ministers had to play the role of fathers and mothers. This was by adopting the families of comrades who went into exile.[76] Furthermore, the church ministers also travelled abroad to lobby support from the international community to impose sanctions on the South African government, in order for the apartheid government to stop with the oppression of black people in South Africa. To mention just a few in this regard it will include the well-respected

70. S. Ndlovu, *The Soweto Uprising: The Road to Democracy in South Africa*, 3 South African Democratic Trust 344 (2011).
71. Kauuova, *supra* n. 70, at 4.
72. *Ibid.*
73. *South African History Online Place of Birth of Nelson Mandela*, www.sahistory.org.za (accessed 5 Jun. 2018).
74. *Ibid.*
75. Coertzen, *supra* n. 68, at 128.
76. *Ibid.*, at 129.

noble peace prize laureate Archbishop Desmond Tutu and Rev. Frank Chikana.[77] They are not the only ones. Many ministers and other church formations established by missionaries played a central role in the liberation of South Africa. Church leaders and ministers were also not spared from the apartheid brutality as some were torched, arrested, harassed, assaulted, and at some point, some of the churches were destroyed.[78] This was attributed to the fact that now churches were seen as a threat by the apartheid government and churches who opposed apartheid had to be eliminated. Religion was used as a vehicle for the emancipation of the black community from the start of the revolution until the end of this period. This was because at some point when most struggle stalwarts like Nelson Mandela, Govan Mbeki and Walter Sisulu were arrested religious leaders were working with the ANC and other liberation movements underground to keep the emancipation of black people a reality because it was in danger of being extinct through their imprisonment.[79] During the imprisonment of various political leaders, religious leaders played a role in ensuring that the struggle did not end. This was through securing funding for political parties, paying visits to political prisoners and going to hospitals to visit the injured in order to keep the struggle alive and relevant at that particular time. Furthermore, religious leaders were actively involved in conducting mass funerals caused by the apartheid government which was instrumental in killing of black people in big numbers, and an example to note is the 16 June 1976 Soweto uprising which cost the lives of thousands of young people.[80]

31. Religious leaders at the dark time of South Africa brought the necessary light using religion in order to keep the country sane and alive. Prayer sessions were held across the length and breadth of South Africa, for the almighty God to intercede on behalf of the African people who were in desperate need of liberation and freedom from their oppressors. This history and role of religion in the social makeup of the South African community can best be described as a period similar to that of the children of Israel who were slaves in Egypt. Moses, who was their Shepherd as anointed by God, managed to take them out of slavery because they were the children or a nation that was favoured by God.[81] The prayer, which religious leaders and the struggle stalwarts engaged in, was that if God can deliver the children of Israel surely He could liberate the people of South Africa from the hands of their oppressors in their country of birth. It is important to bear in mind that it will not be correct to say that the only source for the liberation of South Africa emanates from religion. Other factors and role players were also instrumental in the liberation movement. However, the point is that religion played a significant role when it comes to the emancipation of black people from the hands of the oppressors. Even though religion, which came through other means, was or is considered foreign in the African continent,, it can be said that it is one of those things or aspects that

77. *Archbishop Desmond Tutu Is Awarded the Nobel Peace Prize,* https://www.sahistory.org.za/dated-event/archbishop-desmond-tutu-awarded-nobel-peace-prize (accessed 13 Sep. 2018).
78. Coertzen, *supra* n. 68, at 130.
79. Kauuova, *supra* n. 70, at 5.
80. Ndlovu, *supra* n. 71, at 344.
81. Chapter 3 *Exodus* Verses 11–15. God hears the cries of his children and delivers them from the hands of their oppressor.

played a significant role in the freedom of South Africa. The pluralistic nature of the South African community did not hinder the role religious practices played in the liberation of South Africa. This is because the Muslim religious community and Hindu religious community joined in the struggle with the aim of ensuring that apartheid is defeated. Members from the Muslim and Hindu communities also joined the various struggle organisations like the ANC with the aim of liberating South Africa.[82] These collective efforts from different religious background affirm that religion as a practice is about collective effort and attaining social cohesion through peaceful process irrespective of where one comes from.

32. The religion of the African people or the indigenous people of South Africa is their customary practices and culture. They worship ancestors and God, in the sense that it is believed that one cannot worship God without first having considered the ancestors. Indigenous people hold the view that the ancestors are the connector between people on earth and God.[83] The indigenous cultural practice or way of life was sidelined by the colonisers and mostly seen as barbaric and the coloniser's way of life and law, which is the common law, was imposed at the exclusion of African customary law and culture.[84] More about the African customary law or practices will be discussed extensively later on in this research work. The various tribes in South Africa include the Ndebele community, Sepedi community, Tswana community, Nguni community, North and South Sotho community, Venda community and Tsonga community. All these above-mentioned tribes or communities make the South African community and have their traditional leaders.[85] These tribes or communities are spread out across the country in the nine provinces and their leaders working with the various church leaders and political leaders were instrumental to the emancipation of South Africa. Traditional leaders, making use of traditional healers as religious and cultural practitioners of indigenous systems at some point, challenged the colonisers and won the battle. For example, the fight between the Zulu tribe and the English in Natal in which the Zulu tribe won the battle to keep their land against the land invaders at the time who were from an English background.[86] Furthermore, this affirms that African religion and culture also have a rich history when it comes to the social make-up of South Africa, and it can be said that it was also one of the driving forces in the liberation of South Africa.

33. It is important to note that most if not all of the struggle stalwarts were part of these indigenous communities that make South Africa, and making use of African religion was and is still a way of life in South Africa. Traditional leaders were also instrumental in rallying their people in the indigenous communities, largely rural areas of South Africa, to join and fight the apartheid system. This affirms that

---

82. Ndlovu, *supra* n. 71, at 345.
83. N. Mndende, *An Introduction to African Religion*, 28 Surrender 1 (2012).
84. *Ibid.*, at 2.
85. J. Kiernan, *African Traditional Religions in South Africa*, 15 De Jure 27 (2008).
86. E. Frances, *History of the Zulu War and Its Origins* 12 (Chapman and Hall 1880).

the struggle against apartheid using religion was a joint venture with the aim of liberating the people of South Africa.[87] Traditional leaders and traditional healers who are the spiritual diviners in the African communities formed alliances with church leaders and ministers from the different religious dominations with the purpose of bringing down the apartheid monster of oppression and discrimination.[88] This allegiance is continuing to this very day in the democratic dispensation with the aim of ensuring that the pluralistic nature of the South African society is not undermined, and this body is currently known as the South African Council of Churches. This body consists, or is made up, of different religious leaders in South Africa, who champion the aspect that religious dominations can be diverse in nature, but they have a common element or aspect, which is to establish peace and stability in society, irrespective of the background an individual comes from.[89] The use of religion and other measures finally brought the apartheid regime to its knees and it was an end to this cruel system in 1994, which is the period known as the new dawn and the democratic dispensation of South Africa. From the above assessments, it is very much clear that religion, which is the focus of this book, is central to the liberation of South Africa, without under estimating the importance of other role players like the various political parties, international community and investors. They all played a significant role but religion managed to bring people from diverse background together to fight the common enemy, which was the terrible apartheid system. Focus will now shift to the post-apartheid period and still outlining the relevance and importance of religion in the new dawn, which is the democratic dispensation of South Africa post-1994.

B. Post-Apartheid Period

34. After the defeat of apartheid in South Africa using religion among other things, this was now the time for liberation and freedom. South Africa is now under the rule and control of the Constitution, which is the supreme law of the land.[90] Religion still has relevance and it plays an important role during this period. This is because the struggle against apartheid was fought through religion and won through the practice of religion among other things, which meant that religion remained in the centre stage of everything. Religion by its very nature champions for reconciliation and forgiveness, and in South Africa this was no different. The Truth and Reconciliation Commission (TRC) was formed and chaired by Archbishop Desmond Tutu with the aim of ensuring that the oppressor asks for forgiveness to the victims with the aim of reconciling both black and white people.[91] The purpose was to ensure that both black and white people are given the opportunity to talk about the

---

87. *Ibid.*
88. *Ibid.*, at 13.
89. Mndende, *supra* n. 84, at 3.
90. Section 2 of the Constitution.
91. S. Peter, *Restorative Justice and the South African Truth and Reconciliation Commission*, 32 SA Journal of Philosophy 35 (2006).

painful past they suffered and reconcile them in order to build a new South Africa through the mechanism of forgiveness, which is embedded in religion.[92]

The establishment of the TRC at the dawn of democracy affirms the important role religion has in the history and social aspect of South Africa. Hence one cannot discuss religion from a social aspect while excluding the historical aspect because they both share identical aspects and cannot be separated but can be viewed as hand in glove. Today in South Africa when one looks or assesses the position of religion it can best be described as firmly protected and enshrined in the Constitution and through various statutory provisions. For example, section 15 of the Constitution states that everyone has a right to choose their religion or belief freely.[93] The independence of the individual and religion is protected by the Constitution, and this constitutional provision and other provisions that have been set by the government of the day to ensure the independence of religion will be explored in depth later in the following chapters. However, it is important to note that religious organisations like churches enjoy autonomy and are tax exempted in South Africa as they are considered non-profit organisations who aim to bring the necessary social cohesion for all the people of South Africa.[94]

35. From the diverse religious communities, which South Africa comprises of, and as they are part of the population of South Africa for statistical purposes, it is important to consider the following aspects. The Christian community is the majority in South Africa and consists mostly of people from the indigenous communities, and are largely black people. This does not mean that only black people are Christians because white people are also Christians and this does not mean that all white and black people are Christians. There are exceptions to the general rule or norm while bearing in mind that freedom of religion is guaranteed in South Arica.[95] The Christian community is followed by the Muslim and Hindu religions respectively, who are a minority.[96] The fact that the Christian religion is the one that is dominating in South Africa does not mean that minority religious groups or organisations must be discriminated against or oppressed because religious freedom and practice is not based on numbers but it is based on the fundamental right that is guaranteed by the Constitution. Religious practices and way of life still dominate how people relate in South Africa. Religious and cultural practices are starting to be introduced in schools around South Africa, and this initiative is laying a solid foundation for the development and sustainability of religion and cultural practices in South Africa.

---

92. The TRC was based on the foundation of restorative justice and it is argued that this process in some cases it was not effective because there are still lots of social ills in South Africa, which includes for example racism. *See* Peter, *supra* n. 92 at 36.
93. Section 15 of the Constitution.
94. Section 21 of the Companies Act 68 of 2008.
95. Section 15 of the Constitution.
96. Coertzen, *supra* n. 68, at 130.

III. Conclusion

36. Religious and cultural practices form a part of the hybrid community of South Africa and have been used successfully in bringing the necessary stability and social cohesion that is key to the development of South Africa. Religious and cultural practices have proven to be the connector of different people with different backgrounds to share a common goal, which in the South African context was to defeat the apartheid government. Furthermore, in the new democratic dispensation religion is one of the fundamental rights that is protected by the Constitution and must be respect at all costs owing to its historical roots to the development and history of South Africa. The same can be said about cultural practices in the new constitutional dispensation. In the next chapter, reference will be made to the Constitutional position and recognition of religious practices in South Africa from a legal point of view.

Part I. Legal Framework and Sources

Chapter 1. Constitutional Position of Religion

§1. Introduction

37. Religion and religious practices form a part of the democratic dispensation of South Africa, due to the role religion played in the liberation movement of the country. Today, in the democratic dispensation religious practices are one of the fundamental rights that are protected and enshrined in the Constitution and other pieces of legislation. This without a doubt affirms the importance and independence of religious practices in the new constitutional order. Religion is a way of life or a lifestyle that is embedded in how people relate with each other from the same religious group and beyond in order to reach or connect with other people from other religious affiliations. This important aspect of religious practices being a neutraliser and a connector of different people from different lifestyles is what is championed in the South African jurisdiction through the affirmation or guarantee that is afforded to everyone when it comes to the right or freedom to choose any religion of one's choice without being forced or pushed to do so.[97]

38. Freedom of religion as a point of departure is important to note that it is guaranteed by the Constitution, which is the supreme law of the land. This means that any religion or religious practice found to be contrary to the Constitution will be declared invalid.[98] In this chapter focus will be placed on the position of religious and cultural practices from a constitutional point of view, and this will entail discussing in detail the various constitutional provisions that are aimed at the protection and promotion of religion in South Africa. Various court cases will also be discussed where the right to religion from both a legal and a social perspective are considered.

§2. The Right to Religion and Culture

39. Before getting into the discussion in connection with the constitutional protection of the right to religion and religious practices, it is important to first define what religion is as a starting point. This will enable one to have a comprehensive

97. Section 15 of the Constitution.
98. Section 2 of the Constitution.

understanding of this diverse and yet important concept called religion. It is important to note and acknowledge that religion is not defined in the Constitution, and there is no generally accepted definition that is provided either by the international law or by instruments when it comes to this definition.[99] Religion can be defined as diverse ways of being human and there are many various ways of being religious.[100] All religions are concerned with ways of seeing and being in the world and it is clear from this definition that the value of religion lies with people who are following that religion to observe the various facets that are connected with that particular religion. Three facets can be identified which are connected with religion: first, religion as a belief comes from the aspect of the existence of a deity or the adherence to doctrines and this rests with in most cases of African cultural practices or religion.[101] Second, religion is considered as a way of life and this facet puts emphasis on actions and practices that distinguish the believer from other religious dominations. In this case, for example, a Christian is prohibited to engage in adultery or polygamy as prohibited by the Bible, but this conduct is allowed in cases where a the believer is a member of the African cultural practices or belief.[102] The third and last facet is based on the aspect that religion is an identity, which focuses on the self-worth of an individual who forms part of the religious or cultural group. This means that an individual believer remains an individual despite being a member of a religious or cultural group, and this does not entail that the individual is now non-existent when it comes to the group.[103] Individual identity and freedom is protected by the Constitution as it will be outlined below when the constitutional provisions in connection to religion are discussed.

40. Section 15 of the Constitution states that everyone has a right to choose their own religion.[104] This means that each and every citizen in the Republic will be able to choose his or her religion without being put under pressure, but this also means that to partake in a religious practice is a voluntary process which an individual must exercise. The right to choose one's religious practices in South Africa goes hand in hand with other fundamental rights which are enshrined and protected by the Constitution. All the rights outlined in the Constitution do not operate in isolation but complement each other to a lesser and greater extent. This also applies to the right to choose one's own religion freely without any undue influence and this is affirmed by section 16 of the Constitution. Section 16 of the Constitution states that everyone has a right to freedom of religion and expression.[105] This provision emphasises the above-mentioned point that rights in the Constitution work hand in hand with the aim of enhancing and protecting the interests of members of the community.

99. I. Chipkin & A. Leatt, *Religion and Revival in Post-apartheid South Africa,* 9 RESA 39 (2009).
100. *Ibid.*
101. Mndende, *supra* n. 84, at 3.
102. All customary marriages are polygamous in nature and one husband is permitted to get married to more than one wife at a particular time. *See* Mndende, *supra* n. 84, at 4.
103. Chipkin & Leatt, *supra* n. 100, at 39.
104. Section 15 of the Constitution.
105. Section 16 of the Constitution.

41. This is attributed to the fact that section 15 gives the right to an individual the right to choose his or her own religion and religious practices, and section 16 of the Constitution strengthens this choice which the individual enjoys in connection with choosing their religion and religious practices. By focusing on the freedom and voluntariness of an individual in choosing his or her religious practices without any undue influence, it can be argued or seen as a way of affirming the autonomy of the individual person when it comes to their freedom to religion. The right to choose one's own religion and freedom in relation to that religion was tested in various court cases in South Africa, but only the landmark court cases will be considered. In *Christian Lawyers Association v. Minister of Health*,[106] the facts were in connection with the legality of abortion in South Africa. The argument that was advanced by the Christian Lawyers Association was that abortion is morally wrong and constitutes a violation of the right to life as protected by the Constitution.[107] In response, the government argued that abortion was not contrary to the right to life on the basis that a foetus does not qualify as a human being. Furthermore, abortion was a personal and voluntary choice on the part of an individual. The court held that the Choice on Termination of Pregnancy Act was indeed lawful and did not contravene any provisions as alleged and this application was dismissed by the court.[108]

42. However, in *Hay v. B*,[109] the parents of the child who was in desperate need of a blood transfusion to save his life refused the medical procedure on religious basis.[110] The parents of the child refused the blood transfusion on religious grounds as they were members of Jehovah witness church which prohibits blood transfusions.[111] Since the medical practitioner could not get the consent of the parents, there was nothing that he or she could do to save the life of the child, and if the medical practitioner acted contrary to this it would be seen as a violation of the right to religious practices according to section 15 of the Constitution, which was the basis of the arguments that were raised by the parents of the minor child.[112]

This court case took place while the legal system of South Africa was still developing and in the infancy stages. In the years that followed, there was a development in the law in terms of legislation. For example, the introduction of the Children's Act 38 of 2005, which focuses on the best interests of children irrespective of their background, religion and socio-economic circumstances.[113] The adoption of the Children's Act and the application of section 28 of the Constitution, which deals with the best interests of all children, changed the position of children and their parents especially when it comes to life-threatening situations. This meant that parents could not refuse medical treatment for their children anymore on religious grounds.

106. *Christian Lawyers Association v. Minister of Health* 1998 (11) BCLR 1434 (T), 1998 (4) SA 1113 (T).
107. *Ibid.*
108. *Christian Lawyers Association Case, supra* n. 107, at 1114.
109. *Hay v. B* 2003 (3) SA 492 (W).
110. *Ibid.*, 1114.
111. *Ibid.*
112. *Hay Case, supra* n. 110, at 1115.
113. Section 2 of the Children's Act 38 of 2005.

The medical practitioner can approach the court for permission to administer treatment to that child, because the right to life supersede the right to religion on the basis of public policy. This happened in 2014 when the Centre for Child Law at the University of Pretoria was requested by a medical practitioner to help him approach the court in order to save the life of a child who was in desperate need of blood transfusion and the court granted this order on the basis of section 28 of the Constitution.[114] From these court cases it is very much clear that a balancing exercise is key in connection to the various constitutional rights.

43. It is important to note that in South Africa one cannot discuss the position of religious practices without making reference to cultural practices. This is because as indicated earlier both religion and culture are linked to the belief systems of people who subscribe to either religion or culture in the African sense. Section 30 of the Constitution states that everyone has a right to choose the cultural group or practice of his or her own.[115] Furthermore, this constitutional provision on the right to culture is strengthened by section 31 of the Constitution, which affirms the freedom of everyone when it comes to taking part in a cultural practice or group.[116] Religion and culture are one and the same thing and this is affirmed by the definition of culture. Culture is defined as a way of life that is common to a group of people, a collection of beliefs and attitudes, shared understandings and a pattern of behaviour that allow people to live together in peace but set them apart from other people.[117] It is very much clear that both religion and culture are one because they both have to do with a lifestyle and subscribing to the cultural and religious pattern of a certain group.

44. Owing to the painful history of South Africa as already discussed above, culture and cultural practices are associated with black Africans and governed by African customary law. For a culture or cultural practice to be recognised in South Africa, it must fulfil the following requirements: The custom must have existed for a long time; the community at large must recognise the custom; the custom must be reasonable; and the custom must be consistent and subject to the Constitution and other legislation.[118] These standards as set out by the legislature aim to ensure that culture is not abused, and in particular African customary law because it is largely unwritten. These conditions do not aim to undermine the right to culture of indigenous people but aim to safeguard this right from abuse. Cultural practices could be in most instances associated with indigenous people, but sections 30 and 31 respectively of the Constitution by stating everyone it means that people from different backgrounds and ethnic groups are free to partake in cultural practices of indigenous people.[119] This is because to take part in either culture or religion is a voluntary process and it only lies with the individual. Even though these standards as set out by the Constitution in connection with the recognition of culture are not

114. Section 28 of the Constitution.
115. Section 30 of the Constitution.
116. Section 31 of the Constitution.
117. Mndende, *supra* n. 84, at 6.
118. *Ibid.*, at 7.
119. Section 30 of the Constitution and s. 31 of the Constitution.

explicitly outlined for religion, it can be argued that for a religious group to be protected by the Constitution it is necessary that it meet the standards as outlined above. More about the position of African customary law, which gives rise to culture, will be explored later. The right to choose culture and the freedom that is associated with choosing such a culture are not absolute and this is also the case when it comes to religion and can be limited by section 36 of the Constitution as the general limitation clause.[120]

45. Apart from the rights and freedom to choose religion and culture, other fundamental rights are key as already outlined in realising and protecting religion or cultural practices. Focus will now shift to those fundamental rights with the aim of protecting as well as promoting the rights to religion and cultural practices from a South African point of view.

§3. The Right to Dignity and Equality in Realisation of the Right to Religion and Culture

46. Section 10 of the Constitution states that everyone has inherent dignity and the right to have their dignity respected and protected.[121] Further, it is important to enquire what human dignity means. It is defined as an individual's or a group's feeling, its self-respect and self-worth. Human dignity is concerned with physical and psychological integrity and empowerment, and is harmed or violated by unfair treatment premised upon personal traits or circumstances, which do not relate to individual needs, capacities or merits. The right to human dignity is enhanced by laws, which are sensitive to the needs, capacities and merits of different individuals, taking into account the context of their differences.[122] The right to human dignity is harmed or violated when people are marginalised, ignored or devalued, and it is enhanced when laws recognise the full place of all individuals and groups within society. Religious and cultural practices are protected because they are central to human dignity.[123] Scholars like Lenaghan argue that due to the racial discrimination past of South Africa it is important that human dignity should be highly valued as a constant reminder of the past inequality and discrimination.[124] This is due to the impact of the past inequality and discrimination on the values of indigenous cultural and religious practices. Furthermore, it is important to consider what does equality in the context of religion and cultural practices really entail before linking both the right to human dignity and equality in realising the right to culture and religion.

---

120. Section 36 of the Constitution is the general limitation clause and it will be discussed extensively later on.
121. Section 10 of the Constitution.
122. T. Asad, *Discipline and Reasons of Power in Christianity Formations and Dignity* 34, (John Hopkins University 2006).
123. *Ibid.*, at 35.
124. C. Lenaghan, *Shots in the Streets: Violence and Religion in South Africa, Contemporary South African Debates* 68, (Oxford University Press 2004).

47. Section 9(1) of the Constitution states that everyone is equal before the law and has the right to equal protection and benefit of the law.[125] De Vos and Freedman argue that the right to equality can best be described in two ways. First, the right to equality does not entail a guarantee that all people should be treated identically all the time, regardless of their personal attributes or characteristics, social or economic status. The right should therefore be viewed as entailing more than a formal prohibition against discrimination. Second, the right to equality must guarantee more than equality before the law and must focus on the effects or impact of legal rules or other differentiating treatment on people. The right to equality cannot therefore focus merely on whether two or more people have been treated in an identical manner by the legal rule or by the company or another individual concerned.[126] One can assert that this two-stage enquiry in relation to the meaning of equality is balanced and in line with the principle of substantive equality.[127] For the purposes of this research project and of relevance when discussing the equality clause is section 9(3) of the Constitution which provides that the state may not unfairly discriminate directly or indirectly against anyone on one or more grounds, including race, gender, sexual orientation, age, religion, belief, disability, culture, language and birth.[128] From the above discussion of both the right to dignity and equality, it is important to note that the right to equality plays an important role in the recognition of dignity and both are fundamental. Accordingly, from this assertion, there is a connection between dignity and privacy and this view was confirmed in *Christian Education South Africa v. Minister of Education*.[129] In this case, it was held that equality does not require that we treat everyone equal but that we treat everyone with equal concern and respect irrespective of his or her religious or cultural practices.[130] Goldstone J, in this case, summarised the importance or connection between dignity and equality in the following way:

> At the heart of the prohibition of unfair discrimination lies a recognition that the purpose of our new constitutional and democratic order is the establishment of a society in which all the human beings will be accorded equal dignity and respect regardless of their membership in a particular group.[131]

48. This court case like many others and the constitutional provisions, which aim to champion the realisation of religious and cultural practices, can be seen as a way of remedying the past injustices while at the same time ensuring that the new democratic order upholds the fundamental rights of the people of South Africa. It is also important to consider that dignity is not only defined in different ways, but it also operates in different respects. Dignity can function as a specific right as

---

125. Section 9(1) of the Constitution.
126. P. De Vos and M. Freedman, *South African Constitutional Law* 421, (Oxford University Press, Cape Town 2015).
127. *Ibid.*, at 422.
128. Section 9(3) of the Constitution.
129. *Christian Education South Africa v. Minister of Education* [2000] ZACC 11, 2000 (4) SA 757 (CC); 2000 (10) BCLR 1051 (CC).
130. *Ibid.*
131. *Ibid.*, at 1052.

entrenched in section 10 of the Constitution.[132] Additionally, dignity may determine how the entrenched right like the right to choose religion and cultural practices without any form of discrimination in order to achieve equality can be practiced.[133] This means that dignity can serve as a basis to the realisation of a fundamental right that is upheld and protected by the Constitution, which in this case is the right to choose cultural and religious practices of one's choice. This is often the reason in South Africa that dignity is classified as a value because different provisions in the Bill of Rights must be interpreted in order to promote the values that underlie a democratic society. These values include human dignity, equality and freedom.[134] Without a doubt, it is clear that human dignity lies in the heart of the new democratic order of South Africa.

49. Dignity therefore stands in relation to the application of all other rights and these rights shape our understanding of dignity. For this reason, dignity has particular relevance to our understanding of the right to freedom of religion and culture. Similarly, the right to equality is of particular importance to our appreciation of the right to freedom of religion and culture as outlined. The interrelationship between dignity and equality in many ways can be comparable to the interrelation that exists between religion and culture as explored. For diversity to prosper it is essential that the value of dignity and equality as well as the interrelation between religion and culture are held when addressing issues regarding the manifestation of indigenous culture and religion, in particular.

§4. The Right to Privacy and Right to Religion and Cultural Practices

50. Before discussing the right to privacy in South African law, it is important to consider that the Constitution entrenches this right in section 14.[135] Furthermore, it is necessary to take into consideration that the right to privacy is also upheld and protected by various statutes, including the Protection of Personal Information Act and the National Health Act.[136] More about the role of these two statutes will be explored later when discussing the position of domestic legislations that are in place in relation with the protection of religion and cultural practices.

---

132. Section 10 of the Constitution.
133. Section 15 of the Constitution.
134. Section 2 of the Constitution.
135. Section 14 of the Constitution and s. 2(a) of the Protection of Personal Information Act 4 of 2013 (herewith referred to as Poppy) set out the purpose of this Act by outlining that the aim of this Act is to give effect to the constitutional right to privacy, by safeguarding personal information when processed by a responsible party, subject to justifiable limitations that are aimed at balancing the right to privacy against other rights, particularly the right of access to information and protecting important interests, including free flow of information within the Republic and across international borders.
136. Section 2 of Poppy and s. 14 of the National Health Act 61 of 2003.

I. Common Law Definition of Privacy

51. Under common law, every person has personality rights such as the rights to physical integrity, freedom, reputation, dignity and privacy.[137] According to Neethling, privacy is defined as an individual condition of a person's life, which is characterised by seclusion from the public and publicity.[138] Since privacy is considered as personal facts which a person has determined to exclude from the knowledge of other people, this means that the right to privacy can only be violated when a person learns of the true private facts of another person against his or her will.[139]

52. The right to privacy has long been entrenched in the common law, in which everyone has personality rights which include the right to physical integrity, freedom, reputation, dignity and privacy.[140] This right to privacy has been recognised as an independent personal right that applies to both natural persons and juristic persons, such as companies.[141] The violation of the right to privacy is determined by the court, in terms of the reasonable standard in the interests of the community in order to guard against any form of injustices on an individual or a company.[142] Privacy underpins dignity since a violation of the right to privacy infringes on the right to dignity.[143] This is because these two rights relate to each other, in the sense that the right to privacy ensures that one maintains confidential information that he or she has, and if such information is published in a derogatory manner it will damage the dignity of the person.[144]

53. This rich history, which emanates from common law with regard to the right to privacy and other related rights such as dignity and reputation, clearly affirms Burchell's view that 'the common law of privacy in South Africa still provide the lion's share when it comes to the right to privacy standard'.[145] This is despite the fact that currently in the new constitutional dispensation, the rights to privacy, dignity, reputation and freedom are recognised individually and independently in order to give them more recognition and further interpretation; this was not the position in common law previously as these rights were treated collectively rather than individually.[146]

---

137. J. Neethling, J.M. Potgieter & P.J. Visser, *Neethling's Law of Personality* 31 (Butterworths, Durban 2005).
138. *Ibid.*, at 32.
139. A. Roos, *Personal Data Protection in New Zealand: Lessons for South Africa*, 4 PER/PELJ 191 (2008).
140. *Ibid.*, at 192.
141. Neethling et al., *supra* n. 138, at 37.
142. *Ibid.*, at 38.
143. Section 10 of the Constitution.
144. *C v. Minister of Correctional Services* 1996 (4) SA 292 (T).
145. J.M. Burchell, *Media Freedom of Expression Scores as Strict Liability Receives the Red Card: National Media Ltd v. Bogoshi*, 8 SALJ 1 (1999).
146. Section 14 of the Constitution and s. 12 of the Constitution it protects everyone's right to freedom and security.

II. Legality of the Common Law Right to Privacy in the Constitutional Era

54. Section 8(3) of the Constitution provides that when applying a provision of the Bill of Rights to either a natural or juristic person, a court must apply, or if necessary, develop the common law to the extent that legislation does not give effect to the right under consideration, in order to properly give effect thereto.[147] Furthermore, a court may develop the rules of common law to limit a right, provided the limitation is in line with section 36(1) of the Constitution. Section 39(3) of the Constitution, which deals with the interpretation of the Bill of Rights,[148] states that the Bill of Rights does not deny the existence of any other rights or freedoms that are recognised or conferred by the common law, customary law or legislation to the extent that they are consistent with the Bill of Rights. These constitutional provisions clearly affirm the view that Burchell holds about the common law carrying a lion's share on the right to privacy, as outlined.

55. However, the common law position with regard to the right to privacy still needs the affirmation of the Constitution before it might be used as a point of departure. The common law right to privacy can be limited as well as declared null and void if it is found to be contrary to the provisions of the Constitution.[149] The position is quite clear that although the common law right to privacy in South Africa is fundamental due to its historical scope, it is however subject to the Constitution which is the crucial aspect that Burchell omitted in his assessment of the position of common law right to privacy.

56. The right to privacy has long been protected in our law dating back to the time it was inherited from the Roman law system in which the *actio iniuriarum* was provided as a remedy in cases where there was an invasion to the sanctity of a person's home.[150] This has been the position before the constitutional dispensation came into existence and provided for the right to privacy in section 14 of the Constitution.[151] Two examples of case decisions dealing with the common law right to privacy can be found in the 1900s. First, an English case will be discussed and then a South African decision thereafter.

Tolly v. Fry and Sons Ltd [1931] AC 333

147. Section 8(3) read with s. 8(2) of the Constitution.
148. Section 39(3) of the Constitution.
149. Section 2 of the Constitution.
150. Neethling et al., *supra* n. 138, at 38. The remedy with regard to the *actio iniuriarum*, which owes its historical development to Roman law, was established and used to recover damages (*solatium*) based on the intentional infringements of one of the acknowledged personality rights. In order to succeed with this remedy, an infringement of a legally acknowledged interest of personality is required and such infringement has to be intentional in the process. *See Jackson v. NICRO* 1976 3 SA 1 (A) 11, in which the history and purpose of this Roman remedy was clearly outlined and applied with the aim of making an informed decision in the process.
151. Section 14 of the Constitution.

57. In *Tolly v. Fry and Sons Ltd,*[152] the plaintiff was an amateur golfer and the defendant was a producer of candy. The defendant, in his advertisement of chocolates, showed the plaintiff with a protruding bar of chocolate in his back pocket.[153] This form of advertisement was not done with the consent of the plaintiff and the plaintiff challenged this on the basis that it violated his right to dignity and privacy. Furthermore, the plaintiff argued that such advertisement was done without his consent, and that it had the potential to destroy his golfing career as he was still an amateur, due to the misleading nature of the advertisement.[154] The court found the conduct of the defendant as unlawful and ordered the defendant to compensate the plaintiff for the loss or harm which he had suffered as a result of the advertisement.[155]

***O'Keeffe v. Argus Printing and Publishing Company Ltd* 1954 (3) SA 244 (C)**

58. In *O'Keeffe v. Argus Printing and Publishing Company Ltd,*[156] the plaintiff was a well-known radio personality who consented to the publication of her photograph, taken at the pistol range, to be used for a newspaper article by the defendant company.[157] The defendant company, however, used the photograph of the plaintiff for advertising purposes, which was not what the plaintiff consented to in the first place. Watermeyer AJ held the view that the conduct of the respondent company was unlawful and that the respondent acted contrary to the right to privacy of the plaintiff and was ordered to compensate the plaintiff for the violation of her right to privacy with regard to the derogatory advertisement.[158]

III. The Right to Privacy in Section 14 of the Constitution of the Republic of South Africa, 1996

59. Section 14 of the Constitution states that everyone has the right to privacy, which includes the right not to have their person or their home searched; their property searched; their possession seized; or the privacy of their communications infringed.[159] This provision of the Constitution, which guarantees the right to privacy, is supported by Schoeman who defines privacy as a state or condition of limited access to a person.[160] Furthermore, the right to privacy was explained in *National Media Ltd ao v. Jooste.*[161] Harms AJ explained that the right to privacy

152. *Tolly v. Fry and Sons Ltd* [1931] AC 333; [1931] UKHL 1.
153. *Tolly Case, supra* n. 153, at 334.
154. *Ibid.*
155. *Tolly Case, supra* n. 153, at 335.
156. *O'Keeffe v. Argus Printing and Publishing Company Ltd* 1954 (3) SA 244 (C).
157. *Ibid.*
158. *Ibid.*, at 245.
159. Section 14 of the Constitution.
160. F. Schoeman, *Philosophical Dimensions of Privacy: An Anthology* 5 (Cambridge University Press, Cambridge 1984).
161. *National Media Ltd ao v. Jooste* 1996 (3) SA 262 (A).

can best be described as an individual condition of life characterised from the public and publicity. This condition includes all personal facts which the individual has determined to exclude from the knowledge of other people and in respect of which he or she has wills that they be kept private.[162] *Kennedy and Arnold v. Ireland*[163] takes the discussion of the right to privacy further by stating that:

> ... though not specifically guaranteed by the Constitution, the right to privacy is one of the fundamental personal rights of the citizen which flow from the Christian and democratic nature of the state. The nature of the right to privacy must be such as to ensure the dignity and freedom of an individual in the type of society envisaged by the Constitution, namely a sovereign, independent and democratic society.[164]

60. This reasoning by the court surely envisages the position of the South African jurisprudence, in protecting the right to privacy of an individual from other people in a democratic state. Furthermore, this means the right to privacy of an individual as a member of a religious group is protected and this extends to the religious organisation.

IV. The Limitation of Constitutional Rights

A. Rationale for the Limitation of Rights

61. In a democratic state where the Constitution is the supreme law of the country, it is important that there must be a balance between the rights and duties which citizens enjoy for democracy to function.[165] This will entail limiting the rights of other people in the public's interest through a general limitation clause. Rautenbach and Malherbe argue that there can be no peace and order in any society when people's rights are exercised absolutely without limitations.[166] However, it is important to take into account that the limitation of rights must not be arbitrary, but needs to be in line with the Constitution, taking into account values such as human dignity, equality and freedom.[167] Devenish explains the background of the general limitation clause, and particularly, what it entails and its purpose.[168] He outlines that:

> It is widely accepted in the domestic law of most countries, in international law and according to international and other human rights documents, that only a very limited number of rights, if any [,] are absolute. These include freedom from torture, the abuse and exploitation of children and possibly freedom from

162. *Ibid.*, at 263.
163. *Kennedy and Arnold v. Ireland* [1987] IR 587.
164. *Ibid.*
165. Section 2 of the Constitution.
166. C. Rautenbach & R. Malherbe, *Constitutional Law* 156 (LexisNexis, Cape Town 2015).
167. Section 36(1) of the Constitution.
168. G.E. Devenish, *A Commentary on the South African Constitution* 121 (Butterworths, Durban 1998).

> servitude; freedom of conscience, belief, thought and opinion. The overwhelming majority of human rights and liberties which are of necessity are restricted by inherent duty, which should be perceived as the inextricable counterpart of a corresponding right, to respect the rights of others. The classical example in this instance is that freedom of speech does not allow one person to defame another nor would it [call for a] sanction [to be imposed on] a person shouting fire in a full theatre when there is no fire.[169]

62. This means that there must be a link between a right and the means by which it is limited in order for it to be considered a lawful and justifiable limitation.

In addition, section 38 of the Constitution provides that anyone whose rights have been infringed can seek appropriate relief from a court or tribunal. Sections 36(1) and 38 of the Constitution set out the tone for dealing with the issue of the limitation clause, which is applicable in the relationships between the state and private individuals, as well as in the relationships between private individuals themselves. The next step will be to look into the limitation clause and also to make an enquiry as to whether it is valid or not.

B. The General Limitation Clause in Section 36(1) of the Constitution of the Republic of South Africa, 1996

63. Section 8 of the Constitution is important in describing the relationship between the state and private individuals.[170] This section facilitates a more realistic way of dealing with the rights of all people and balancing them with the obligations of the state. For example, section 14 of the Constitution does not grant people an absolute right as the state can violate the right to privacy of the individual with the aim of protecting the public's interests.

64. Similar to any other rights contained in the Constitution, the right to privacy can be limited according to section 36(1) thereof. This section provides the general limitation clause and states that the rights in the Bill of Rights may be limited only in terms of the law of general application to the extent that the limitation is reasonable and justifiable in an open and democratic society based on human dignity, equality and freedom. This limitation ought to be effected while taking into account factors such as the nature of the right. This means the purpose of the right; the importance of the purpose of the limitation, which has to do with the reason behind limiting the right; the nature and extent of the limitation, which has to do with an enquiry (which is normally done by the court) into looking at the way the limitation

---

169. *Ibid*

170. Section 8(1) of the Constitution states that the Bill of Rights applies to all laws. The Bill of Rights binds all the branches and spheres of government bodies. This means that it must be followed by: (a) the legislatures, which is a body that makes law; (b) the executive, which is the body that carries out law; (c) the judiciary, which are the courts. Section 8(2) of the Constitution states that the Bill of Rights applies to all matters between citizens and the government. This means it applies vertically between citizens and the government. The Bill of Rights also applies horizontally in matters which arise among people or businesses.

took place; the relation between the limitation and its purpose, meaning that there must be a balance between the infringement of a right and its purpose in order for the limitation to be justifiable; and the less restrictive means to achieve the purpose, which means that the limitation must achieve benefits which are proportional to the impact of that limitation.[171]

65. However, it is important to note that in *National Coalition for Gay and Lesbian Equality and Another v. Minister of Justice and Others,*[172] the court held that the limitation of gay rights on moral grounds was not proportional to the benefits that were aimed to be achieved. In this case the court held that, the enforcement of private moral views of a section of the community, which is based to a large extent on prejudice, cannot qualify to be a legitimate purpose.[173] There is nothing, in the proportionality enquiry, to weigh against the extent of the limitation and its harmful impact on gays. In this case the court stated that there is no justification for the limitation of gay rights.[174]

66. The operation of section 36 of the Constitution can be illustrated using the example of a medical practitioner who discloses the health status of a patient who has acquired the Ebola virus to other people. In this instance, the disclosure can be justified as acting in the public interest in order to curb the spread of the virus. Furthermore, an equitable balance exists between the purpose of the disclosure and the limitation of the right to privacy, given the dangerous nature of the Ebola virus. Thus, such a disclosure is aimed at protecting the public's interest and ensuring that the particular patient is placed in quarantine. This will not amount to a violation of the right to privacy of the patient but can be considered a protected disclosure, according to section 36(1) of the Constitution.

67. In South African law there are court cases in which the role and legitimacy of section 36 of the Constitution have been described in detail. For example, in *S v. Manamela* the court stated that:

> In essence, the court must engage in a balancing exercise and arrive at a global judgment on proportionality and not adhere mechanically to a sequential check-list. As a general rule, the more serious the impact of the measure on the right, the more persuasive or compelling the justification must be. Unfortunately, the question is one of degree to be assessed in the concrete legislative

---

171. Section 36(1)(a)–(e) of the Constitution further argues that there is no hard and fast rule for determining whether the purpose of the limitation is considered constitutionally legitimate. However, broadly speaking the constitutional values of openness, democracy, freedom, equality and dignity will play a role in determining what a legitimate purpose is. This shows that courts will judge each case based on its own merits and decide what constitutes a legitimate purpose. Without a doubt, this is the correct approach which needs to be adopted in order to ensure that the law is in line with the public interest and that the public interest is upheld at all times.

172. *National Coalition for Gay and Lesbian Equality and Another v. Minister of Justice and Others,* (CCT11/98) [1998] ZACC 15; 1999 (1) SA 6; 1998 (12) BCLR 1517 (9 Oct. 1998).

173. *Ibid.*, at para. 37.

174. *Ibid.*, at para. 38.

and social setting of the measure, paying due regard to the means which are realistically available in our country at this stage, but without losing sight of the ultimate values to be protected. Each particular infringement of a right has different implications in an open and democratic society based on dignity, equality and freedom. There can be accordingly no absolute standard for determining reasonableness.[175]

68. However, the issue of protected disclosure in our law is a complicated matter and needs to be judged based on the facts of each case. Our courts have rejected a number of cases on the basis that the disclosure of the health status of a person is unlawful and does not amount to protected disclosure, and thus not justified in section 36 of the Constitution. For example, in *NM v. Smith*,[176] in which the Human Immunodeficiency Virus (HIV) status of three people was published in a book without their consent, the court rejected the argument raised by the respondent that the HIV status of the three applicants was known by the public, as the court held that such disclosure was not in the public's interest and thus did not amount to a protected disclosure.[177]

69. In *Maje v. Botswana Life Insurance*,[178] the plaintiff sued the defendant company for publishing her photograph in the company's newsletter and thereby depicting her as being HIV positive. The defendant company argued that it was in the public's interest to make such a disclosure, and the court dismissed such reasoning on the basis that the publication was offensive and had caused her humiliation.[179] The conduct of the defendant company did not amount to a protected disclosure and was not in the public's interest, but it amounted to an actionable offence.[180] Based on the above court cases it is clear that for a disclosure to be protected and justified, it must be in the public's interest and thus it can be argued that such a disclosure

---

175. *S v. Manamela and Another* 2000 (3) SA 1 (CC), 'the court stated that the limitation of constitutional rights for a purpose that is reasonable and necessary in a democratic society involves weighing up of competing values and ultimately an assessment based on proportionality. This is implicit in the provisions of S 33(1)(c) of the Interim Constitution. The fact that different rights have different implications for democracy, and in the case of our Constitution, for an open and democratic society based on freedom and equality, means that there is no absolute standard which can be laid down for determining reasonableness and necessity. Principles can be established, but the application of those principles to particular circumstances can only be done on the case-by-case basis. This is inherent in the requirement of proportionality, which calls for the balancing of different interests. In the balancing process, the relevant considerations will include the nature of the right that is limited, and its importance to an open and democratic society based on freedom and equality, the purpose for which the right is limited and the importance of that purpose to such a society, the extent of the limitation, its efficacy, and particularly where the limitation has to be necessary, whether the desired ends could reasonably be achieved through other means less damaging to the right in question. In the process, regard must be had to the provisions of section 33(1)(c) of the Interim Constitution, and the underlying values of the Constitution, bearing in mind that as a Canadian Judge has said, the role of the court is not to second-guess the wisdom of policy choices made by legislators'.
176. *NM v. Smith* Case, *supra* n. 1 at 250.
177. *Ibid.*
178. *Maje v. Botswana Life Insurance* [2001] 2 BLR 626.
179. *Ibid.*
180. *Ibid.*, at 627.

must serve a certain purpose, which is in the public's interest before it can be considered justifiable according to section 36 of the Constitution.

70. In the context of religion and culture, which are the central themes of this study, it is important to note that the rights to religion and cultural practices are not absolute for an individual or group as protected by the Constitution. This means that the right to religion and culture can be limited if it is in the public interests to do so according to section 36(1) of the Constitution.[181] The right to privacy of a religious or cultural group can be limited if it is found to be reasonable and justifiable in an open and democratic society, based on human dignity, equality and freedom. In South Africa the right to religion and religious practices was limited, for example, in the case of the Jehovah witness church on the stance of the refusal to take blood transfusion based on religious grounds. This was found to be contrary to the right to life and contrary to the public interest and hence the limitation on the right to religion as per section 15 of the Constitution.[182] It can be argued that the limitation clause is a neutraliser in keeping harmony between the rights and the limitations of rights as contained in the Constitution.

§5. CONCLUSION

71. The constitutional position in South Africa in relation to religion and cultural practices is without a doubt firmly entrenched and protected based on the various provisions that were explored above. This means that despite the diversity of the South African population religious and cultural practices are encouraged and promoted because religion and culture is one of those practices that are necessary and important to bring and achieve the social cohesion. It was also outlined in the chapter that despite the autonomy of an individual to choose religion and culture and freedom that come with such choice, these rights are not absolute and can be limited according to section 36 of the Constitution, which is the general limitation clause. The limitation of these rights and other fundamental rights must not be viewed as discriminatory or arbitrary to religion and culture, but the limitation itself aims to ensure that these rights are not used adversely or abused at the expense of other rights and other people, bearing in mind the pluralistic society such as South Africa.

The focus of the work will now be shifted to various legislations which are entrenched in South African law, with an aim of promoting and protecting the rights to religion and culture among other things.

181. Section 36(1) of the Constitution.
182. Section 15 of the Constitution.

Chapter 2. Legislation Aimed at the Promotion of Religion and Cultural Practices

§1. INTRODUCTION

72. The Constitution was adopted in 1996 as the supreme law of the land, which has a Bill of Rights that contains fundamental rights like the right to choose and freedom in relation to religion and culture among other things. Furthermore, in ensuring that these fundamental rights are realised section 7(3) of the Constitution mandates the government to respect, protect, promote and fulfil the rights in the Bill of Rights.[183] Religious and cultural groups also have an obligation to ensure that when they exercise their rights to choose and freedom in relation to religion and culture they comply with the conditions that have been set by the Constitution.[184]

73. After the constitutional enactment ensuring the protection of religious and cultural practices, there were questions in relation to the implications of this new position. This resulted in the formulation of the South African Charter of Religious Rights and Freedoms as one of the codes that aim to regulate religious and cultural practices from a legislative point of view. However, it is important to note that the background for the formulation of this piece of legislation emanates from Albie Sachs, a struggle stalwart who also became one of the justices in the Constitutional Court in the new dawn of democracy. Sachs wrote during the drafting period of the Constitution in 1990 in connection with the position of religious and cultural practices in South Africa:

> Ideally in South Africa, all religious organisations and persons concerned with the study of religion would get together and draft a charter of religious rights and responsibilities – it would be up to the participants themselves to define what they consider to be their fundamental rights.[185]

An in-depth discussion of this charter as to how it was formed, the purpose of the charter and how does it regulate religious and cultural practices in connection with the Constitution as the supreme law in the land will be considered.

§2. THE FORMATION OF THE SOUTH AFRICAN CHARTER OF RELIGIOUS RIGHTS AND FREEDOMS

74. This charter on Religious Rights and Freedom was drafted over a number of years, and it was a continuous process by a number of people from diverse backgrounds who make the South African community or society. This included academics, religious leaders, government commissioners, human rights groups and

183. Section 7(3) of the Constitution.
184. Section 36 of the Constitution.
185. A. Sachs, *The Soft Vengeance of a Freedom Fighter* 3 (Amazon 1990).

organisations, the international community, legal experts, media groups, and major religious and cultural groups in South Africa were consulted with.

75. After all the work was done by these various participants, then the charter was endorsed in 2010 in the presence of the then Deputy Chief Justice of South Africa, Dikgang Moseneke, who was of the view that the charter represented the diversity and religious as well as cultural freedom of the people of South Africa.[186] At the occasion, ninety-one leaders who represented different religious and cultural groups were present in order to affirm the importance of the charter when it comes to the regulation of religion and culture. Furthermore, different religious and cultural groups were present at the occasion to show solidarity to this charter which aimed to strengthen the independence and freedom of religious and cultural practices. The total number of people who were signatories for the adoption of this charter to be law in South Africa was estimated to be 10.5 million.[187] This shows that the various religious and cultural communities in South Africa welcomed the charter because it was viewed as one of those pieces of legislation that aimed to promote the independence and freedom of religious and cultural practices in South Africa. Currently, the charter is available in six of the eleven official languages of South Africa, which include English, Afrikaans, Zulu, Xhosa, Sotho and Tswana.[188] This means that work needs to be done in ensuring that the charter is written in all the eleven official languages of South Africa to ensure that it is available and accessible to all people, and language must not be a barrier when it comes to understanding and interpretation of the charter. After the public endorsement of the charter, a South African Council for the Protection and Promotion of Religious Rights and Freedoms was established to oversee the process of the charter to be formally adopted as law in South Africa.

76. The discussion about the aim and purpose of this charter, which ensures the right to choose and freedom in connection to religion and culture, will now follow.

§3. Purpose, Aim and Objective of the South African Charter of Religious Rights and Freedoms

77. The South African Charter of Religious Rights and Freedoms defines freedom, rights and responsibilities of religious and cultural beliefs and the relationship between the government of South Africa and its citizens.[189] The Bill of Rights affirms that everyone has the right to freedom of religion and culture, while section 234 of the charter makes provision for civil organisations like churches and cultural groups to draw up charters of rights and responsibilities, which may then be enacted by parliament.[190] This charter on religious rights and freedoms is the first of its kind

186. Coertzen, *supra* n. 68, at 130.
187. *Ibid.*, at 131.
188. *Ibid.*, at 132.
189. Section 1 of the South African Charter of Religious Rights and Freedom, 2010.
190. Section 234 of the South African Charter of Religious Rights and Freedom.

in South Africa and apart from addressing freedoms and rights of religion over and against the government, the charter is very useful and important for the relationship between the different religious and cultural groups in South Africa, irrespective of their diversity. The charter aims to help the different religious and cultural groups in South Africa. It is not the purpose of this charter to bring one religious belief or cultural practice in South Africa, but to strengthen diversity and independence of religious and cultural practices in South Africa.[191]

78. The charter defines the rights and freedoms that each religion or culture in the Republic can claim while at the same time working together with other religions and cultural groups in the public interest and for the common good of the entire population of South Africa, which is pluralistic in nature. The charter also serves as a useful tool in ensuring that religious and cultural groups are in a position to determine their own self-identity in terms of rights and freedoms. It is important to note that if religious and cultural groups do not make use of this tool as provided by the charter, they will find that their rights and freedoms will be determined by the courts, which can be a very difficult exercise due to the expensive nature of litigation in South Africa.[192] Even if parliament decides not to enact the charter into law, religious and cultural organisations can always make it part of their own body of rules and regulations, which then will have to be taken into account by the courts in case of a dispute.

79. The charter consists of a Preamble of eight articles, which outline the purpose of the charter. This is then followed by the charter itself, comprised of twelve articles with sub-divisions, stating the religious rights and freedoms of religious people and communities in South Africa. Furthermore, the charter states what freedom of religion means to religious believers and religious organisations in South Africa within a South African context, as well as the specific rights, responsibilities and freedoms that are associated with freedom of religion.[193] These include the following, among others:

(a) The right to gather to observe religious beliefs.
(b) Freedom of expression regarding religion.
(c) The right of citizens to change their faith.
(d) The right of persons to be educated in their faith.
(e) The right of citizens to educate their children in accordance with their philosophical and religious convictions.
(f) The right to refuse to perform certain duties or assist in activities that violate their religious belief.
(g) The rights of religions to institutional freedom.[194]

---

191. Section 1 of the South African Charter of Religious Rights and Freedom.
192. Coertzen, *supra* n. 68, at 133.
193. Section 2 of the South African Charter of Religious Rights and Freedom.
194. *Ibid.*

80. All these rights and freedoms, which are outlined in the charter, will be extensively discussed in the following chapters with the aim of showing the interrelatedness between the Constitution and the charter. This is because the aim of both these legal instruments is to ensure that the right to choose, and freedom to practice, a religion or culture is protected and promoted in the new constitutional order of South Africa.

§4. Conclusion

81. In conclusion, from the above assessment it is very much clear that religious and cultural practices are both constitutionally and statutorily entrenched, irrespective of the diversity of the South African society. A piece of legislation like the South African charter on religious rights and freedom aims to champion the right to freedom of religion and cultural practices in South Africa. This is while at the same time ensuring that different religious and cultural organisations work together with the purpose of protecting the freedom of religion and culture and the independence that is associated with it.

Chapter 3. Typology of the South African System

§1. Introduction

82. South Africa is a multicultural and pluralistic community with people who belong to different cultural and religious beliefs or groups. It is important to outline at this stage particularly when it has become necessary to discuss the different religious and cultural beliefs that make the South African society that all of them are equal. This means that there is no religion or cultural practice or group that dominates the other but all are equal and important before the law and the broader South African community, and this is evident from the different provisions of the Constitution that aim to strengthen and protect the independence as well as freedom of religious and cultural practices.

83. 84. In this chapter the focus will be on the classifications, outlining the different religious groups and cultural practices in South Africa in their diversity. It is important to note that emphasis will be on the African cultural practice or culture in this chapter because this forms a part of the belief of the indigenous people of South Africa. This cultural system was sidelined by the colonisers of South Africa and the effects are still felt to this very day because African customary law, which regulates the cultural practice of the indigenous people, is under-developed and not implemented to its full capacity in South Africa.

§2. Different Religious Dominations in South Africa

I. Christianity

84. Christianity is the religion of the people who follow Jesus Christ as their Saviour and Lord who together with the Father and the Holy Spirit is the True God. Out of love for the world, the Father sent the Son to the world in order to save it from sin.[195] The Son died for the sin of the world, and rose from the dead after three days of his death. As the Lord who raised from the dead, Christ is the head of the whole of creation, which is the Kingdom of God.[196] Christ was at the same time also given as the head of the church, his body, the fullness of Him that fills everything in everybody. They who believed in Him are His church and at the same time members of His Kingdom, called to witness to the world in every aspect of their life about Jesus Christ, the Lord of the whole of creation.[197] For Christians, the Bible is the Word of God which guides them in their religious and moral life in all spheres of life. It is important to note that Christianity reached South Africa very early in the Western Calendar.[198] The reformed religion came to the southern tip of

195. Holy Bible 2 Corinthians Ch 5: Verses 14–15.
196. Holy Bible Hebrews Ch 2: Verses 14.
197. R. Elphick, *Christianity in South Africa: A Political, Social and Cultural History* 11 (University of California Press 1997).
198. *Ibid.*, at 12.

Africa from Western Europe in 1652, through the establishment of a refreshment post for their ships at the Cape of Good Hope by the Dutch East Indian Company.[199] Today the majority of the population of South Africa, both black and white, follow the Christian faith as a way of life.[200]

II. Muslim

85. Muslim religion is a minority religion in the South African community as the number of people who follow the Muslim religion are estimated to be 1.5% of the total population of South Africa.[201] When tracing the historical background of the Muslim religion in the South African context, it is important to bear in mind that these were the Indians who were brought to South Africa as slaves in 1652 to work in the sugar plantations in Natal.[202] The people who subscribe to the Muslim religion are mostly in the Indian and coloured communities of South Africa, and it is important to note that they do not isolate themselves from other religious dominations; this is because inter-faith is championed.[203] Muslims use the *Qur'an* as their book of faith and believe in Mohammed as their Saviour and that of the world.[204] Even though Muslims are a minority in South Africa, their religious rights and freedoms are firmly entrenched in the Constitution. This is evident from the fact that during the recognised periods of worship like the month of Ramadan, people who form part of this religion are given the space to pray without any form of discrimination, in the workplace or schools.[205]

III. Hindu Religion

86. Hindu religion is a minority religious group in South Africa, which consists of about 1.1% of the population of South Africa.[206] Hindus arrived in South Africa during the British rule of colonisers in which they were promised good jobs and income by the then British government with the security that they will be allowed to practice their religion without disturbance.[207] This was not the case as Hindus were discriminated and often denied the opportunity to practice their religion while working in the sugar plantations of Natal. In the Hindu religion, in general, worship of Mother Goddess Parvati and Lord Sri Shiva is more common among Hindus of South Indian origin. Hindus of North Indian origin worship Rama, or in the case of

199. *Ibid.*, at 12.
200. *Ibid.*, at 13.
201. A. Taybob, *Islamic Resurgence in South Africa: The Muslim Youth Movement*, 12 Journal of Religion in Africa 104 (2014).
202. *Ibid.*
203. *Ibid.*, at 105.
204. *Ibid.*
205. *Ibid.*, at 106.
206. B. Brain, *Setting Down Roots: Indian Migrants in South Africa, 1890–1911* 12 (Wits University Press 1990).
207. *Ibid.*

Gujaratis, Vishnu.[208] However, it is important to note that there are many similarities among Hindus in South Africa, which ultimately make the Hindu community as a whole.

87. Deepavali is an important festival for the Hindu community in South Africa and is a major attraction of the city of Durban, where different festivals and other festivities are held. The city of Durban is home to the majority of people who follow the Hindu faith and way of life, and that is the reason why this city serves as a hub for the Hindu festival, in most cases.[209]

IV. Jewish Religion

88. The Jewish Religion in South Africa came as a result of the British colonisers in South Africa in 1669, and this was due to the fact that they came for economic opportunities in South Africa at the time, in particular with the discovery of diamond in Johannesburg.[210] Jewish religion is one of the minority religions in South Africa, and the people who subscribe to this form of religion are mainly stationed in the Western Cape province of South Africa.[211] The Jews, like the Christians and Muslims, believe that God created everything and that he is the Lord of the Universe. They further believe that they have an agreement with God as stated in the Bible in which God makes a promise to Abraham and his descendants that he will be their God and they will be his people.[212]

89. There are two Jewish groups: Orthodox and Reformed Jews. Orthodox Jews are the ones who follow the law strictly without any compromise and only worship in Hebrew.[213] While the Reformed Jews, who came into existence in 1800, apply the law with much flexibility, and not strictly. They also worship in Hebrew and make use of their indigenous language when they worship.[214] It is important to note that like all other minority religions in South Africa, the Jews experienced a measure of discrimination, but their discrimination was less hostile and cruel as compared to other minority religious groups like Muslims.[215] This was attributed to the fact that Jews belonged to a white privileged class who looked similar to the oppressors at that time in South Africa.[216]

208. *Ibid.*, at 13.
209. *Ibid.*
210. F.A. Hugh, *South African Jews and Apartheid*, 34 Patterns of Prejudice 23 (2000).
211. *Ibid.*
212. *Ibid.*, at 24.
213. *Ibid.*
214. *Ibid.*, at 25.
215. *Ibid.*
216. *Ibid.*, at 26.

V. African Traditional Religion

90. African beliefs and practices emanate from all aspects of life at the level of both the individual and society. One does not convert into this religion. Everyone in the African society is believed to be born into the religion and thus becomes an active member or participant of a spiritual journey, which originates from the spiritual world and eventually goes back to the spiritual world.[217] It is argued that African religion includes fundamental beliefs, which are common to all Africans, despite the fact that Africa consists of a multitude of nations, with complex cultures and languages.[218] Because of these commonalities based on religion, it is asserted that it must always be spoken in the singular. However, Kiernan holds a different view to the above assertions on the basis that African religion is pluralistic and must be considered like that and not in the singular. This is because it denotes all the indigenous religions in South Africa – not only the faiths of the various black African people but also faiths of people like the San and the Khoi.[219] The term 'indigenous religions' would of course not only apply to South Africa, but it would go beyond and also include the whole of the continent of Africa due to its diverse nature.

91. It is believed that African beliefs and practices originated from the spiritual world, and they were handed down to the physical world by the forbearers by word of mouth, by practical participation in ritual activities and by the day-to-day teachings of the elderly to the children.[220] There is no founder for African religion. This religion was revealed to the first generation, who were directly created by the supernatural power or God. This supernatural being gave to the first generation all laws and taboos on how to live in harmony with God, with other people and nature.[221] People were instructed to marry each other for reproductive purposes, and parents were tasked with the duty to lead their offsprings by example in a manner that will promote good moral values.[222] When the first generation died, their spirits joined the spiritual world where the Creator lives. This marked the beginning of ancestors who became the messengers of the Creator and the supervisors of the physical world.[223] The ancestors looked after the welfare of the living mainly through the elderly, who taught the youth orally and through rituals.[224] In this way, the religion was passed on from one generation to another generation. From time to time, the elders also revealed themselves to the living of all ages through dreams and visions.[225]

---

217. Mndende, *supra* n. 84, at 40.
218. *Ibid.*
219. *Ibid.*, at 41.
220. *Ibid.*
221. *Ibid.*, at 42.
222. *Ibid.*
223. *Ibid.*, at 43.
224. *Ibid.*
225. *Ibid.*, at 44.

92. African religion can be summarised in five elements according to the assertions that are raised by Oosthuizen. That African religion is a belief in a supreme being; belief in lesser Gods; belief in spirits; ancestral beliefs and the practice of magic and medicine each with its own cult.[226] According to Oosthuizen, these core elements relate with each other because African thinking and practice are always in terms of groups. Everything is part of a group and each individual only makes sense in as far as it is connected with everything else.[227] This would mean that the physical and the spiritual world are very closely connected in African religion.

§3. Conclusion

93. In conclusion, it is important to note that South Africa is made up of people with different religious and cultural beliefs, who make up the South African community, which is generally regarded as pluralistic in nature. Focus will now shift to the freedom and independence of religion and religious practices in South Africa on the individual, collective and organisational basis.

226. *Ibid.*
227. *Ibid.*, at 45.

Part II. Religious Freedom in General

Chapter 1. Individual and Collective Religious Freedom

§1. Introduction

94. Individual and collective religious freedom will be discussed jointly in this chapter. This is because both individual and collective religious freedom cannot be separated based on the following reasons. An individual does not practice a religious or cultural practice alone or in isolation with other people; t for that religious practice or culture to be recognised, it must be done in the collective with other people. This means that the individual freedom, which a person enjoys when it comes to religious and cultural freedom, forms a part of the collective in the practice of that particular religion or culture. Based on these reasons, this chapter will highlight the importance of religious freedom from an individual point of view, which in turn will result to a collective religious freedom, which the particular individual enjoys as protected by the Constitution.

95. Constitutional provisions, which aim to promote the freedom of religion for both an individual and the collective, will be discussed in this chapter. Furthermore, the extent to which a right protects the individual and collective freedom when it comes to religion and cultural practices is explored. The views of various authors in relation to freedom of religion for both an individual and collective are outlined for a comprehensive understanding of the constitutional protection of freedom to religion on both an individual and collective basis.

§2. Constitutional Protection of the Right to Religious Freedom on an Individual and Collective Basis

96. Despite the constitutional protection of religious and cultural freedom in South Africa, it is still difficult to simply define the freedom of religion and culture as a constitutional right. This is because it is much more of a concept that needs to be circumscribed. It is also a concept that can continue to develop in future due to the changing needs of the society. Section 15(1) of the Constitution provides that everyone has the right to freedom of conscience, religion, thought, belief and opinion.[228] It is understood that groups that observe distinctive cultural and religious practices are also guaranteed this freedom of religion and culture on a collective basis. However, they must ensure that their personal laws and customs do not violate the Constitution.[229]

97. This means that primarily it remains the task of the individual and religious groups of South Africa to identify those rights in a way that is consistent with their own religious identity and within the ambit of the Constitution and the laws of the country all the time. If the religions and religious people of South Africa do not accept this role and task it will be taken away from them by government, courts and society and it will be done in a way which will not necessarily further freedom of religion, and this can be returning to the previous era of colonialism and apartheid.[230] In the era of colonialism and apartheid religious freedom was not in existence as people were prohibited at some point to take part in a religious practice of their choice, and let alone they could not be in a position to choose the religion or cultural practice of their choice. Briefly, this means that in the new constitutional dispensation in fulfilling their task, churches and cultural groups must make sure of their deep roots and identity and make sure of the rights and obligations that spring from those roots and identity. They must also take note of the Bill of Rights as contained in the second chapter of the Constitution, as well as the Acts of parliaments, which further describe the content and application of the rights concerned.[231]

§3. The Extent of the Right to Freedom of Religion on an Individual and Collective Basis

98. While section 15(1) of the Constitution guarantees the right of every person to belief or hold an opinion,[232] subsection 15(3) of the Constitution draws a distinction between freedom of belief, on the one hand, and the right to practice such belief or religion on the other.[233] This section indirectly restricts the right of any person to practice his or her religion or culture. Section 15(3) (a) of the Constitution states that this provision does not prevent legislation recognising (i) marriages concluded under any tradition, or a system of religious, personal or family law; or (ii) systems of personal and family law under any tradition, or adhered to by persons professing a particular religion.[234] Section 15(1) (a) of the Constitution can best be described as a negative right as opposed to a positive right on the basis that there is no compulsion on the part of the legislature to legally recognise religious practices and religious law. Instead, the legislature has a choice not to pass any law at all that will recognise religious practices or law.

228. Section 15(1) of the Constitution.
229. Section 36 of the Constitution.
230. Mndende, *supra* n. 84, at 43.
231. Section 2 of the Constitution.
232. Section 15(1) of the Constitution.
233. Section 15(3) of the Constitution.
234. Section 15(3) (a) of the Constitution.

99. It is argued that the constitutional protection of freedom of conscience, religion, thought, belief and opinion only protects the religious belief, but it does not recognise the observance of the rules of religion, which is religious laws. What is actually not protected by this provision is the practice associated with the recognition of religious law. In most instances, belief and practice cannot be separated because belief leads to the practice based on that belief. It is absurd to have any other interpretation of section 15 of the Constitution other than to conclude that this provision, read with sections 15(2) and (3) of the Constitution, guarantees both the freedom of belief and the freedom to practice that religion.[235] This conclusion is based on two reasons: section 15(3) of the Constitution automatically prohibits any religious law or practice of religion that violates any provision in the Constitution. Although this provision seems unnecessary since it is a repetition of the supremacy clause of the Constitution as per section 2, its purpose seems to be to ensure that the legislature will not codify unconstitutional religious law.[236] The point is that this provision obviously does not affect the constitutional recognition of any religious belief or practice of religious law that is not contrary to the Constitution.

100. Furthermore, section 15(3) of the Constitution also seems unnecessary because it does not assist in the protection of religious freedom, but simply provides that the legislature may not be prevented by any other law from recognising systems of religious or personal law. That is, the correct interpretation of section 15 of the Constitution is that freedom of religious beliefs and freedom to practice any law associated with any such religion or belief should be regarded as protected by the Constitution.[237] Consequently, all aspects of religious law, such as Christian laws and Jewish laws among others, currently form a part or should be regarded as forming a part of the South African legal system where they are not inconsistent with the Constitution.[238] When interpreting the common law, courts should interpret section 15 of the Constitution to include both the right to freedom of belief and the right to practice that belief.[239] Any other interpretation will defeat the objectives of this provision. The current attitudes of the courts when interpreting section 15 of the Constitution is considered below. It will be exposed that courts are not in favour of adopting a liberal interpretation by protecting both belief and practice of that belief according to the principles of the relevant personal laws, even if those personal laws do not violate the Bill of Rights.

235. Section 15(2) and (3) of the Constitution.
236. Section 15(3) of the Constitution.
237. Section 15(1) of the Constitution.
238. Section 36 of the Constitution.
239. Section 39 of the Constitution.

§4. The Attitudes and Conduct of the South African Courts While Interpreting Freedom of Religion on an Individual and Collective Basis

101. If the above interpretation to freedom of religion in South Africa is correct, why do South African courts continue to refuse to legally recognise and enforce religious laws, in particular, to legally recognise religious marriages? The interpretation clause in the Constitution according to section 39(1)[240] obliges courts to promote the values that underlie an open and democratic society based on human dignity, equality and freedom and to promote the spirit, purport and objects of the Bill of Rights. For example, African and Muslim religions as well as customary law do not share the same prestige and protection that is afforded to civil law in matters that concern the protection of human rights. Cleary the courts have come to realise the inconsistencies in their application of the Bill of Rights pertaining to personal laws which include religious laws. Despite this realisation, the courts have constantly refused to legalise religious marriages. In *Ryland v. Edros*,[241] the court had to consider whether public policy depended on the views of the majority Christian religion under the current constitutional dispensation. The court in this case concluded by outlining that it is quite inimical to all the values of the new South Africa that the courts should only brand a Muslim marriage contract as offensive to public policy. It is offensive and a regress to the democratic South Africa, which is diverse and pluralistic in nature.[242]

102. Despite this observation, the courts still refused to give legal recognition to Islamic marriages. Similarly, in holding that the consequences of any contract resulting from an invalid Muslim marriage may be recognised under certain circumstances, the court in *Amod v. Khan*[243] held that the new Constitution ushered in a new value system based on equality, religious freedom and dignity as opposed to the *boni mores* of society.[244] In this case, Mahomed CJ held that:

> the insistence that the duty of support which such serious de facto monogamous marriage imposes on the husband is not worthy of protection, can only be justified on the basis that the only duty of support which the law will protect in such circumstances is a duty following from a marriage solemnised and recognised by one faith or philosophy to the exclusion of others. This is an untenable basis for the determination of the *boni mores* of society. It is inconsistent with the new ethos of tolerance, pluralism and religious freedom which had consolidated itself in the community even before the formal adoption of the interim Constitution on 22 December 1993.[245]

---

240. Section 39(1) of the Constitution.
241. *Ryland v. Edros* 1997 (2) SA 690 (CC).
242. *Ibid.*
243. *Amod v. Khan* 1947 (2) SA 432 (N).
244. *Ibid.*, at 433.
245. *Ibid.*

103. Furthermore, in this case the court expressly refused to recognise Islamic marriages. The reasons why courts in South Africa do not recognise and enforce religious marriages and laws were revealed in *Ismail v. Ismail.*[246] In this case, the court had to decide whether the consequences of a Muslim marriage, which is still legally unrecognised because it is regarded as potentially polygamous and therefore contrary to public policy, could be legally enforced upon, divorce.[247] Judge Van Reenen in this case held that:

> The marriage law of the State is rooted in Christianity whereas the union of plaintiff and defendant is based on a non-Christian faith. At the advent of Protestantism, marriage was secularised and passed under the control of the State. However, although as a result of the advent of Protestantism marriage was secularised, the countries of the Reformed Church retained the idea of its being of divine institution in its general origin.[248]

104. It is important to note that this case was decided ten years before the adoption of the Constitution and the Bill of Rights in South Africa, yet the courts have consistently refused to recognise religious marriage law other than Christian marriages, even in the new constitutional dispensation. Immediately after the Interim Constitution came into operation, the court in *Marais v. The Master* held that potentially polygamous marriages might contravene the gender equality principle on the following basis.[249] Apart from that, the principle of gender equality embodied in sections 8(2) and 119(3) in the Constitution and constitutional principles stated in section 232(4) may well lead to the conclusion that polygamous marriages are as unacceptable to the *boni mores* of the new South Africa as they were to the old.[250] Another problem as outlined above is that in *Amod v. Khan* the court consciously refused to deal with the issue of the validity of a Muslim religious marriage under the current constitutional dispensation, arguing that the duty of spouse support does not depend on the validity of a marriage.[251] The court held that, for the purposes of the dependant's action, the decisive issue is not whether the dependant concerned was or was not lawfully married to the deceased, but whether or not the deceased was under a legal duty to support the dependant in a relationship, which deserved recognition and protection as per common law.[252]

105. From the above assessments and court cases it is clear that courts have an important role to fulfil in order to ensure that fundamental rights are upheld and protected. This is unfortunately not yet the case as outlined, particularly when dealing with freedom of religion on an individual and collective basis. Focus from a constitutional perspective will now shift to a statutory law perspective when it comes to the position of freedom of religion on an individual and collective basis.

---

246. *Ismail v. Ismail* 1983 1 SA 1006 (W).
247. *Ibid.*
248. *Ibid.*, at 1007.
249. *Marais v. The Master* 1984 (4) SH 288 (D).
250. *Ibid.*
251. *Amod v. Khan, supra* n. 1 at 434.
252. *Ibid.*

§5. The Position of the Charter of Religious Rights and Freedom in Individual and Collective Freedom to Religion

106. The charter, as it was discussed above, as one of the instruments in South Africa that serves to promote and protect freedom of religion in line with the Constitution also puts provisions in place that ensure that the freedom of religion on an individual and collective basis is upheld. This includes making the following provisions without getting into the detail:

(a) The right to believe or not to believe.
(b) The obligations of the state with regard to religious rights.
(c) The right to observe and exercise one's religion.
(d) The right to education consistent with one's religious convictions.
(e) The right to maintain particular matrimonial, family and personal legal traditions.
(f) The right to institutional freedom.
(g) The rights and obligations of religion with regard to the laws of the land.
(h) The right of religion to freedom of expression.
(i) The right of religion to freedom of association.
(j) The right of religion to freedom of propagation.
(k) The right to religious dignity.
(l) The right of religion to solicit, receive, manage and spend voluntary financial and other forms of support and contributions.
(m) The right of religion to conduct upliftment and charity work in the community and to establish, maintain and contribute to charity and welfare associations, and solicit, manage, distribute and spend funds for this purpose.

107. All of the above in fact describes what freedom of religion on an individual and collective basis entails. It can also be said in the words of John Witte as he argues that freedom of religion includes:

(1) freedom of conscience;
(2) the free exercise of religion;
(3) religious pluralism;
(4) religious equality;
(5) the separation of church and state; and
(6) the disestablishment of religion by the state.[253]

108. From the above provisions it is apparent and clear that religious freedom on an individual and collective basis is firmly entrenched in the charter in conjunction with the Constitution as the supreme law of the land in South Africa. However, as already pointed out, from the perspective of the constitutional law and protection in relation to freedom of religion there is a problem with regard to the interpretation

253. Section 2 of the South African Charter of Religious Rights and Freedom.

and application of this provision. A liberal approach in connection with the interpretation and application of freedom of religion on an individual and collective basis is essential. It should not rely on *boni mores* as a limiting factor because the only way a fundamental right can be limited in a just and fair manner is through the application of section 36 of the Constitution as the general limitation clause.[254]

§6. Conclusion

109. Considering all the issues discussed above, one is bound to conclude that the South African courts have failed to promote the values that underlie an open and democratic society based on human dignity, equality and freedom.[255] Furthermore, the courts have failed to promote the spirit, purport and objects of the Bill of Rights by not recognising that religious personal law have become part of the mainstream legal system in South Africa. The courts have failed to protect the right of different people to practice their religion by not recognising that the right to freedom of religion includes not only the mental aspects of belief but also freedom to practice and have legal protection of the rules of that religion.

110. By favouring the rules of the Christian faith, the courts have unfairly discriminated against other religions. To rectify this mistake, and to deal with such problems in a free and democratic society based on freedom and fundamental human rights, the legislature must develop laws to reflect legal diversity and legal pluralism as contemplated in section 15 of the Constitution. Different laws have different functions to fulfil, functions for which they are best suited, and this view is supported because it includes or accommodates for diversity, which is the South African position. Other laws need not of necessity be in conflict with state laws; they may complement state laws and direct the development of state laws in harmony with basic values underlying non-state legal and social systems. Clearly, there is a need for recognition of legal pluralism concerning matters of private law. This would comply with the right to freedom of religion as guaranteed in the Constitution, considering the fact that everyone has a right to follow a religion of his or her choice, either as an individual or as a collective.

254. Section 36 of the Constitution.
255. Section 2 of the Constitution.

Chapter 2. Organisational Religious Freedom

§1. Introduction

111. Freedom of religion and culture as protected and enshrined in both the Constitution and legislation means that different religious organisations as well as cultural organisations have the freedom to practice and uphold their religion and culture. These organisations in the South African context refer to churches, synagogues and mosques due to the diverse and pluralistic nature of the South African community. Furthermore, it will also refer to the cultural groups or associations of the different indigenous communities and societies. The freedom to practice religion and cultural practices that is vested with the churches and cultural associations means that these organisations are free to determine how their churches and cultural groups must operate in realising their faith or belief in the process. However, this must be done in such a way that it does not infringe any rights or is not contrary to the Constitution as the supreme law of the land.

112. In this chapter focus will be placed on how faith-based organisations in South Africa are free to determine their administrative position in realising individual and collective rights to freedom of religion and culture.

§2. The South African Charter of Religious Rights and Freedoms

113. Section 9 of the charter provides that religious organisations are subject to the rule of law in South Africa, despite the freedom that is guaranteed to them for religion and cultural practice.[256] The charter accepts the reality that religious organisations exist and function within society, are therefore a part of society, and that in principle they are therefore subject to the law. As a matter of fact, a religious organisation carries a particular responsibility to teach its members good citizenship and the importance of observing the law in the interest of society as a whole for the common good.[257] A religious organisation may also not act in a way that is blatantly illegal, for example to force its members to perform acts or rituals that are physically harmful or may damage or destroy the property of others or to force minors to marry. This will be found to be contrary to the Constitution and such practices will be declared null and void to such an extent that criminal prosecution will have to be carried out.[258]

114. However, if a religious organisation, while exercising the rights guaranteed in the charter in a lawful way, feels it cannot in good conscience observe a particular law, it must be able to justify itself on the ground that it is inconsistent with its beliefs and convictions as guaranteed by the Constitution.[259] Sometimes a law may

256. Section 9 of the South African Charter of Religious Rights and Freedom.
257. *Ibid.*
258. Section 2 of the Constitution.
259. Section 15 of the Constitution.

even contain exemptions on religious grounds, precisely to accommodate those who disagree based on their beliefs or convictions. An example is the Civil Union Act, which provides that religious marriage officers must not be compelled to solemnise marriages between persons of same sex.[260] The approach followed in this provision does not imply that a religious institution or its leaders may break the law. The right to not observe a particular law, with which one disagrees, is not a licence for illegal behaviour.

115. Similarly, it is necessary to recall that diversity depends upon difference being respected and that law should not be used as a means of homogenising society. There are those who wish everyone to believe as they do on matters where we should allow differences of opinion and this is mostly the case in relation to terror attacks like the Boko Haram group in Nigeria.[261] Maintaining a space for difference of beliefs – sometimes on matters that are deeply held by those who believe otherwise, particularly in relation to political issues on the day and on matters like sexual orientation, gender rights or religious beliefs – however, is one of the marks of a mature and respectful society and a recognition of law understood within its proper bounds. Due to the importance of diversity in relation to freedom of persons, organisations and society itself, the onus should be on those who wish to challenge the presumption of diversity to show legal regulation and restriction is necessary and in the public interest. The hurdle for such a rebuttal should be significant since the dignity and freedom of difference is what is at stake.

§3. Independence of Religious Organisations in South Africa

116. Section 9(2) of the charter deals with the duty of the state to respect, protect, promote and facilitate the right to religious freedom of organisations in South Africa as protected by the Constitution.[262] The state must recognise the authority of religious organisations over their own affairs, and the state must govern justly, constructively and impartially in the interest of everybody. Recognition of a legal presumption in favour of diversity and the limits of law in relation to internal matters of religion alongside generally applicable principles of procedural fairness and natural justice are important to achieve justice, fairness and respect for difference within a constitutional order such as South Africa.[263] Developing legal recognition of how these principles interact is an aspect of the evolution of both legal and societal maturity and a rejection of the domination of the past.

260. Section 33 of the Civil Union Act 2004.
261. *Nigeria's Boko Haram*, https://www.bbc.com/news/world-africa-31784538 (accessed 9 Aug. 2018).
262. Section 9(2) of the South African Charter of Religious Rights and Freedom.
263. *Ibid.*

117. Furthermore, section 9(3) of the charter deals with the internal religious matters which are free from the state and the courts in as far as religious organisations are concerned.[264] In this aspect, the charter puts emphasis on the important consequence of the independence of religious organisations, in that the state, including the courts, may not regulate or prescribe matters concerning religious confessions, doctrines and ordinances. This actually reflects the long-standing principle in the law that the state should not make laws on religious matters, and courts should not decide cases of religious confessions, doctrines and ordinances.[265] This is because matters of religious doctrine, confessions and ordinances are not matters over which courts may rightly claim jurisdiction, as they have no expertise in relation to religious revelations. In addition, should courts delve into this area, they would diminish the independence that religious organisations are free and independent from the state interference. This will mean that there will be an overreach on the part of the courts because they would enter into an area of religious dogma, which is something beyond the proper purview of the law.

§4. Assessment of Religious Organisation's Freedom and Independence in South Africa

I. Christian Faith

118. Despite the challenges, which were outlined above when it comes to the interpretation of the freedom to religion by the courts in South Africa from a constitutional point of view. It is noted that churches are still able to develop their own rules and procedures in order to realise and enjoy the freedom of religion, which is guaranteed by the Constitution and the charter as outlined. The Christian faith is the majority in South Africa and comprises of different churches or organisations with different distinctive features and practices. However, despite their differences they share a common element or factor, which is they are all the followers of Christ who is the Saviour of humankind.[266] The head of the church will normally be the reverend or a pastor and will be assisted by a group of elders, which is normally called the church council. The church council will play a very important role when it comes to the day-to-day operations of the church and advice the minister or pastor on key strategic decisions which need to be taken or followed by the church for it to grow and be sustainable.

119. Because the Christian faith is the largest in South Africa, reference will be made in this study to one church organisation with the aim of showing the independence and freedom, which religious organisations enjoy in South Africa from the perspective of both a constitutional and statutory law. For example, the Methodist Church of Southern Africa is one of the churches that came to South Africa through

264. Section 9(3) of the South African Charter of Religious Rights and Freedom.
265. *Ibid.*
266. Holy Bible 2 Corinthians Ch 5: Verses 14.

the missionary work and intervention during the colonial era. This church as a religious organisation enjoys freedom and independence, which is upheld and protected by both the Constitution and the charter. The Methodist Church has its own Constitution as a church, which is Laws and Disciplines of the Methodist Church of Southern Africa (L&D). L&D prescribes how the church must be run on a daily basis and how congregants must behave in church and what they must wear as a way of representing the broader church.[267] Furthermore, L&D prescribes how office bearers are appointed and elected in the church and outline how these office bearers must execute their role and functions in the process.[268]

120. The development of an instrument like the L&D affirms that religious organisations are autonomous in South Africa and independent from state interference in realising a fundamental right such as freedom to religion and cultural practices. The same can be said with other churches in South Africa like Roman Catholic Church, Lutheran and Dutch Reformed Church among others. They are in a position to develop their own internal rules and processes in such a way that it will govern how they operate in South Africa in harmony and peace with other faith-based organisations.[269] This religious freedom, which the Christian faith enjoys despite their majority status in South Africa, does not mean that they are given a blanket clause to develop rules that are contrary to the rule of law in South Africa or to divide society by formulating discrimination rules. In such cases, the Constitution as the supreme law of the land will be the one that reigns supreme and such rule or practice of the religious organisation will be declared void.[270]

II. Muslim Faith

121. When it comes to the Muslim faith or religion, the mosques are considered as independent religious organisations and have rules and standards, which form part of the Muslim religion and tradition, which in a sense sets them apart from other religious organisations. In the mosques, normally a prayer leader is a person that is well respected among the congregants, and he leads a productive and honest lifestyle. Members of the congregants will vote for the prayer leader.[271] Since a mosque is regarded as a place of worship, cleanliness and respect is championed among congregants at all times with the aim of keeping the peace and stability in the mosque. Proper dress code in which long clothes that do not reveal any part of the body is prescribed, and women to further cover their heads is encouraged because anything that is contrary to this will be viewed as disrespectful.[272] As per

267. Z. Siwa, *The Laws and Disciplines of the Methodist Church of Southern Africa* 2 (Methodist Publishing House 12 ed. 2016).
268. *Ibid.*, at 3.
269. Section 15 of the Constitution.
270. Section 2 of the Constitution.
271. Taybob, *supra* n. 202, at 110.
272. *Ibid.*

the sitting arrangement in the mosque, women will sit behind men and that is the tradition of the Muslim religion since ancient times.[273]

III. Hindu Faith

122. Just like other faith-based organisations, Hindus have they own rules and traditions when it comes to observing their faith. The Hindus make use of the temple as a place of worship or assembly and study among other things.[274] The head priest, as the spiritual leader, will be the one who leads the prayers in most cases and when it comes to the sitting arrangement men and women would sit in separate areas. Before entering the temple, it is a standard practice that all congregants must remove their shoes as a sign of showing respect and honour in the temple.[275]

123. When it comes to reform, Hindu faith, which came into existence from the 1,800, holds a different view from the standard practice of the ancient Hindu-based doctrines and practices.[276] According to the reformed Hindu practice, they have introduced musical instruments and a choir to assist with the fellowship in the temple. Furthermore, the practice and rule that congregants when they enter into the temple must remove their shoes has been removed and is no longer in operation, and the rule that men and women must sit in separate places does no longer apply, which was the position in the ancient Hindu religion.[277] With the reformed Hindu faith, it is clear that the position in terms of their doctrine and rules, they follow a format that is similar to that of a church which is a place of worship for the Christian faith.

IV. Jewish Faith

124. In the case of the Jews, they make use of a synagogue as a place of worship to God. Like other religious groups, the Jewish religion is also autonomous when it comes to formulating rules and regulations in connection with how a member or congregant of the Jewish religion must behave.[278] Most of these rules or regulations emanate from the Jewish traditions or the way of life, for example if a boy child has been born, then such a child must go for circumcision on the eighth day of his birth for him to be welcomed to the Jewish community and religion.[279] This is attributed to the fact that it is in line with the convenant made with God and the people of Israel.

273. *Ibid.*, at 111.
274. Brain, *supra* n. 207, at 13.
275. Ibid., at 14.
276. *Ibid.*
277. *Ibid.*
278. Hugh, *supra* n. 211, at 30.
279. *Ibid.*

125. The Jews when they enter the synagogue there are also rules and procedures which they need to observe as a sign of showing respect and honour in the synagogue. For example, the congregants must wear a kippah, which traditionally only men are permitted to wear, but due to the reform, women can also wear it.[280] In addition, a prayer leader is also in charge in leading the congregants during prayer sessions as a Shepard of God.[281]

V. African Customary Faith

126. When it comes to African faith there is a common element between all indigenous groups or cultural affiliations. The most common denominator is the fact that there is a belief in the ancestors who are the messengers of God.[282] In the African faith ancestors play an important role in the lives of the indigenous community to such an extent that great respect is bestowed on the graveside. People must wear in an appropriate manner when they enter the graveyard; for example, men must wear a jacket and women must wear a dress or skirt but cannot wear a trouser, as that will be viewed as disrespectful and contrary to the public interests.[283] Furthermore, it is a standard practice among the indigenous communities that one cannot speak ill about the dead because they play an important role when it comes to the development and up keep of society.[284]

127. In terms of African faith it is important to bear in mind that the practice of appeasing the ancestors is key and this includes the ceremony of slaughtering a cow, goat, sheep and chicken as well as brewing African beer as all these will serve as sacrifices to the ancestors.[285] Sacrifices are made to pray for rain or to welcome a new season as a way of expressing gratitude, or requesting for intervention to the ancestors and God for prosperity and development in society. When performing African cultural practices or religion, normally these ceremonies will be performed at a certain place which is earmarked as sacred by the family or community which is similar to churches in the Christian faith or a synagogue in the Jewish faith.[286] The spiritual leader in the African culture will normally be a traditional healer who is the mouthpiece of the ancestors, and in some instances, the eldest member of the family can lead this ceremony. This can include the uncle or aunt who both play a vital role in African culture due to their bloodline and position in the cultural context.[287] This means that it is not everyone who can play the role of a spiritual leader when it comes to the aspect of African religion.

---

280. *Ibid.*, at 31.
281. *Ibid.*
282. Mndende, *supra* n. 84, at 18.
283. *Ibid.*
284. *Ibid.*, at 19.
285. *Ibid.*
286. *Ibid.*, at 20.
287. *Ibid.*

128. There are also processes and rules that need to be observed when it comes to practising African religion with the aim of ensuring that this religion is maintained and preserved for generations to come. For example, initiation of young boys is a cultural practice that is observed to mean that one moves from a stage of being a boy to the stage of being a man. During this practice only old men are permitted to perform and guide these young boys, and this practice happens in the mountains during the winter season and no woman is permitted to enter or be part of this practice. As such an interference by a woman during this cultural practice will be a taboo and a violation of this ancient and prestigious cultural practice of the indigenous community.

§5. Conclusion

129. Without a doubt in South Africa freedom of religion from an individual, collective and organisational perspective is upheld and protected by the Constitution and statutory instruments. However, there is still a problem with the interpretation of this fundamental right by the courts as outlined, but this does not take away the autonomy and independence which religious and cultural organisations enjoy as outlined above. The autonomy of the various religious and cultural organisations in South Africa, which make up the pluralistic society, can formulate rules and procedures that govern their operations with the aim of realising their right to freedom of religion and independence. This is not an absolute power, in the sense that these religious and cultural organisations can't do as they please and violate the rights of other people. Such conduct will be viewed as unlawful and void in terms of the Constitution as the supreme law of the land.

130. Religious organisations, despite their independence and freedom, are still subject to the rule of law and must comply and uphold the supremacy of the Constitution in their internal regulations and processes. Regarding the independence and freedom of religion and practice, from an organisational point of view, attention will now be on the recognition and position of religious societies in South Africa in the next chapters.

Part III. Legal Status of Religious Communities

Chapter 1. Formal Status and Autonomy of Religious Communities

§1. Introduction

131. In South Africa, various religious and cultural practices are recognised and upheld, which make up the South African community. As already outlined in the previous chapters, freedom of religion and culture are firmly entrenched and protected in South Africa from a number of vantage points, and this includes from an individual, collective and organisational perspective. The question that now arises is does this freedom and independence to choose religion and culture stretch beyond the above-mentioned constitutional perspectives to such an extent that it gives rise to the protection and promotion of a religious and cultural community in existence in South Africa.

132. This chapter aims to interrogate if there is a recognition of religious and cultural communities in South Africa and if so to what extent, bearing in mind the freedom that is afforded to religion and culture from an individual, collective and organisational point of view. Furthermore, it will be argued whether the freedom to choose and practice religion and culture from these different perspectives ultimately gives rise to the promotion and development of the various religious and cultural communities, and if so, are religious and cultural communities afforded the necessary legal status and protection by the Constitution. Furthermore, this chapter aims to interrogate the legal position and independence of religious and cultural communities in South Africa. The legal status and independence of religious and cultural communities cannot be separated because they relate with each other to a lesser or greater extent.

§2. Constitutional Position on Religious Communities

133. Section 31(1) of the Constitution states that persons belonging to a cultural, religious or linguistic community may not be denied the right, with the other members of that community: (a) to enjoy their culture, practice their religion and use their language; (b) to form, join and maintain cultural, religious and linguistic associations and other organs of civil society.[288] Furthermore, section 31(2) of the Constitution states that the rights mentioned in section 31(1) above may not be exercised in a manner that is found to be inconsistent with the provisions of the Bill of Rights.[289]

134. From the above assessment of this constitutional provision, it is clear that religious and cultural organisations can indeed form a part of a religious or cultural community and the common denominator in the formulation of that community or society will be the religion or culture. However, in the formulation of religious or cultural community through religion or culture it is important to bear in mind that such a religious or cultural community is still subject to the rule of law and the supremacy of the Constitution in South Africa. This means that despite the autonomy to formulate their own rules and procedures from an organisational point of view as well as from a communal point of view, for example, in setting out standards of living for the members of their community they are still subject to the rule of law. What this means is that a religious or cultural community cannot exercise its rights and powers in a manner that it undermines the Constitution to such an extent that the rights of other people who do not form part of that religious or cultural community are compromised. Hence section 31(2) of the Constitution serves as a cautionary clause to section 31(1) of the Constitution, which encourages religious and cultural communities to be formed.

135. It is important for the reader to bear in mind that despite the recognition and encouragement by the Constitution on the formulation of religious and cultural communities, it does not mean that these communities operate in isolation from the state. They are autonomous to such an extent to realise and uphold their religion and culture as protected by the Constitution.[290] This means that religious and cultural communities are not independent states or communities in South Africa who can operate in parallel to the state or to even compete with the state, but these religious and cultural communities form part of the government, and in executing their daily role are required to champion the interests of the government. The recognition of religious and cultural communities in South Africa, with their diversity, aims to bring and achieve the necessary social cohesion without paying attention to the difference or diversity that are identified in the various religious and cultural communities. Religious and cultural communities play an important role in building the society of South Africa because among other things, which the law cannot be, they are able to instil values and good manners on people to relate with each other in a respectful and dignified manner. They teach members values like respect, tolerance, compassion and solidarity, which form part of the African concept of *Ubuntu, botho*, as one of the values that underpin our Constitutional democracy.[291]

---

288. Section 31(1) of the Constitution.
289. Section 31(2) of the Constitution.
290. Section 31(1) of the Constitution.
291. G.J. Van Niekerk & L. Wildenboer, *The Origins of South African Law* 177 (UNISA 2010).

136. In the African culture, maintaining peace and social relations is important for the survival and growth of society in which emphasis is not placed on the individual but on the broader community as a whole.[292] This principle is embedded in the African concept of *Ubuntu,* which is a normative source of indigenous law and implies coexistence of human rights and freedoms, on the one hand, and collective duties, on the other hand. It places emphasis on the role of the individual as a member of the community.[293] Furthermore, *Ubuntu* implies a spirit of humanness and includes virtues such as compassion, forgiveness and human dignity.[294] *Ubuntu*, as an indigenous law source, was recognised in *S v. Makwanyane*,[295] where the Constitutional Court was of the view that in interpreting the Bill of Rights, indigenous values ought to be put to effect, and thus develop a specific South African understanding of human rights and indigenous law as a system of law in our country.

137. Religious and cultural communities in addition to instilling values and good behaviour on their members are also important in other factors. They hold a position of trust and confidence among their members. This means that they can be in a position to reach to their members with efficiency and help their members to deal with other personal and spiritual issues, which in most instances can fall beyond the powers of the state.[296] This relationship of trust cannot be underestimated because it is influential and gives power to the religious and cultural organisations. That is the reason why it is said that religious and cultural communities are obliged to promote the interest of the state to their members because they hold the position of trust and confidence.

138. Reference will now be made to the various religious and cultural communities that are recognised in South Africa in their diversity. Furthermore, the role and practice of these religious and cultural communities will be explored, and how they use their position of trust from the point of a religious or cultural community with the aim of ensuring that democracy in South Africa is strengthened and upheld will be interrogated.

§3. Recognised Religious and Cultural Communities in South Africa

139. South Africa is a multi-faceted and rainbow nation due to the various religious and cultural communities that make up the South African population. For this discussion, reference will not be made to all religious and cultural communities that are recognised in South Africa, but the discussion will be restricted to two of these religious and cultural communities. The reason for this limitation is on the basis that all religious and cultural communities share a common element, which is to bring

292. C. Gibbs, *Recent Developments Regarding South African Common Law and Customary Law of Succession* 23 (University of Pretoria Press 2015).
293. Van Niekerk and Wildenboer, *supra* n. 292, at 177.
294. *Ibid.*, at 178.
295. *S v. Makwanyane* 1995 (6) BCLR 6655 (CC).
296. *A Gross Abuse of a Position of Trust,* https://www.churchtimes.co.uk/articles/2016/9-december/comment/opinion/tackling-abuse-on-pitch-and-pulpit (accessed 17 Aug. 2018).

the necessary social cohesion through the process of instilling discipline and respect among its members and the broader community.

140. The Christian community, which is the leading religious domination in South Africa, will be discussed, and it will be outlined how this religious community share similarity with other religious communities that are recognised in South Africa. The cultural community of the indigenous people of South Africa who are governed by African customary law will be discussed, and the discussion about the different tribes or clans in terms of sharing the same identity like *Ubuntu* will be explored in the process.

I. The Christian Community

141. When it comes to the Christian community, it is important to note that South Africans are a very religious people. The question that arises is what is the influence and role of Christians in the South African society? In answering this question it is important to consider this important aspect because trust plays an important role in a democratic society as it consolidates the society and its institutions.[297] Christians are organised in churches as denominations and congregations, and churches appear consistently to enjoy the highest level of trust across the country, with four out of five South Africans indicating that they strongly trust churches.[298] Institutions that are at the bottom of the relative levels of trust in South Africa are the police and labour unions due to the levels of corruption and fraud, which is dominating particularly in state institutions.[299]

142. This places the church and its membership in an important position to have an influence on the South African society. The majority of the population of South Africans believe that there is a God, most members of the Christian community engage in prayers twice or three times a day on average and this is a practice that is instilled in young children and the broader Christian community. What is also championed in this community is honesty in dealing with the state, and this relates to tax compliance and accessing of social grants. In the case of tax compliance the Christian community is expected to encourage their members that in addition to paying their offerings to church as stated in the Holy Bible, they are still obliged to honour their commitment to the state and pay their taxes on time and with diligence.[300] Failure to do so will amount to a serious violation of both state laws and the Christian way of life in which honesty and integrity is central and fundamental. Furthermore, the Christian community also discourages its members from engaging in illegal activities of defrauding the state like accessing social grants even though they do no need it as this is an un-Christ like conduct and is infringing the laws of

297. *Churches and Government Likely to Abuse Trust,* https://www.benkhumalo-seegelken.de/suedafrika-aktuell/1753-church-and-government/ (accessed 30 Sep. 2018).
298. *Ibid.*
299. *Ibid.*
300. W.J. Schoeman, *South African Religious Demography: The 2013 General Household Survey* 4 Theological Studies 2 (2017).

the land.[301] These two examples show and confirm the relationship between the state and the Christian community in working together for ensuring that the necessary social cohesion is being maintained.

143. Christians hold strong orthodox views in relation to the Christian belief or doctrine, and they believe in God and in the Bible as the literal word of God. Jesus is the solution to all of the world's problems.[302] The Christian community has conservative views on support for capital punishment and oppose abortion. In this case, for instance, when it comes to capital punishment the state has taken the position of the Christian community on the basis that only God is the one that gives and takes away life and no man is empowered to take life.[303] This was deliberated in the landmark court case of *S v. Makwanyane*, post-democracy in which a discussion about the abolishment of capital punishment was considered and abandoned.[304] In this case, the highest court of the land held that capital punishment was inhumane, degrading, cruel and contrary to the provisions of the Bill of Rights. This practice was declared unconstitutional in South Africa because it was not in line with the spirit, purpose and objects of the Constitution.[305]

144. This court case and the abolishment of this practice in the South African context affirmed the relationship between the state and the Christian community and with the other religious communities in ensuring that capital punishment is abolished in the new democratic dispensation. The aim in this case was to ensure that the rule of law is upheld in South Africa and that the value of upholding and not taking life from the perspective of a religious community is realised at the same time. However, not all the time will the state and the Christian community agree on everything, for example as already mentioned abortion is prohibited in the Christian community but it is a practice that is recognised as lawful in South Africa.[306] Section 10 of the Constitution states that everyone has inherent dignity and the right to have their dignity respected and protected.[307] Furthermore, it is important to enquire what human dignity means. It is defined as how an individual or group feels, its self-respect and self-worth. Human dignity is concerned with physical and psychological integrity and empowerment and is harmed or violated by unfair treatment premised upon personal traits or circumstances, which do not relate to individual needs, capacities or merits. The right to human dignity is enhanced by laws, which are sensitive to the needs, capacities and merits of different individuals, taking into account the context of their differences.[308] The right to human dignity is harmed or violated when people are marginalised, ignored or devalued, and it is enhanced when laws recognise the full place of all individuals and groups within society.

---

301. *Ibid.*, at 3.
302. Holy Bible Hebrews Ch 2: Verses 14.
303. *S v. Makwanyane, supra* n. 296, at 6655.
304. *Ibid.*, at 6656.
305. *Ibid.*
306. Section 2 Choice on Termination of Pregnancy Act 92 of 1997.
307. Section 10 of the Constitution.
308. Schoeman, *supra* n. 301, at 4.

145. The legalisation of a practice like abortion on the part of the state is to realise the fundamental right, which is the right to human dignity, and bodily integrity, which a person enjoys in a democratic society. However, the state still encourages good behaviour and practices among its citizens, while at the same time bearing in mind that, for example, a young child can be raped and be pregnant and must be afforded the opportunity to terminate the pregnancy if they so choose. The socio-economic needs or circumstances can also be a factor when it comes to the termination of pregnancy and the right of the woman who chooses to do so must be respected and upheld as protected by the Constitution.[309] Another important aspect to be considered when it comes to the termination of pregnancy for its legalisation in South Africa is that it was attributed to the fact that many people, in particular young girls, were opting for illegal abortions which was not safe and this resulted in millions of lives lost in the process.[310] The state had to act in the public interest and to stop this practice from escalating and this resulted in the adoption of the Choice on Termination of Pregnancy Act in South Africa, with the aim of legalising adoption. It is clear that when acting in the public interest the state can differ with a religious community about a particular point of view and while at the same time ensuring that the rule of law is upheld and protected.

146. The foundation of Christian communities when dealing with the state is honesty and integrity with the aim of ensuring that the necessary social cohesion is upheld and achieved. The Christian community also instils values and principles to their members as to how they are expected to relate with each other and other people on a daily basis. These include the Ten Commandments in the Bible, which read as follows: You shall not have any other gods before me; you shall not make any idols; you shall not take the name of the Lord your God in vain; remember the Sabbath day, to keep it holy; honour your father and mother; you shall not murder; you shall not commit adultery; you shall not steal; you shall not bear false witness and you shall not covet.[311]

147. These values form part of the Christian community and other religious communities. It can be argued that these values like not to steal, lie, commit murder among others form part of the criminal justice system of South Africa and a person who commits these offences can either be imprisoned or fined. This is in addition to being condemned from a religious community on the basis that a person who does not uphold and respect the Ten Commandments is a sinner who needs the grace of God to change his or her ways. A religious community and the state work together in this context to ensure that the religious convictions are upheld, on the one hand, and the rule of law is protected, on the other hand, because the Constitution is the supreme law of the land. Furthermore, the Christian community also instils good conduct and good manners on children who must be taught to have discipline and respect for elders, and this is done in most cases by punishing the children when

309. Section 10 of the Constitution.
310. Schoeman, *supra* n. 301, at 5.
311. Holy Bible Exodus Ch 31: Verse 18.

they defy the Christian laws or values that are communicated in the Bible.[312] This practice or value of punishment through corporal punishment used to be the way of also maintaining discipline and order in schools; however, the laws have since changed to such an extent that corporal punishment is considered a crime in South Africa today.[313]

148. Meaning both the state and the Christian community and other religious communities are no longer singing from the same hymnbook when it comes to the topic of corporal punishment. A teacher or parent in South Africa today cannot effect corporal punishment on a child without being dismissed or suspended from employment as a teacher or arrested or warned as a parent in some cases.[314] This is attributed to the fact that currently in South Africa corporal punishment is abolished and considered a crime and teachers and parents are encouraged to adopt other disciplinary measures to punish their children other than to go for corporal punishment, in order to avoid being in trouble with the law.

149. The values and principles of the Christian community is without a doubt similar to the other religious communities because the fundamental value that is key in all religious affiliations and communities is honesty and respect. This is key in relation to how people in the religious community relate with each other and other people, and it is important to note that religious communities are independent and autonomous on the basis of the trust and confidence they share with their members as already outlined. However, the autonomy and independence of religious communities is to a lesser or greater extent limited because it is still bound by the laws and procedures of the state and cannot encourage their members to engage in practices that are illegal and unlawful because this will be contrary to the public interest and unlawful. Religious communities are statutorily bound to ensure that in realising their mission and objectives they promote the interests and position of the state. This is attributed to the fact that a religious community occupies a strategic position of trust with its members, and this relationship cannot be underestimated at all by the state. Focus will now shift to the cultural community in South Africa that is subject to customary law and how it operates in the new democratic dispensation where human rights are upheld and protected.

II. Cultural Community under African Customary Law in New Democratic South Africa

150. The painful past of South Africa during the apartheid and colonial era has left a devastating effect on the application and development of cultural communities who are governed by African customary law. This was because African customary law was viewed as inferior to the common law and this resulted in cultural communities to be seen as non-existent or irrelevant because the legal system that

312. *Ibid.*
313. Section 10 of the South African Schools Act 1996.
314. *Ibid.*

governed them was not recognised.[315] This left most, if not all, of the various cultural communities in South Africa under-developed and poor, which is currently the *status quo* in the rural parts or provinces of South Africa post the democratic dispensation in South Africa.[316]

151. The democratic government with the adoption of the Constitution as the supreme law of the land had a duty to remedy the injustices of the past, which is to ensure that African customary law is developed and applied, to such an extent that it is put on the same level with the common law. This meant that African customary law must apply in parallel with the common law and does not need the common law to validate it, which was the position during the colonial and apartheid era period. Now only the Constitution can validate and confirm this legal system. The aim was to ensure that cultural communities are also protected and are independent through the application of a recognised legal system, which in this case is African customary law. Cultural communities are under the rule and governance of traditional leaders, who assume the role on the basis of birth, and in the past only male children could assume this role; however, this position has since changed because it was found to be discriminating against women on the basis of gender. This means that today a woman can be a leader of a traditional community in South Africa with the aim of ensuring peace and stability in that particular society.[317] It is believed that since traditional leadership is acquired by birth, a traditional leader is the one who shares some links or connection with ancestors for the development and prosperity of that cultural community. This link or connection with the ancestors makes the traditional leader a spiritual leader of the community.[318]

152. The traditional leader runs the traditional community in consultation with the traditional council, which is normally a group of elders who play the role of advisor and counsellor for the traditional leader. This is to safeguard the traditional leader from engaging in arbitrary decisions, which can harm the community.[319] Cultural communities are governed by African customary law, which also outlines the roles and powers of a traditional leader in the community because the interest of all the members of the community are key rather than individual interests.[320] The role and functions of a traditional leader in a cultural community is to provide support and care to members of the community; to allocate land to members of the community for residential and agricultural purposes; to provide safety in case of war; to provide or mete out the administration of justice through indigenous courts in case there is a dispute between members of the community and to ultimately ensure that there is peace and stability in the entire community.[321] The traditional leader will be executing all these functions under African customary law, and the traditional leader will ensure that discipline and order are achieved to such an extent that corporal

315. Mndende, *supra* n. 84, at 23.
316. *Ibid.*, at 24.
317. Bekker & Rautenbach, *supra* n. 18, at 48.
318. *Ibid.*
319. *Ibid.*, at 199.
320. *Ibid.*
321. *Ibid.*, at 200.

punishment and fines can be imposed in cases where a crime is committed, such as stock theft or a delict like adultery.[322] However, if a traditional leader engages on corporal punishment to such an extent that the wrong doer is injured, it will mean that such a traditional leader must be charged with assault and imprisoned on the basis that corporal punishment has been abolished in South Africa.[323] A similar case was reported in 2015 of *S v. Dalindyebo* in which the Pondo king was engaged in activities of assault, attempted murder and arson in meting out punishment to one of his subjects and was sentenced to twelve years of imprisonment for such illegal acts.[324] This court case and the assertions raised above do confirm that even though a cultural community is recognised in South Africa, and operating under African customary law is not without a doubt an independent community that operates parallel with the state, such a cultural community is subject to the Constitution and the rule of law.

153. The legal subjects or people that belong to a cultural community are bound by the authority of the traditional leader who can be either a king or a queen, depending on the clan and the circumstances of each clan or cultural community. The customs and traditions of that particular community will govern their daily lives and activities, which will promote their way of life and well-being. The customs and traditions of a cultural community will include encouraging group interests rather than individual needs on the basis that the needs of a group build a community for the betterment of tomorrow.[325] Respect and tolerance for elders and other people in general, which is key for the well-being of a particular individual and the broader community, is fundamental and encouraged. This goes with the aspect on the observance and recognition of cultural practices such as appeasing the ancestors and praying for rain among other things.[326] Dressing in indigenous clothes like lion skin among other things for men, and to engage in lawful customary marriages which can be either monogamous or polygamous in nature with the aim of procreation in a lawful and honourable manner to ensure that growth and sustainability in a cultural community are endorsed.[327] One of the distinguishing features of cultural communities in South Africa apart from other communities is the principle of restorative justice. This means that members of the community who are in dispute with each other will have to negotiate and speak with each other to resolve their disputes in an amicable manner. The traditional leader who will play a vital role in resolving the dispute of the parties will champion the same approach because no retribution or prisons are condoned or permitted in cultural communities.[328]

154. Even though traditional communities find themselves under traditional rule and administration, they are autonomous and independent. However, these communities and the traditional leaders are assisted by the local government through the

322. *Ibid.*
323. Section 10 of the South African Schools Act.
324. *S v. Dalindyebo* (090/2015) [2015] ZASCA 144 (1 Oct. 2015).
325. Mndende, *supra* n. 84, at 24.
326. *Ibid.*, at 25.
327. *Ibid.*
328. Bekker & Rautenbach, *supra* n. 18, at 201.

various municipalities spread out in the country, with the aim of ensuring that service delivery, such as the provision of water and sanitation, is provided to the people.[329] This means that there is co-operation between traditional leaders and the state when it comes to the running of traditional communities, particularly in the area of service delivery and administration. This collaborative approach is something that has been exploited by both traditional leaders and the state because traditional leaders would edge and encourage members of their society to vote for a particular political party on the basis that it is working for the needs of the community. In return, the political party that has been voted in power can return the favour through the provision of gifts and donations to the relevant traditional leader and his or her community.[330] Furthermore, the relationship between traditional leaders and government has been strengthened today in South Africa because all traditional leaders receive some form of remuneration from the state on the basis of their position. This form of benefits and privileges make or put traditional leaders in a position where they are able to promote the interests of the state to all the members of their community because they are in a position of trust and power with their subjects in the community. A similar stance that was identified when discussing the position of a religious community above. The fact that a traditional community shares some links or benefits as provided by the Constitution does not mean that such a community can do as it pleases. It must ensure to promote the spirit, purpose and objective of the Bill of Rights in South Africa, meaning that the rule of law must be upheld and protected at all costs. For instance, traditional leaders who promote or engage in illegal activities by violating the rights of people in their communities under the pretence of culture will be prosecuted by the state. This was the case where traditional leaders are seen as oppressing and discriminating against women based on gender or in cases where the children are abused by making them to be married to old men. This is based on forced marriages; these practices are contrary to human rights and not in line with the provisions of the Constitution.[331]

155. Traditional communities are also connected with each other and other communities through the principle of *Ubuntu,* which is embedded in all traditional communities, irrespective of the clan or language a person comes from or speaks. In all indigenous communities the values of respect, honesty, sharing, caring, love, compassion and solidarity are key. This is attributed to the fact that focus on a group and not an individual is championed, and it is important when it comes to traditional communities on the basis that a group is the one that makes and develops a community rather than a single individual who must become part of a group. For example, the principle of sharing and group emphasis in a traditional community is found where the eldest member of a family is working and the other members are unemployed. The eldest member of the family will ensure that he or she shares his or her salary with the rest of the members of the family with the aim of empowering those who are unemployed.[332] Meaning, that the interests of the member of

329. Section 150 of the Constitution.
330. Section 151 of the Constitution.
331. Section 2 of the Constitution.
332. Bekker & Rautenbach, *supra* n. 18, at 201.

the family, in this case as the eldest that works, are not taken into account, rather the interests of the entire family members for the purposes of their survival and development for the continued existence of the family and the community in turn.[333] Group interests are affirmed rather than the individual interests, which are in most instances seen to be selfish and not a way of life when it comes to a traditional community. In this, it is without a doubt affirmed that sharing is caring, and it is practiced on a daily basis in traditional communities.

§4. CONCLUSION

156. South Africa recognises religious and traditional communities based on their religion or cultural affiliations. These religious and traditional communities are autonomous and independent, but their independence and autonomy are subject to the rule of law and the Constitution as the supreme law. Religious and cultural communities cannot do or say something that is contrary to the Constitution or which will violate human rights and even to put the state in danger. The next chapter will focus on the fundamental rights, which religious communities enjoy in South Africa as protected by the Constitution.

333. *Ibid.*, at 202.

Chapter 2. Fundamental Rights and Religious Communities

§1. Introduction

157. Since religious and cultural communities are part of the population and state of South Africa, it means that these religious and cultural communities are under the supremacy and rule of the Constitution. Religious and cultural communities are entitled to the various fundamental rights, which are enshrined and protected by the Constitution, in addition to the various rights and responsibilities that emanate from that particular religious or cultural community.

158. The provision of fundamental rights by the state to members of the religious and cultural communities emanates from the responsibility the state owes to the citizens of South Africa with the aim of ensuring that they lead a productive and dignified life. Furthermore, the state has an obligation in ensuring that it realises these rights based on the reasonable available resources at its disposal. Various fundamental rights are provided by the Constitution in chapter two, which is the Bill of Rights, and a discussion about these rights as well as the application and limitation of the fundamental rights that are protected in the Bill of Rights will be explored. From the above assessment, it is clear that religious and cultural communities enjoy fundamental rights like any other citizen of the Republic, and this is because a person who is part of a religious or cultural community does not cease to be a citizen of the Republic of South Africa by voluntarily being part of a religious or cultural community.

§2. Chapter Two of the Bill of Rights of the Constitution of the Republic of South Africa, 1996

159. The Bill of Rights is found in chapter two of the Constitution and section 7(1) of the Constitution states that the Bill of Rights is the cornerstone of democracy in South Africa. It enshrines the rights of all people in our country and affirms the democratic values of human dignity, equality and freedom.[334] Section 7(2) of the Constitution provides for the realisation of the rights contained in the Bill of Rights by providing that the state must respect, protect, promote and fulfil the rights in the Bill of Rights.[335] Furthermore, section 7(3) of the Constitution states that the rights in the Bill of Rights are subject to the limitation clause as stated in section 36(1) or elsewhere in the Bill of Rights.[336] The question that now arises is who does the Bill of Rights apply to, or to put it differently, the question should be who is bound by the provisions of the Bill of Rights as it enshrines fundamental rights to all the people of South Africa, irrespective of their origin, colour, religion and culture.

---

334. Section 7(1) of the Constitution.
335. Section 7(2) of the Constitution.
336. Section 7(3) of the Constitution.

160. Section 8(1) of the Constitution states that the Bill of Rights applies to all law and binds the legislature, executive, judiciary and all organs of state.[337] Furthermore, section 8(2) of the Constitution provides that a provision in the Bill of Rights binds a natural or a juristic person if and to the extent that it is applicable, taking into account the nature of the right and the nature of any duty imposed by that right.[338] For the application and implementation of the fundamental rights that are contained in the Bill of Rights, section 8(3) of the Constitution states that when applying a provision of a Bill of Rights to a natural or juristic person in terms of subsection (2), a court: (a) in order to give effect to a right in the Bill of Rights must apply, or if necessary develop, the common law to the extent that legislation does not give effect to that right; and (b) may develop rules of common law to limit the right, provided that the limitation is in accordance with section 36(1).[339] Lastly, when it comes to the application of the rights contained in the Bill of Rights, section 8(4) of the Constitution states that a juristic person is entitled to the rights in the Bill of Rights to the extent required by the nature of the rights and the nature of that juristic person.[340] It is important to note that the Bill of Rights makes provision to a number of fundamental rights, which cannot be discussed in detail for this study. Apart from the fundamental right, which is the right to freedom to choose religion or culture on the part of an individual or group as already discussed, the discussion here will be limited to the central and core fundamental right, which is close to the hearts of all the people of South Africa, irrespective of their religion, culture, colour or creed, and it is the right to access healthcare services. The reason for limiting the discussion to only the right to access healthcare services is because it is one of the precious and safeguarded rights due to its importance in preserving and maintaining life. It is dear to both religious and cultural communities in South Africa, despite the fact that religious and cultural communities do hold the view that life is given by God and he alone can preserve and take it. This point is not disputed at all and the discussion on the right to access healthcare must be seen as a way the government relates with both religious and cultural communities in fulfilling fundamental rights.

161. Section 27(1) of the Constitution guarantees everyone the right to access healthcare services, which include reproductive healthcare, and that no one may be refused emergency medical treatment.[341] The government has a duty to fulfil this obligation to provide healthcare services to the people based on reasonable availability of resources, including the private healthcare sector, to a greater or lesser extent, particularly in emergency medical treatment.[342] This means that if healthcare practitioners deny people medical treatment or emergency medical treatment, their actions would amount to a violation of the right to access healthcare as protected by the Constitution.

---

337. Section 8(1) of the Constitution.
338. Section 8(2) of the Constitution.
339. Section 8(3) of the Constitution.
340. Section 8(4) of the Constitution.
341. Section 27(1) of the Constitution.
342. Section 27(2) of the Constitution.

162. Violating the protection of socio-economic rights, enshrined in the Constitution, goes against the purpose for which these rights were designed because chapter two of the Constitution is aimed at addressing the imbalances and injustices of the past. Further, to unfairly and unjustifiably deny or violate socio-economic rights constitutes a contravention of section 7(2) of the Constitution, which states that the government must respect, promote, protect and fulfil the rights in the Bill of Rights.[343] This provision clearly outlines the duty of government not only to refrain from interfering with the enjoyment of these rights, but also to promote, protect and enhance their realisation as part of its constitutional mandate.

163. The debate as to whether socio-economic rights are justiciable or not has long been settled in the South African context, evident from landmark court judgments, such as *Soobramoney v. Minister of Health (KwaZulu-Natal).*[344] This case will be discussed later in this chapter.

164. In this chapter, the focus will not be on the justifiability of socio-economic rights, but rather on the role of the courts in protecting socio-economic rights such as the right to access healthcare services. Brand argues that socio-economic rights can be enforced by the courts in two ways: first, through their lawmaking powers of interpreting legislation and developing the rules of common law, and second, the courts ought to adjudicate constitutional and other challenges experienced in state procedures that are intended to advance socio-economic rights.[345] The interpretation stated by Brand on the role of the courts in enforcing socio-economic rights is in line with sections 8(2)–8(3) of the Constitution, when it relates to the application and enforcement of the rights in the Bill of Rights.[346]

165. Treatment Action Campaign v. The Minister of Health[347] serves as a fundamental and instrumental case to the provision of section 27 of the Constitution. In this case, the applicants sought an order against the Minister of Health and the healthcare authorities of all nine provinces, except the Western Cape. The order requested was to compel the Minister and all healthcare authorities in the eight provinces to implement an effective national programme to prevent or reduce mother-to-child transmission of the HIV and Acquired Immune Deficiency Syndrome (AIDS) virus, which would include the dispensing of Nevirapine to pregnant women who are HIV positive, and to their babies. Nevirapine is described as an antiretroviral drug that can prevent or inhibit the spread of HIV.[348] The applicants in this case described to the court the alarming rate of mother-to-child transmission and the impressive reduction that would be achieved by the administration of the

343. Section 7(2) of the Constitution and *See* B. Bekink, *Principles of South African Constitutional Law* 81 (LexisNexis, Johannesburg 2012).
344. *Soobramoney v. Minister of Health, Kwazulu Natal* 1998 (1) SA 765 (CC).
345. D. Brand *Introduction to Socio-Economic Rights in the South African Constitution in Brand D and Heyns C Socio-Economic Rights in South Africa* 1 (Butterworths, Durban 2005).
346. Section 8(2) of the Constitution.
347. *Treatment Action Campaign (TAC) v. The Minister of Health* 2002 (4) BCLR 356 (T).
348. *Ibid.*, at 357.

drug.[349] It was claimed in court that a single dose of the drug can reduce this form of transmission by 50%, but this claim was disputed by the respondents who maintained that the success rate was probably considerably lower.[350] The respondents further in general contended that immediate implementation of antiretroviral treatment in these cases would be impossible because of insufficient financial resources.[351]

166. Judge Botha concluded, with regard to section 27 of the Constitution, that a countrywide prevention programme of this kind is an ineluctable obligation of the state.[352] Moreover, Nevirapine would be affordable if the programme is properly planned. What was required in the opinion of the court was a plan striving towards a comprehensive coverage and implementation of Nevirapine. The court accordingly made an order to this effect.[353] The court also made it clear that to order an immediate roll-out of the drug would be impossible and not practical, and as a result of this view, it ordered the respondents to report back to it within three months regarding the status of planning for the implementation of the roll-out of the drug. In this case, the court ruled in favour of the applicants.[354]

167. This decision of the high court was confirmed by the Constitutional Court,[355] as it reasoned that government policy, which provides for the availability of the Nevirapine drug in only certain areas of public hospitals, was inflexible and unconstitutional. Government was ordered to roll-out the administration of Nevirapine nationally for preventing mother-to-child transmission of HIV.[356]

168. Furthermore, section 27(2) of the Constitution provides that government must take legislative and other measures within its reasonable resources to achieve the progressive realisation of each of these rights.[357] What constitutes 'reasonable resources' will be determined on the basis of the facts of each case. Despite the fact that the right to access healthcare services is constitutionally enshrined, it is still important to note that considerable inequalities exist with regard to access to healthcare services. These inequalities are largely caused by the discrepancies found in healthcare resources and barriers to access health, such as vast distances, high travelling costs from rural areas, long queues and disempowered patients. As already discussed colonial past had a negative effect on the rural landscapes of South Africa not only from a religious and cultural point of view but also from an access to healthcare perspective.[358] For example, in *Soobramoney v. Minister of Health*

349. *Ibid.*
350. *Ibid.*, at 358.
351. *Ibid.*, at 360.
352. *Ibid.*, at 361.
353. *Ibid.*, at 362.
354. *Ibid.*, at 364.
355. *Ibid.*, at 365.
356. *Ibid.*
357. Section 27(2) of the Constitution.
358. M. Chersich, *Inequalities in Access to Health Care in South Africa*, 32 JPHP 3 (2011).

(KwaZulu-Natal),[359] the court dealt with the issue of unavailability of healthcare services to a patient who was suffering from kidney failure, but had been denied treatment in the relevant public hospital. The conduct of the hospital was found to be justified due to the lack of resources and in line with the provisions of the Constitution.[360]

§3. Public and Private Healthcare in South Africa

169. In South Africa, similar to most developing countries, the provision of healthcare services is divided into two sectors, namely the private healthcare sector and the public healthcare sector. In South Africa, these two systems of healthcare treatment were established during the apartheid era and are still sustained in the modern-day democratic South Africa.[361] In order to contextualise the two systems of healthcare, it is necessary to provide a brief history relating to how these two systems of healthcare were formed during the apartheid era. During apartheid, the quality and nature of healthcare services were based on three factors, namely race, location and income.[362] Homelands were formed with the aim of formally separating people in different areas, and thus the same principle was applied to healthcare services, which were inadequately and poorly administered to the majority of black people in the homelands.[363]

170. During apartheid, the church played a significant role in the administration of healthcare services to the majority of the people, due to the failure of the then government in providing healthcare services to the people in the public sector setting.[364] This shows that religion did not only play a major role in the political struggle of South Africa but went beyond in such a way that healthcare services were provided by religious groups at some point in South Africa. The formation of the private sector came into existence as a result of the economic downturn and the pressure which was placed on the government by the medical industry, resulting in the government having to regulate the healthcare sector.[365] Through the regulation of the healthcare sector, the private healthcare sector rose as an institution of healthcare services. How healthcare is administered in both the private and public sectors will be discussed now.

---

359. *Soobramoney v. Minister of Health, supra* n. 345, at 765.
360. *Ibid.*, at 766.
361. S. Beck, *Health Policy, Health Services, and Cancer Pain Management in the New South Africa*, 17 JPSM 17 (1999).
362. D. McIntyre, *The Health and Health System of South Africa: Historical Roots of Current Public Health Challenges*, 374 Lancet 817 (2009).
363. S.M. Tollman & K. Kautzky, *A Perspective on Primary Health Care in South Africa* 2 SAHR 20 (2008).
364. *Ibid.*, at 21.
365. C.D. Naylor *Private Medicine and the Privatization of Health Care in South Africa* 11 Social Science & Medicine 1153 (1998).

I. Private Healthcare Services

171. It is estimated that only 15% of the South African population has access to private hospital treatment.[366] In private hospitals, for example, cancer and other health-related treatments are found to be effective and functional.[367]Such efficiency and effectiveness are achieved by means of readily available drugs and the presence of state of the art medical equipment.[368] Private healthcare system is accessed by those with access to insurance or medical aid funding, or those who can afford to pay for it directly from their own pockets.[369] Private healthcare treatment, compared with the public healthcare treatment accounts for the largest share of total healthcare financing, which comprises of medical aid schemes and private out-of-pocket payments made by patients.[370] Private healthcare treatment is generally viewed as the best available, and it is seen to be the most effective with regard to cancer treatment and other chronic medical conditions.[371]

172. However, there are challenges with regard to private healthcare system, as it is generally expensive to access, and thus is not easily accessible to the indigent who mostly are part of the religious and cultural communities of South Africa who need the quality of care offered in private institutions.[372] As a consequence, some medical aid schemes are not willing to pay for cancer treatment and other chronic medical conditions in private institutions, due to the high cost associated with treatment in the private healthcare system. One example of this situation is the case of a woman who died in 2007 of breast cancer because her medical fund, Global Health, refused to pay for her treatment.[373]

---

366. *Ibid.*, at 1153.
367. *Ibid.*, 1154.
368. *Ibid.*
369. Chersich, *supra* n. 359, at 102.
370. J. Akazili & J. Ataguba, *Health Care Financing in South Africa: Moving Towards Universal Coverage*, 28 CEM 75 (2010).
371. *Ibid.*, at 76.
372. *Ibid.*, at 77.
373. *Ibid.*, at 78. The Competition Tribunal commented on the state of healthcare in South Africa, and how the private healthcare system has become a lucrative business. The provision of adequate healthcare to all the citizens of the country is clearly an important plank in the government's efforts to tackle poverty and inequality. High and middle income earning South Africans (and this would include a significant proportion of those in employment) receive healthcare through South Africa's sophisticated private healthcare system comprising the full gamut of general practitioners, specialists, hospitals and pharmacies. Private healthcare system is funded by an array of medical schemes serviced by the administration companies, data processing companies and managed care companies that are an integral part of South Africa's sophisticated 'first world' private healthcare system. However, majority of the population (and this includes a significant number of those in the lower reaches of formal employment) rely on the public healthcare system for meeting its needs. The reality, and possibly the only agreed certainty in the fraught debate surrounding the provision of healthcare in South Africa, is that the private healthcare system, and most notably, although not exclusively, the private hospital network, is characterised by significantly excess capacity while the public healthcare system is simultaneously resource-constrained and increasingly unable to cope with the demands which it faces. A major thrust of the government's efforts to improve healthcare provisioning is thus to utilise the excess capacity in the private healthcare system, to reduce the

173. Section 27 of the Constitution also binds private hospitals to provide medical treatment to patients during an emergency situation.[374] In addition, section 5 of the National Health Act supports the position that both private and public healthcare establishments cannot deny anyone emergency medical treatment.[375] Pieterse argues that the constitutional obligation placed on both private and public hospitals to provide emergency medical treatment is a direct and immediate entitlement.[376] The question as to what constitutes emergency treatment has been left open by the Constitution. This has had serious financial implications on private hospitals as they are obliged to render emergency medical treatment to people who cannot guarantee the necessary payment.[377] Since the question of what constitutes emergency medical treatment has been left open by the Constitution, it is assumed to be a question of fact, which will be decided on a case-by-case basis. However, in the landmark case of *Soobramoney v. Minister of Health (KwaZulu-Natal)*,[378] discussed in detail below, the court defined emergency medical treatment as a 'sudden catastrophe which calls for immediate attention'.[379] The broad and flexible nature of this definition is supported because it caters for all forms of healthcare emergencies without excluding the other.

174. Due to the seriousness of the obligation to provide emergency medical treatment, and the inconsistencies of some private hospitals in implementing and realising this socio-economic right to healthcare services, specific legislation was enacted, which deserves to be mentioned. Item 3(b) of the Schedule to the Promotion of Equality and Prevention of Unfair Discrimination Act 4 of 2000 reinforces the obligation imposed on private hospitals to render emergency treatment to a patient by prohibiting the discriminatory refusal of emergency treatment.[380] As a consequence, a person living with cancer in an emergency situation may be admitted to a private hospital to receive treatment, and unfortunately, that particular hospital will be permitted to bill the patient for the services which they have rendered without taking into account whether the patient can afford the bill or not.

175. The obligation of private hospitals to admit patients in emergency situations is further stressed by the Hospital Association of South Africa; this association has incorporated a draft code of ethics for private hospitals across the country.[381] In the draft code of ethics for private hospitals, emergency treatment is defined as

---

demands on the public healthcare system, to move a strata of those presently reliant on public healthcare system over to the private healthcare system.

374. Section 27 of the Constitution.

375. Section 5 of the National Health Act.

376. M. Pieterse, *Enforcing the Right Not to Be Refused Emergency Medical Treatment: Towards Appropriate Relief*, 18 Stellenbosch Law Review 75 (2007).

377. M. Pieterse, *Legislative and Executive Translation of the Right to Have Access to Health Care Services*, 10 Law Democracy and Development 243 (2010).

378. *Soobramoney v. Minister of Health, supra* n. 345, at 766.

379. *Ibid.*, at para. 36.

380. Item 3(b) of the Schedule to the Promotion of Equality and Prevention of Unfair Discrimination Act 4 of 2000.

381. Proposed Code of Ethics for Private Hospitals – Hospital Association of South Africa *Private Hospital Review 2008: Examination of Factors Impacting on Private Hospitals* 67 HASA 44 (2008).

'treatment that is necessary to stabilise an emergency medical condition'.[382] The code requires private healthcare establishments to render such treatment ethically without any form of discrimination and subject to reasonable compensation. This definition of emergency medical treatment as provided for by the Hospital Association of South Africa is in line with the definition that was provided by the court in the *Soobramoney* case, because both definitions cover all forms of medical emergencies.[383]

II. Public Healthcare Services

176. Public healthcare treatment in South Africa is administered to 75% of the population, which is largely located in urban and rural areas, including religious and cultural communities.[384] The public healthcare treatment of cancer and other related illnesses is characterised by inadequate treatment facilities, less or unavailable drugs, overcrowding of patients, unqualified staff and abuse of patients by staff.[385] The public healthcare treatment system is often seen as the exact opposite of private healthcare treatment in the country. For example, persons living with breast cancer have to wait for weeks before undergoing surgery in the Groote Schuur Hospital situated in Cape Town,[386] due to the unavailability of funds from the National Department of Health. Not enough staff is available and in addition to this, defective, as well as old equipment, is used to assist patients.[387] Currently, in Durban there has been a shortage of oncologists to an extent that one oncologist is responsible for thousands of cancer patients, resulting in poor healthcare afforded to persons living with cancer in need of urgent healthcare.[388] National Department of Health has admitted that there is a shortage of oncologists across the country, and this is surely a call for concern when it comes to the treatment of cancer and other chronic conditions in South Africa.[389] If this problem is not resolved as soon as possible it means that millions of persons living with cancer in South Africa, who rely on public healthcare system, will either lose their lives or will be denied access to healthcare due to a shortage of oncologists in the country. This will be contrary to the fundamental right to healthcare as enshrined in the Constitution. These obstacles, in addition to the physical pain that persons living with cancer experience, cause a lot of distress for them and their families, and are unfortunately beyond their control.[390] Furthermore, this lack of healthcare services leaves people with only one hope, which is prayer to God for a divine intervention.

382. Clause 1 of the Proposed Code of Ethics for Private Hospitals 67.
383. *Soobramoney v. Minister of Health, supra* n. 345, at para. 36.
384. Falkson G and Beck S *Prevalence and Management of Cancer Pain in South Africa* 94 Pain 76 (2001).
385. Chersich, *supra* n. 1 at 103.
386. Falkson and Beck, *supra* n. 1 at 76.
387. Falkson and Beck, *supra* n. 1 at 77.
388. *Ibid.*
389. Chersich, *supra* n. 1 at 104.
390. Patijn J *High Prevalence of Pain in Patients with Cancer in a Large Population-Based Study in the Netherlands* 132 Pain 312 (2007).

177. Proper measures need to be taken to avoid or prevent the government from disregarding the realisation of the right to access healthcare services for all South Africans. However, this has proven to be an impossible task in the case of cervical cancer, where the screening process has already perpetuated and reinforced inequality with regard to the availability of such facilities.[391] This is because cancer-screening facilities are available only to a select few in South Africa. For instance, women in rural areas who live far away from clinics are deprived of this service, while those located in cities and urban areas, situated closer to hospitals and clinics, will have better access to these healthcare services.

178. Around 3,800 women, who mostly rely on public healthcare treatment, die of cervical cancer every year in South Africa.[392] This is a high percentage, which points to a potential problem in the public healthcare system. The Cancer Association of South Africa has stressed the importance of making screening programmes widely available for women for purposes of early detection of any cancer in order to reduce the number of deaths caused by cancer.[393] Patients making use of public hospitals may compel the hospital to provide treatment to them, but this may be limited, as the state will provide the treatment based on its available resources in accordance with section 27 of the Constitution.

§4. *Soobramoney v. Minister of Health (Kwazulu-Natal)* 1998 (1) SA 765 (CC)

179. This case involved a patient who was in need of dialysis treatment on a weekly basis, as he was suffering from kidney failure and needed treatment in order to survive.[394] The state argued that there were not enough facilities to offer this treatment as it was expensive.[395] The state, however, would provide treatment based on available resources in order to preserve the life of the patient in question.[396] The court further reasoned that the decision of the Minister of Health was made in good faith and through rational decision-making, found to be in the best interest of the public, and therefore did not warrant any interference by the court.[397] The court ruled in favour of the Minister and refused to compel the hospital to act in the manner that the patient anticipated, but it confirmed that the decision of the hospital in question was based on its available resources in its provision of treatment, and as expected according to section 27 of the Constitution.[398]

---

391. Denny L, Kuhn L and Batra P *Utilisation and Outcomes of Cervical Cancer Prevention Services Among HIV-Infected Women in Cape Town* 100 SAMJ 39 (2010).
392. Express C *More Services Needed to Fight Cancer* 23 UFS 8 (2008).
393. *Ibid.*, at 9.
394. *Soobramoney v. Minister of Health, supra* n. 345, at 1697.
395. *Ibid.*
396. *Soobramoney v. Minister of Health, supra* n. 345, at 1698.
397. *Soobramoney v. Minister of Health, supra* n. 345, at 1699.
398. *Ibid.* In *Government of Republic of South Africa and others v. Grootboom*, (CCT11/00) [2000] ZACC 19; 2001 (1) SA 46; 2000 (11) BCLR 1169 (4 Oct. 2000), the facts involve the eviction of informal residents of a privately owned land which was reserved for the construction of low-cost

180. Based on the provision of section 27(2) of the Constitution and the *Soobramoney* case, one can deduce that the right to access healthcare services, as guaranteed in section 27(1) of the Constitution, is not absolute, in the sense that it is dependent on the availability of reasonable resources.[399] As to what really constitutes reasonable resources surely needs a precise definition. The vagueness of reasonableness in this context entails that a patient's fate in a public healthcare establishment would depend on the available resources, without the hospital being held liable for a lack thereof. The public healthcare establishment may argue that it acted within the limits of its reasonable resources to save the life of the patient. The issue of reasonable resources in the provision of healthcare services is a form of a qualified step to the provision of healthcare services, requiring closer investigation as to what it entails, and the effect it has on the overall right to healthcare services as provided for in the Constitution.

181. The first step in addressing the vagueness of the concept of reasonable resources as stated in section 27 of the Constitution is to accept the position that was followed in *Government of Republic of South Africa and others v. Grootboom.*[400] In this case, the standard of reasonableness was adopted to test the relevant legislation regarding its fulfilment or realisation of socio-economic rights. The case is relevant for this discussion since the right to have access to healthcare services is also a socio-economic right and forms the point of discussion at present. According to the court, any measures which are adopted with the aim of fulfilling socio-economic rights ought to encompass a standard that is reasonable in both its conceptualisation and implementation.[401] This simply means that there must be a clear allocation of tasks and responsibilities to different stakeholders in government, and appropriate financial and human resources must be made available in order to realise socio-economic rights. This reasoning of the court regarding the duty of government to fulfil socio-economic rights is in line with the International Covenant on Economic, Social and Cultural Rights (ICESCR), of which South Africa is a member.[402] This international human rights instrument states that in order for a state party to be able to attribute its failure to meet at least its minimum

---

houses. The residents opposed their eviction by the government on the basis that the government was obliged to provide them with alternative or temporary shelter as enshrined by the Constitution, until they acquired permanent accommodation with their offsprings. The court held that the government was obliged to provide temporary or alternative shelter, on demand, to children and parents and in instances where parents are unable to accommodate their own children. The court further reasoned that this duty by the government to provide shelter to the people is independent and the state must take reasonable legislative and other measures within its disposal to realise this goal. In this case the court concluded by outlining that the state was obliged to provide temporary accommodation to the evicted dwellers without taking into account the aspect of availability of resources; this was due to the fact that there was a crisis and an emergency. The appeal court, however, was in disagreement with the court of *a quo* when it came to this decision. *Government of the Republic of SA v. Grootboom, ibid.*, at 1169.

399. Section 36(1) of the Constitution.
400. *Government of the Republic of SA v. Grootboom, supra* n. 399, at 1169.
401. *Ibid.*
402. ICESCR Adopted and opened for signature, ratification and accession by General Assembly Resolution 2200 A (XXI) of 16 Dec. 1966 and came into force on 3 Jan. 1976 in accordance with Article 27 of the UN.

core obligations to a lack of resources, it must demonstrate that every effort was made to use all resources that are at its disposition to satisfy, as a matter of priority, those minimum obligations.

182. The most important aspect which the court highlighted in the *Soobramoney* case is that legislation or policy which is aimed at the realisation of socio-economic rights must be transparent and communicated effectively to all the parties concerned, in order for it to be effectively implemented without confusion. However, this is unfortunately the current state of affairs particularly in South Africa's public healthcare system.[403] Pieterse argues that the non-fulfilment of the right to access healthcare services for patients in public healthcare system is not only attributed to inadequate legislation and healthcare policy, but it is also due to failures to translate and interpret the right to its correct meaning and aim.[404] Pieterse's view is correct, as this problem has arisen in the context of determining what constitutes reasonable resources as provided for in section 27(2) of the Constitution.[405]

183. The major challenge in both private and public hospitals in South Africa is the shortage of qualified staff. This challenge results in a frustration of the relationships between patients and medical personnel.[406] In the context of cancer and other chronic medical conditions, the shortage of staff will result in the unavailability of necessary resources and support to deal with the demands of the oncology unit.[407] The problem of shortage of staff is not only a challenge in South Africa, between the private and public healthcare centres, but affects the entire African continent, and it has become a major factor in crippling healthcare systems on the continent.[408]

184. In Africa, 10,000 medical professionals graduate each year, but within five years, about half of them leave for Europe in search of greener pastures.[409] Most African graduates leave their countries due to low salaries, restricted career development options and work overload, which is a common feature in the African healthcare systems.[410] Change is urgently required, which may be done by providing career counselling in medical schools and professional institutions, as well as the provision of mentorship programmes to students. Offering incentives to generate an interest and commitment among young medical professionals in order for them to stay in Africa and not go abroad is also a mechanism that may possibly alleviate the problem.[411]

403. *Government of the Republic of SA v. Grootboom, supra* n. 399, at 1170.
404. Pieterse, *supra* n. 378, at 251.
405. *Ibid.,* at 252.
406. M. Venter, C. Venter & K. Botha, *Cancer Treatment in South Africa: A Narrative Literature Review* 22 JPA 462 (2012).
407. *Ibid.*, at 463.
408. *Ibid.*
409. I.F. Adewole, *Challenges and Opportunities in Cancer Control in Africa: A Perspective from the African Organisation for Research and Training in Cancer* 14 Series 147 (2013).
410. *Ibid.*.
411. *Ibid.*, at 148.

§5. Conclusion

185. From the above assessment, it is clear that members of a religious or cultural community too suffer from different chronic diseases like cancer, and they experience challenges in accessing healthcare services. Access to healthcare services as protected and enshrined in the Constitution is key to the survival of religious and cultural communities as well as the broader public of South Africa. This is despite believing in God or ancestors, the right to healthcare services as protected in the Constitution cannot be underestimated but must be championed and realised for the continued existence and health of the South African population. Furthermore, government must work hard to ensure that this right is realised and due to the current state of public healthcare system in South Africa, a collaboration between the private and public healthcare sectors is encouraged. This will be seen as the first step in addressing the challenges that are faced by the public healthcare sector, and it will be an opportunity for the private and public healthcare sectors to learn from each other. The aim in this learning experience will be to enhance the healthcare system of South Africa, which without a doubt will be in the public interests.

Chapter 3. Contractual Religious Freedom

§1. Introduction

186. The practice and freedom of religion or culture is a voluntary process in the sense that the individual is the one that chooses to follow or to subscribe to a particular religion or culture on their own volition. This means that one cannot be pushed or forced to take part in a particular religion or culture against their will in South Africa because this will be viewed as contrary to the freedom to choose a religion or culture of one's choice as expressly guaranteed and protected by the Constitution.

187. Since the practice and participation of religion and culture is a voluntary process, which the individual person expresses to a particular religion or culture, the question that now arises is whether or not a contract exists between the religious freedom of an individual and a particular religion or culture? If there is such a contract, then it is considered as the horizontal application of rights or an equal relationship as per the Constitution. Furthermore, it can be asked if there is a relationship between a religious or cultural society and the state, it can be classified as the vertical application of rights and is normally an unequal relationship between the state and its citizens. The aim of this chapter will be to explore the relationship between an individual and a religious society and to further explore the relationship between a religious community and the state in line with the Constitution as the supreme law of the land.

§2. Individual and Collective Contractual Religious or Cultural Freedom

188. An individual and the collective, which normally is a group or community, enjoy the freedom to choose their own religious or cultural beliefs as protected by the Constitution.[412] However, does this freedom to choose religion or cultural beliefs on the part of an individual or collective amount to a contract with that religion or cultural belief, and the answer to this question is yes on the following basis. The fact that religious and cultural beliefs are entered into voluntarily without any pressure on the part of an individual or collective creates a form of a contract between the religious or cultural belief and the particular individual or group. For example, a person who chooses to subscribe to a certain religious and cultural belief also agrees to the personal laws that come with those religious or cultural beliefs.[413] This means that an individual or the group take the responsibility to uphold and protect the personal laws or codes which stem from that particular religion or cultural

412. Section 15 of the Constitution.
413. *Personal Law,* https://www.merriam-webster.com/dictionary/personal%20law (accessed 11 Jun. 2018).

belief while at the same time the religious or cultural organisation aims to uplift and make the individual or group to grow spiritually and to uphold their human dignity in the process.[414]

189. One can say that the relationship or contract that exists between a religious or cultural belief with an individual or group is reciprocal in nature and this emanates from the fact that the individual or group is bound to uphold the dignity and reputation of that religious or cultural belief. In turn, the religious or cultural belief has the responsibility to not discriminate or violate the rights of an individual or group as their members.[415] This means that, for example, a religious or cultural belief which an individual or group has chosen voluntary cannot discriminate that individual on the basis that he is gay, as this type of conduct by either the religious or cultural belief will be contrary to public policy, and it will be in breach of contract between the individual concerned and that religious or cultural belief.[416] In South Africa it is important to mention that there has been lots of disputes which are in connection with the discrimination of congregants who are either gay or lesbian and who have been excluded by the church on the basis of their sexual orientation. This conduct by the various churches was found to be discriminatory and in breach of contractual religious freedom between an individual and the church and in most cases the church was called on to correct and stop engaging in practices that are discriminatory and contrary to the provisions of the Constitution as the supreme law of the land.[417] The same can be said about an individual or group that belongs to a religious or cultural belief and engages in conduct that tarnish or put the religious or cultural belief into disrepute. In this case the conduct of either the individual or collective will be discouraged and in some instances the individual or group concerned will be put through disciplinary proceedings with the aim of restoring the integrity of that particular religious or cultural belief.[418]

190. The relationship between an individual and a group with a religious or cultural affiliation is based on a horizontal application as provided by section 8(1) of the Constitution.[419] This type of relationship is equal between the parties and the same can be said with the contractual relationship between an individual and a religious or cultural affiliation in terms of exercising the freedom to religion and culture. An individual can hold a religious or cultural affiliation accountable for acting contrary to the law and *vice versa*, in cases where the terms of their contractual freedom to religion or culture are not upheld but violated. By outlining that this is an equal relationship it means that both an individual and a religious or cultural affiliation are on equal footing, and no one exercises more power on the other but are equal in the exercise of their role and function in realising this contractual freedom to religion as enshrined and protected by the Constitution.

414. Section 10 of the Constitution.
415. Section 9 of the Constitution.
416. Section 9(1) of the Constitution.
417. *Ibid.*
418. *Personal Law*, https://www.merriam-webster.com/dictionary/personal%20law (accessed 12 Jun. 2018).
419. Section 8(1) of the Constitution.

191. Voluntariness lies at the heart of the contract between the religious or cultural affiliation and the individual or group. This in the sense means that in realising the contractual freedom to religion or culture the state cannot compel its citizens to follow a certain religion or culture because this will be contrary to the requirements for a valid religious or cultural freedom as provided by the Constitution.[420] Furthermore, this implies that the state must embrace diversity and the various religious and cultural practices in South Africa without interfering or prescribing the type of religion or culture the people must follow. It must be left to an individual or group to exercise their right. Where it is found that the state has entered into a space of pushing people to a certain religion or culture at the exclusion of other religions or culture, such conduct will be a violation of contractual freedom to religion and a violation of the right to freedom to choose religion or culture freely as protected by the Constitution.[421]

192. The same applies to parents. They cannot compel their children to follow a certain religion or culture, but they can play a role in ensuring that their children grow up in a particular religion or culture. This does not mean that the children are bound to follow the religion or culture in which they grew up, or be compelled to follow it once they are old. Children once they reach the age of majority, which is 18 years in South Africa, they are considered to be old and have the liberty to choose the religion or culture they would want to be associated with.[422] As long as that particular child is now a major and there is no other factor that serves as a deterrence towards affecting the mental capacity of that child, then such a child is also entitled to voluntarily choose his or her own religion or culture without the interference of parents or community.[423] In turn, the child is expected to abide by the rules or procedures of that particular religion or culture he or she has chosen, without undue influence or pressure. Mental illness, intoxication and provocation are some of the factors which are used to determine the capacity or state of mind of an individual and to measure if a person is able to draw a distinction between right and wrong.[424] These factors will aid when determining the voluntariness of a person to subscribe to a particular religion or culture.

193. Schools and workplaces are also duty bound to respect the voluntary aspect of an individual in relation to contractual freedom to religion or culture. Meaning a school or workplace cannot compel a student or an employee to subscribe to a particular culture or religion that is dominating in that particular school or workplace. Space must be provided for diversity and tolerance in the workplace or school with the aim of ensuring that the contractual religious freedom which an individual has chosen is not violated as protected by the Constitution.[425] Reasonable accommodation is one of the factors that needs to be adopted in the workplace and schools with the purpose of ensuring that diversity and tolerance are embraced in such a way that

---

420. Section 15(1) of the Constitution.
421. *Ibid.*
422. Section 12 Children's Act.
423. Section 12(1) Children's Act.
424. Section 14 Children's Act.
425. Section 15(1) of the Constitution.

no religious or cultural practice is discriminated against or undermined irrespective of its popularity and dominance in South Africa.[426] The principle of reasonable accommodation is more of a labour law concept and its meaning, purpose and value in the South African labour context will be explored extensively later on when discussing whether ministers qualify as employees in South Africa.

§3. The Relationship Between the State and Religious or Cultural Affiliations in Realising Contractual Freedom to Religion

194. The state has an important role to play in the relationship between religious or cultural affiliations and individuals or groups, despite the fact that such a relationship is private and voluntary in nature. Government's role in this relationship is based on the principle of the vertical application as provided by section 8(1) of the Constitution which is an unequal relationship.[427] This relationship is unequal on the basis that the state is the one that exercises more power on the individual and the individual is the one that must comply with the rules and procedures that have been set by the government in this regard.[428] An example of this form of relationship is the paying of taxes and an individual who defaults on paying taxes will be held liable for non-compliance, and it is a criminal offence not only in South Africa but across the globe.

195. In this unequal relationship the role of government is that of regulator and overseer, in how religious and cultural affiliations relate with their members. This means that the government ensures that people are not abused or degraded under the pretence of religion or culture because it has been proven on a number of occasions that culture and religion can be used to the detriment of people. The same can be said in South Africa in which the Commission for the Promotion and Protection of the Rights of Cultural, Religious and Linguistic Communities (CRL) has on a number of occasions exposed the inhumane and degrading practices of charismatic churches in South Africa, in which people are expected to eat grass and snakes, as part of religion.[429] Such practices are seen as contrary to human rights, such as the right to dignity among other things and pastors who were caught subjecting their congregants to these practices were arrested and prosecuted for violating the laws of the Republic.[430]

426. Section 15(1) of the Employment Equity Act 55 of 1998 deals with reasonable accommodation and encourages employers to embrace diversity in the workplace with the aim of protecting the rights of people, including the right to choose religion or culture.
427. Section 8(1) of the Constitution.
428. Section 8(2) of the Constitution.
429. Section 181 of the Constitution provides for the establishment of the CRL Commission which is a chapter nine institution and draws its mandate from the Constitution as the supreme law and this commission aims to safeguard the abuse of religious and cultural practices to the detriment of the people of South Africa.
430. Section 10 of the Constitution.

196. Currently, in South Africa as one of the measures to ensuring that religion and cultural practices are not abused which has been the case of late. The state through the CRL makes sure that churches and cultural institutions comply with the law. Furthermore, churches and cultural organisations are now expected to disclose their finances, church practices and culture with the aim of ensuring that the people of South Africa are not subjected to unnecessary abuse under the pretence of religion or culture which has unfortunately been the norm. The regulatory conduct of the state in this context is very much clear that it aims to uphold and protect the contractual freedom to religion which exists between an individual or a group and a particular religion or culture, while at the same time without any abuse being the order of the day.

197. Regulation on the part of the state does not take away the power or autonomy of an individual or group and the particular religion or culture, but it aims to ensure that the constitutional provision to freedom of religion is protected and preserved and that the rule of law is upheld all the time. This is attributed to the fact that the Constitution is the supreme law of the land and any law or conduct found to be contrary to the Constitution must be declared null and void.[431] However, the regulatory and supervisory role on the part of the state is not the only factor that is necessary for the preservation of the contractual freedom to religion or culture. The state has a working relationship with religious and cultural organisations by providing them with the necessary support and finances among other things. Furthermore, the state regards religious and cultural organisations as non-profit organisations and provide tax exemptions to these organisations in the interest of the public of South Africa and also to realise the freedom to religion or culture as protected.[432] In addition to these concessions by the state on religious and cultural organisations, these organisations are also duty bound to ensure that they comply with the rule of law and that they promote the government and the supremacy of the Constitution.

198. This means that the relationship between an individual and a particular religious or cultural group is reciprocal, and the same can be said about the relationship between the state and a religious or cultural organisation. For example, the state cannot prescribe the internal processes of a religious or cultural organisation, and a religious or cultural organisation cannot engage in conduct that put the state at risk or violates the human rights of its members, as this conduct will be contrary to the law and breach of the relationship of the two organisations.[433] The state and religious or cultural organisations work hand in glove with the aim of ensuring that the contractual freedom to religion an individual or group enjoys is realised and protected.

199. Since the state is also bound by the Constitution it ensures that it does not put in place policies or structures that serve as deterrence on people and religious or

431. Section 2 of the Constitution.
432. Section 21 of the Companies Act.
433. Section 8(1) of the Constitution.

cultural organisations to realise the contractual freedom to religion or culture. Government is duty bound to take its mandate from the Constitution and not exercise its powers in an arbitrary manner in its role of overseeing and regulating in this important relationship between an individual or group and a particular religion or culture. One can also argue that the overseeing role of government can be limited to a lesser or greater extent by section 36(1) of the Constitution, in cases where it is found that government is abusing its position.[434]

§4. Conclusion

200. In conclusion, from the observations above it is clear that the contractual freedom to religion or culture emanates from the principle of voluntariness. One must subscribe to a particular culture or religion freely without undue pressure or fear from anyone or the state for that matter. This chapter and the previous chapter were based on the domestic position of South African law, when it comes to the position and championing of the right to freedom of religion and culture. In the following chapters focus will shift to the international perspectives when it comes to religious law.

434. Section 36(1) of the Constitution.

Part IV. International, Transnational, Regional Effects on Religious Communities

Chapter 1. International Law Effects on Religious Communities

§1. Introduction

201. From the position of domestic law in the recognition of religious communities to the international perspective in the recognition and upliftment of religious communities in the international platform. Since the local position in South Africa has been explicitly outlined in terms of giving recognition to religious communities and they are regarded as independent communities, which enjoy autonomy from state intervention or interference in terms of their day-to-day operations and personal laws. However, as it was discussed, this independence is subject to the rule of law and the supremacy of the Constitution.

202. In this chapter, the focus will be on the role of international law in relation to the recognition of religious communities in ensuring that these religious communities are protected and preserved apart from the domestic laws that are in place as per each jurisdiction. Reference will be made to the relationship between international law and religion as a first point of departure and then the role of international law as well as international tribunals like the International Criminal Court (ICC) play in this particular sphere. Various court cases and also legal instruments on an international scale will be explored in this chapter, with the aim of highlighting the importance and role of international law towards religious communities.

§2. Relationship Between Religion and International Law

203. There is a close relationship between religion and international law because both of them aim to achieve social cohesion in different parts of the world. That is the reason why it can be argued in this case that religion has never been completely exiled from international law, but has always been part of international law venture.[435] Indeed, religion has played and continues to play a significant role in the

435. D. Evans, *Religion and International Law: Friends or Foes?*, 32 SOAS 638 (2010).

evolution of international law even though this relationship is often perceived to be complex and controversial for different reasons. On the one hand, the controversy surrounding this relationship may be attributed to the apparent differences in the nature of religion (sacred) and that of international law (secular).[436] Evans argues that the place of religion in the international legal system, or any legal system that purports to be secular, is likely to be controversial and complex.[437] On the other hand, religion and international law are identical in other ways. Generally, both religion and international law are important social phenomena that relate respectively to fundamental social issues in human society, which have often stimulated passionate disagreement about their proper content and functions.[438]

204. Both can be seen as systems of social ordering, as ethical or normative regimes or semi-autonomous social fields. Also, both religion and international law can be politicised and manipulated by elites, which adds further to the complexity of their relationship.[439] Owing to this complexity, the relationship between religion and international law can be analysed from different perspectives, depending on one's objective into the enquiry between religion and international law.[440] Based on the close relationship that exists between religion and international law, the question is whether religion and international law are friends or foes, and this question will be answered by the discussion of the various theories that aim to describe best the relationship that exists between religion and international law in ensuring that social cohesion is attained universally. One of the points that emerge is that there exists an old and ongoing relationship between religion and international law, which would be very difficult, if not impossible, to dissolve.[441] The strings of that romance are however in the hands of people, as human agents for both religion and international law, and they can be pulled for better or for worse.[442] This is where both a religious community and international law agencies need to pull together to ensure that social cohesion is maintained and preserved. Neither the separationist theory, which argues that religion and international law must be separated, nor the accomodationist theory, which is based on the idea that religion and international law must work together and if there is a conflict between the two then a compromise needs to be achieved, brings the best out of this relationship between religion and international law.[443] The strict adherence to the separationist theory portrays an adversarial relationship between religion and international law, or even promotes such a relationship and it closes the eyes of society to the positive contributions that religion has made, and could still make, to enhance and legitimise modern international law.[444]

---

436. *Ibid.*, at 638.
437. *Ibid.*, at 639.
438. *Ibid.*
439. *Ibid.*, at 640.
440. *Ibid.*
441. F.A. Boyle, *Foundations of World Order: The Legalistic Approach to International Relations 1898–1922* 1 (Duke University Press 1999).
442. *Ibid.*, at 2.
443. *Ibid.* .
444. *Ibid.*, at 3.

205. The accomodationist theory portrays a harmonistic picture between religion and international law, but its idealistic tendency may lead society too naively to disregard areas of tension which need careful consideration and resolution for the benefit of our common humanity.[445] This leads to the conclusion that the relationship between religion and international law should be objectively perceived as double-edged and capable of being either amicable or adversarial, depending on what society and human agents of the relationship between religion and international law want it to be. Meaning that the relationship between religion and international law is not always smooth sailing, but there are conflicts between the two systems and dialogue is key in ensuring that the said conflicts are resolved through a process of accommodation and negotiation among other things.

206. On the one hand, the importance of international law will continue to be felt as the world becomes globally smaller and nation states and people become more interdependent.[446] On the other hand, the fact that strong religious convictions exist throughout the world means that religion will continue to have significant impact on international law in each of their identified levels of interaction. The growing interest and wave of academic writing and court cases on the subject is a strong indication in this regard and is often ignored or overlooked at the international level.[447] Baderin argues that if international law is to achieve its aim of developing a legal framework that emphasises humanity and dignity in today's world, then international lawyers and academics can no longer afford to ignore the importance that religion plays for many people and many societies, simply because it can be predicted that it will continue to play that role in future.[448]

207. It would be for the benefit of both religion and international law and for the enhancement of humanity, if all, as human agents of both systems, strive to promote a friendly and harmonistic relationship between religion and international law rather than a belligerent or adversarial one.[449] This can be achieved by not downplaying the areas of difference between the two systems, but rather accentuating those areas of common ground and there must be a commitment to resolving possible areas of tension with good faith and with the principle that everything can be discussed.[450]

208. The principle of good faith is one of the implied terms found in all modern-day contracts, including the contract of employment between the employer and employee. According to the implied term of good faith, parties to the contract are

445. *Ibid.*
446. *Ibid.*
447. *Ibid.*, at 4.
448. S. Baderin, *From International Law of Christianity to the International Law of the Global Citizen* 3 JHIL 183 (2001).
449. *Ibid.*
450. *Ibid.*, at 184.

bound by everything which good faith reasonably and equitably demands.[451] This generally means that the parties to a contract will be acting in good faith and will not without reasonable or proper cause conduct themselves in a way that will destroy or seriously damage the relationship of trust and confidence between them.[452] Focus will now shift to the position of international law in relation to the protection and recognition of religious communities.

§3. The Position of International Law on Religious Freedom

209. International law just like domestic law has taken steps to guarantee the broader right of freedom to religion. The Universal Declaration of Human Rights (UDHR) states that all individuals have a right to freedom of thought, conscience and religion, including the freedom to change his or her religion or belief, and freedom, either alone or in community with others and in public or private, to manifest his or her religion or belief in teaching, practice, worship and observance.[453] This statement was incorporated into Article 18 of the International Covenant on Civil and Political Rights (ICCPR), which champions for the freedom of religion of an individual and collective and promotes the independence of religious communities in the international context.[454]

210. Though not embodied explicitly in other areas of international law, this right to free exercise of religion has been incorporated by reference into both the Genocide Convention and the Refugee Convention. Both Conventions include religious minorities among the kinds of groups that are protected by the convention. For example, the Genocide Convention criminalises acts committed with the intention to destroy, in whole or in part, a religious group or community.[455] While the Refugee Convention defines refugee as anyone who owing to a well-founded fear of being persecuted for reasons of religion is outside the country of his or her nationality or is unwilling to avail himself to the protection of that country.[456] While the Genocide and Refugee Convention protect minority groups, the 1970 United Nations Educational, Scientific and Cultural Organisation (UNESCO) Convention protects religious property through the protection of cultural heritage.[457] It is important to note that though the right to freedom of religion is codified in international conventions, it is unclear how this obligation is enforced through international courts.

---

451. M. Van Jaarsveld, *An Employees Contractual Obligation to Promote Harmonious Relationships in the Workplace – When are the Stakes too High? Some Pointers from the Judiciary*, 19 SA Mercantile Law Journal 205 (2007).
452. *Ibid.*
453. Article 18 UN UDHR 1948.
454. Article 18 ICCPR 1976.
455. Article 13 Genocide Convention 1948.
456. Article 2 Refugee Convention 1951.
457. Article 28 UNESCO Convention 1970.

211. For example, the Permanent Court of International Justice (PCIJ) dealt with the question of minority rights three times, in the case concerning the rights of minorities in Upper Silesia in 1928, in the advisory opinion concerning Greco-Bulgarian Communities in 1930 and in the court's advisory opinion concerning minority schools in Albania, issued in 1935.[458] In all three cases, the PCIJ established the obligation of countries to ensure de facto equality for minority religious communities. This de facto right included the right of a community to maintain its cultural identity, language and religion and to educate its children in its own schools.[459] In the case concerning the Rights of Minorities in Upper Silesia in 1928, the PCIJ was called to adjudicate a dispute between Poland and Germany arising out of a provision in the German-Polish convention which allowed minority religious communities in both countries to choose the language of instruction for their children.[460] The convention specifically applied to religious and linguistic minorities in Upper Silesia, a piece of territory divided between the two countries. The dispute arose because Polish authorities decided that around 7,000 children who had been enrolled in German-language schools in Upper Silesia were invalidly registered as religious or linguistic minorities and therefore had no right to attend schools established for German students under the convention.[461] The PCIJ dealt with this case under the Geneva Convention, which guaranteed every person the right to declare their affiliation to a linguistic or religious group. This is because the individual had the right to declare or not his or her affiliation to a minority religious or linguistic group, and his or her government did not have the right to review this declaration.[462] This standard was not designed to allow anyone to claim affiliation with any group solely to get a benefit. Instead, whether an individual belonged to a minority community and therefore is entitled to a benefit under law is a question of fact and not intention.[463] While a government may require an individual to provide some kind of proof of his stated affiliation, a government may not decide for itself which community an individual belongs to. The PCIJ applied this standard to Polish government's actions, saying that the protections of the Geneva Conventions were designed to avoid any kind of internal discrepancies within a government about the identification and classification of an individual.[464]

212. The concern was that this kind of discrepancy would lead to an individual being deprived of a benefit that he was entitled to as a member of a minority group. While this prohibition may lead to some individuals unjustly benefiting from a system designed to protect minorities, a government has to take this risk.[465] By reviewing and questioning the decisions made by individual parents to enrol their children

458. C.F. Amerasinghe, *The Permanent Court of International Justice and the International Court of Justice: Contentious Jurisdiction* 46 (International law E Books 2008).
459. *Ibid.*
460. *Ibid.*, at 47.
461. *Ibid.*
462. *Ibid.*
463. *Ibid.*, at 48.
464. *Ibid.*
465. *Ibid.*, at 49.

in minority schools, the Polish government violated the Geneva Convention's mandate.[466] In reaching this decision, the PCIJ treated religion as both a personal and collective right. Under the PCIJ's reasoning, religious minorities have the right to educate their children as they see fit. As such, minority communities have the right to establish institutions like schools and places of worship and this right is a collective one. In contrast, the right to freedom of religion is a personal one, since individuals have the right to decide which group they belong to. The individual, not the community or the government, has the right to be identified as a member of a minority community and to choose where and how his or her children are educated.[467] In this case and the other cases mentioned above that are dealt by the PCIJ in connection with the protection of religious communities in the international context, the PCIJ adopted a hybrid conceptualisation of the right to freedom of religion.[468] The court upheld this model of minority rights in 1930, in its advisory opinion concerning the rights of minority groups in Green and Bulgaria.[469] There, the court considered the obligations imposed by the two states by a bilateral reciprocal emigration treaty. As part of its decision, the court held that this communal right to religion, minority communities had the right to maintain their traditions and way of life.[470]

213. Unlike the PCIJ, the International Court of Justice (ICJ) has no explicit mandate to consider and adjudicate questions related to minority treaties or treaties related to cultural rights. As a result, the ICJ has seen a limited number of cases related to cultural rights and adopted a narrow approach when it comes to the interpretation of that particular right to religious communities.[471] Even though the ICJ has a narrow approach and limited jurisdiction when it comes to the recognition and promotion of religious communities, the role of both the PCIJ and ICJ can be summarised as follows as international organs which aim to protect religious communities. Though the PCIJ and ICJ have only considered the rights of minority religious groups in a handful of contexts, both courts have suggested that the right to freedom of religion is both an individual right based on an individual's conscience and a communal right, honoured as a way to protect a population's traditions and a way of life.[472] However, unlike its predecessor, the ICJ's rulings on this question has been narrower and to the detriment of religious communities in the international context. As such, the ICJ's jurisprudence has provided little actual protection to minority religious groups around the world, including those in states under the ICJ's jurisdiction. Litigation centred on protecting the cultural and religious rights of minority populations has largely taken place in domestic courts, meaning minority populations remain vulnerable legally.[473]

466. *Ibid.*
467. *Ibid.*
468. *Ibid.*, at 50.
469. *Ibid.*
470. *Ibid.*, 51.
471. G. Zyberi, *The International Court of Justice and the Rights of Peoples and Minorities*, 14 Series 2 (2013).
472. *Ibid.*
473. *Ibid.*, at 3.

§4. Religious Rights and Communal Protection by International Criminal Law

214. Unlike PCIJ and ICJ, the cases before international criminal tribunals, which touch on the rights of religious minorities, focus almost exclusively on the rights of minority communities to their religious cultural symbols, rather than the rights of the individual members of these minority communities.[474] The International Criminal Tribunal for the former Yugoslavia (ICTY) and ICC have both focused exclusively on the destruction of religious property and sacred sites during armed conflicts. In part, this focus is driven by the jurisdiction of the two tribunals – both tribunals assign criminal penalties to individuals who violate the laws of war and commit crimes of genocide or crimes against humanity. As such, cases before these bodies focus almost exclusively on acts committed by an individual or group of individuals against a group.[475] The cases before ICTY that relate to the rights of religious minority communities have focused almost exclusively on the allegations of plunder and excessive destruction of property, in violation of the laws and customs of war and the Geneva Conventions.[476]

215. Because of ICTY's role in prosecuting autocracies committed during armed conflict in the Balkans in the 1990s, the cases heard by the tribunal relating to minority rights focused primarily on the destruction of cultural and religious property. However, by saying that the intentional destruction of cultural and religious property constituted a violation of the laws and customs of war, the tribunal also articulated a version of minority rights based on the rights of a community to access houses of worship and its sacred shrines.[477] This right is identical to the one articulated by the ICJ in its advisory role relating to the wall on Palestinian territory. ICTY's jurisprudence combined with its advisory opinion from the ICJ suggest that the right of a minority population to preserve and access their sacred sites is protected through the protection of the sites themselves.[478] Like ICTY, ICC has only considered minority rights in the context of the prohibition against the international destruction of cultural and religious property. The ICC has considered this issue once in *Prosecutor v. Ahmad Al Faqi Al Mahdi.* In this case the court considered charges against A, an Islamic cleric who worked with Al Qaeda of the Islamic Maghreb (AQIM) in Mali to specifically target religious practices deemed contrary to AQIM teachings.[479] As part of this campaign A and others started to destroy religious sites of cultural and historical importance around Mali, including mausoleums and shrines.[480] The court held that this destruction constituted a war crime

474. A. Birdsall, *The International Criminal Tribunal for the Former Yugoslavia – Towards a More Just Order?*, 8 PCD 3 (2006).
475. *Ibid.*
476. *Ibid.*, at 4.
477. *Ibid.*
478. *Ibid.*, at 5.
479. *Prosecutor v. Ahmad Al Faqi Al Mahdi* 48 (ICC, 2016).
480. *Ibid.*

under Article 8(2) (e) (ii) of the Rome Statute, which prohibits attacks against civilian targets during armed conflicts.[481] The court also held that the attacks violated both the Geneva Convention and the Second Protocol to the Hague Convention, and as such constituted a criminal act within the court's jurisdiction.[482]

216. The court ultimately sentenced A to nine years in prison for the destruction of these monuments, holding that these crimes were committed with the intent to impose a specific religion on the population of Timbuktu through armed conflict.[483] The religious sites targeted were specifically protected by the UNESCO Convention and they were destroyed because of their cultural significance to the local people. As such A not only intended to traumatise the local population through a campaign of terror but he also intended to rob the population of religious and cultural symbols vital to its identity.[484] Despite ICC's holding in *Prosecutor*, similar cases heard by the ICTY is limited specifically to the destruction of religious monuments during war time; however, the court's rhetoric in the case suggests a broader conception of the protection of religious rights.[485] Like ICTY, the court focuses specifically on the right of a population to its cultural and historic sites, especially to shrines that have been part of the area's tradition and history for generations. However, the court also suggests that these acts become more serious when committed as part of a systematic programme of imposing religious beliefs on a population. This raises the possibility that, in the future, the court will be more willing to look into a country's practices relating to its treatment of minority religious groups.[486] However, the ICC's holding in *Prosecutor* is limited in a few important ways, which again suggest that the vast majority of protections afforded to religious minorities come from domestic rather than international law. First, *Prosecutor* rests with Article 8(2) (e) of the Rome Statute, which defines war crimes, meaning its holding is limited to the context of armed conflict. Therefore, *Prosecutor* cannot be used to target practices by world leaders during peacetime.[487] Second, the ICC's jurisdiction itself is limited by the Rome Statute and by the fact that it has a limited enforcement mechanism. So even if the ICC chooses to enquire more deeply into a state's policies with respect to its minority groups, it is unclear whether this enquiry will actually be meaningful or function as a way to protect minority populations.[488] Finally *Prosecutor* is currently the only case in which the ICC has charged and convicted someone for the destruction of cultural property, and it is unclear how and if the ICC's jurisprudence in this area will evolve in the future.

---

481. *Ibid.*
482. *Ibid.*, at 49.
483. *Ibid.*
484. *Ibid.*
485. *Ibid.*, at 50.
486. *Ibid.*
487. *Ibid.*
488. *Ibid.*, at 51.

§5. CONCLUSION

217. Though the UDHR and the International Convention on Civil and Political Rights articulate the right to freedom of religion as a deeply personal one, international tribunals, including the ICC, ICJ and ICTY have all conceptualised this right primarily as a communal one. These international tribunals have opted overwhelmingly to protect freedom of religion by protecting the rights of minority communities to educate their children and maintain access to their sacred sites. As a result, international law does very little work towards protecting the right to freely exercise of religion for minority communities outside the context of an armed conflict.

218. In part, this is due to the jurisdictional constraints placed on international tribunals. It is rare for a question concerning free exercise of religion to be heard by a body like the ICC or the ICJ outside the context of armed conflict, or at least hostility. This lack of protection is also driven by the conceptualisation of religion as a communal right, since it requires only that a community be allowed to exist, not that the individual members of the community be able to participate fully and freely in the political process. As a result, international law is fundamentally weak in the area of protecting and enforcing the ICCPR's mandate that every individual has the right to freedom of conscience, belief and culture.

Chapter 2. Transnational Law Effects on Religious Communities

§1. Introduction

219. International business transactions occur within a web of legal frameworks. Some are national, like the legal systems of the parties' countries, while others are international, like the legal systems created by various treaties or the rules of customary international law. Yet the division between national and international should not be overstated. In fact, as we shall see that international law penetrates and influences national systems in a variety of ways, while national laws and practices shape international law. To avoid misleading distinctions, Professor Philip Jessup (later a judge on the ICJ) proposed the phrase 'transnational law'. Jessup defined this phrase 'to include all law which regulates actions or events that transcend national frontiers. Both public and private international law are included, as are other rules which do not wholly fit into such standard categories'.[489] In the best of all possible intellectual worlds, the student of international business transactions would already have acquired a familiarity with international law and its relation to domestic legal systems in a course like Public International Law or Transnational Law. However, realism and experience suggest that this is not always the case. This chapter therefore introduces some of the concepts an international business lawyer must understand.

§2. Does Transnational Law Represent a New and Independent Legal Order

220. Transnational Law can be defined as the law of non-state governance systems. The definition suggests both commonalities and differences between 'transnational law' as a distinct legal field and conventional legal fields derived from the legal orders of nation states.[490] Like domestic law fields, transnational law includes both a constitutional element (a basic set of presumptive and supreme organising principles and rules), a substantive element (implementing the constituted system) and a process element (rules for the development of substantive rules and dispute resolution.[491] Unlike domestic legal orders, transnational law covers a wide number of distinct governance communities existing simultaneously and organised beyond the rule that imposes power of states. Moreover, these governance communities are not necessarily organised in the same way as states with a population and a defined geographic territory and an institutional framework exercising plenary authority.[492] Rather, transnational law communities may be understood as functionally differentiated communities organised for mutual benefit for specific objectives. They can include groups, institutions and networks. Transnational law can thus be understood as the study of the system of principles and rules applied in lieu of or in addition to

489. P. Jessup, *Transnational Law* 8 IGLP 1 (2012).
490. *Ibid.*, at 2.
491. *Ibid.*, at 3.
492. *Ibid.*, at 3.

the domestically germane law of a state, or of the community of states, to the relationships among persons and institutions.

§3. What Persons or Entities Control the Creation and Regulation of Transnational Law

221. No one person, entity or single institution controls transnational lawmaking. Transnational law is not dependant on a single lawgiver or regulator. Transnational law is thus the study of law that does not belong to, or can be controlled, by any single system of domestic or international law, as both have been traditionally constituted.[493] Diffusion of regulatory authority is one key to understanding the structure of Transnational lawmaking. Another key is functional differentiation of authority among a wide variety of political and non-political communities. The system of hierarchical and vertically integrated regulatory systems grounded on the state as the pinnacle of lawmaking and on the community of states as the disciplinary mechanism for relations among states has been augmented by regulatory systems covering matters beyond the reach of any single state.[494] Transnational law starts from the premise that law and lawmaking are no longer the exclusive preserve of political states, or of the community of states. Transnational law provides that political communities no longer hold a monopoly on lawmaking and that law systems are no longer grounded on an identity between law (or regulatory authority) and the state (and the community of states). Autonomous supra-national actors, private global actors and communities grouped by function or affinity can, to some extent, produce regulatory systems understood to be transnational, and this without a doubt includes both religious and cultural communities.

§4. Is Transnational Law a Single or Unitary System of Laws, Rules and Principles

222. Transnational law is not a unitary system of laws and rules analogous to the legal structure of a state, or the treaty and custom structure of the law of nations. Transnational law is an amalgam of rule systems, of hard and soft laws, that are limited in scope, but effective within the scope of the authority of the rulemaking community.[495] The key characteristic of transnational sub-systems is their functional limits. Like classical federal systems, all transnational sub-systems are based on grants of limited and specific authority. These limits are defined sometimes by commercial law, investment, human rights and sometimes by other factors such as shared belief, citizenship and in particular political communities and religion or culture in some instances.[496] The aggregation of sub-systems together constitutes the

493. *Ibid.*, at 4.
494. *Ibid.*
495. H.E. Foley, *Incorporation, Multiple Incorporation and the Conflict of Laws*, 42 Harvard Law Review 516 (2000).
496. *Ibid.*

field of Transnational law. However, Transnational law as a whole is grounded in certain principles and rules that form the basic focus of any study in this field. The foundational premise rests on acceptance of the existence independent of the control or authority of any one state or of the community of states of a system of non-national, supra-national or multi-national principles and rules applicable, in accordance with its own terms and logic, to public and private actors, natural and juridical persons.[497]

§5. What Are the Principal Features of the Study of Transnational Law

223. There are four characteristics that form the basis of the study of the grounding rules and principles of Transnational law and lawmaking: (1) scope of authority, (2) institutional autonomy, (3) regulatory authority and (4) effectiveness of power to settle disputes.[498] These have a constitutional element, the organising principles, that give the regulatory community form and set its organisational boundaries. These include the Constitution of a government apparatus and the rules for the operation of the governance with power vested in this organisation. They also have substantive and process elements.[499] These include the rules, laws and other norms that are produced or administered by the community and the process rules through which they are applied, enforced, constructed and interpreted. This is where religious and cultural communities fit in perfectly because they are independent and autonomous in the sense that they are in a position to determine their own rules and procedures. However, in the South African context this independence must be exercised in line with the prescripts of the Constitution, which is the supreme law of the land.

224. From the study and discussion of transnational law which in a nutshell deals with the personal laws of religious and cultural communities, in this case in the South African context. These personal laws aim to strengthen and build the religious and cultural communities from state inference. An example of legal rules or principles that can be viewed as Transnational law in this case is the L&D as outlined above, and as already discussed that such personal laws must be in line with the provisions of the Constitution.

§6. Conclusion

225. The independence and autonomy of religious and cultural organisations to establish rules and procedures in connection with their day-to-day operation and governance form part of Transnational law. The power of these religious and cultural bodies to establish their own rules and procedures forms part of the freedom

497. *Ibid.*, at 517.
498. *Ibid.*
499. *Ibid.*, at 518.

to choose religion or culture as protected by the Constitution. Furthermore, it is clear that Transnational law will strengthen democracy in the case of the diverse nature of the South African community and it is also subject to the Constitution.

Chapter 3. Regional Law Effects on Religious Communities

§1. Introduction

226. While the United Nations (UN) operates at an international level, there are also regional institutions that aim to promote human rights. These institutions produce regional treaties and declarations, for example Regional Human Rights Institutions. The European Commission (EU) on Human Rights and the European Court of Human Rights are responsible for applying the European Convention on Human Rights. The Inter-American Commission on Human Rights and the Inter-American Court of Human Rights apply the American Convention on Human Rights. The reports, recommendations and decisions that come from these bodies are very persuasive, and they have been referred to by domestic courts in interpreting various human rights provisions. In this chapter, we will focus on the African system. African Union (AU) is Africa's principal organisation for the promotion of socio-economic integration of the continent. The AU grew out of the Organisation for African Unity (OAU). The OAU was a regional body with multiple objectives, including ridding Africa of the legacy of colonisation and apartheid, as well as fostering social and economic co-operation among African states. In 1999, the Heads of States of the OAU issued the Sirte Declaration that called for the creation of the AU.[500] The AU's role in the African continent is to accelerate the political and socio-economic integration of the continent; to encourage international co-operation, taking into account the UN Charter and the UDHR; to promote and protect human rights in accordance with the African Charter and other relevant human rights instruments; to work with relevant international partners in doing away with preventable diseases and in promoting good health on the continent.[501] More about the role and importance of the AU will be outlined in this chapter and other legal instruments and organisations that aim to build and enhance regional law will be discussed.

§2. African Commission on Human and Peoples' Rights

227. The African Commission on Human and Peoples' Rights is the monitoring body for the African Charter on Human and Peoples' Rights. The African Commission aims to do the following in its monitoring process or system in the continent of Africa:

(a) To receive reports from state parties, interstate reports and individual complaints where there are serious abuses.
(b) It has the power to investigate abuses.

---

500. B. Andemicael, *The OAU and the UN: Relations Between the Organisation of African Unity and the United Nations* 34 (Africana, New York 1976).
501. *Ibid.*, at 38.

(c) It can bring these abuses to the attention of the Assembly of Heads of State and Government of the AU. After being informed of abuses, the Assembly may then request the Commission to make an in-depth study, and to present a report with conclusions and recommendations.[502]

228. It is clear that the African Commission on Human and Peoples' Rights in its monitoring process ensures that fundamental rights of people which include the right to choose religion and culture freely are protected from a Transnational law perspective. If it does indeed find some form of abuse it will escalate it to the mother body which is the AU to ensure that the abuse is corrected and the rule of law is upheld in the continent of Africa.

§3. Regional Agreements

I. The African Charter on Human and Peoples' Rights (African Charter)

229. The primary regional instrument for South Africa relevant to economic, social and cultural rights is the African Charter on Human and Peoples' Rights (African Charter). South Africa became a party to the African Charter in 1996. The African Charter includes the 'right to enjoy the best attainable state of physical and mental health' in Article 16. The Article goes on to provide: 'State Parties to the present Charter shall take the necessary measures to protect the health of their people and to ensure that they receive medical attention when they are sick.'[503]

230. It is important to note that when reading the African Charter progressive realisation and available resources are not mentioned. This suggests that the obligations of the state parties to fulfil a right and not just the minimum essential level of the right are immediate.[504] However, this was not the interpretation given to Article 16 by the African Commission in *Purohit and Moore v. The Gambia.*[505] Among other rights, this case also concerned the right to health for mental health patients. The Commission noted the dissimilarity between the wording of the right to health in the charter and the ICESCR, but interpreted the charter in line with the ICESCR. It read into Article 16 the qualifiers of available resources and progressive realisation.

231. Based on the problems that were encountered in the above-mentioned court case, it resulted in the development and drafting of the Abuja Declaration. The Abuja Declaration of 2001 was drawn up by the Heads of State and Government of the OAU (now AU). The declaration considered the HIV/AIDS pandemic: 'As a state of emergency in the continent. To this end, all tariff and economic barriers to

502. R. Murry, *Human Rights in Africa: From the OAU to the African Union* 5 (Cambridge University Press, Cambridge 2005).
503. Article 16 African Charter on Human and People's Rights 1986.
504. Section 27 of the Constitution.
505. *Purohit and Moore v. The Gambia [Communication 241/2001].*

access to funding of AIDS-related activities should be lifted.' The states committed themselves to: take all necessary measures to ensure that the needed resources are made available; mobilise all the human, material and financial resources required to provide care, support and quality treatment; and enact and use appropriate legislation and international trade regulations to ensure the availability of drugs at affordable prices.[506] From the drafting of this declaration and its adoption in the African continent it can be said that it was a vehicle for ensuring that a fundamental right which in this case was access to healthcare services is realised through making use of Transnational law. One can thus argue that Transnational law had a positive effect on the continent of Africa in fighting the HIV/AIDS pandemic due to co-operation and commitments of Member States.

II. African Court on Human and Peoples' Rights

232. In June 1998, the Protocol to the African Charter on Human and Peoples' Rights on the Establishment of the African Court on Human and Peoples' Rights (ACHPR) (the Protocol) was adopted by the OAU, the predecessor of the AU. Civil society and Non-governmental Organisations (NGOs) played an important role in the drafting and adoption of this Protocol. More than five years after the adoption of the Protocol, fifteen ratifications were achieved, and the Protocol took effect on 15 January 2004.[507]

233. At the time of writing, the ACHPR was in the process of being set up, despite a decision that was taken at an AU summit meeting in 2004 to merge the ACHPR with the African Court of Justice that has yet to be established. The merger raises complexities that threatened to delay the establishment of the ACHPR, but the AU decided that setting up the court is a priority, and it should not be delayed.[508] While human rights activists welcome the move to create the ACHPR, there are some concerns regarding the proposed manner of operation. For example, there are no clear rules as to how the ACHPR and the African Commission are to work in relation to each other, especially regarding the procedure for referring cases from the African Commission to the ACHPR. Another important concern is that the Protocol restricts access to the ACHPR by individuals and NGOs. State parties are required to sign a declaration that states that they recognise the competence of the ACHPR to receive complaints from NGOs.[509] If a state has not done this, the ACHPR will not be able to hear complaints against that state that are brought by NGOs. Further, only NGOs that have an observer status with the African Commission may file cases before the ACHPR.

234. Without a doubt, the establishment of the ACHPR is a good move for the continent of Africa, but it presents challenges as already outlined in ensuring that

506. Article 2 of the Abuja Declaration of 2001.
507. Article 2 of the ACHPR 2004.
508. Article 5 of the ACHPR.
509. Article 6 of the ACHPR.

fundamental rights of people are protected through easy access to the court as a first step. However, bearing in mind the colonial past of the entire continent of Africa, gradual progress is eminent and the same can be said with how the ACHPR operates; it is bound to change in the interest of all the people of Africa.

III. New Partnership for Africa's Development

235. New Partnership for Africa's Development (NEPAD) was adopted at the 37th session of the Assembly of Heads of State and Government in July 2001 in Lusaka, Zambia. NEPAD is an integrated development plan to address key social, economic and political priorities. It is meant to develop values and monitor their implementation within the AU to: promote accelerated growth and sustainable development; eradicate widespread and severe poverty; and halt the marginalisation of Africa in the globalisation process.[510]

236. Policy reform in the human development sector is a priority, and it focuses on health together with other areas, such as religion and cultural practices. Some of the health objectives of NEPAD are to:

(a) strengthen programmes for containing communicable diseases;
(b) achieve secure health systems that meet needs and support disease control effectively; and
(c) successfully reduce the burden of disease on the poorest people in Africa. NEPAD's action plans include strengthening Africa's participation in processes aimed at procuring affordable drugs and exploring the use of alternative delivery systems for essential drugs and supplies.[511]

237. NEPAD as a policy reform document aims to work hand in hand with the Abuja Declaration or even to strengthen it to a lesser or greater extent. This is because both legal instruments aim to ensure that access to healthcare services and other socio-economic rights are realised in the African continent with the necessary urgency.

IV. African Peer Review Mechanism

238. In July 2002, the African Peer Review Mechanism (APRM) was established at the 6th summit of the Heads of State and Government Implementation Committee of NEPAD. The APRM is a self-monitoring mechanism designed to track the progress and performance of Member States in their quest for democracy, human rights and good governance. The APRM's main aim is to foster the adoption of appropriate laws policies, standards and practices that lead to political stability

510. Article 1 of the NEPAD 2001.
511. Article 2 of the NEPAD.

and economic growth and integration.[512] The APRM does this through: sharing experiences; reinforcing successful and best practices; and reviewing the performance of particular countries – one of the four focus areas for review is socio-economic development.[513]

239. The APRM depends on Member States voluntarily agreeing to be reviewed. South Africa was one of the first countries to agree, and this process began in late 2005. The peers or fellow heads of states receive reports prepared by experts and make recommendations. However, they have no mandate to punish Member States that do not follow these recommendations.[514] However, there is pressure on participating states to achieve positive reports from the review team as the APRM influences donor organisations and countries on which countries should be favoured as funding recipients. This is a strong incentive for states to conform with NEPAD's values.[515]

V. Southern African Development Community

240. The Southern African Development Community (SADC), previously the Southern African Development Coordination Conference, was formed in April 1980, by nine southern African countries: Angola, Botswana, Lesotho, Malawi, Mozambique, Swaziland, Tanzania, Zambia and Zimbabwe. It has now expanded to a membership of fourteen countries, including South Africa.[516] One of SADC's aims is: 'to achieve development and economic growth, alleviate poverty, enhance the standard and quality of life of the peoples of Southern Africa and support the socially disadvantaged through regional integration'.[517] In 1997 SADC adopted the SADC HIV/AIDS and Employment Code. This Code has been used by NGOs to try to persuade governments to adopt legislation that is similar to the Code. The SADC Protocol on Health aims to improve the health of the population through close co-operation in the region, recognising that this is: 'essential for the effective control of communicable diseases, non-communicable disease and for addressing common health concerns'.[518] The Protocol aims to: promote activities that have the potential to improve health, such as the development of healthcare personnel and facilities; encourage co-operation with international partners; and help develop common strategies to address the needs of vulnerable groups.[519]

241. In August 2000, SADC approved the Principles to Guide Negotiations with Pharmaceutical Companies on provision of Drugs for Treatment of HIV/AIDS and related conditions in SADC Countries. SADC was making efforts to implement its

512. Article 2 of the APRM 2002.
513. Article 2 of the APRM.
514. Article 3 of the APRM.
515. Article 4 of the APRM.
516. Article 1 of the SADC 1980.
517. Article 2 of the SADC.
518. Article 3 of the SADC.
519. Article 4 of the SADC.

HIV and AIDS Strategic Programme and Plan of Action 2003–2007, and this plan has been implemented in 2019 in order to reach more people.

§4. Conclusion

242. In conclusion, it is very much clear that from a continental perspective Africa is making serious inroads in ensuring that fundamental rights are upheld and protected. This is through the adoption of various legal instruments as discussed and also the various tribunals or forums that implement and protect the said rights as enshrined in the various declarations as part of Transnational law. Despite some of the challenges, Africa is without a doubt in the right direction and Transnational law makes it easy for all the fifty-four countries in Africa to co-operate and work together as a unit, with the aim of bringing the necessary social cohesion to the people of Africa. Focus will now be on exploring the relationship between religion and politics in South Africa, bearing in mind the close relationship that is shared by both religion and politics in the struggle against apartheid.

Part V. Religion and Politics

Chapter 1. Influence of Religion in Politics

§1. Introductions

243. The history of religion and politics in South Africa is fundamental and important. This can be attributed to the important role religion played in the development of political consciousness in South Africa with the sole purpose of ensuring that apartheid comes to an end. Religion was also used as one of the mechanisms to oppress the majority of black people, by imposing a foreign religion on local people and forcing them to abandon their own religion and culture in the process. This meant that for the liberation of the black majority in South Africa, it was without a doubt necessary to make use of the very same religion to end the evil apartheid era.

244. Religion provided the necessary space and resources through the various church organisations in South Africa with the sole purpose of ensuring that the people of South Africa are free from political oppression. This is the reason why in this chapter it is argued that religion is fundamental in the democratic dispensation of South Africa. The role and importance of religion will be discussed in the chapter with the sole aim of outlining the historical links or importance which are shared by both religion and politics in South Africa today, post-1994.

§2. The Role of Religion in Politics

245. The role of religion in politics is a topic that has gained enormous significance in light of the events of 11 September 2001. It has left almost everyone thinking that religion in politics is probably negative, harmful and something to be avoided, especially the events in New York and Washington exemplify the negative role of religion while operating in the political sphere.[520] I say this without presuming any particular upswing by the dominant discourse. I fear that the discourse that we are engaged in currently has the potential to create an atmosphere in which religion is simply disqualified from any role in politics. This is very difficult for generations of activists who found a comfortable intersection between religion and

520. *September 11 Terror Attacks Fast Facts*, https://edition.cnn.com/2013/07/27/us/september-11-anniversary-fast-facts/index.html (accessed 2 Aug. 1988).

politics.[521] Therefore, focus will be on experiences of religion and politics in South Africa not as a representative of a political party, but as a co-religionist. It is very difficult to accept that religion should be disqualified from any political role, even potentially. It is very clear that the role of religion in politics up to 1994 was fairly undisputed.[522] On the one hand, one finds the NG Kerk, theologically justifying apartheid, helping to present the apartheid state as Christian, but creating no distinction between what it saw as Christian values and the need for the apartheid state to be the abode of these particular values.[523] This was the major form of discourse on religion and politics. Against this, across racial, religious and denominational lines, there were three broad responses to the apartheid state.

246. First, it was the accomodationist response, which consisted largely of traditional orthodox and conservative elements, not very active in opposing apartheid but simply carrying on with life. Second, it was the response of those who represented and adopted religious discourse to challenge apartheid, reaching a high point with the development of a liberation theology and where religious figures became active partners in the struggle, making contributions that often ranged from simple prayer right up to martyrdom. Third, and lastly, the response to the apartheid state in the pre-1994 period was far more fundamentalist.[524] It was a hybrid fundamentalism rather, sometimes operating in the realm of traditional orthodoxy and at other times reflecting aggressive evangelism and an aggressive relationship with God as the antidote to evil that goes on in society. This strand within the challenging discourse argued that what we are busy with is not the formation of a democratic South Africa, but an Islamic or theocratic state.[525] They thus put the margins a little bit further than what the broad spectrum of people in South Africa were busy with. The liberation theology discourse was in a sense hegemonic. It had powerful charismatic advocates, a coaching theology and popular appeal, but its projection was often disproportionate to its actual size. In the Muslim community the few acted as a fig-leave for the many. A similar situation was often projected in Christian theology as well, as if all Christians were active in the anti-apartheid struggle.[526]

247. The truth is that those who were opponents of apartheid hid their marginal status because of the need to create the illusion of a mass popular theological uprising against apartheid. This is the first instance where forgiveness should be asked for in advance. The post-apartheid, post-1994 construction of a democratic state effectively displaced the focused role of religion in politics so prevalent in the apartheid era at the intellectual, theological and philosophical level.[527] It was replaced

521. H. Bredekamp & R. Ross, *Missions and Christianity in South African History* 14 (Wits University Press 1995).
522. *Ibid.*, at 15.
523. W. Saayman, *Being Missionary, Being Human: An Overview of Dutch Reformed Mission* 12 (Cluster Publications 2007).
524. *Ibid.*, at 13.
525. J. Kruger, *The Human Search for Meaning: A Multi-Religious Introduction of the Religions of Humankind* 42 (National Book Printers 1996).
526. *Ibid.*, at 43.
527. *Ibid.*, at 44.

by ambiguity, confusion and demobilisation of religious forces and of the religious discourse in politics. The key debates that emerged at the time centred on the new role of the church or of religion in post-apartheid South Africa.[528] Some church leaders felt that now that the real political leaders were free, the priests should return to the parish. The church should now get on with the business of religion and politicians should take on the business of politics.

248. Without a doubt the model-building that was done up to 1994 eroded in that moment of ambiguity, confusion and demobilisation. The impact of this ambiguity resulted in the return of the dichotomy between religion and politics. Church activists were recalled to the parish, resulting in an abdication by intellectuals from the project to transform South Africa. This abdication was a product of the discourse at the time of 1994, which still prevails until today.[529] This discourse centred on the question of the essential nature of the democratic state in the post-1994 period. The answer to that question, which is now the dominant and hegemonic position, is that the essence of democracy is opposition; if you don't have an opposition, you don't have a democracy. The alternative to the same question, which is far less than hegemonic, is that the essence of democracy in a post-apartheid, post-polarised, unequal, damaged society has to be transformation.[530]

249. It is not suggested that there must be a creation of three walls between the two options: either choose or oppose transformation. The argument is that the consequence of the idea of opposition as the central thesis of democracy has resulted in the pronouncement that the church has only one role politically, and that is to be the watchdogs of society.[531] One can understand where the dominant discourse comes from because it speaks to the sentiments of a group who have enjoyed all the power and privilege up to 1994, and who are now set up as the opposition. They therefore see opposition as the essence of democracy.[532] It speaks to the discourse of a group who understood that quick transformation was to be the essence of a democratic project to which they had to contribute, and therefore, it was not comfortable to say that the essence of democracy is transformation. It was far better to say that the essence of democracy is opposition, with all its racial assumptions, and that a black government should not be trusted.[533] It is argued that the Constitution was precisely crafted in a way to facilitate the ongoing contribution of religion to politics. When the Constitution was conceived and when it came to discussion on the relationship between state and religion, there were four options.

250. There was the option that we should craft an atheistic state, hostile to religion, a state that had nothing to do with religion or did not encourage it. This option

528. Saayman, *supra* n. 524, at 15.
529. Bredekamp & Ross, *supra* n. 522, at 16.
530. *Ibid.*
531. *Ibid.*, at 17.
532. Saayman, *supra* n. 524, at 16.
533. *Ibid.*

would have been betrayal because the very nature of South African society is fundamentally religious.[534] It would also betray the entire role that religion played in bringing us to freedom. Therefore, this option never really became an issue of discussion in our work towards our new Constitution.[535] Then there were obviously those who thought that maybe this is the moment to craft a theocratic state. South Africa has its own problems in terms of its diversity and it was not necessary to encourage this kind of discourse because it simply involved too many complications.[536] The third option was to construct a secular state where religion operated in the private sphere, where individuals believe what they choose, as long as there is tolerance of the religious sentiments of other people.[537] The option that was eventually adopted and that's placed in the Constitution is one of the secular state, but with an active role for religion, that guarantees freedom and even guarantees access to resources in order to make sure that religions are equitably represented.[538]

251. For example, town planning should make sure that religious sites are made available equitably, the airwaves and television time have to be proportionally allocated, etc.[539] The Constitution guarantees equality of status for all religions, but for proportion of facility, i.e., you don't have Muslims and Christians sharing equal time on radio since their demands are different. What is disappointing is that the broader religious community did not take up this discourse and therefore the fourth position was not debated.[540] Important questions concerning challenges and responsibilities were not raised. What we had was religion claiming its rights. The question to be asked is whether the religious establishments fulfilled their responsibility because the Constitution did not simply guarantee religious freedom; it also offered an opportunity for religious activism.

252. No one seems to be crafting a theology of transformation to guide and bring the people of South Africa together, as was the case with the theology of liberation in the apartheid years. A theology of transformation, for example, should be able to help in influencing and guiding the debate on whether South Africa is indeed a country of two nations, one, rich and white and the other, poor and black.[541] In this way, it isn't a matter of simple political discourse about what exists or not, or whether the former President of South Africa Thabo Mbeki is right or wrong. There must be some objective value to ensure that we are guided by in our scriptures, so that we speak with conscience and not simply with the pragmatism of the

---

534. T. Masuku, *Prophetic Mission of Faith Communities During Apartheid South Africa, 1948–1994: An Agenda for a Prophetic Mission Praxis in the Democratic South Africa*, 42 Missionalia 4 (2014).
535. *Ibid.*
536. *Ibid.*, at 5.
537. *Ibid.*
538. Section 15(1) of the Constitution.
539. Saayman, *supra* n. 524, at 17.
540. *Ibid.*
541. Masuku, *supra* n. 535, at 6.

moment.[542] We thus begin to spell out the complementary roles of politics and religion, without holding that religious activists should be politicians, or that politicians should be religious activists, or that they all have the same role. Politicians in South Africa often operate on a pragmatic basis in which tasks are mediated with the idea and the identification of one country and two nations. Religion asserts values; it asks questions of right and wrong, of practical implementation, of fulfilling God's command.[543]

253. Religion, for example, consults scripture on our role in persuading those who have to share and those who have not to be patient until the sharing starts taking effect. The absence of a theology of transformation creates an intellectual gap in society.[544]

254. The role of religion in addressing equality and freedom in South Africa in the current democratic dispensation is something that cannot be underestimated, but is very crucial. Religion has to confront its own ideological debates and try to fashion a broad consensus and coherence that is once again hegemonic.[545] Not all religious people were against apartheid, but those who were managed to occupy a position of hegemony in the 1993/1994 period. Therefore, it stands to reason that not all religious institutions will embrace transformation, but those that do must become hegemonic. To achieve this religion must overcome ideological divides and confront those within the broader religious community because the dominant discourse is still the traditional conservative orthodox one.[546]

255. Raymond Williams is correct in saying that orthodoxy cannot push back boundaries; it merely confirms.[547] Religion need to go beyond practicing a discourse that emerges in vicious form from time to time. Fundamentalist discourse avoids the real challenges of society by operating at the soft underbelly of an emerging democracy, taking on issues such as abortion, homosexuality, prostitution and crime to name but a few. Antonio Gramsci explains that transitions are the most painful and dangerous periods in the life of a nation because the old is not yet dead and the new has not yet been born.[548] When you have the old and the new coexisting so uncomfortably in society, you always have a powder keg. The discourse in religion that only seeks to ignite that powder keg does not confront the need to be the best undertakers or the worst of the old, and the midwives for the best of the new. South Africa need to bring about a discourse that hopefully starts to position religion as the undertaker of the worst and the midwife of the best. South Africa may not be able to convert everyone to this discourse, which some call post-modernist, others progressive. The point is that a new discourse must be hegemonic, due to the pluralistic nature of the South African community.

---

542. *Ibid.*
543. *Ibid.*, at 7.
544. *Ibid.*
545. *Ibid.*, at 8.
546. *Ibid.*
547. R. Williams, *Keywords: A Vocabulary of Culture and Society* 16 (Fontana 1976).
548. A. Gramsci, *Architect of a New Politics* 157 (Louisiana University Press 1990).

§3. CONCLUSION

256. From the discussion above it is clear that religion and politics are central to the new democratic dispensation of South Africa. Furthermore, as outlined above in relation to the important role of religion in society, it is very important to note that religion in its efforts of bringing social cohesion must also play a central role to the transformation debate of South Africa. This with the aim of ensuring that equality and freedom is also attained by religious organisations, as protected by the Constitution and bearing in mind the painful past which South Africa comes from, which is something that cannot be underestimated in the new democratic order.

Chapter 2. Political Influence in Religion

§1. INTRODUCTION

257. The close relationship between politics and religion in South Africa is one of the factors that shows political influence on religion to a lesser or greater extent. Political leaders in the past resorted to religion or churches for political support and liberation, bearing in mind that most political leaders have a religious background which was the position even during the apartheid era. Political leaders can influence religious communities and churches by providing donations and often visiting these churches or communities with the aim of growing political power and support in that particular community or church.

258. Politicians will go to the various churches with the aim of pulling the church members to vote for them during elections. This means that they will go and influence church members of a certain religious domination either to put them in power or to keep them in power come election period. Most of the people in South Africa who are part of religious dominations are poor and socio-economically disadvantaged and these politicians have exploited them to such an extent that they will promise service delivery to these vulnerable people before elections, but after elections do the opposite. This has resulted in serious consequences for politicians in recent times in a sense that violent service delivery protests around South Africa are taking full steam against politicians who lie to the general masses and use religion to gain cheap political points. The political influence on religion can best be described as a double-edged sword in which if this relationship is handled with care and honesty it can be prosperous to the entire community and politicians, in order to attain social cohesion. However, if this relationship is abused, which is often the case, then it will be detrimental, resulting in some cases lives being lost during service delivery protests. In addition, members of the religious community will lose trust in the government and not engage in democratic processes like going to vote and this without a doubt will be detrimental to the young democracy of South Africa, which cannot be separated from the importance of religion.

§2. THE INTERSECTION BETWEEN RELIGION AND POLITICS

259. Focus will now be on the intersection of politics and religion through the lens of African traditional religion. African traditional religion which is most widely practiced among the indigenous people of South Africa is rich in diversity.[549] All religious dominations worship a supreme being, and in this case is the ancestors. This belief system accepts all other religions as equally valid constructions of

549. Mndende, *supra* n. 84, at 16.

knowledge and does not pretend to hold all the answers to human problems. Other views are considered, and often adopted, making African traditional religion a dynamic institution.[550]

260. Because African traditional religion believes that all authority eventually comes from the spirit world, any expression of power necessarily has religious implications. This principle manifested itself in several ways in the traditional Zulu culture, and it continues to surface in the current politics of Africa.[551] Religion was used to explain political failures and occupied the place of the political system in stateless societies. It also explains the tension over land ownership in modern southern Africa. Religion was employed in pre-colonial southern Africa to conceptualise the divide between good and evil in society.[552] Evil typically centred on the misuse of power. In societies where political institutions have abused the authority vested in them, religious discourse can be seen as an attempted remedy by means of reordering of power.[553] Religion was used to explain political misfortune in the community. Ngwabi Bhebe proposes that some Shona groups blamed their downfall on a former ruler who had abandoned traditional beliefs.[554] The ancestors sought to punish him and thus allowed the Zulu conquerors to triumph. In the same manner, when the king of Ndebele land, Lobengula, failed to hold off white invaders in the late nineteenth century, it was said that he received messages from Mwari, the Shona deity, claiming that the colonisers were actually Mwari's sons sent to punish Lobengula for excessive raiding of the Shona people.[555] In more recent times, the link between spiritual evil and political failure is still evident. A popular story in the Democratic Republic of Congo constructs an underground world where demons construe evil plans leading to the destruction of Africa. Using religious imagery, devilish creatures are depicted in modern situations, even with an underground international airport for all sorts of sorcerers and magicians flying in and out, and with various evil characters equated to World Bank and International Monetary Fund (IMF) agents.[556] Thus, a religious story is used to explain disillusionment over domestic political and economic processes and international political development and equate political failure with evil.

261. Similarly, religion often replaced central authority in early stateless societies. Religious structures and hierarchies translated into political organisers. This is known as authority over the spiritual world and thus translated into authority over people, making religion an outstanding means of instrumental sing political

550. *Ibid.*
551. *Ibid.*, at 17.
552. *Ibid.*
553. *Ibid.*, at 18.
554. B. Ngwabi, *The Historical Dimensions of Democracy and Human Rights in Zimbabwe* 16 (Mambo Press 2001).
555. *Ibid.*
556. *Ibid.*

power.[557] The use of lineage and clans, which continues today in indigenous communities, concentrates authority in the most senior living descendant. Different religious sects may also function as political factions, sects may counterbalance central authority, and by developing an ideology and relating particular actions and mundane practices within a wider set of meanings leading to a more honourable and dignified complexion to social conduct. Here, religion offers an alternative to the typical political process, and still achieves the same end, a more organised and accessible society.[558]

262. With this strong foundation in religion, different members of traditional African societies were able to manipulate religion for political purposes. Bhebe points out that when Zulu King Mzilikazi conquered the Shona in southern Zimbabwe he adopted many of their religious beliefs in order to integrate the Shona into Ndebele land peacefully for the political benefit of his kingdom.[559] Simultaneously, the Shona religious authorities were trying to curry favour with the new rulers by accepting them as God-sent.[560] With the beginnings of Christianity in Africa, and the creation of African Independent Churches, this same theme was evident. Africans appropriated certain aspects of Christianity that seemed politically beneficial yet maintained their traditional ritual practices. When the Christian missionaries gained strength during the colonial period, however, Africans allowed for more Christian customs in order to negotiate with the white conquerors.[561] In this way religion was used as a means to political stability and prosperity. Likewise, religious communities acted as targets for political mobilisation. By declaring authority from a supreme being, African rulers could demand taxes, labour or participation in the army.[562]

263. The most drastic mixing of religion and politics occurs when the line between the two is erased entirely. The belief that everything derives power from the spiritual world allows religion to actually usurp politics. Politicians or tribal rulers then come to rely on religion for their very legitimacy.[563] Religion determines power relationships and commercial ties, and thus it becomes the main institution in society. When Ndebele King Mzilikazi moved his Zulu followers out of South Africa and into Zimbabwe, conquering and incorporating groups as he went, he eventually found himself in an unfamiliar territory. This threatened his power because, as the king, he was responsible for appealing to his ancestors for rain for his people. However, his careful study of the clouds in South Africa was useless in Zimbabwe.[564] The Shona, whom he conquered in Zimbabwe, were experts in the Zimbabwean clouds and told Mzilikazi that if he appeased Mwari (the Shona deity) with black oxen, tobacco and other useful things, he could cause a downpour even

557. *Ibid.*, at 17.
558. *Ibid.*
559. *Ibid.*
560. *Ibid.*, at 18.
561. *Ibid.*
562. *Ibid.*, at 19.
563. *Ibid.*
564. Mndende, *supra* n. 84, at 18.

though there might be a drought.[565] Mzilikazi thus adopted the Shona belief in Mwari to maintain his political power. Moreover, a succession crisis followed Mzilikazi death, and a council was chosen to select the next king. Their leader, Mncumbata, appealed to Mwari for help, who directed him to Lobengula, Mzilikazi's son. Thus, the new king remained to some extent dependent on Mwari for his power to rule the Ndebele and their subjects. Lobengula's authority rested on the Ndebele's belief in Mwari, the traditional Shona God.[566]

264. Religion is able to manipulate politics because traditional African religion incorporates the spiritual world to such a high degree in everyday life. The ancestors are just as important as the people of the human world are. In this sense, the evolving political language of Africa tends to regard politics as a metaphor for movements in a spirit world rather than *vice versa.* Because the ancestors are the cause of evil and suffering, they also hold the key to peace and happiness.[567] All religious, and therefore political, communication is directed towards this end. Thus, the only way to attain political power is through the ancestors.

265. Recent religious revival in Africa demonstrates the prevalence of political and religious interdependence. At a time when political failure is widespread, coupled with the disillusionment of World Bank and IMF programmes for development, rekindled religious sentiment reflects disappointment with the current system of governance. Because religion locates evil in society, and allocates power accordingly, ambiguity concerning the morality of the state logically breeds religious excitement.[568] The same applies to nations with strong governments like South Africa. Here, religious movements challenge the very basis of legitimacy of states which operate through institutions and norms of governance originally created in colonial times. A considerable factor in the anger over forced removals under the apartheid regime stems from traditional religious beliefs. On the Zulu homestead, a clan's land is regarded as sacred. Ancestors are buried there, and it is also the site of important rituals. By forcing people to leave their lands, the central beliefs of African traditional religion were placed under direct attack by the state.[569]

266. Before the bureaucratic institutions established by colonial powers took form, religious specialists played an integral role of checking the chief or central authority figure to guard against corruption and tyranny. This concept has not faded with modernisation and education, as many Africans today continue to hold beliefs derived from traditional cosmologies which they apply to their everyday activities. In the contemporary African mind, concepts concerning ancestral interference with daily life still hold a considerable amount of influence. Instead of trying to eliminate or disregard such customs, international donors and developers must work within this unique African framework to find solutions to the continent's problems.

565. *Ibid.*
566. *Ibid.*, at 19.
567. *Ibid.*
568. *Ibid.*, at 20.
569. *Ibid.*, at 21.

§3. The Importance of Religion and Politics from a Perspective of Politicians in South Africa

267. 'Politics can do with the influence of the church.' Mmusi Maimane, the leader of Democratic Alliance (DA), which is the official opposition in South Africa, said this during his Good Friday address at the Faith Ways Bible Church in Soweto: 'You might be thinking why politicians are coming to churches, that it must be elections,' he said, 'I hold the belief that politics can do with the influence of the church. I still hold the belief that even those who founded the ANC were committed Christians.'[570]

268. He told congregants to 'look forward', and they murmured in agreement. 'It is time for us to say that we know where we come from. We are going to cross this Red Sea.'[571] Earlier on that day, former President Jacob Zuma called on Christians to pray for politicians, as the devil was trying to interfere with their work.[572]

269. He spoke to thousands of members of the Universal Church of the Kingdom of God at Ellis Park Stadium in Johannesburg. Zuma called on the congregants to pray for members of parliament to behave and conduct themselves in a respectful manner. He professed his love for the church and said the atmosphere in the stadium made him want to attend church regularly.[573] In KwaZulu-Natal, Premier Senzo Mchunu said the building of a stronger province would not succeed without religious leaders' involvement.[574] Senzo argued that churches must remain focused on their work of healing the nation and fighting injustices at all times.[575]

270. A prominent aspect in the speech of Senzo was when he said churches should criticise government when they felt it was going astray and failing to meet the expectations of society. He invited religious leaders to pray for a prosperous province.[576] Furthermore, Archbishop Thabo Makgoba of Cape Town called on government officials 'to tell the truth' about the Gupta family's alleged influence on matters of state capture. The Dominican Order of Catholic priests asked Public Protector Thuli Madonsela to investigate the Gupta's alleged involvement in ministerial appointments.[577] These few recent incidents in the South African position affirm that a strong relationship exists between religion and politics, and it is something that cannot be underestimated at all. Religion is instrumental and fundamental in the development of the new South Africa. This is affirmed by the participation of various political leaders in churches across the length and breadth of South Africa, to

570. *Why Is the Church a Non-player in Politics*, https://mg.co.za/article/2016-03-23-why-the-church-is-a-nonplayer-politically (accessed 14 Aug. 2018).
571. *Ibid.*
572. *Ibid.*
573. *South African Political Parties Go to Church*, https://www.news24.com/elections/news/sa-political-parties-go-to-church-20140420 (accessed 16 Aug. 2018).
574. *Ibid.*
575. *Ibid.*
576. *Ibid.*
577. *Ibid.*

seek spiritual intervention in cases where political solutions are not available in pursuit of building a prosperous South Africa.

§4. CONCLUSION

271. Religion and politics are dependent on each other and must work together in order to achieve the desired society in South Africa. This is the reason why politicians often visit churches across the country with the aim of growing their spirituality and also to seek for the intervention of the church in some instances to resolve a societal problem. This is taking place because both politicians and churches strive for the best interests of the people of South Africa, and to achieve this they must work together even though at times they might encounter some conflicts between them.

Chapter 3. Interaction Between Religion and State on a Political and Legal Level

§1. INTRODUCTION

272. The rich history that is shared by both religion and politics during the colonial and apartheid era has had a profound effect on the development and influence of the South African legal system. This in a nutshell means that basically the supreme law of the land in the form of the Constitution had to be developed to embrace diversity and to accommodate religion and religious communities. Law had to ensure that there is equality in South Africa, without any form of discrimination and maltreatment on all the inhabitants of South Africa, which was unfortunately the position in the past.

273. However, it is important to note that even though politics and religion had a profound effect on the development of the South Africa legal system, this does not mean that the judiciary has lost its independence and impartiality. The South African judiciary as one arm of the government is independent from political and religious interference, and it is only accountable to the Constitution as the supreme law of the land. One can argue that the close relationship between politics, religion and law from a historical perspective is aimed at ensuring that social cohesion is achieved and maintained, and that law is used as an instrument to determine the process or manner in which this social cohesion is achieved in a way that is fair and just in the interest of ensuring that no human rights are sacrificed at the expense of the other. This chapter explores as to how law, religion and politics operate with the aim of ensuring that social cohesion is maintained and how is the balancing exercise between the three spheres achieved in the interest of the religious community in this context.

§2. RELIGION AND LAW IN SOUTH AFRICA

274. It is clear that in South Africa, like in the many places in which modernity was introduced through colonialism and missions, modernisation was not accompanied by an increasing level of atheism.[578] Quite the contrary; Christianity is a significant part of both our colonial and apartheid histories. It is clear that in terms of the fates of personal faith, South Africa has been a deeply religious country – religious in African customary terms, in Christianity and a significant minority of other religious traditions – particularly Islam, Hinduism and Judaism.[579] Religion has also been very significant to public and political life. The apartheid regime politicised Christianity in a number of ways through the National Party's close association with the Dutch Reformed Church, through apartheid's political theology of race and through Christian national education. Opposition to apartheid also drew heavily on

578. Mndende, *supra* n. 84, at 18.
579. *Ibid.*

Christianity. The development of liberation theology in the 1970s gave a platform to the Christian and Muslim leaders in the 1980s when they became politically active as the state cracked down on the leadership of civic and political organisations.[580]

275. The negotiations that framed South Africa's new democracy also paid attention to the place religion does, and ought to, take in the country. The negotiators acknowledged the importance of religion to the majority of South Africans and the variety of religious traditions in the country. The 1996 Constitution includes a strong right to freedom of religion, and the possibility of the presence of religion in state institutions under conditions of fairness and neutrality. Various forms for customary religious marriages are also provided for.[581] The drafters of the Constitution were also very clear to exclude reference to Christianity in the Preamble, and emphasised the importance of state neutrality towards various religions in law. During important state events, religious leaders of a variety of traditions play a role. At the same time, the TRC and political rhetoric often draw on the legacy of liberation theology.

§3. Politics and Law in South Africa

276. Politicians and the government of the day in South Africa also subscribe to the Constitution as the supreme law of the land. This means that the leader of a political party, for example the ANC president in South Africa, is also the president of South Africa. In taking the oath of office the president pledges that he or she will uphold the oath of office and respect the rule of law of the Republic and also encourage people to follow the rule of law.[582] Failure by the president to uphold this oath of office will constitute a gross violation of the Constitution, and he or she will be impeached from office on the basis of his or her failure to uphold his or her oath of office as the first citizen of the Republic.[583] This implies that the president as the first citizen is the one that must be the first defender of the Constitution and lead by example in ensuring that the rule of law is upheld and protected all the time.

277. Political parties in South Africa that exist because it is a multi-party democracy are also bound by the rule of law of the Republic. This means that the various political parties and their leaders cannot engage in practices which will put the Republic in danger such as to incite violence or criminality as this will be contrary to the laws of the Republic.[584] Such conduct by political parties and their leaders will be condemned as criminality, and they could face arrest and prosecutions in some instances for defying the law and aiming to bring chaos in the Republic of South Africa. Just like church organisations, political parties are also autonomous

580. *Ibid.*, at 19.
581. Section 15 of the Constitution.
582. Section 83 of the Constitution.
583. Section 89 of the Constitution.
584. Section 44 of the Constitution.

in the sense that they can develop their own Constitution or Code of Conduct which their members are expected to subscribe to it. However, such Constitution or Code of Conduct must not be contrary to the Constitution of the Republic in such a way that it is found to be discriminatory against certain groups of people. If this is found to be the case then such a Constitution or Code of Conduct will be found to be void and not in the interest of the administration of justice.[585]

278. Political parties and leaders are also bound by the court decisions or judgments that are given by the courts, and must ensure that they are implemented and not defied by them as they can face penalties such as contempt of court.[586] This means that political parties and their leaders must be exemplary all the time when it comes to ensuring that the rule of law is upheld and protected. Political leaders must encourage their members to be citizens that respect the rule of law because failure to do so has severe consequences for those leaders that aim to defy the law.[587] Such a duty rests with political parties due to the close relationship they enjoy with members of the public on the ground, and they are duty bound to ensure that their members are disciplined and are always on the right side of the law.

279. Political parties just like church organisations in South Africa, despite being autonomous, also face internal or external legal problems that threaten their existence or affect their reputation in South Africa. For example, political parties of recent times in South Africa have resorted to courts to resolve their internal disputes or for the court to assist them with the interpretation of their internal Code of Conduct or Constitution because such legal instruments will all the time be tested by the Constitution of the Republic as the supreme law.[588] This affirms the role and importance of the judiciary in South Africa because they not only provide legal certainty but go beyond to bring the necessary political certainty with the involvement and interpretation of the law. It is one of the major reasons why political parties and their leaders in South Africa must ensure that the rule of law is protected and upheld at all costs because law plays an important part in ensuring that even political problems which do not have political solutions are resolved through legal intervention and interpretation.

280. Politicians and political parties are not exempted from the rule of law. They are in fact the people that must guard and protect the rule of law. This will be possible by encouraging their members and the general public about the role and importance of the rule of law in the current democratic dispensation of South Africa. The future of the South African democracy lies with a collaboration between political, religious and law, working together with the aim of ensuring that social cohesion is achieved through embracing diversity and respecting the Constitution as the supreme law post-colonial and apartheid South Africa.

585. Section 2 of the Constitution.
586. Section 41 of the Constitution.
587. Section 42 of the Constitution.
588. Section 2 of the Constitution.

§4. CONCLUSION

281. South Africa as a pluralistic society requires a joint venture between law, religion and politics in ensuring that societal peace and growth are maintained at all costs. Furthermore, since law, religion and politics have an important role to play in a unique manner with the purpose of ensuring that the democracy of South Africa grows and succeeds on a daily basis, it is something that cannot be underestimated and must be protected. If there is a conflict between politics and religion which is often the case, law will come in with the role of a regulator and protector, to aim to resolve the said conflict in the public interest. While bearing in mind that religious and political institutions are autonomous and independent.

Part VI. Labour Law Within Religious Communities

Chapter 1. Scope of Application of Labour Law

§1. INTRODUCTION

282. The Constitution of South Africa was adopted on 10 May 1996 and came into effect on 4 February 1997. This means that the scope of application of the Constitution of South Africa is very vast, and it is not restricted to religious freedom and culture which is the focus in this regard. However, the constitutional application extends to the relationship between an employer and employee with the sole aim of safeguarding a fundamental right which is enjoyed by everyone, which is the right to choose an occupation or trade freely as protected by the Constitution.[589] This means that the Constitution aims to play a role of ensuring that there is fair labour practices that are taking place in the labour market, without any form of exploitation and discrimination.

283. The role of the Constitution in the labour market and the various statutory provisions which deal with labour matters will be explored in this chapter, with the aim of exposing the importance and significance of the labour market that if it is unregulated it can result in dire consequences.

§2. LABOUR RIGHTS IN THE CONSTITUTION

284. The Constitution contains a Bill of Rights, which enshrines the rights of all South Africans, and reference will now only be made to labour rights as the theme of this chapter. The following labour rights are enshrined in the Constitution:

(a) everyone has the right to fair labour practices;
(b) every worker has the right to form and join a trade union and to participate in the union's activities;
(c) every worker has the right to strike;
(d) every employer has the right to form and join an employers' organisation and to participate in the activities of the organisation; and

589. Section 22 of the Constitution.

(e) every trade union, employers' organisation and employer has the right to engage in collective bargaining.[590]

285. All these rights set the tone in relation to the employer and employee relationship and it is very important that they are upheld and protected.

§3. Labour Regulation

I. The History of Labour Law

286. Prior to the discovery of gold and diamonds in South Africa, the economy could by and large be described as agrarian, with the main economic activity being agriculture. The relationship between employers and domestic workers and farm workers was governed by various Acts, including the Master and Servants Act 15 of 1856.[591] At this time, the employment relationship was regarded as being a 'master and servant' relationship.

287. When gold and diamonds were discovered, mining activity in South Africa rapidly increased. The booming mining industry brought with it an influx of labour and workers to the mines. The first South African trade union, the Carpenters' and Joiners' Union, was established in 1881, largely to protect the interests of skilled foreign workers working on the mines.[592]

288. As the mining industry developed, the difference in political power between whites and blacks became entrenched as trade unions, catering largely for white workers, mobilised increasingly on the basis of race. In 1911, the Mines and Works Act was passed which reserved various types of work for white workers only.[593] This era was very turbulent and a number of strikes, with the aim of securing the position of white workers on the mines, took place.

289. After the general strike of 1914, martial law was declared and trade union leaders were deported from South Africa. The 'labour peace' which ensued was short-lived as the circumstances of the mines worsened due to the economic depression, a large foreign debt and the rising costs of living.[594] The mines responded by restructuring. This led to a number of white workers being retrenched, which in turn led to the abolition of the ratio between skilled white workers and unskilled black workers on the mines.[595]

590. Section 22 of the Constitution.
591. Section 1 of Master and Servants Act 15 of 1856.
592. Section 2 of Carpenters and Joiners Union Act 1881.
593. Section 2 of the Mines and Works Act 1911.
594. P. Delius, *The History of Migrant Labour in South Africa (1800–2014)*, 17 SHSA 30 (2017).
595. *Ibid.*, at 17.

290. This situation gave rise to the 1922 strike, one of the watershed moments in South African labour history. The result of this strike was the passing of the Industrial Conciliation Act in 1924.[596] This Act was the direct forefather of the Industrial Conciliation Act of 1956, which was later renamed as the Labour Relations Act (LRA) of 1956. This Act accorded recognition to trade unions representing white workers, while a separate system for black workers was created.[597]

291. In 1948, the National Party came into power, which saw the birth of Afrikaner Nationalism. In 1950, the Suppression of Communism Act was passed which led to the large-scale repression of union activists. The 1950s was a turbulent time in the political history of South Africa. This turbulence led to the amendment of the 1956 Industrial Conciliation Act, in order to ensure tougher controls over black workers.[598]

292. The Wiehahn Commission of Inquiry was established in 1979 to investigate the labour situation in South Africa. The resultant report of the Commission went on to change the face of South African labour relations and labour law. The most consequential recommendation made by the Commission was the extension of freedom of association to cover all persons, irrespective of race or sex. The result was that trade unions representing black workers were now able to make use of the machinery of the LRA of 1956.[599]

293. The period between 1991 and 1994 saw the birth of the new democratic South Africa. In 1994, the Interim Constitution, Act 200 of 1993, came into effect. The Act totally changed the constitutional basis of the South African legal system and it became clear that the LRA of 1956 was not in line with the new constitutional order.

294. In 1994, the Department of Labour appointed a ministerial legal task team to draft new labour legislation and the LRA 66 of 1995 came into effect on 11 November 1996. The Act heralded a new era in South African labour law.[600]

II. The Sources of Labour Law

295. The sources of South African labour law include:

(a) legislation;
(b) judicial precedent (judicial decisions), including arbitration awards;
(c) collective agreements;
(d) common law; and

596. *Ibid.*, at 18.
597. *Ibid.*, at 18.
598. *Ibid.*, at 19.
599. Section 1 of the LRA.
600. *Ibid.*

(e) custom and legal writings.[601]

296. Since the democratisation of South Africa after April 1994, the country's labour law was among the first areas of law to be reformed. The main employment law statutes of South Africa are the following:

(a) the Labour Relations Act 66 of 1995 (LRA);
(b) the Basic Conditions of Employment Act 75 of 1997 (BCEA);
(c) the Employment Equity Act 55 of 1998 (EEA);
(d) the Skills Development Act 97 of 1998 (SDA);
(e) the Unemployment Insurance Act 30 of 1996 (UIA);
(f) the Occupational Health and Safety Act 85 of 1993 (OHSA); and
(g) the Compensation for Occupational Injuries and Diseases Act 130 of 1993 (COIDA).

297. Employment protection legislation applies to all employees who ordinarily work in South Africa. Therefore, the legislation also covers employees who work partly outside South Africa and partly inside South Africa. It also applies regardless of the stated governing law of any employment contract or the nationalities of either the employee or the employer.[602] It is not possible for an employee to contract out of statutory employment protection unless the legislation specifically permits it and only to the extent permissible in terms of the legislation. In many cases, the legislation is supported by codes of practice which may be statutory codes of practice drawn up by the National Economic Development and Labour Council or non-statutory codes of practice issued by the Commission for Conciliation, Mediation & Arbitration (CCMA).[603] These codes of practice, although often merely providing guidelines and accordingly not always being of direct legal effect, are taken into account by the labour courts in deciding whether or not an employer has breached statutory employment regulations. Additionally, there are numerous laws implementing health and safety regulations.

298. Unlike the law in certain other countries, collective agreements are normally legally enforceable between employers and trade unions. The LRA 66 of 1995 ('the LRA') supports the primacy of collective agreements and emphasises the need for organised labour and business to regulate their relationship by entering into collective agreements, which bind the employer, the union's members and, where the union represents more than 50% of the employees in a workplace, and if such intent is stated, non-union members in the workplace.[604]

---

601. M. Gragon, *Essential Labour Law* 218 (Butterworths, Durban 2016).
602. *Ibid.*, at 218.
603. *Ibid.*, at 219.
604. Section 2 of the LRA.

III. Contract of Employment

299. When discussing the employment contract the starting point should be that a written contract of employment is not strictly a necessary requirement for the validity of an employment relationship.[605]

300. However, the Basic Conditions of Employment Act (BCEA) compels an employer to give an employee a host of prescribed employment details in writing when they start working with that employer.[606] This is to ensure certainty between the employer and employee, and it attempts to protect the 'vulnerable' employee against the employer who holds the purse strings in the event of any disagreement as to the details.

301. It therefore makes more sense to enter into a detailed contract of employment upfront with the employee, and thereby bind him or her to the company's disciplinary codes and the like at the same time, rather than to provide a list of prescribed details in a rather ad hoc or non-contextualised manner. A comprehensive contract also has the benefit of being signed by both parties and is therefore legally binding.[607] The essence of an employment relationship is in the details of its contract regarding commencement, currency and termination since in most cases the document will 'speak for itself'.

IV. Prescribed Written Particulars

302. The contract of employment between the employer and the employee is something that is going to be explored in detail because it is very important to know what is in the contract of employment, for purposes of determining the intention of the parties and the scope of their employment contract.

303. Terms and conditions of a contract of employment are found in section 29 of the BCEA. They are:

(a) *The name and address of the employer* – These may seem like obvious details, but many an employee claim has been thwarted either by insufficient or incorrect details in this regard (e.g., regarding the identity of the employer party, especially in a group or division of companies, and in enforcing a judgment).
(b) *The employee's name* – Again apparently an obvious requirement but it provides certainty as to the parties.
(c) *The employee's occupation or a brief description of the work for which the employee was employed* – Certainty on this issue is required for reasons relating to advancement and remuneration, giving instructions which are reasonable to that post, performance, discipline and the like.

605. Gragon, *supra* n. 602, at 219.
606. Section 10 of the Basic Conditions of Employment Act 75 of 1997.
607. Gragon, *supra* n. 602, at 220.

(d) *The employee's place of work* – If the employee is required or permitted to work at various places, an indication of this. The employee should agree at the outset of the employment relationship that he or she would be required to travel, work at various stations, failing which he or she may not easily be compelled to do so later without an agreement. The employer may have to remunerate the employee additionally for undertaking work at different places, and could not simply compel the employee to do so if it has not been agreed initially in the contract.
(e) *The date of commencement of employment* – This is important for the calculation of benefits such as leave, recognition of length of service and the benefits attached to it – e.g., severance pay in a retrenchment.
(f) *The employee's ordinary work days and work hours* – The BCEA sets minimum standards (more specifically, maximum hours and days that may be worked) in this regard, which may not be changed in most cases even with an employee's consent.
(g) *The employee's wage or the rate and method of calculating wages* – This has undoubtedly been included for certainty, consistency and protection for the employee, in addition to ensuring that the employee is fairly remunerated for leave and for overtime among other things.
(h) *The overtime rate* – It must be clearly communicated.
(i) *Any other payments in cash or kind to which the employee may be entitled (and for payments in kind, the value of those payments)* – Again it is important for calculating the monetary value and benefits to which an employee would be entitled during leave, on termination (e.g., severance pay on retrenchment is calculated at one week's remuneration per completed year of service, not one week's base salary). Remuneration includes the value of a number of benefits.
(j) *The date when remuneration will be paid* – Promotes certainty and consistency.
(k) *Details of any deductions that will be made from the remuneration* – In addition to reasons of certainty, the nature and amount of deductions that can be made from an employee's remuneration is strictly regulated by the BCEA.
(l) *The amounts of leave which can or must be taken* – It will suffice to state that the BCEA strictly regulates when and how much leave may and shall be taken. This relates not only to annual leave but also to sick, maternity and family responsibility leave in respect of which there are legislated minimums that may not be contravened.
(m) *The period of notice* – This too is strictly regulated by the BCEA – the quantity and manner of giving notice of termination of a contract of employment is normally three months, except in exceptional circumstances.
(n) *A description of any bargaining council or sectoral determination that covers the employer's business* – This too is in protection of employee rights since it is often the case that terms and conditions of employment more favourable to the BCEA are collectively negotiated within a bargaining council. The bargaining council standards will then take precedence over those set out in the BCEA.
(o) *Any period of employment with a previous employer that counts towards the employee's period of employment.* For example, the length of service with a seller who transfers its business to the purchaser is obliged to recognise the

employee's previous length of service, for the purpose of calculating benefits, such as severance packages, long service awards and the like.
(p) *A list of any other documents that form part of the contract – (and an indication of where a copy of these documents may be obtained).* This would relate to policies and procedures, and the like, of the employer. It again provides certainty to the parties of the employment relationship.[608]

304. If and when any of the above details change, the employee is to be notified of the change and be given a copy of the change by the employer which is in line with fair labour practices of South Africa.

V. Employment Contracts

305. South African labour force has grown and based on this growth and expansion two labour contracts are in existence. Fixed-term contracts, the duration of the contract is clearly specified between the parties. The contract will endure for the specified period, or upon the happening of a particular event or until a particular task is completed.[609] Unless otherwise agreed, such a contract cannot be terminated during its current form without good cause, unless the parties have agreed otherwise. If after the contract has lapsed and the employee remains in the employment of the employer, the contract may be tacitly renewed, provided that it is consistent with the parties' conduct.[610] The LRA expressly provides that non-renewal of a fixed-term contract is equivalent to a dismissal in circumstances where the employee expected the employer to renew it on the same or similar terms but the employer either failed to renew the contract at all or offered to renew it on less-favourable terms.[611]

306. In the case of contract to indefinite period, the duration of the contract is not specified by the parties. The contract will endure until:

(a) it is terminated by agreement;
(b) one of the parties gives the contractually specified or reasonable notice to the other;
(c) either party elects to terminate on fundamental breach;
(d) on retirement of the employee at the agreed age, normally 65 years;
(e) summary termination of the employee;
(f) death of either party; and
(g) insolvency of the employer.[612]

608. Section 29 of the Basic Conditions of Employment Act.
609. Gragon, *supra* n. 602, at 220.
610. *Ibid.*, at 221.
611. Section 186 of the LRA.
612. Gragon, *supra* n. 602, at 221.

VI. Probation

307. It is quite common that employers engage employees for a probationary period, which may be negotiated and stipulated in the contract of employment.[613] After expiry of the probationary period, the employer is entitled to decide whether to retain the services of the employee on a permanent basis. The Code of Good Conduct, contained in the LRA, expressly provides for probationary periods for employees. The Code stipulates that the probationary period must be reasonable given the circumstances of the job and the time it takes to determine the employee's suitability for the job. The probationary period can be extended, in suitable circumstances.[614]

308. The Code also states that following completion of the probationary period, probationary employees should not be dismissed unless they have been given appropriate remedial treatment, and they have been allowed a reasonable period for improvement but have failed to improve their performance. If the employer fails to counsel the probationary employee and thereafter fails to confirm the employee's employment, such termination will amount to a dismissal.[615]

VII. Termination of the Contract of Employment

309. The contract of employment can be terminated on the following grounds in the South African labour market:

(a) on expiration of the agreed period of employment;
(b) on completion of the specified task;
(c) by notice duly given by either party;
(d) by summary termination in the event of a material breach on the part of either party;
(e) by repudiation;
(f) by mutual agreement;
(g) by death of either party;
(h) by the insolvency of the employer; and
(i) by the supervening impossibility of performance, where either party becomes permanently unable to perform his or her obligations in terms of the contract.[616]

310. The contract may not be terminated in the absence of a justified reason.

The LRA expressly recognises the following grounds for termination of the employment contract:

(a) misconduct on the part of the employee;

---

613. Section 186 of the LRA.
614. Section 187 of the LRA.
615. Section 186 of the LRA.
616. Section 188 of the LRA.

(b) the employee's poor work performance and/or incapacity;
(c) the operational requirements of the employer.[617]

311. The concept of unfair dismissal is a right created by statute and contained in the LRA. Chapter VIII of the LRA, along with an accompanying code of practice, has made a large contribution to systematising and clarifying this important area of South African employment law. Unfair dismissals now fall into four categories:

(a) automatically unfair dismissals;
(b) dismissals which an employer cannot justify on the basis of:
 (i) the employee's conduct;
 (ii) the employee's capacity; or
(c) the operational requirements of the business.[618]

312. Besides these substantive grounds, dismissal will also be unfair where it is not effected in accordance with a fair procedure. Meaning that the procedure followed must be fair and just through giving the concerned employee an opportunity to make the necessary representations and state his or her side of the story.

313. The meaning of dismissal has been extended by statute to include:

(a) a failure to renew a fixed-term contract when there was a reasonable expectation of renewal;
(b) a failure to allow an employee to resume work after maternity leave;
(c) selective employment following a collective dismissal; and
(d) a situation where an employee terminates the contract because the employer has made continued employment intolerable (constructive dismissal).[619]

314. A dismissal will be automatically unfair if it is found to comply with the following grounds:

(a) offends against an employee's rights to freedom of association;
(b) is by reason of an employee's refusal to do protected strikers' work;
(c) is effected to compel an employee to accept a demand relating to employment;
(d) constitutes victimisation for exercising any rights under the statute;
(e) is by reason of an employee's pregnancy; and
(f) constitutes discrimination on some arbitrary ground.[620]

315. An employee's rights emanate from either statute or common law. While there does exist certain residual common law rights arising out of breach of contract which may be proceeded with by an employee to the high court of South

617. Section 188A of the LRA.
618. *Ibid.*
619. *Ibid.*
620. *Ibid.*

Africa, the exercise of these rights is rare and the relevance to liability issues extremely limited. By far the vast majority of cases arise out of a breach of statutory rights and obligations.

316. A dismissal that is not automatically unfair is deemed unfair unless the employer proves that the dismissal is for a fair reason either related to the employee's conduct or capacity or based on the operational requirements of the employer.[621] In addition, the employer must prove that the dismissal was effected in accordance with a fair procedure. Accordingly, even if the dismissal is proven to be related to the employee's conduct, capacity or based on the employer's operational requirements (i.e., substantively fair), it will nevertheless still be unfair (procedurally unfair) if the employer has not followed a fair procedure, as already emphasised above.

§4. Remedies Available

317. An aggrieved employee who has been either dismissed or discriminated against has the following reliefs against the employer. The commissioner may:

(a) order the employer to reinstate the employee from any date not earlier than the date of dismissal;
(b) order the employer to re-employ the employee, either in the work in which the employee was employed before the dismissal or in other reasonably suitable work on any terms and from any date not earlier than the date of dismissal; or
(c) order the employer to pay compensation to the employee.[622]

§5. Conclusion

318. From the above assessment it is clear that the general background of the South African labour market is inclusive and that it aims to put both the employer and the employee on equal footing. This means that equity in the workplace is championed in the interest of ensuring that even employees who find themselves in an uncompromising position are still in a position to claim their rights to fair labour practices. The employer is expected not to abuse his or her position of power but must act in the interests of both the business and the employees, while being guided by the principle of equity and not disregard the rule of law which might yield dangerous results.

621. Section 188 of the LRA.
622. Section 188A of the LRA.

Chapter 2. Religious Ministers and Labour Law

§1. Introduction

319. In this chapter the focus will be on whether religious ministers or pastors qualify as employees in South African labour law. This has been one of the contentious issues in the South African labour market on the basis of the fact that religious ministers owe their allegiance to God and not the church because they are called by God. Furthermore, to interrogate whether religious ministers who conclude contracts of employment with the various churches they serve: does this give rise to the employer and employee relationship, while bearing in mind that ministers are called and are serving God.

320. In South Africa there is uncertainty as to whether religious ministers qualify as employees in terms of the BCEA.[623] Due to this uncertainty various labour legislations and court cases will be discussed to provide clarity and certainty in this matter. Religious ministers play an important role in society in general and their contribution cannot be underestimated because they play a leading role in ensuring that social cohesion is achieved.

§2. The Position of Religious Ministers When it Comes to Employment in South Africa

321. As already discussed in the introduction, in the South African context there is uncertainty when it comes to the employment status of religious ministers. This is attributed to the fact that whether they are employees or not and if so who employs them, is it either the church or God? Various views in the labour market will be discussed with the aim of getting the necessary clarity and position as to the employment status of religious ministers in the South African jurisdiction.

322. Who is an employee in terms of South African labour law? An employee is a person who works under the care and control of an employer in rendering his or her services and in return to be remunerated by the employer for services rendered.[624] As to whether a minister in a church qualifies as an employee is a question that will be discussed below.

323. Although there are no court cases in South Africa in which the church was held vicariously liable for the acts of its priests or ministers, there are judgments in other areas in which the question of whether a minister was an employee or not was decided. For example, in *Schreuder v. NG Kerk Wilgespruit,* the Labour Court found that the minister was an employee of the congregation of the Dutch Reformed

623. Section 2 of the Basic Conditions of Employment Act.
624. *Ibid.*

Church where he worked because a letter of calling (*beroepsbrief*) in which the rights and duties of both parties were set out created a contract of employment.[625]

324. The agreement was that the minister would receive a wage (*traktament*) in exchange for his services to the congregation, which was also regarded as an indication that this was an employment contract.[626] The court also took the provisions of the *Kerkorde*', which *inter alia* contains rules and regulations for the employment of minsters and the termination of their services into account, in church. From the *Kerkorde* it is clear that the minster is regarded as an employee of the church, since it prescribes the requirements for a fair dismissal in terms of the LRA.[627] The court in this case accepted that the minister served more than one employer entity within the church group, namely the congregation, *die ring* and the *synod,* and thus was an employee and all laws in South Africa that deal with the employer and employee relationship were applicable to this contract and the conduct of the church in dismissing the minister was found to amount to breach of contract.[628]

325. In contrast to the *Schreuder* case the Labour Court found in *Church of the Province of Southern Africa, Diocese of Cape Town v. CCMA*, which also dealt with a claim for unfair dismissal, that a priest in the Anglican Church could not be regarded as an employee.[629] The court based its finding on the fact that the parties did not intend concluding an employment contract, and that there was no contract of employment that outlined the terms and conditions between the parties. In a nutshell the court held that the intention of the parties was not clear and thus no contract of employment was in place and the priest in this case did not qualify as an employee.[630]

326. Another interesting judgment to this legal uncertainty with the position of ministers is the court case of *Diocese of Cape Town v. CCMA.* In this case, the court accepted the church's evidence that it only provided the space and opportunity for the priest to fulfil his calling to work for God.[631] The court held that 'the relationship between a church and a minister is not apt, in the absence of clear indications of a contrary intention in the document, to be regulated by a contract of service'.[632] The court emphasised the importance of a contract and it must be clear in the sense that the intentions of the parties must be made known. In a similar vein, in *Salvation Army (South African Territory) v. Minister of Labour*, the court granted a

625. *Schreuder v. NG Kerk Wilgespruit* 1999 20 ILJ 1936 (LC).
626. *Ibid.*
627. *Ibid.*, at 1937.
628. *Ibid.*
629. *Church of the Province of Southern Africa Diocese of Cape Town v. Commission for Conciliation Mediation and Arbitration and Others* (C619/2000) [2001] ZALC 141 (7 September 2001).
630. *Ibid.*
631. *Diocese of Cape Town v. CCMA* 2002 3 SA 385 (LC).
632. *Diocese of Cape Town v. CCMA, supra* n. 1 at 385.

declaratory order to the effect that officers (preachers) of the Salvation Army are not its employees and that the Salvation Army is thus not bound by labour legislation.[633]

327. The court accepted evidence that the Salvation Army only provides the space for the officer to answer his call of duty to God and that the officer is only given guidance on administrative tasks, not on how God is to be served.[634] The fact that no remuneration is paid, but only a living allowance, and that there was no guarantee that this would be paid, also indicated to the court that there was no employment relationship.[635] What is clear from the above is that the stance of the South African courts was that in the light of the perceived special calling of a priest, he will not be regarded as an employee of the church unless it is clear that there was indeed an intention to create such a relationship.[636]

328. Since these judgments were handed down, South African courts have been prepared to hold that even though a contract specifies that it is not an employment contract, the court could, by taking the reality of the relationship into consideration, decide that a contract of employment exists between the parties based on the intention between the parties.

329. Another interesting case to note in the establishment of the relationship between a church and a minister is *Nkosinami Nkomonde v. the Diocese of Pretoria (Anglican Church of Southern Africa).*[637] The issue at the heart of this matter was whether or not priests, elders, ministers and pastors are employees of the church. If they are employees of the church, in whose name do they practice?[638] This issue has come before the CCMA and the Labour Court in various guises with various decisions. As the commissioner wrote in his award. '... an Anglican Priest working at a Diocese and being paid a stipend is performing work and placing his labour potential at the disposal of the Church. His duties are clearly defined, the times at which he must perform certain of his duties are set and he is capable of being performance managed by his seniors.'[639]

330. According to section 200A of the LRA, such a person can be presumed to be an employee. Indeed, the award continues, 'If one approaches the matter from an atheistic point of view then what is left behind is quite obviously an employment relationship.'[640]

---

633. *Salvation Army (South African Territory) v. Minister of Labour* (J 464/02) [2004] ZALC 64 (2 Sep. 2004).
634. *Ibid.*
635. *Salvation Army (South African Territory) v. Minister of Labour, supra* n. 634, at 65.
636. *Ibid.*
637. *Nkosinami Nkomonde v. the Diocese of Pretoria (Anglican Church of Southern Africa)* 2015 4 SA 399 (LC).
638. *Ibid.*
639. *Nkosinami Nkomonde v. the Diocese of Pretoria, supra n. 638, at 400.*
640. *Ibid.*

331. Against this argument the commissioner then writes:

> *I believe it is necessary to consider a fundamental concept of what a Christian Church is and where it allegedly derives its authority from. The Theocratic and hierarchical structure is as follows:*
> *At the top is 'God', then Jesus, then the Angels and then one drops down to the Earthly realm were one has the Church, a hierarchical structure within that particular Church, which cascades down to the Priest at the bottom.*[641] *All structures below God derive their authority, power, wisdom and decision-making ability from God. As such, even the Priest at the bottom end of the hierarchy operates under the power of God's spirit and for God's purposes for the people on earth.*[642]
> *The bottom line is that the Priest works for God (according to the Anglican faith) and that any arrangements, including financial payments are simply assistance by the Church to allow the Priest to fulfil his relationship and 'calling' with God to the fullest extent possible.*[643]

332. The commissioner, then went on to rule that as a consequence the church cannot allow an organisation of the state such as the CCMA to tell it whom it may appoint as priests or who it should or should not address. That's a decision left to God.[644]

333. The commissioner ruled that the applicant, in this matter was a priest of the Anglican Church against his dismissal by the church, fell outside the ambit of the LRA. Therefore, the CCMA did not have the authority to consider the matter. In short only God can determine this matter through those who manage the affairs of the church on behalf of God.[645]

334. South African courts have in recent years progressively extended protection against unfair dismissal to categories of persons not previously regarded as employees. Courts interpreted labour legislation in light of the Constitution to include persons with illegal and invalid contracts, as well as persons who were described as independent contractors. Despite this development, the LAC in *Church of the Universal Kingdom of God v. Myeni* held that the minister of religion could not be regarded as an employee because in the court's view the parties had no intention to create legal relations.[646] This judgment can be criticised on the ground that the court regarded the form of agreement, which described the service of the minister of religion as voluntary, as the only determining factor and did not take the

641. *Ibid.*
642. *Nkosinami Nkomonde v. the Diocese of Pretoria, supra n. 638, at 401.*
643. *Ibid.*
644. *Nkosinami Nkomonde v. the Diocese of Pretoria, supra n. 638, at 402.*
645. *Ibid.*
646. *Universal Church of the Kingdom of God v. Myeni and Others* (DA 3/14) [2015] ZALAC 31; [2015] 9 BLLR 918 (LAC); (2015) 36 ILJ 2832 (LAC).

de facto relationship between the parties, which pointed to an employment relationship, into account.[647] The LAC insisted that there must be a contract for the minister of religion to be regarded as an employee, while the Constitution and the LRA do not require a contract.[648] Reliance on English jurisprudence by the LAC is further not appropriate because of certain peculiarities regarding the history of churches in England, which are not shared by South African churches. The LAC did not take the unequal bargaining position of ministers of religion, nor the relevant constitutional rights of the parties, into account. The spiritual dimension of the minister's service seems to be decisive in the court's decision, although the misconduct was not a spiritual matter falling within the domain of the church's own decision-making powers. This matter should have been dealt with by civil courts since there was no danger of doctrinal entanglement.[649]

§3. CONCLUSION

335. In conclusion, from the discussion above and the reasoning of the courts in the above-mentioned court cases, it is clear that in South African law whether a religious minister qualifies as an employee or not is a question of fact. This means that the intention of the parties must be made clear from the beginning of the relationship in terms of a written contract and the contract of service must state out clearly the rights and responsibilities of the religious minister as well as that of the church as an employer.

336. If there is no form of contract or remuneration as well as the terms of service stated and the religious minister is only awarded an allowance by the church, this means that the religious minister will not be considered an employee because he or she would not fit in the requirements of an employee. In such instances as it was alluded above the church would be considered a body that is used to help the religious minister to realise his or her calling to God, and this will not constitute a contract of employment between the religious minister and the respective church. In a nutshell, this means that for a religious minister to be considered an employee and for the church to be held liable on the basis of all labour regulations in South Africa, which are applicable to all employees, must be fulfilled and upheld.

647. *Ibid.*
648. *Ibid.*, at 2833.
649. *Universal Church of the Kingdom of God v. Myeni and Others*, *supra* n. 647, at 2834.

Chapter 3. Other Church Employees and Labour Law

§1. Introduction

337. The contract of employment between the employer and employee is key to be able to distinguish one as an employee or a volunteer of a particular organisation or church in this particular context. In this chapter, based on the background and arguments that were raised in the previous chapter, it is important to draw a distinction between employees of the church and volunteers of the church who normally render their services for free and will not qualify as employees of the church. The aim in this chapter is to ensure or distinguish if indeed church employees or volunteers are protected by the various labour laws in South Africa and if so to what extent.

338. Furthermore, in most instances the church congregants who are normally old and unemployed would render their services to the church for the betterment and prosperity of that church on a voluntary basis. Will this type of service by the congregants of a particular church amount to a contract of employment, thus giving rise to the employer and employee relationship? This is a question of fact and will be considered in this chapter with the aim of ensuring that people who render their services to the church as either employees or volunteers are protected from any form of abuse. The various employees of the church who play a supporting role in addition to the volunteers of the church will be considered because they qualify as employees in South Africa.

§2. Church Employees in South Africa

339. Churches in South Africa are non-profit organisations and are autonomous in the sense that they are independent bodies, and in most instances, some churches have an administrative and clergy offices.[650] Normally, the clergy office will consist of a minister who can be supported by other clergy officials, such as the Bible woman and a deacon who are normally the people who will assist the minister to execute his or her role in terms of spiritual and evangelical works.[651] Furthermore, in the office of the clergy there are preachers of the word of God and these are lay preachers who also assist the minister with the preaching and other evangelical works which are in connection with the church.[652] The Bible women and deacons are people just like ministers who are called from the higher place to serve in the house of the Lord, and these people are not required to work or get formal employment outside the church. This is the position in the Methodist Church and both the Bible woman and deacon will receive a salary or allowance on a monthly basis and get other employment benefits such as medical aid and provident fund.[653] This

650. Siwa, *supra* n. 268, at 3.
651. *Ibid.*
652. *Ibid.*, at 4.
653. *Ibid.*

means that both a Bible woman and deacon are employees of the church and there is an employment contract between them and the church in the sense that they will report to the relevant church minister, who will also be responsible for allocating work for them in terms of the church operations.[654]

340. Bible women and deacons are also expected to retire at the given time just like ministers and give other people an opportunity to lead the church for the purposes of its sustenance and growth for generations to come.[655] Deacons and Bible women can also be dismissed from their role in cases where it is found that they acted with misconduct and in violation of the rules and procedures of the church.[656] The same could be said of a minister once it has been established that he or she is an employee and is involved in conduct that puts the image of the church into disrepute. In such instances the minister will have to be disciplined and in other instances he or she can be dismissed by the church.[657] From the role and assessment of the position of both the Bible woman and deacon it is clear that they are employees of the church in the office of the clergy, and they are protected by the various labour law provisions in South Africa and in their employment contract such provisions must be incorporated. This means that the church cannot dismiss them unfairly or discriminate against them as this will be contrary to the provisions of labour law and most importantly in violation of the provisions of the Bill of Rights as per section 9 of the Constitution.[658] When it comes to lay preachers this is not the case as these are people who also say that they are called to serve in the house of the Lord and in return for having rendered all the services as per the office of the clergy they do not receive any remuneration or donation.[659] Lay preachers have employment outside the church and for them to be preachers of the word of God is a special gift on its own in the sense that it is not everyone who can be a preacher and for them to execute this role is without a doubt an honour which no money or gratitude can express or equate.[660] From the role of lay preachers in the church it is clear that these are people who are congregants of the church and play an important role in ensuring that the Gospel of God is spread out, and they are not employees of the church. This means that lay preachers are not employees of the church and are not protected by any labour laws in South Africa.[661] Lay preachers hold the view that their reward is not from this earth but it is from a higher place which is in heaven, and that reward awaits them when they leave this earth.[662] From the office of the clergy it is clear that it is not everyone who is in that office qualifies to be an employee as it was outlined above. However, all the participants in that office share a common goal or vision that the word of God must reach the four corners of the world as per their respective callings in the ministry of Christ.

654. *Ibid.*, 5.
655. *Ibid.*
656. *Ibid.*, at 6.
657. *Ibid.*
658. Section 9 of the Constitution.
659. Siwa, *supra* n. 268, at 6.
660. *Ibid.*, at 7.
661. *Ibid.*
662. *Ibid.*, at 8.

341. Another arm of operations in churches today in South Africa is the administrative office and in this office the people who work in it play a supporting role to the clergy. The personnel in this office will include the secretary to the minister and other administrative staff. In most cases the secretary and the administrative personnel will work in the office on a full-time basis and be provided with a contract of employment in relation to how they are going to be remunerated and their working conditions.[663] These people will support the clergy and ensure that the church administration is handled with care and efficiency in the interest of the church. Furthermore, the administrative personnel will report to the minister and can also be disciplined by the church in case it is found that there is a violation of the Code of Conduct or such a personnel has acted in an improper manner.[664] Meaning that the administrative personnel in churches are employees and are covered by the labour laws of South Africa against any form of discrimination or unfair dismissals which will be contrary to the interests of the public. However, it is important to also note that the administrative personnel of a church can be volunteers who are also members of that particular church and this will be a different story and position. Meaning that these administrative personnel will not qualify as employees of the church and will not receive any form of labour protection but will be volunteers who have an aim of ensuring that they work towards the development of their church. This form of conduct will be considered as good will, and it will not constitute employment.[665]

342. Furthermore, apart from the office of either the clergy or administrative personnel there are also other office bearers who work in church, including the caretaker and the gardener who are in charge with ensuring that the church premises are kept clean and safe.[666] In most cases the church caretaker and gardener will have an employment contract with the church and do not have to be a church member of that particular church domination. This means that they are only rendering a service to the church which is the upkeep of the church and its premises and to be remunerated for the services rendered.[667] Caretakers and gardeners will normally report to the office of the minister who will act on behalf of the church and ensure that they render their services as per their employment contract.[668] It is possible that in this case some church members of the various dominations can delegate this role to their members due to financial constraints and in such a case there will not be any contract of employment and the services rendered will be a voluntary initiative which aims to benefit the church.[669] In this case there will not be any employment contract in place and both the caretaker and the gardener will not be remunerated because they will be acting in the interest of the church. Under these circumstances the church will provide some kind of a gift or gratitude to the various congregants

663. *Ibid.*
664. *Ibid.*, at 9.
665. *Ibid.*
666. *Ibid.*, at 10.
667. *Ibid.*
668. *Ibid.*, at 11.
669. *Ibid.*

who have rendered their services as required by the church.[670] This gift will not amount to remuneration because it will create an impression that a contract of employment exists between them and the church. When it comes to the employment of church employees, the facts and circumstances of each case need to be taken into account to determine if we are dealing with an employee or volunteer in church because these are two different things.

§3. Conclusion

343. To qualify as employees of the church, they need to comply with the provisions of the Labour laws in South Africa and must have an employment contract with the church for the services to be rendered. People who are volunteers in church with the aim of assisting in church and who are thanked for the services do not qualify as employees and cannot claim employment benefits to that effect. In a nutshell the intention of both the church and the various personnel must be made clear when it comes to the obligations they intend to create in order for an employment contract to be created.

670. *Ibid.*, at 12.

Part VII. Religious Communities and Protection of the Individual

344. In this part, the discussion focuses on the protection of fundamental rights which an individual enjoys as a member of a religious or cultural group. However, it is important for the reader to note that some of these fundamental rights which an individual enjoys as a member of a group have been extensively discussed in the text earlier, and it will not be necessary to repeat in this chapter. The fundamental rights which an individual as a member of a religious or cultural community, which have been discussed already, include the right to privacy; freedom of expression; due process norms and the various religious groups, bearing in mind the pluralistic nature of the South African community; and the issue of unfair discrimination both from the constitutional and legislative point of view.

345. The topics that will be covered in this part are the following: freedom to marry; professional secrecy and medical deontology will be discussed under one heading due to the fact that they interrelate and have to do with the employment or profession which a member of a religious community holds in society and lastly there will be a discussion on penal law and religion as the concluding chapter on the topic of the role of an individual in a religious or cultural community.

Chapter 1. Freedom to Marry

§1. Introduction

346. In the new South Africa where the Constitution is the supreme law of the land, it means that people's fundamental rights are entrenched and protected by the Constitution and this includes the freedom of an individual person to marry. An individual of a certain religious group or culture can get married to another person who is from another religious or cultural group, and they can conclude into either a civil marriage, Muslim marriage or an African customary law marriage. This is to affirm and show the diversity of the population of South Africa, and that today black and white people irrespective of their religion or culture can marry each other. It is important to note that this was not the position during the colonial and apartheid era, in which only whites would be married to each other and the same would apply with black people. Mixed racial marriages were not permitted during the apartheid era.

347. However, in the new South Africa the position without a doubt has changed to such an extent that the freedom to marry is left to the individuals to choose as they please irrespective of their religion or cultural background. Despite this wonderful progress in the development of the South African law in recognition and affirmation of the dignity of people to have the freedom to choose who they want to marry, there are challenges. Some of these challenges will be discussed in this chapter with the aim of exposing this hindrance as well as coming up with solutions in the process in order to ensure that the freedom to marry which an individual enjoys is protected and promoted.

§2. Failure to Recognise Religious Marriages

348. South Africa's Constitution allows for laws to be passed that recognise religious and traditional marriages as well as personal laws and family laws. These are systems that regulate marriage, divorce, matters affecting children, including the important aspect of guardianship, custody, access, maintenance and succession as well as inheritance.[671] This is a serious departure from the colonial and apartheid period as already discussed above, when South Africa's common law definition was restrictive in such a way that marriage was defined as a union between one man and one woman. In other words, a legal marriage in South Africa was characterised by heteronormativity and monogamy.[672] This was at the exclusion of same-sex marriages and polygamous marriages and this affected African customary marriages, Muslim marriages, Jewish and Hindu marriages, which are all polygamous in nature and the freedom to marry an individual was compromised.[673]

671. C. Rautenbach, *Gender Equality and Religious Family Laws in South Africa*, 3 Queensland UTLJ 168 (2003).
672. *Ibid.*
673. *Ibid.*, at 169.

349. The situation changed when the Constitution was adopted and, for example, there was an adoption of the Recognition of Customary Marriages Act, which afforded African customary marriages full legal recognition.[674] Furthermore, there was an adoption of the Civil Union Act which made marriages between same-sex couples legal and enforceable in South Africa.[675] These two pieces of legislations aimed to promote the freedom of an individual to marry. without paying attention to his or her religion, culture, gender or sexual orientation which is key in a constitutional democracy.

350. However, the above cannot be said about polygynous marriages such as Muslim marriages, Hindu marriages and Jewish marriages because they have not yet been afforded legal recognition. This failure on the part of the legislature can be said to be tantamount to saying that religious marriages are inferior and less deserving of legal protection and enforcement.[676] Furthermore, this failure infringes the rights to human dignity, freedom of religion and equality of spouses in religious marriages and the children born within those marriages.[677] It can be argued that also this failure restricts the freedom of an individual to exercise their right to choose whom they want to marry, and this cannot be permitted in the new South Africa.

351. Women are particularly badly affected by this non-recognition, for example a Muslim wife who has substantially a small estate compared to her husband. She could be left financially destitute should the marriage end in a divorce.[678] Yet she has no recourse under South African law because her marriage is not legally recognised, but given putative recognition like a civil marriage.[679] Furthermore, Muslim women are also at a disadvantage because polygyny is not regulated. This means that men can enter into subsequent marriages without their knowledge and this is very unfair as well on the basis of equality.[680] Women also have difficulty accessing a Muslim divorce even though their right to divorce is recognised under Islamic law. This has resulted in many women being held hostage in unwanted marriages, which again affects the freedom to marry as protected by the Constitution.[681] This means that people who are married under religious marriages or law are at a serious disadvantage.

352. Government is called upon to give attention to the plight which the Muslim community and other minority religious groups experience, which is without a doubt unjust. The non-recognition of these unions not only is it an infringement to their right to freedom of religion and belief but also it could be taken further to say

---

674. Section 1 of the Recognition of Customary Marriages Act 120 of 1998.
675. Section 1 of the Civil Union Act.
676. Rautenbach, *supra* n. 672, at 169.
677. Section 10 of the Constitution.
678. F. Cachalia, *The Future of Muslim Law in South Africa*, 12 OP 22 (1991).
679. *Ibid.*
680. *Ibid.*, at 23.
681. *Ibid.*

that it is a violation of the human dignity of the people who are part of these minority religious groups.[682] The freedom to marry, which each and every person enjoys in the Republic, is compromised when it comes to these minority religious groups, and it is unjust because it is not in line with section 36 of the Constitution which is the limitation clause.[683] Furthermore, the non-recognition of Muslim marriages and other minority religions means that people who are members of these religious groups at some point will desist to conclude into such marriages or union. However, it will not be on the basis of their free will but on the fact that they want to avoid the consequences which come with the non-recognition of such unions. This is unfortunate and requires urgent attention of the legislature. Freedom to religion or culture goes hand in hand with the freedom to marry and they speak one and the same thing with the purpose of affirming the independence and autonomy of an individual in the new South Africa.

§3. Conclusion

353. In conclusion, freedom to marry is central in a democratic dispensation and it is something that the government must never underestimate and must ensure that it is protected to avoid the injustices of the past. Government, and in particular the legislature, is called upon to act with the necessary urgency in giving recognition to Muslim marriages and other minority religious marriages, and as it was discussed the failure to do so is in actual fact a direct violation of the freedom to religion and freedom to marry, which are without a doubt central pillars of building social cohesion.

682. Section 10 of the Constitution.
683. Section 36 of the Constitution.

Chapter 2. Professional Secrecy and Medical Deontology

§1. Introduction

354. Members or individuals of either a religious congregation or cultural group apart from their beliefs and values belong to different professions. This means that an individual as a member of a religious or cultural group makes contact with different people or groups as part of his or her employment among other things. What happens if there is a dispute between the profession of an individual and their culture or religion at a particular point in time. This conflict of interests will be explored in this chapter.

355. Professional secrecy and medical deontology are discussed in this chapter as one because they both have to do with a professional life of a member of a religious or cultural group. Due to this connection between the two, they will be discussed in this chapter together with the aim of showing or drawing a distinction between a religious life of an individual and the professional life of that particular individual. Furthermore, it will be explored as to how a balance could be maintained between the professional life and a religious life of an individual, to avoid the conflict between the two as mentioned.

§2. Professional Secrecy

356. As already outlined above members of a religious or cultural group can occupy very high positions or professions based on their level of education or talents. It is important to note that freedom of trade or occupation is guaranteed in South Africa by the Constitution and this includes members of religious and cultural dominations.[684] This means that religious and cultural organisations have individuals in the church who are qualified as lawyers, accountants, medical doctors, nurses and teachers to name but a few. The professions that are listed above and those which are not mentioned have their professional standards and procedures, which an individual irrespective of his or her religion or culture must comply with in order to be admitted to that profession. For example, if an individual church member is admitted as an attorney in South Africa it means that all of his or her professional conduct will be tested and regulated by the law society of that particular area and will be expected to take an oath.[685] Taking an oath is ensuring that individual attorneys are to uphold and protect their profession and that they will owe their allegiance to the profession they have chosen. This will mean that the individual concerned is expected to protect and uphold the integrity of that profession, irrespective of his or her religious beliefs or convictions.[686]

684. Section 22 of the Constitution.
685. Section 11 the Attorney's Act.
686. Section 12 the Attorney's Act.

357. This means that they are expected to act in the interest of their profession based on the work they do, and in most instances it will be in the public interest to act in that particular manner, and hence the religious or cultural convictions of that particular individual will take a back seat or be limited, and it will be justified.[687] In cases where the individual professional decides to act contrary to the prescribed procedures and rules of his or her profession – which in this case as an attorney with the aim of promoting and protecting his or her religious or cultural beliefs – it can lead to a disciplinary hearing and dismissal of that particular individual on the basis that he or she has disregarded his or her oath of office, which constitutes gross negligence.

358. Furthermore, an individual member of a religious or cultural group must be in a position to draw a distinction between his or her professional life and his or her religious life. This is very much fundamental because a religious or cultural life of an individual is subjective in nature and based on principles of love, caring and compassion and while the professional life is the exact opposite. Professional life is an objective process in which no feelings are applicable but only the correct interpretations and applications of the laws and standards are supreme.[688] Maintaining and drawing a distinction between the professional and religious or cultural life of an individual is key because the two are different and must be kept separate in most instances to avoid conflict because in most instances an individual might be conflicted between his or her professional life and their religion or culture.

359. Drawing a distinction between professional life and religious or cultural life of an individual will help the particular professional to maintain professional privilege and this means that communication between that professional and his or her clients is privileged.[689] Meaning that a professional is prevented to disclose communication he or she had with clients to anyone, and the courts cannot compel the said professional to disclose such information because it will be based on the fact that it is privileged and must be protected.[690] If that professional discloses the said information it will result in dire consequences for the said professional in the sense that he or she can be dismissed for breaching this privileged information and in some instances clients can institute a civil claim against the said professional for the violation of their right to privacy based on the unwarranted disclosure.[691] This means that individual professionals must exercise the necessary caution and ensure that privileged information that emanates from their professional capacity is not disclosed in their private or religious spaces, even though such information adversely affects their private or religious life.

687. Section 36 of the Constitution.
688. Section 2 of the Attorney's Act.
689. Section 20 of the Attorney's Act.
690. Section 21 of the Attorney's Act.
691. Section 25 the Attorney's Act.

§3. Medical Deontology

360. Medical deontology has to do with an aspect of maximum professionalism that is expected on the part of a medical doctor in ensuring that the patient is treated successfully.[692] A great deal of professional ethics is expected of a medical doctor who in this case can also be a member of a religious or cultural group, and the medical practitioner must act in the best way possible as prescribed by their profession in treating the patient.[693] This will entail giving the patient the prescribed medication, maintaining the diagnosis and treatment of the patient private and avoid making conversations that will endanger the health and well-being of that particular patient.[694] In a nutshell a medical practitioner is expected to provide a great deal of support and professionalism and act ethically in the process because there is a thin line between religion and ethics, but in this instance the medical doctor is bound by the ethics of the profession and not that of his or her religious convictions.[695]

361. In cases where it is found that a medical doctor has acted contrary to his or her oath of office and followed his or her conscience, which can have a religious or cultural component, the medical practitioner will be subjected to disciplinary proceedings by the regulatory body which is the Health Professions Council of South Africa (HPCSA). In some instances he or she will be suspended or dismissed from practising medicine.[696] This is because the HPCSA acts as a watchdog against medical practitioners who abuse their position against vulnerable members of the community.

§4. Conclusion

362. From the above it is very much clear that professionalism is key in maintaining a working balance between religion and professional life. These two roles must not be mixed because if they do then a conflict will arise and it will be without a doubt to the disadvantage of the individual member of that religious domination or cultural group. Furthermore, religious and cultural groups must encourage their members to pursue various career opportunities, which in return will strengthen and boost their existence as well as sustenance on the basis of the support which these professionals will bring to the church. Furthermore, it will be argued in the chapter to come that direct financing of religious organisations or communities depends on active members of that particular religion or culture taking part in active economic opportunities.

---

692. Section 26(1) the Attorney's Act.
693. W. Mary, *Medical Ethics Made Easy*, 21 Professional Care Management 81 (2016).
694. *Ibid.*, at 22.
695. *Ibid.*, at 23.
696. *HPCSA Regulatory Body of South Africa's Medical Practitioners*, https://citizen.co.za/news/south-africa/1991655/health-professions-council-struggles-in-fight-against-corruption/ (accessed 17 Aug. 2018).

Chapter 3. Penal Law and Religion

§1. Introduction

363. South Africa is a young democracy under the rule and supremacy of the Constitution, which took over the role from parliament which used to pass arbitrary laws to the detriment of the majority of black people. Despite the infancy, South African Constitution has made great strides in the protection of fundamental rights which all the people of South Africa enjoy today. However, the same cannot be said about religion and culture in South Africa because they are ancient and form part and parcel of the daily lives of the people of South Africa and the broader continent of Africa.

364. This chapter will look into areas where the religious and cultural beliefs of people are compromised or violated by others. If indeed it is found that there is such a violation of either religion or culture what would be the mitigating factor to address the injustice in terms of the law. While bearing in mind that freedom of religion and culture is firmly protected by the Constitution and if there is a threat to this particular right then the law must provide the necessary remedies to address this injustice.

§2. Penal Law and Religion

365. Throughout this research project, the importance of the role of religion and culture in South Africa is emphasised, and it can be said that they both form part and parcel of the lives of all people in the Republic. In addition, remembering the role religion played in the liberation of South Africa during the dark era of apartheid and how currently in the new democratic dispensation religion plays a major role in bringing the necessary social cohesion.[697] This significant and important role that is played by religion on the daily lives of the people of South Africa is the major reason why today freedom of religion and culture is one of the fundamental rights in the Constitution.[698] This affirms and recognises the role religion has played and is still playing to this very day, which is something that cannot be underestimated without a doubt. The fact that religion is fully protected by the Constitution and other pieces of legislation means that it is significant in society and any threat to this freedom to religion will amount to a threat to the Constitution. Freedom of religion and culture means the liberty to choose without any fear or favour, and this must be done in a way that does not undermine or discriminate against the religion or culture of other people.[699] This means that freedom of expression in connection with a religion or culture does not mean that it is a freedom to promote one religion

697. J.C.W. Van Rooyen, *Dignity, Religion and Freedom of Expression in South Africa*, 12 OR 1 (2011).
698. Section 15(1) of the Constitution.
699. Van Rooyen, *supra* n. 698, at 2.

or culture at the expense of the other.[700] Freedom of expression in this case means the freedom to exercise and enjoy your cultural and religious practices in line with the rule of law.

366. South Africa being a pluralistic society, diversity is the order of the day and must be embraced. This means that the majority religious dominations cannot discriminate or dominate the minority, but all religious or cultural groups must coexist and the element of equality and fairness must be present in this particular space.[701] Furthermore, this means that hate speech, which is directed at destroying the image and reputation of one religion by the other, is not condoned in South Africa, and it will be contrary to the provisions of freedom of expression.[702] Hate speech or slander also violates fundamental rights which include in particular human dignity and freedom of religion, and this can result in the guilty party being held liable for damages by a civil court.[703] Furthermore, hate speech can in some cases lead to criminal prosecution because it can result in people engaging in violent activities in some instances with the aim of protecting their religious or cultural belief.[704] This shows that when there are cases of hate speech among different religions or cultural groups, it is important to resort to law-enforcement agencies first because failure to do so can result in people taking the law into their own hands, resulting in severe consequences.

367. Law as a regulatory mechanism, in this case the Constitution as the supreme law, must intervene and ensure that all religious and cultural organisations are protected and maintained. There must not be any form of bias or favouring one at the expense of the other, but the rule of law must be the dominating one all the time. Furthermore, law has indeed a role to play as a mediator in cases where religious and cultural organisations are at loggerheads with each other and bring the necessary stability and order, with the aim of ensuring that the rule of law remains supreme all the time. The role of the law as a mediator is key in order to prevent and stop religious wars from taking place in the Republic, which is unfortunately the order of the day in some African countries.

§3. CONCLUSION

368. Tolerance and respect for the rule of law and other religious and cultural groups is expected from each and every member of South Africa with the aim of ensuring the democracy of South Africa is protected and upheld at all costs. Failure to respect the religious and cultural affiliations of other people will result in dire consequences, which can either lead to a civil or a criminal claim in the process. This must be avoided at all costs and a great deal of tolerance for diversity is key

700. Section 16(1) of the Constitution.
701. Van Rooyen, *supra* n. 698, at 3.
702. Section 16(1) of the Constitution.
703. Van Rooyen, *supra* n. 698, at 4.
704. *Ibid.*, at 5.

in ensuring that different religious and cultural dominations coexist in South Africa because they all play an important role in the building of the new South Africa.

Part VIII. Church Financing

Chapter 1. Direct and Indirect Financing of Religious Communities

§1. Introduction

369. Religious communities for their survival and continuous existence need the necessary financial support from within the religious community and also from external people or donors. This must be done within the parameters of the law of the Republic and there must not be any form of criminality during this process of raising funds.

370. Religious communities like any autonomous institutions have needs and obligations to fulfil and raising funds is one of the major factors that contribute towards the existence of that religious community or organisation. This is one of the major reasons that religious organisations are not required to pay tax in South Africa, unlike other countries such as Germany and Denmark.[705] It can be said that the reason why religious communities or organisations are not taxed in South Africa is because of the rich history and role religion has played in the democratic dispensation of South Africa, and furthermore not to put financial burden on them which can adversely affect their existence. This chapter will focus on how religious communities receive or raise funds directly or indirectly with the sole purpose of keeping their ministry alive in their respective communities and also in the interest of their members.

§2. Direct Financing of Religious Communities

371. Section 11 of the South African Charter of Religious Rights and Freedom states that like all voluntary associations, a religious institution has the right to perform different actions to further its objectives and to receive, manage and spend money for that purpose.[706] During this initiative, religious institutions are expected to act as responsible bodies to govern themselves in financially sound and responsible ways and to act in the best interests of their members.[707] Direct financing

705. M. Valova, *Financing Churches and Religious Societies in the 21st Century*, 322 RCVP 146 (2010).
706. Section 11 of the South African Charter of Religious Rights and Freedom.
707. *Ibid.*

entails a process in which within the members of that particular religious community or institution different fund-raising initiatives are made in order to keep the ministry going. Direct financing of religious institutions can include members taking the initiatives by paying their offerings monthly to the church; by contributing to church collections; by paying their affiliation or registration fees annually; by participating in thanksgiving services, which is normally annual.[708] These are some of the recognised fundamental practices which are directly contributing to the financial position of a religious community. This is one of the ways in which members of that particular religious community or institution are taking charge in building as well as preserving their religious community by ensuring that it is financially sound to meet its obligations either on a monthly or daily basis.

372. Furthermore, direct financing from the members of the religious community or organisation can emanate from engaging in different fund-raising initiatives with the aim of ensuring that the religious ministry is financially afloat and sound. This can include to make talents by inviting other churches to their church and raise funds and while at the same time building relations with other religious groups.[709] Organising choir competitions, concerts and sell tickets to members of that church community and the broader community. During the concert different beverages can be sold to the people who are coming for the concert and this will in the process raise the necessary funds that are required by the church.[710] Direct financing is a way of showing commitment by the members of that religious community to their ministry, and they are expected to lead by example and preserve the ministry for the next generation. Furthermore, direct financing can be seen as a way in which the children of God are showing and expressing gratitude to God for all the blessings and mercies he has showed them.[711] In a nutshell, direct financing of a religious community or organisation can be described as a covenant the children of God have with their Creator and is something that is fundamental. All the funds that are raised either directly or indirectly as will be discussed below must be directed towards the ministry of God in that particular religious community or organisation. In cases where this is not the case, then in most instances the people who are involved with church administration which include the minister can be charged and convicted for fraud and corruption by the law-enforcement agencies in South Africa.[712] This affirms that despite the autonomy of a religious community or organisation it is still subject to the rule of law and must not act contrary to the law.

373. Religious communities or organisations must account for their financial position to their members monthly and annually. This will affirm that it is transparent and ensure that it records as well as audits their financial statements in order to give confidence to their members who are its direct funders for its existence.[713] Furthermore, in South Africa religious communities or organisations are expected to

708. Siwa, *supra* n. 268, at 85.
709. *Ibid.*, at 86.
710. *Ibid.*
711. Holy Bible 2 Corinthians Ch 9: Verses 7.
712. Section 12 of the South African Charter of Religious Rights and Freedom.
713. Siwa, *supra* n. 268, at 87.

submit their financial statements to the South African Revenue Services annually even though they are tax exempted and this is to guard against money laundering and other criminality which can take place in religious communities or organisations.[714] The primary funders of a religious community or organisation are its members, and this gives rise to direct financing of that particular ministry and hence religious communities or organisations are all the time in need of more members with the sole aim of growing their ministry and funding among other things.

§3. Indirect Financing of a Religious Community

374. Indirect financing of a religious community is the opposite of direct financing. Indirect financing has to do with sourcing out funds from external people or businesses who are not part of that religious community or organisation, but must share the same goal in most instances.[715] Indirect financing of a religious community in most cases has to do with the church leadership engaging with sponsors to support the various initiatives or projects of that particular religious community or organisation. This can include looking for sponsors from the local business people; political parties and in some instances to international bodies for funding.[716] In return the sponsors can ask the religious organisation or community to account as to how they utilised the funds and in some instances particularly businesses can request the religious community or organisation to advertise its services to their members in return to the sponsorship they have given them.[717] This will be a reciprocal relationship in which both parties are benefiting from each other, while at the same time building the necessary social cohesion and stability.

375. When it comes to indirect financing of religious communities or organisations it is important to note that an external body must not compel or make a religious community or organisation to compromise its values and the rule of law, to such an extent that it is now seen to promote some form of criminality which is contrary to the law. For example, a religious community or organisation to be used to safe keep proceeds of illegal activity in return for the donation it received will be contrary to the rule of law.[718] Furthermore, religious communities or organisations cannot accept a donation from a business or political party on condition that it is to defame or discriminate against other religious communities or organisations, as this type of conduct will be seen as contrary to public policy and destroying social cohesion.

376. From the above assessment of indirect financing of religious communities or organisations it is clear that for this to happen in order to sustain the ministry of the church it must be a reciprocal relationship based on sound legal principles. This

714. Section 13 of the South African Charter of Religious Rights and Freedom.
715. Valova, *supra* n. 706, at 147.
716. Siwa, *supra* n. 268, at 88.
717. *Ibid.*, at 89.
718. Section 13 of the South African Charter of Religious Rights and Freedom.

means that donors of that religious community or organisation must not impose conditions that will compromise the values and ethos of that religious community, which can result in an act of crime being committed in the process. Religious communities or organisations must not compromise their values in return for donations because if they do then it will be to their detriment and downfall in the process. Religious communities or organisations as important bodies in societies are vehicles of social cohesion and nation building, and they must uphold the rule of law and their values or morality even if it means that they are not to receive external donations or funding.

§4. CONCLUSION

377. In conclusion, from the discussion above it is very much clear that church funding either direct or indirect is fundamental for the continuous existence of a religious community or organisation. However, this does not mean that religious communities or organisations can engage in illegal activities to obtain church funding but are bound by their internal processes as well as the rule of law in South Africa which sets and guides them in relation to how they are expected to get funding. In cases where they are found to act contrary to those processes and laws they will face the full might of the law, which can put their existence in serious risk or threat, if necessary care is not exercised.

Part IX. Education

Chapter 1. Religious Education in Public and Private Schools

§1. Introduction

378. In this chapter emphasis will be placed on the education system of South Africa with specific reference to religious education included in the school curriculum. South African education system just like that of other countries in the world is divided into two parts or sections, that is, public and private schools. This is similar to the public and private healthcare sectors in South Africa which was discussed in the earlier chapter, when emphasising recognised fundamental rights. A discussion will be drawn in this particular instance as to how both private and public schools operate in South Africa in fulfilling their constitutional obligations of giving adequate access to education as per the Constitution.[719]

379. Furthermore, section 7 of the South African Charter of Religious Rights and Freedom states that everyone has the right to be educated or to educate their children or to have them educated in accordance with their religious or philosophical convictions.[720] This means parents have the right to choose the form of education for their children and schools are delegated the parental authority when they take on the task of education, because the state has no prior right to educate the young especially on religious matters which is something people will exercise independently and freely without the interference of the state.[721]

§2. Religious Education in Public Schools

380. Public schools in South Africa are considered to be in a position similar to the public hospitals as discussed above, because this system of education is faced with lots of challenges which include lack of resources; poor infrastructure; fewer committed teachers and overcrowding in most instances.[722] It is stated in the South

719. Section 29(1) of the Constitution states that everyone has a right to a basic education, including adult basic education, and to further education, which the state through reasonable measures must make progressively available and accessible.
720. Section 7(1) of the South African Charter of Religious Rights and Freedom.
721. Section 15(1) of the Constitution.
722. *Education in South Africa a System in Crisis*, https://city-press.news24.com/News/education-in-south-africa-a-system-in-crisis-20160531 (accessed 19 Oct. 2018).

African Constitution that each and every one or child in this instance has a right to access to quality education.[723] This means that public schools are the ones that are fulfilling this particular role for the people of South Africa, and most if not all public schools are no-fee schools. This simply means that they are fully funded by the government.[724] Public schools apart from offering mainstream educational curriculum to the children, for example subjects like Mathematics, English, and Biology to name but a few, they also offer religious education, for example Biblical studies.[725] Furthermore, each and every morning in most if not all public schools children are expected to gather at the school hall or ground for their daily prayers and for the school announcements and this is one practice that forms part and parcel of religious education in public schools.[726]

381. Public schools are playing a very vital role when it comes to the education system of all the children of South Africa, and they must be supported and protected to ensure that they are sustained. Freedom of religion includes the principle that every person has the right to receive and provide religious education, training and instruction, even in public schools.[727] As in all education, the state may choose to subsidise this education, which is the case in public schools by even inviting church ministers to come and minister to the children who belong to a certain religious belief about the importance of that religion or belief.[728] This is another example of the approach in South Africa in which the government is not prohibited to be involved in religious matters, but it has a positive duty to promote and facilitate the exercise of the right to religious freedom as provided by the Constitution.[729] The only condition that is imposed is when the state does engage with religion, for example to subsidise religious education and training, it has to act impartially and may not favour anyone or discriminate against anyone. This will entail the government to consult with all religions in that particular area or community to ensure credibility of programmes and to minimise any fears from religious communities.[730] This means that care should also be given to the principles of accommodation and alternative delivery so that diverse means are not confused with the ends the state wishes to achieve.

382. People and children in particular carry their convictions with them, also in public places like schools, and must be offered an opportunity and space to observe their religion. Logically, this means ethos will reflect the beliefs or convictions of the majority in the school. However, this ethos must be practised in an equitable, free, voluntary and non-discriminatory way. The rights of minorities must be respected and no one in school may be victimised or discriminated against because

723. *Ibid.*
724. *Ibid.*
725. *Ibid.*
726. Section 15(1) of the Constitution.
727. Section 29(1) of the Constitution.
728. Section 7(2) of the South African Charter of Religious Rights and Freedom.
729. Section 7(1) of the South African Charter of Religious Rights and Freedom.
730. Section 7(2) of the South African Charter of Religious Rights and Freedom.

their beliefs or convictions differ from those of the majority.[731] The law challenges public schools to develop mechanisms and processes to ensure that these conditions are met, with the aim of ensuring that children from different religious backgrounds are accommodated and not discriminated against on the basis of their religion. This means that public schools must embrace the diversity of the South African community and not see it as an excuse to favour one group at the expense of the other.

§3. Religious Education in Private Schools

383. Private schools are accessed by children whose parents are able to access this form of education in South Africa, which is independent education from the state. Private education is normally considered to be superior to the public education system because of adequate resources; limited number of children and committed educators as well as good infrastructure as compared to the public education system.[732] Children who access private education just like their peers in public schools in addition to their mainstream educational curriculum they are also offered religious studies and they will also have their morning or afternoon devotions to enable them to stay in touch with their religious convictions even if they are in school.[733] Furthermore, parents will choose to educate their children in a private school that will promote, or is closely linked with, their religious beliefs or convictions. Section 7(3) of the South African Charter of Religious Rights and Freedoms states the right to establish private educational institutions on the basis of a particular religion or philosophy or faith, and the right to observe and impact religious convictions.[734] Such a private institution may refuse to promote, teach or practice other beliefs or convictions. Parents apply to and contract with such an institution on a voluntary basis and it should be accepted that they cannot then object on behalf of their children to the religious activities of the institution.[735]

384. To allow an objector to change a school effectively subordinates religious rights in association with individual rights. The right to exit exists, and it is preferable as a last resort should accommodation become impractical or impossible.[736] There is no right to obliterate the belief frameworks of a single ethos school because a learner may object to the religious or philosophical viewpoint of a school. It is viewed that the alternative approaches that misapply appropriate accommodation as a misunderstanding of how the principles of reconciling different beliefs should operate properly alongside respect for diversity and accommodation.[737]

731. *Ibid.*
732. *Why Private Schools Are an Excellent Investment Opportunity in South Africa*, https://businesstech.co.za/news/business/189369/why-private-schools-are-an-excellent-investment-opportunity-in-sa/ (accessed 19 Oct. 2018).
733. *Ibid.*
734. Section 7(3) of the South African Charter of Religious Rights and Freedom.
735. *Ibid.*
736. Section 7(4) of the South African Charter of Religious Rights and Freedom.
737. *Ibid.*

385. Parents who take their children to private schools to access private education and religious education are bound by the code or practices of that private school, and they must ensure that they comply with them. This means that parents and their children who have subscribed to those rules and practices of the private school voluntarily without any undue influence or pressure are expected to uphold those rules and procedures at all costs. They cannot impose new rules or procedures in the name of religion, unless of course it is found that the code or laws that are in place in that private school are discriminatory in nature or promote violence. Then in such cases they will be declared to be null and void.[738] If this is not the case then the parents or the child involved can elect either to uphold those rules or to exit the school because they have joined the school on a voluntary basis and without undue pressure.

§4. Conclusion

386. It is clear that both private and public education in South Africa are offered in two different environments, but when it comes to religious education both systems aim to ensure that children are educated based on their religious convictions in that particular school without imposing or discriminating against other religious dominations. Furthermore, both private and public schools in their administration of religious education promote diversity and cannot promote propaganda and violence among other things because this will be viewed as contrary to the Constitution which is the supreme law of the land. Based on this assessment it can thus be said that both public and private schools when it comes to religious education share a common goal of using religion as a mechanism to tolerate diversity and to bring the necessary social cohesion to the South African community.

738. Section 7(5) of the South African Charter of Religious Rights and Freedom.

Chapter 2. Religious Schools

§1. Introduction

387. The rich history and foundation of religion in the South African community is something that has been stressed and outlined throughout this book. This means that religion is something that cannot be underestimated in South Africa and that is why it is firmly entrenched and protected by the Constitution. Religion due to its history and position in South Africa and across the globe is a system that has grown and it is now independent from law and political inference but is still subject to the rule of law in South Africa, which is the Constitution.

388. Religion as an independent system is a specialised area of expertise for people who are followers of that religion and those who show interest in that religion. This means that religion apart from serving as a faith to those people who follow it, it has also become a career or a field of study to other people. Due to this development in religion and the pluralistic society of South Africa a need has arisen for the establishment of religious schools with the aim of teaching the theological perspective and understanding of religion or faith. Various religious schools have been established which aim to serve as training institutions for the various religious groups in South Africa, in their diversity.

§2. Religious Schools in South Africa

389. Religious schools which are also called parochial schools or sectarian schools are schools that are aimed at teaching the religious viewpoints of a particular faith or religion.[739] The pluralistic nature of South Africa and the recognition of the different religious beliefs has resulted in different religious schools being developed as well as preserved as one of the ways to ensure that religion or belief remains relevant and in existence in South Africa.[740] The majority of the population of South Africa are Christians as already outlined and this has resulted in the development of Christian religious schools, for example Methodist Church Religious School of South Africa and the Presbyterian Religious School of South Africa, among others.[741] These Christian religious schools teach the Christian doctrines of their respective churches and also cater as training grounds for up and coming prospective ministers and lay preachers of that particular church.[742] Furthermore, in addition to their teaching of a specific indoctrination of a certain church the candidates in these seminary schools are also taught aspects of church administration, and how they are expected to carry themselves in society as well as within the church.[743] This

739. F.T. Hodgson, *Religion and Culture in Public Education in South Africa* 184 (BEHB Printers 2010).
740. *Ibid.*, at 185.
741. Siwa, *supra* n. 268, at 99.
742. *Ibid.*
743. *Ibid.*, at 100.

in a certain way can be seen as a way of preserving the religion or religious convictions because the candidates in these seminary schools are taught the doctrine of the church which can include aspects of history, procedure and processes of the church and the social position of the church. It is something that must be preserved and applied after they qualify.[744]

390. Candidates in seminary schools can spend about three to four years doing their diploma and after this period serve about two to three years of probation before being ordained as ministers.[745] However, this is different from preachers and other candidates like deacons who will be ordained after two years of completing either a certificate or diploma in that particular domination.[746] Furthermore, the people who train these candidates in these seminary schools will be ministers of that particular religious domination, and they must have served for a long time in the church and have the necessary experience and knowledge to impart to the young ones who are in seminary school.[747] Seminary schools in offering religious education are still expected to comply with the rule of law in such a way that they cannot discriminate against other people in need of that type of education or to promote things like violence or propaganda against another religion or group.[748] If this is found to be the case then such a seminary school will be either suspended or closed because it will be promoting illegality which is not in the public interest.

391. Other religious groups which are the minority religious dominations like the Muslim and Jewish also have their religious schools that are in operation in South Africa.[749] Just like the Christian faith or religion, the Muslim; Hindu and Jewish religious schools are teaching the Muslim, Hindu and the Jewish doctrine to its candidates. Normally the people who will be attending these religious schools will either be a leader in that domination or a member with the aim of ensuring that they have a better understanding about their religion or faith.[750] The same standards and procedures that are found in Christian religious seminaries are applicable to the Muslim, Hindu and Jewish religious schools, that is promoting peace and reconciliation without pushing for violence and discrimination. Because engaging in such conducts or practices will be found to be contrary to the Constitution and the rule of law.[751]

392. In the context of African customary law as a religious and cultural practice of the indigenous people of South Africa there is a religious school as well for the people who practice this system.[752] However, not everyone can go to this religious school which is called an initiation school, but only an individual who has a special

744. *Ibid.*
745. *Ibid.* at 101.
746. *Ibid.*
747. *Ibid.*, at 102.
748. Section 2 of the Constitution.
749. Siwa, *supra* n. 268, at 99.
750. Hugh, *supra* n. 211, at 70.
751. *Ibid.*, at 71.
752. Mndende, *supra* n. 84, at 50.

gift of seeing things of the future and who has been chosen by the ancestors for this type of training.[753] The individual chosen to go for an initiation training will be a traditional healer after having gone through an extensive training of about six to twelve months, learning the African way of life as well as how to heal people using different African ways and herbs.[754] Initiation schools for traditional healers are also seen as one of the mechanisms for the preservation of African customary law and practices, and it is a very important and significant process. Initiation schools are also bound by the rule of law and must comply with the law at all costs and not promote any form of illegality or discrimination in their training and practice.

§3. Conclusion

393. Religious schools are important for the survival of the various religious dominations in South Africa because they are the foundation for preparing the next generation of people in a particular religious domination. The teachings of these religious schools must be in line with the purpose and supremacy of the Constitution, in such a way that they must not promote discrimination or violence against each other or members of other religious dominations. If this is found to be the case, then freedom of religion and culture will be compromised and there will be some form of consequences in place as laid down by the law.

753. *Ibid.*, at 51.
754. *Ibid.*, at 52.

Selected Bibliography

Articles

Adewole, I.F. 'Challenges and Opportunities in Cancer Control in Africa: A Perspective from the African Organisation for Research and Training in Cancer'. Series 14 (2013): 147.

Akazili, J. & J. Ataguba. 'Health Care Financing in South Africa: Moving Towards Universal Coverage'. CEM 28 (2010): 75.

Baderin, S. 'From International Law of Christianity to the International Law of the Global Citizen'. JHIL 3 (2001): 183.

Beck, S. 'Health Policy, Health Services, and Cancer Pain Management in the New South Africa'. JPSM 17 (1999): 17.

Birdsall, A. *The International Criminal Tribunal for the Former Yugoslavia–Towards a More Just Order?* PCD 8 (2006): 3.

Buscher, B. 'The Political Economy of Africa's Natural Resources and the Great Financial Crisis'. TESG 26 (2011): 137.

Burchell, J.M. 'Media Freedom of Expression Scores as Strict Liability Receives the Red Card: *National Media Ltd v. Bogoshi*'. SALJ 8 (1999): 1.

Cachalia, F. 'The Future of Muslim Law in South Africa'. OP 12 (1991): 22.

Chersich, M. 'Inequalities in Access to Health Care in South Africa'. JPHP 32 (2011): 3.

Chipkin, I. and A. Leatt. 'Religion and Revival in Post-apartheid South Africa'. RESA 9 (2009): 39.

Coertzen, P. 'Constitution, Charter and Religions in South Africa'. AHRLJ 14 (2014): 127.

Delius, P. 'The History of Migrant Labour in South Africa (1800–2014)'. SHSA 17 (2017): 30.

Denny, L., L. Kuhn & P. Batra. 'Utilisation and Outcomes of Cervical Cancer Prevention Services Among HIV-Infected Women in Cape Town'. SAMJ 100 (2010): 39.

De Lille, P. 'The Western Cape Government: History and Regional Development Profile for the City of Cape Town'. WCG 3 (2018): 14.

Evans, D. 'Religion and International Law: Friends or Foes?'. SOAS 32 (2010): 638.

Express, C. 'More Services Needed to Fight Cancer'. UFS 23 (2008): 8.

Falkson, G. & S. Beck. 'Prevalence and Management of Cancer Pain in South Africa'. *Pain* 94 (2001): 76.

Selected Bibliography

Fourie, C. 'Why Is Johannesburg the Economic Hub of South Africa: A City of Opportunities for Entrepreneurs'. *Randburg Sun* 3 (2017): 4.

Foley, H.E. 'Incorporation, Multiple Incorporation and the Conflict of Laws'. *Harvard Law Review* 42 (2000): 516.

Hugh, F.A. 'South African Jews and Apartheid'. *Patterns of Prejudice* 34 (2000): 23.

Jessup, P. 'Transnational Law'. IGLP 8 (2012): 1.

Kauuova, W.R. 'The Religions of Southern Africa: A Synopsis of Their Basic Beliefs'. *Institute for Reformed Studies* 325 (1997): 3.

Kiernan, J. 'African Traditional Religions in South Africa'. *De Jure* 15 (2008): 27.

Lyratzopoulos. G., G. Abel & C. Saunders. 'What Explains Worse Patient Experience in London? Evidence from Secondary Analysis of the Persons Living with Cancer Experience Survey'. BMJ OPEN 4 (2014): 1.

Ma, K. 'Control and Conscience: Positivist Approaches to Religion in India and South Africa'. SOAS LJ 260 (2015): 261.

Mary, W. 'Medical Ethics Made Easy'. *Professional Care Management* 21 (2016): 81.

Marrison, J. 'Nelson Mandela, South Africa's First Black President, Dies at the Age of 95'. *The Washington Times* 5 (2015): 14.

Masuku, T. 'Prophetic Mission of Faith Communities During Apartheid South Africa, 1948–1994: An Agenda for a Prophetic Mission Praxis in the Democratic South Africa'. *Missionalia* 42 (2014): 4.

McIntyre, D. 'The Health and Health System of South Africa: Historical Roots of Current Public Health Challenges'. *Lancet* 374 (2009): 817.

Mndende, N. 'An Introduction to African Religion'. *Surrender* 28 (2012): 1.

Morris, J. 'Northern Province and the Karoo Rock in South Africa'. *Journal of Human Evolution* 55 (2016): 11.

Naylor, C.D. 'Private Medicine and the Privatization of Health Care in South Africa'. *Social Science & Medicine* 11 (1998): 1153.

Ndlovu, S. 'The Soweto Uprising: The Road to Democracy in South Africa'. *South African Democratic Trust* 3 (2011): 344.

Nkosi, J. 'Investment Potential of the Platinum Province: Areas of Investment That Are Expected to Grow Fast Include Fuel Cell Technology, Machinery for Mining, Energy Manufacturing and Component Production'. *Global Africa* 4 (2017): 12.

Patijn, J. 'High Prevalence of Pain in Patients with Cancer in a Large Population-Based Study in the Netherlands'. 132 *Pain* 312 (2007).

Peter, S. 'Restorative Justice and the South African Truth and Reconciliation Commission'. *SA Journal of Philosophy* 32 (2006): 35.

Pieterse, M. 'Enforcing the Right Not to Be Refused Emergency Medical Treatment: Towards Appropriate Relief'. Stellenbosch Law Review 18 (2007): 75.

Pieterse, M. 'Legislative and Executive Translation of the Right to Have Access to Health Care Services'. Law Democracy and Development 10 (2010): 243.

Proposed Code of Ethics for Private Hospitals – Hospital Association of South Africa. 'Private Hospital Review 2008: Examination of Factors Impacting on Private Hospitals'. HASA 67 (2008): 44.

Rautenbach, C. 'Gender Equality and Religious Family Laws in South Africa'. *Queensland UTLJ* 3 (2003): 168.
Reagan, T. 'South African Sign Language and Language in Education Policy in South Africa'. *Linguistics* 38 (2008): 166.
Roos, A. 'Personal Data Protection in New Zealand: Lessons for South Africa'. PER/PELJ 4 (2008): 191.
Schoeman, W.J. 'South African Religious Demography: The 2013 General Household Survey'. *Theological Studies* 4 (2017): 2.
Taybob, A. 'Islamic Resurgence in South Africa: The Muslim Youth Movement'. *Journal of Religion in Africa* 12 (2014): 104.
Tollman, S.M. & K. Kautzky. 'A Perspective on Primary Health Care in South Africa'. SAHR 2 (2008): 20.
Van Jaarsveld, M. 'An Employees Contractual Obligation to Promote Harmonious Relationships in the Workplace-When are the Stakes too High? Some Pointers from the Judiciary'. SA Mercantile Law Journal 19 (2007): 205.
Valova, M. 'Financing Churches and Religious Societies in the 21st Century'. RCVP 322 (2010): 146.
Van Rooyen, J.C.W. 'Dignity, Religion and Freedom of Expression in South Africa' OR 12 (2011): 1.
Venter, M., C. Venter & K. Botha. 'Cancer Treatment in South Africa: A Narrative Literature Review'. JPA 22 (2012): 462.
Zyberi, G. 'The International Court of Justice and the Rights of Peoples and Minorities'. *Series* 14 (2013): 2.

Books

Amerasinghe, C.F. *The Permanent Court of International Justice and the International Court of Justice: Contentious Jurisdiction* (Hague: International Law E Books 2008), 46.
Andemicael, B. *The OAU and the UN: Relations Between the Organisation of African Unity and the United Nations* (New York: Africana 1976), 34.
Asad, T. *Discipline and Reasons of Power in Christianity Formations and Dignity* (Baltimore: John Hopkins University 2006), 34.
Bekker, J. & C. Rautenbach. *Introduction to Legal Pluralism* (4th ed., Durban: LexisNexis 2015), 12.
Bekink, B. *Principles of South African Constitutional Law* (Johannesburg: LexisNexis 2012), 81.
Brain, B. *Setting Down Roots: Indian Migrants in South Africa, 1890–1911* (Johannesburg: Wits University Press 1990), 12.
Brand, D. *Introduction to Socio-Economic Rights in the South African Constitution in Brand D and Heyns C Socio-Economic Rights in South Africa* (Durban: Butterworths 2005), 1.
Bredekamp, H. & R. Ross. *Missions and Christianity in South African History* (Johannesburg: Wits University Press 1995), 14.
Boyle, F.A. *Foundations of World Order: The Legalistic Approach to International Relations 1898–1922* (Durham: Duke University Press 1999), 1.
De Vos, P. & M. Freedman. *South African Constitutional Law* (Cape Town: Oxford University Press 2015), 421.

Selected Bibliography

Devenish, G.E. *A Commentary on the South African Constitution* (Durban: Butterworths 1998), 121.

Elphick, R. *Christianity in South Africa: A Political, Social and Cultural History* (Los Angeles: University of California Press 1997), 11.

Frances, E. *History of the Zulu War and Its Origins* (Pretoria: Chapman and Hall 1880), 12.

Gramsci, A. *Architect of a New Politics* (Lafayette: Louisiana University Press 1990), 157.

Gragon, M. *Essential Labour Law* (Durban: Butterworths 2016), 218.

Hodgson, F.T. *Religion and Culture in Public Education in South Africa* 184 (Pretoria, BEHB Printers 2010).

Kruger, J. 'The Human Search for Meaning. A Multi-religious Introduction of the Religions of Humankind'. (Goodwood: National Book Printers 1996), 42.

Lenaghan, C. *Shots in the Streets: Violence and Religion in South Africa, Contemporary South African Debates* (Oxford: Oxford University Press 2004), 68.

Murry, R. *Human Rights in Africa: From the OAU to the African Union* 5 (Cambridge: Cambridge University Press 2005), 5.

Neethling, J., J.M. Potgieter & P.J. Visser. *Neethling's Law of Personality* (Durban: Butterworths 2005), 31.

Ngwabi, B. *The Historical Dimensions of Democracy and Human Rights in Zimbabwe* (Mambo Press 2001), 16.

Rautenbach, C. & R. Malherbe. *Constitutional Law* (Cape Town: LexisNexis 2015), 156.

Richardson, F. *The Orange Free State and Inhabitants* (11th ed., Cambridge: Cambridge University Press 1999), 5.

Sachs, A. *The Soft Vengeance of a Freedom Fighter* (Amazon 1990), 3.

Saayman, W. *Being Missionary, Being Human: An Overview of Dutch Reformed Mission* (Pietermaritzburg: Cluster Publications 2007), 12.

Siwa, Z. *The Laws and Disciplines of the Methodist Church of Southern Africa* (12th ed., Johannesburg: Methodist Publishing House 2016), 2.

Schoeman, F. *Philosophical Dimensions of Privacy: An Anthology* (Cambridge: Cambridge University Press 1984), 4.

Van Niekerk, G.J. and L. Wildenboer. *The Origins of South African Law* (Pretoria: UNISA 2010), 177.

Williams, R. *Keywords: A Vocabulary of Culture and Society* (London, Fontana 1976), 16.

Websites

Archbishop Desmond Tutu is Awarded the Nobel Peace Prize, https://www.sahistory.org.za/dated-event/archbishop-desmond-tutu-awarded-nobel-peace-prize (accessed 13 September 2018).

A Warm Welcome to Kwazulu Natal in South Africa, http://www.zulu.org.za/about/key-facts/welcome (accessed 16 August 2018).

A Gross Abuse of a Position of Trust https://www.churchtimes.co.uk/articles/2016/9-december/comment/opinion/tackling-abuse-on-pitch-and-pulpit (accessed 17 August 2018).

Churches and Government Likely to Abuse Trust https://www.benkhumalo-seegelken.de/suedafrika-aktuell/1753-church-and-government/ (accessed 30 September 2018).

Education in South Africa a System in Crisis https://city-press.news24.com/News/education-in-south-africa-a-system-in-crisis-20160531 (accessed 19 October 2018).

HPCSA Regulatory Body of South Africa's Medical Practitioners https://citizen.co.za/news/south-africa/1991655/health-professions-council-struggles-in-fight-against-corruption/ (accessed 17 August 2018).

Mpumalanga and the Big Five Animals, https://www.mpumalanga-info.co.za/provinces/article/378/big-five-animals-in-south-africa (accessed 17 August 2018).

Nigeria's Boko Haram https://www.bbc.com/news/world-africa-31784538 (accessed 9 August 2018).

Personal Law https://www.merriam-webster.com/dictionary/personal%20law (accessed 11 June 2018).

September 11 Terror Attacks Fast Facts https://edition.cnn.com/2013/07/27/us/september-11-anniversary-fast-facts/index.html (accessed 2 August 2018).

Statistics of the Population of South Africa, http://www.statssa.gov.za/publications/P0302/P03022018.pdf (accessed 11 July 2018).

South African History Online www.sahistory.org.za (accessed 5 June 2018).

South African History Online Place of Birth of Nelson Mandela www.sahistory-.org.za (accessed 5 June 2018).

South African Political Parties Go to Church https://www.news24.com/elections/news/sa-political-parties-go-to-church-20140420 (accessed 16 August 2018).

The Great Limpopo Province, History of the Province http://www.dolimpopo.com/limpopo-province (accessed 6 July 2018).

Why Private Schools are an Excellent Investment Opportunity in South Africa https://businesstech.co.za/news/business/189369/why-private-schools-are-an-excellent-investment-opportunity-in-sa/ (accessed 19 October 2018).

Why Is the Church a Non-Player in Politics https://mg.co.za/article/2016-03-23-why-the-church-is-a-nonplayer-politically (accessed 14 August 2018).

Selected Bibliography

Table of Cases

Amod v. Khan 1947 (2) SA 432 (N).
C v. Minister of Correctional Services 1996 (4) SA 292 (T).
Christian Lawyers Association v. Minister of Health 1998 (11) BCLR 1434 (T); 1998 (4) SA 1113 (T).
Christian Education South Africa v. Minister of Education [2000] ZACC 11; 2000 (4) SA 757 (CC); 2000 (10) BCLR 1051 (CC).
Church of the Province of Southern Africa Diocese of Cape Town v. Commission for Conciliation Mediation and Arbitration and Others (C619/2000) [2001] ZALC 141 (7 September 2001).
Diocese of Cape Town v. CCMA 2002 3 SA 385 (LC).
Government of Republic of South Africa and others v. Grootboom, (CCT11/00) [2000] ZACC 19; 2001 (1) SA 46; 2000 (11) BCLR 1169 (4 October 2000).
Hay v. B 2003 (3) SA 492 (W).
Ismail v. Ismail 1983 1 SA 1006 (W).
Jackson v. NICRO 1976 3 SA 1 (A) 11.
Kennedy and Arnold v. Ireland [1987] IR 587.
Maje v. Botswana Life Insurance [2001] 2 BLR 626.
Marais v. The Master 1984 (4) SH 288 (D).
National Media Ltd ao v. Jooste 1996 (3) SA 262 (A).
National Coalition for Gay and Lesbian Equality and Another v. Minister of Justice and Others, (CCT11/98) [1998] ZACC 15; 1999 (1) SA 6; 1998 (12) BCLR 1517 (9 October 1998).
Nkosinami Nkomonde v. the Diocese of Pretoria (Anglican Church of Southern Africa) 2015 4 SA 399 (LC).
O'Keeffe v. Argus Printing and Publishing Company Ltd 1954 (3) SA 244 (C).
Prosecutor v. Ahmad Al Faqi Al Mahdi (ICC, 2016).
Purohit and Moore v. The Gambia [Communication 241/2001].
Ryland v. Edros 1997 (2) SA 690 (CC).
S v. Dalindyebo (090/2015) [2015] ZASCA 144 (1 October 2015).
S v. Manamela and Another 2000 (3) SA 1 (CC).
S v. Makwanyane 1995 (6) BCLR 6655 (CC).
Salvation Army (South African Territory) v. Minister of Labour (J 464/02) [2004] ZALC 64 (2 September 2004).
Schreuder v. NG Kerk Wilgespruit 1999 20 ILJ 1936 (LC).
Soobramoney v. Minister of Health, Kwazulu Natal 1998 (1) SA 765 (CC).
Tolly v. Fry and Sons Ltd [1931] AC 333; [1931] UKHL 1.

Table of Cases

Treatment Action Campaign v. The Minister of Health 2002 (4) BCLR 356 (T).
Universal Church of the *Kingdom of God v. Myeni* and Others (DA 3/14) [2015] ZALAC 31; [2015] 9 BLLR 918 (LAC); (2015) 36 ILJ 2832 (LAC).

Table of Legislation

African Charter on Human and People's Rights 1986.
Basic Conditions of Employment Act 75 of 1997.
Constitution of the Republic of South Africa 1996.
Civil Union Act 2004.
Genocide Convention 1948.
International Covenant on Economic, Social and Cultural Rights (CESCR) Adopted and opened for signature, ratification and accession by General Assembly Resolution 2200 A (XXI) of 16 December 1966 and came into force on 3 January 1976 in accordance with Article 27 of the United Nations.
International Covenant on Civil and Political Rights 1976.
The Recognition of Customary Marriages Act 120 of 1998.
The Protection of Personal Information Act 4 of 2013.
The Refugee Convention 1951.
The Master and Servants Act 15 of 1856.
The Carpenters and Joiners Union Act 1881.
The Mines and Works Act 1911.
The Choice on Termination of Pregnancy Act 92 of 1997.
The Abuja Declaration of 2001.
The African Court on Human and Peoples Rights 2004.
The African Peer Review Mechanism (APRM) 2002.
The Attorney's Act of 1979.
The Advocate's Act 74 of 1964.
The Employment Equity Act 55 of 1998.
The New Partnership for Africa's Development (NEPAD) 2001.
The Right of Appearance in Court Act 62 of 1995.
The Southern African Development Community (SADC) 1980.
The South African Schools Act 1996.
The Schedule to the Promotion of Equality and Prevention of Unfair Discrimination Act 4 of 2000.
The Companies Act 68 of 2008.
The Children's Act 38 of 2005.
The National Health Act 61 of 2003.
The South African Charter of Religious Rights and Freedom 2010.
UN Universal Declaration of Human Rights 1948.
UNESCO Convention 1970.

Index

The numbers here refer to paragraph numbers.